Mycotoxins

Mycotoxins

From field to feed

Edited by

Regiane R. Santos

BRILL | WAGENINGEN ACADEMIC

Cover illustration: Joan Miró, The Farm (1921-1922). Oil on canvas, 123.8 × 141.3 × 3.3 cm (overall), 138.4 × 155.9 × 7.6 cm (framed). National Gallery of Art, Washington DC, USA. © Successió Miró, c/o Pictoright Amsterdam, 2024.

Library of Congress Cataloging-in-Publication Data

Names: Santos, Regiane R. editor
Title: Mycotoxins: from field to feed / edited by Regiane R. Santos.
Description: Leiden ; Boston : Brill/Wageningen Academic, [2025] | Series: Food and feed safety, 3050-5240 ; vol.1 | Includes bibliographical references and index.
Identifiers: LCCN 2024059160 (print) | LCCN 2024059161 (ebook) | ISBN 9789004696310 hardback acid-free paper | ISBN 9789004712720 e-book
Subjects: LCSH: Mycotoxins--Physiological effect | Livestock--Diseases | Livestock--Effect of contaminated sediments on
Classification: LCC QP632.M9 M923 2025 (print) | LCC QP632.M9 (ebook) | DDC 615.9/5295--dc23/eng/20250212
LC record available at https://lccn.loc.gov/2024059160
LC ebook record available at https://lccn.loc.gov/2024059161

Typeface for the Latin, Greek, and Cyrillic scripts: "Brill". See and download: brill.com/brill-typeface.

ISBN 978-90-04-69631-0 (hardback)
ISBN 978-90-04-71272-0 (e-book)
DOI 10.1163/9789004712720

This book is printed on acid-free paper and produced in a sustainable manner.

Contents

Figures and Tables

Figures

Tables

Abbreviations

15-ADON	15-acetyldeoxynivalenol
2D	two-dimensional
3+15-ADON	3+15-acetyldeoxynivalenol
3-ADON	3-acetyldeoxynivalenol
3D	Three-dimensional
3R principles	Replacement, Reduction, and Refinement
4-NIV	4-acetylnivalenol
ABC	Adenosine triphosphate-binding cassette
AchE	Acetyl cholinesterase
ADFI	Average daily feed intake
ADG	Average daily gain
ADME	Absorption, Distribution, Metabolism, and Excretion
ADON	Acetyldeoxynivalenol
ADWG	Average daily weight gain
AF	Aflatoxin
AFB_1	Aflatoxin B_1
AFB_2	Aflatoxin B_2
AFBO	AFB_1-8,9-*exo*-epoxide
AFG_1	Aflatoxin G_1
AFG_2	Aflatoxin G_2
AFL	Aflatoxicol
*afl*D	Nor~1 gene
*afl*R	Aflatoxin pathway regulator
AFM_1	Aflatoxin M_1
AFM_2	Aflatoxin M_2
AFP_1	Aflatoxin P_1
AFs	Aflatoxins
AHR	Aryl hydrocarbon receptor
AID	Apparent Ileal Digestibility
AL	Aluminium
ALP	Alkaline phosphatase
ALT	Alanine aminotransferase
AME	Alternariol monomethyl ether
AMH	Anti-Müllerian hormone
AML-12	Hepatocytes from the normal liver of a 3-month old mouse
$AnAFB_1$-KLH	Anaphylatoxin B_1 conjugated with keyhole limpet hemocyanin
ANVISA	Brazilian Health Regulatory Agency (Agência Nacional de Vigilância Sanitária)
ANZFA	Australia New Zealand Food Authority
AOH	Alternariol

AST	Aspartate aminotransferase
ATP	Adenosine triphosphate
ATTD	Apparent Total Tract Digestibility
a_w	Water activity
BCa	Adjusted bootstrap percentile
BCAs	Biocontrol agents
BCRP	Breast cancer resistance protein
BEA	Beauvericin
BK channels	Calcium-activated potassium channels
BME-UV1	Bovine mammary epithelial cells
BVDV	Bovine viral diarrhoea virus
BW	Body weight
BWG	Body weight gain
Caco-2	Immortalized cell line of human colorectal adenocarcinoma cells
CAR	Constitutive androstane receptor
CAST	Council for Agricultural Science and Technology
CAT	Catalase
CC5	Hazardous concentrations for 5 % of a population
CCFAC	Codex Committee on Food Additives and Contaminants
CCs	Cumulus cells
$CD2^+$	Cluster of differentiation 2
$CD21^+$	Complemente receptor 2
$CD4^+$	T helper cells
$CD8^+$	Cytotoxic T cells
CEC	Cation exchange capacity
CerS	Ceramide synthases
CFR	Code of Federal Regulations
CIT	Citrinin
CO_2	Carbon dioxide
COCs	Cumulus-oocyte complexes
CONTAM	Contaminant
CP	Crude protein
CPA	Cyclopiazonic acid
CSA	Climate-smart agriculture
CV	Coefficient of variance
CVM	Centre for Veterinary Medicine
CYP	Cytochrome p450
DAFF	South African Department of Agriculture, Forestry and Fisheries
DAS	Diacetoxyscirpenol
DDGS	Distillers' dried grains and with solubles
de-DON	Deepoxy-deoxynivalenol
DLLME	Dispersive liquid-liquid microextraction

DM	Dry matter
DMI	Dry matter intake
DNA	Deoxyribonucleic acid
DOM-1	Deepoxy-deoxynivalenol
DON	Deoxynivalenol
DON-3G	DON-3-Glucoside
DON-3S	Deoxynivalenol-3-sulphate
DTT	Dichlorodiphenyltrichloroethane
EAs	Ergot alkaloids
EC	European commission
EC_{50}	Concentration that gives half-maximal response
ECB	European corn borer
ECVAM	European Centre for the Validation of Alternative Methods
EE	Ether extract
EFCOVAL	European Food Consumption Validation
EFSA	European Food Safety Authority
EFX	Enrofloxacin
ELEM	Equine leukoencephalomalacia
ELISA	Enzyme-linked immunosorbent assay
ENNA	Enniatin A
ENNA1	Enniatin A1
ENNB	Enniatin B
ENNB1	Enniatin B1
ENNB2	Enniatin B2
ENNB3	Enniatin B3
ENNB4	Enniatin B4
ENNs	Enniatins
ER	Endoplasmic reticulum
ESCs	Embryonic stem cells
EST	Embryonic stem cell test
EU	European Union
FA	Fumonisin A
FA_1	Fumonisin A_1
FA_2	Fumonisin A_2
FAO	Food and Agriculture Organization of United Nations
FAP	Feed additive petition
FB	Fumonisin B
FB_1	Fumonisin B_1
FB_2	Fumonisin B_2
FB_3	Fumonisin B_3
FB_4	Fumonisin B_4
FC	Fumonisin C

FCR	Feed conversion ratio
FD	Fluorescence detection
FDA	Food and Drug Administration
Fe	Iron
FEB	Fusarium Ear Blight
FEEDAP	EFSA panel on Additives and Products or Substances used in Animal Feed
FER	Fusarium ear blight/rot
FGF	Fibroblast growth factor
*Fg*SC	*Fusarium graminearum* species complex
FHB	Fusarium head blight
FI	Feed intake
FP	Fumonisin P
FUM	Fumonisins
GABA	Gamma-aminobutyric acid
GAP	Good Agricultural Practice
GC	Gas chromatography
GC	Granulosa cell (only for Chapter 14)
GER	Gibberella Ear Rot
GES-1	Normal human gastric epithelial cells
GGT	Gamma glutamyl transferase
GI	Gastrointestinal
GIT	Gastrointestinal tract
GM	Genetically modified
GPx	Glutathione peroxidase
GRAS	Generally recognized as safe
GSH	Glutathione
GST isoenzymes α	Glutathione S-transferase α
GST isoenzymes π	Glutathione S-transferase π
GST	Glutathione S-transferase
HACCP	Hazard Analysis Critical Control Point
HBGV	Health-Based Guidance Values
hCG	Human chorionic gonadotropin
HCT116	Cells from the colon of an adult male with colon cancer
hCYP1A2	Cytochrome P450 1A2
HepG2	Liver cells from a hepatocellular carcinoma of a 15-year-old human male
HFB_1	Hydrolysed fumonisin B_1
HK-2	Proximal tubular cell line derived from normal adult male kidney
HPLC	High performance liquid chromatography
hPXR	human pregnane X receptor
HRMS	High-resolution mass spectrometry
HT29	Cell line from a white, female colorectal adenocarcinoma patient
HT29-MTX	HT29 cell line adapted with MTX

HuMIX	Human-microbial crosstalk model
IACUC	Institutional Animal Care and Use Committee
IARC	International Agency for Research on Cancer
IBV	Infectious bronchitis vaccine
IC_{50}	Half-maximal inhibitory concentration
ICM	Inner cell mass
IEBC	Intestinal explant barrier chip
IFIF	International Feed Industry Federation
IgA	Immunoglobulin A
IGF-1	Insulin growth factor-1
IgG	Immunoglobulin G
IL	Interleukin
IPCC	Intergovernmental Panel on Climate Change
IPCS	International Programme on Chemical Safety
IPEC-1	Intestinal porcine epithelial cells 1
IPEC-J2	Intestinal porcine enterocyte line
IPM	Integrated pest management
iPSCs	Induced pluripotent stem cells
IUPAC	International Union of Applied and Pure Chemistry
IVF	*In vitro* fertilisation
IVM	*In vitro* maturation
J774A.1	Cells from the ascites of an adult female mouse with reticulum cell sarcoma
JECFA	Joint FAO/WHO Expert Committee on Food Additives
K_{cat}	Catalytic efficiency
KLH	Keyhole limpet hemocyanin
K_m	Michaelis constant
LAB	Lactic acid bacteria
LAMIC	Laboratory of Mycotoxicological Analysis
LC_{50}	Concentration of inhaled chemical at which 50% of the test animals died
LC-MS/MS	Liquid chromatography tandem mass spectrometry
LD_{50}	Concentration of ingested/absorbed chemical at which 50% of the test animals died
LDH	Lactate dehydrogenase
LLC-PK1	Epithelial kidney cells from the kidney of a male 3-4 week-old pig
LLE	Liquid-liquid extraction
LOAEL	Lowest-Observed Adverse Effect Level
LOD	Limit of detection
LOQ	Limit of quantification
LPO	Lipid peroxidase
LPO	Lipid peroxidation
MAP kinase	Mitogen activated protein kinases
MDCK	Madin-Darby canine kidney cells

MDR1	Multidrug resistance protein 1
MDZ	Midazolam
$MHCII^+$	Major histocompatibility complex II
MII	Metaphase II
MON	Moniliformin
MPA	Mycophenolic acid
mRNA	Messenger ribonucleic acid
MRP1	Multidrug resistance associated protein 1
MRP2	Multidrug resistance associated protein 2
MS/MS	Tandem mass spectrometry
M-SHIME	SHIME including porcine mucin gels in small glass tubes
MTX	Methotrexate
MUC2	Mucin 2
N	Nitrogen
NA	Not applicable
NAT	N-acetyl transferase
NCor	Nuclear receptor corepressor
NDF	Neutral detergent fibre
NDV	Newcastle disease vaccine
NF-κB	Nuclear factor kappa-light-chain-enhancer of activated B cells
NH3-M	Ammonia nitrogen
NIV	Nivalenol
NIV-3G	Nivalenol-3-Glucoside
NO	Nitric oxide
NOAA	National Oceanic and Atmospheric Administration
NOAEL	No-Observed Adverse Effect Level
NPY	Neuropeptide Y
NR	Not reported
Nrf2	Nuclear factor erythroid 2-related factor 2
NSAIDs	Non-steroidal anti-inflammatory drugs
OECD	Organization for Economic Co-operation and Development
OOC	Organ-on-chip (OOC)
OTA	Ochratoxin A
OTAα	Ochratoxin α
OTB	Ochratoxin B
OTC	Ochratoxin C
OTSC	Office of the Texas State Chemist
PAT	Patulin
PBMC	Human peripheral blood mononuclear cells
PBS	Phosphate buffer saline
P-gp	P-glycoprotein
$pHFB_1$	partially Hydrolysed fumonisin B_1

PHO-A	Phomopsin A
PHO-C	Phomopsin C
PHO-E	Phomopsin E
PHOs	Phomopsins
PMSG	Pregnant mare serum gonadotropin
PMTDI	Provisional maximum tolerable daily intake
PPAR	Peroxisome proliferator-activated receptor
ppb	parts per billion
PPE	Porcine pulmonary oedema
ppm	parts per million
PRRSV	Porcine reproductive and respiratory syndrome virus
PXR	Pregnane X receptor
PYY	Peptide YY
QTLs	Quantitative trait loci
QuEChERS	Quick, easy, cheap, effective, rugged, and safe
RASFF	Rapid Alert System for Food and Feed
RNA	Ribonucleic acid
RNS	Reactive nitrogen species
ROS	Reactive oxygen species
Sa	Sphinganine
Sa/So	Sphinganine: Sphingosine ratio
Sf-9 cell line	Cell line derived from pupa ovarian tissue of a fall armyworm
SGR	Specific growth rates
SHIME	Simulator of the Human Intestinal Microbial Ecosystem
SK-N-MC	Neuroblastoma cell line
SLAM	Simultaneous localisation and mapping
SLE	Solid-liquid extraction
So	Sphingosine
SOD	Superoxide dismutase
SPE	Solid-phase extraction
ST	Sulfotransferase
T:O ratio	Tetrahedral:Octahedral sheets ratio
TDI	Tolerable dose intake
TeA	Tenuazonic acid
TEER	Transepithelial electrical resistance
TLC	Thin-layer chromatography
T_{max}	Time to peak drug concentration
TMR	Total mixed ration
TNFα	Tumor necrosis factor α
TOF	Time-of-flight
tRNA	Transfer ribonucleic acid
Twin-SHIME	SHIME with two parallel running reactors

UDP	Uridine-5-diphosphate
UGT	Uridine diphosphoglucuronosyl transferase
UHPLC-MS/MS	Ultra-high-performance liquid chromatography-tandem mass spectrometry
UPLC-Q/TOF-MS	Ultra-performance liquid chromatography-quadrupole/time-of-flight hybrid mass spectrometry
UV	ultraviolet detection
Vd	Volume of distribution
VFAs	Volatile fatty acids
V_{max}	Maximum velocity of an enzymatic reaction
VTG	Vitellogenin
WHO	World Health Organization
ZAN	Zearalanone
ZEN	Zearalenone
ZEN-14G	Zearalenone-14-O-glucoside
ZEN-14S	Zearalenone-14-sulfate
ZEN-16G	Zearalenone-16-O-glucoside
ZGA	Zygotic genome activation
α-ZAL	α-zearalanol
α-ZEL	α-zearalenol
α-ZEL/β-ZEL PO	ratio α-zearalenol:β-zearalenol *per os* (after oral administration)
β-ZAL	β-zearalanol
β-ZEL	β-zearalenol

CHAPTER 1

Introduction to mycotoxins

R.R. Santos

Schothorst Feed Research, Lelystad, the Netherlands
rsantos@schothorst.nl

1 Mycotoxins

The word mycotoxin is derived from the Greek words *mykes*, which means 'fungus', and *toxikos*, which means 'poison' (Bianchini and Bullerman, 2014); therefore the word mycotoxin means 'fungus toxin' or 'fungus poison' (Bullerman, 2003). Mycotoxins produced by filamentous fungi are secondary metabolites (termed by A. Kossel in 1891) that may be responsible for the defence and survival of the organism (Thirumurugan et al., 2018). Because of their ubiquitous presence in nature, and their ability to exert both protective and harmful effects, historically, mycotoxins have been considered both medicinal and toxic. As long as 5,000 years ago, the biological activity of ergot was known in China and was used to control haemorrhages, while during the Middle Ages the first reported case of ergotism (also called St Anthony's Fire) was in France and was characterised by symptoms of hallucinations and gangrene; it resulted in the death of more than 60,000 people after they had eaten bread produced with ergot-contaminated rye (Mostrom, 2016; van Dongen and de Groot, 1995). Other reported outbreaks of major mycotoxin intoxication from 1881 to 1910 included 'cardiac beriberi' in Asia, which was caused by the consumption of rice infected with *Penicillum citreonigrum* and produced the mycotoxin citreoviridin, resulting in *shoshin-kakke* (heart-attacking paralysis). In 1927, pigs fed barley from the United States of America (USA) presented with symptoms of disease including vomiting; the animals refused the feed, which was later analysed and found to contain *Fusarium*. Additionally, the 'alimentary toxic aleukia' in the former Soviet Union in the 1940s was caused by *Fusarium* toxins in cereals retained on the field over winter, affecting both humans and animals (see Review from Pitt and Miller, 2017). More recently, anti-cancer and insecticide properties were observed in the so-called emerging mycotoxin beauvericin (BEA), but its toxic effect on reproductive cells was also reported (see Review from Caloni et al., 2020). Mycotoxins were not recognised as a threat to human and animal health until 1960 with the outbreak of Turkey X disease in the United Kingdom, which resulted in the discovery of aflatoxins (AFs). In this disease outbreak, more than 100,000 turkey poults and other young farm animals were lost because of a toxic substance in groundnut meal (ingredient in the feed ration) imported from Bra-

 | DOI: 10.1163/9789004712720_2

zil that was heavily contaminated with *Aspergillus flavus* and *A. parasiticus*, which produce one of the most common mycotoxins in the feed industry, aflatoxin (Bianchini and Bullerman, 2014; Fujimoto, 2011). Following the discovery of AFs and their carcinogenic effect in humans, the subsequent investigation of novel mycotoxins led to the identification of over 100 toxigenic fungi and over 300 mycotoxins worldwide, with additional mycotoxins still being isolated and identified (Groopman and Kensler, 2005). The most common mycotoxins that have been linked to toxic syndromes in livestock and/or humans include AFs, trichothecenes (deoxynivalenol – DON and T-2/HT-2 toxins), fumonisins (FUM), zearalenone (ZEN), ochratoxin A (OTA), and ergot alkaloids (EAs), but the isolation and characterisation of 'new' mycotoxins (also called *emerging*, although more correctly denominated *neglected* mycotoxins) indicate that human and animals are daily exposed to a mixture of these natural contaminants, resulting in mycotoxicosis.

The term mycotoxicosis was first used in 1952 (Pitt, 2013) and is characterised by a non-infectious disease with symptoms that can be observed in animals after ingestion of mycotoxin-contaminated feed. Exposure to individual or combined mycotoxins in adequate quantities may cause mycotoxicosis, which may affect various health processes depending on the target organs of the specific mycotoxin (Peraica, 2016), and may be acute and/or chronic. Exposure to very high amounts of mycotoxins over a short period results in acute toxicity, whereas chronic mycotoxicosis occurs due to long-term exposure to lower amounts of certain mycotoxins and affects animals over a long period, during which symptoms are often not noted or only related to impaired production performance (Peraica, 2016).

2 An undeniable reality: the impossibility of a free-mycotoxin diet

The manufacture of high-quality feed, combining optimum nutrient levels and safe ingredients, is a precondition for animal health as well as for animal-derived products entering the food chain (Santos, 2021). The selection of feed ingredients will vary depending on availability; for example, in the Americas, maize and soybean meal are the main energy and protein sources used in animal feed, whereas in Asia and Europe feed composition is more complex and typically includes many other grains and cereals, as well as their by-products. To provide an idea of the annual feed output, it was estimated that in 2023, there would be a worldwide production of over 1 billion metric ton of compound feed for livestock (IFIF, 2023) and around 35 million metric tons for pets (Statista, 2023). However, irrespective of the origin of the nutrient source, such as grains, cereals and forages, it is not feasible to completely prevent the field contamination of crops by fungal species that produce mycotoxins. As shown in Chapter 2, mycotoxins are structurally diverse and are produced by a wide range of fungi able to contaminate all types of feedstuffs. Mycotoxin production will depend on the fungi species, geographical area, climate, storage conditions and many other factors that will be discussed in this book. It is worth noting that a single fungal species may synthesise different mycotoxins, and different fungal species can synthesise the same mycotoxin. More importantly, due to the variety and inclusion levels of contaminated feedstuffs, mycotoxin co-contamination is more frequent than single contamination, raising concerns about the effects of combined mycotoxins. Table 1.1 demonstrates that the dietary composition may differ across animal species and even within the same category of animals, depending on the availability of feed components. For instance, the diets for broiler chickens are designed to have comparable quantities of major nutrients even though with different ingredients, as seen by the examples of growing diets provided in Table 1.1. Nevertheless, most feed ingredients contribute to mycotoxin contamination, hence making it impractical to consistently provide a free-mycotoxin diet, or even a diet contaminated with a single mycotoxin. Figure 1.1 shows illustrative photographs of the main contaminated feedstuffs used in the preparation of the diets for Table 1.1, along with their corresponding concentrations of mycotoxins.

It is important to mention that mycotoxins are often present in diets at very low concentrations, which can be measured as parts per million (ppm or mg/kg) or parts per billion (ppb or µg/kg), except when outbreaks of highly contaminated diets are reported. Furthermore, the mycotoxins DON, T-2, ZEN, FUM, and OTA, which are routinely analysed in diets together with AFB_1, are not the only contaminants found in feed samples. The contaminated matrices often contain the modified (DON-3-Glucoside; DON-3G) and the acetylated (3+15-acetyl-DON; 3+15-ADON) forms of DON, as well as the emerging or neglected mycotoxins enniatins (ENNs), alternariol (AOH), alternariol monomethyl ether (AME) and BEA (Table 1.1 and Figure 1.1).

Chapter 2 presents comprehensive information on the modified, acetylated and emerging mycotoxins. In brief, the modified and acetylated forms of DON have the ability to enhance the damage caused by the original mycotoxin, depending on the specific animal species that are exposed to these mycotoxins (see Chapters 4 and 5 for more information). Furthermore, dietary exposure to a mixture of mycotoxins results not only in interactions with these contaminants, but also with other contaminants present in the feed, and even feed constituents (Chapter 3). For instance, the production performance of broiler chickens was impaired after exposure to a maize-based diet contaminated with approx. 3.5 mg/kg DON (Santos and Molist, 2020), but a much lower DON concentration (2.3 mg/kg) was capable of impairing performance when the broiler chickens were fed a viscous wheat-based diet rich in non-starch polysaccharides (Santos and van Eerden, 2021).

TABLE 1.1 Examples of swine and poultry diets multi-contaminated with mycotoxins

	Swine[a]	Broiler chickens[b]	Broiler chickens[c]
Main feed ingredients (%)			
Maize	40.0	3.3	45.0
Soybean meal	3.0	20.0	26.4
Wheat	12.1	55.0	21.2
Wheat middlings	7.2	–	–
Rye	–	7.5	–
Rapeseed meal	–	2.5	–
Sunflower seed meal	10.0	2.6	–
Soybean hulls	7.5	–	–
Sugar beet pulp	5.0	–	–
Linseed	3.7	–	–
Potato	2.2	–	–
Main nutrients (g/kg)			
Dry matter	878	882	878
Energy (kcal/kg)	2269	2950	3000
Crude protein	150	196	206
Crude fat	47	77	67
Crude fibre	70	28	21
Mycotoxins (µg/kg)			
DON	255	2360	3220
DON-3-Glucoside (DON-3G)	–	1670	625
3+15-Acetyl-DON (3+15-ADON)	–	–	59.5
T-2	–	–	–
ZEN	318	–	583
FUM	84	–	42
OTA	–	–	1.6
Enniatins (ENNs)	37.5	45.4	–
Alternariol (AOH)	29.5	4.2	–
Alternariol monomethyl ether (AME)	68.6	2.1	–
Beauvericin (BEA)	27.7	–	–

a Benthem de Grave et al. (2021) – lactation feed for sows.
b Santos and van Eerden (2021) – grower period feed (day 14-28).
c Santos and Molist (2020) – grower period feed (day 14-28).
Analysed mycotoxins with their respective limit of quantification (LOQ): Aflatoxin B_1 (1 µg/kg), Aflatoxin B_2 (1 µg/kg), Aflatoxin G_1 (1 µg/kg), Aflatoxin G_2 (1 µg/kg), Alternariol (2 µg/kg), Alternariol monomethyl ether (2 µg/kg), Beauvericin (5 µg/kg), Citrinin (10 µg/kg), Cytochalasin E (2 µg/kg), Deoxynivalenol (20 µg/kg), 3 + 15 Acetyl-deoxynivalenol (20 µg/kg), Deoxynivalenol-3-glucoside (20 µg/kg), Diacetoxyscirpenol (5 µg/kg), Enniatin A (5 µg/kg), Enniatin A1 (5 µg/kg), Enniatin B (5 µg/kg), Enniatin B1 (5 µg/kg), Fumonisins B1 + B2 (20 µg/kg), Moniliformin (5 µg/kg), Nivalenol (50 µg/kg), Ochratoxin A (1 µg/kg), Roquefortine C (5 µg/kg), Sterigmatocystin (1 µg/kg), T-2/HT-2 toxin (10 µg/kg), and Zearalenone (15 µg/kg).

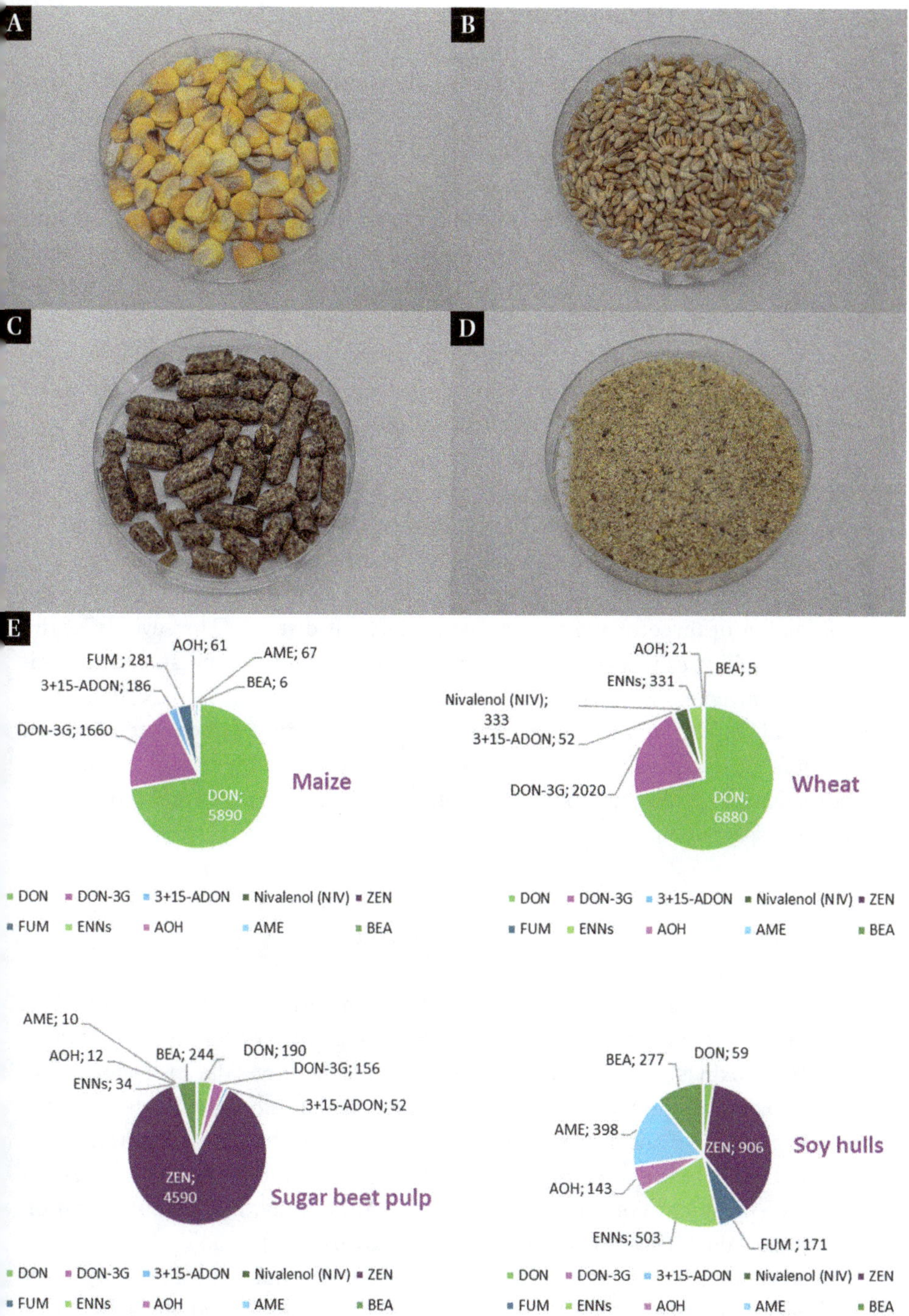

IGURE 1.1 Samples of some contaminated ingredients. A: Maize; B: Wheat; C: Sugar beet pulp; D: Soy hulls; and E: their mycotoxin levels.
PICTURES: ANJA JONGEJAN

3 Main mycotoxins affecting animal health and performance

Mycotoxins can be classified as vomitoxin, hepatotoxins, nephrotoxins, neurotoxins, etc. Some mycotoxins, such as AFs, are carcinogenic and mutagenic, whereas ZEN is an oestrogen-like mycotoxin. Exposure to high levels of these mycotoxins will result in mycotoxicosis-related symptoms. Studies focused on the single-mycotoxin effect are of great importance in order to elucidate the toxicokinetics and toxicodynamic of each mycotoxin, as well as the target organs and species. Such information provides the evidence with which to introduce regulations and recommendations on maximal levels of these contaminants in feedstuffs and final diets. However, with the available knowledge that mycotoxin co-contamination is more frequent than single contamination, it is inevitable that questions regarding the effects of combined mycotoxins will arise.

Table 1.2 provides a summary of the most studied mycotoxins and their main effects after acute exposure. However, the main determinants for the clinical effects, as described in Chapters 4-9, are species-specific and depend on the animal's susceptibility. Factors influencing vulnerability are low or high oral bioavailability, the capacity to degrade mycotoxins into less or more toxic metabolites, or the combination of mycotoxin exposure with infectious diseases. The severity of the symptoms will vary with animal species, age, sex, exposure time and concentration of the mycotoxins in the diet. Mycotoxin exposure, in most cases, will result in subclinical conditions and impaired productivity in swine, poultry, cattle and fish (Chapters 4-7), rendering mycotoxin exposure an underestimated problem. Most reports related to pets and horses focus on clinical cases of intoxication (Chapters 8 and 9, respectively).

4 Detecting and understanding mycotoxins

It was previously estimated that > 70% of feed samples are contaminated with at least two mycotoxins, and the chance of finding an ingredient contaminated with a single mycotoxin is small (Fumagalli et al., 2021). Furthermore, the level of exposure to mycotoxins is influenced by changes in diet. Specifically, when pets consume diets that contain plant-derived ingredients, their exposure to mycotoxins increases (Chapter 8). On the other hand, alternative protein sources like insect larvae in poultry diets are not considered significant sources of mycotoxin contamination because they do not seem to accumulate these toxins (Chapter 10).

Advances in detection techniques have resulted in the quantification of a greater number of mycotoxins, including their modified forms and metabolites. However, the heterogeneous distribution of mycotoxins, combined with the low concentrations (measured in μg/kg or mg/kg) commonly found in feed compounds, poses a difficulty in conducting a proper analysis. Strict sampling plans, which consist

TABLE 1.2 Main mycotoxins and their symptoms in different animal species

Mycotoxins	Main symptoms
DON	Swine: vomiting, feed refusal, intestinal damage, diarrhoea, immunosuppression and inflammation depending on the dose Poultry: gizzard erosion, intestinal damage, immunosuppression, inflammation Ruminants: gastrointestinal disorders, soft stool, immunosuppression, inflammation Fish: feed refusal Pets: vomiting, feed refusal, intestinal damage, diarrhoea, immunosuppression and inflammation depending on the dose Horses: feed refusal
T-2	Swine: vomiting, stomach ulcers, bloody diarrhoea, dermatitis Poultry: oral lesions, gizzard erosion, immunosuppression, inflammation Ruminants: gastrointestinal lesions, ruminal ulcers, intestinal bleeding, irritation in the upper respiratory tract Pets: pancytopenia in cats Horses: oral lesions, pancytopenia, dermatitis
FUM	Swine: pulmonary oedema Poultry: intestinal damage Ruminants: impaired liver function in calves Fish: immunotoxic effects Pets: vomiting, anorexia, liver failure, icterus and enlarged liver Horses: leukoencephalomalacia
ZEN	Swine: reproductive disorders, swelling and reddening of the vulva, increase in the uterus weight, prolonged cycle, abortion Poultry (broiler breeders): decreased egg production, oviduct inflammation, and reduced comb and testicular weight in males Ruminants: reproductive disorders, enlargement of vulva, uterus and mammary glands, ovarian cysts, vulvovaginitis, atrophy of testis and ovaries, abortion Fish: body deformation, intestinal damage Pets: oviduct and uterine hyperplasia and oedema, immunotoxic effects Horses: no significant adverse effects reported
AFs	Swine: liver inflammation, fibrosis, and degeneration or hepatocytes, pale liver Poultry: liver steatosis resulting in enlargement, pale liver and pale egg yolk Ruminants: very high levels needed to cause a decrease in milk yield Fish: body discoloration Pets: central nervous system disturbances, depression, stupor, seizures, liver damage, and coma Horses: liver damage
OTA	Swine: dehydration, kidney necrosis, intestinal congestion, discolouration of kidney and liver Poultry: dehydration, discolouration of kidney and liver Ruminants: dehydration in goats, which are more sensitive than cows Fish: kidney and liver damage Pets: dehydration, kidney failure, neurotoxicity and teratogenic effects, immunotoxic effects, bloody diarrhoea, dehydration, and anorexia Horses: kidney lesions, probably in older horses

TABLE 1.2 Main mycotoxins and their symptoms in different animal species (*cont.*)

Mycotoxins	Main symptoms
CIT	Swine: kidney failure Poultry: swollen liver, pancreas and kidneys, intestinal congestion Horses: kidney lesions
EAs	Swine: reduced or no milk production (agalactia) Poultry: poor feathering, respiratory difficulties, gangrene in the legs, beaks and crests, and ovarian regression Ruminants: gangrene of peripheral parts of the extremities, uterus prolapse and abortion, hyperthermic syndrome Horses: fescue foot (a gangrenous condition of the claws), prolonged gestation, abortion, thickened or retained placenta, decreased milk production or even agalactia

DON: deoxynivalenol; FUM: fumonisins; ZEN: zearalenone; AFs: aflatoxins; OTA: ochratoxin A; CIT: citrinin; EAs: ergot alkaloids
For more details and the effects of other mycotoxins, see Chapters 4 to 9.

of accurate and precise sampling, sample preparation and adequate analysis, are necessary to overcome this challenge. Chapter 11 provides a summary of these steps and emphasises practical measures to prevent the rejection of high-quality feedstuffs and diets, and the acceptance of low-quality ones. In addition to detecting mycotoxins in feedstuffs and animal final diets, biomonitoring involves the detection of mycotoxins in biological samples including tissues, fluids, urine and faeces. Determining the mycotoxin levels in biological samples is mandatory for the registration of a mycotoxin decontaminant, but biomonitoring also serves as a valuable tool for understanding the exposure of a group of animals to multiple mycotoxins. Chapter 12 details the methods of analysis and types of biological samples used for mycotoxin detection. The biological matrix is one of the factors influencing the accuracy of the analysis methods. As an example, milk samples from sows fed diets contaminated with 118 or 312 μg/kg ZEN were analysed for the presence of ZEN and its main metabolites α-zearalenol (α-ZEL) and β-zearalenol (β-ZEL) (Table 1.3; unpublished data). The analysis was performed using three different methods: enzyme-linked immunosorbent assay (ELISA) and liquid chromatography tandem mass spectrometry (LC-MS/MS) with and without a solid-phase extraction (SPE) clean-up step. The only method that properly measured the milk levels of ZEN and its metabolites was the LC-MS/MS coupled with the clean-up step using SPE. LC-MS/MS alone was unable to identify any mycotoxins, resulting in an underestimate of the level of exposure. The ELISA method was unable to differentiate between ZEN and its metabolite β-ZEL. Instead, it measured both ZEN and β-ZEL together at much greater levels, namely 2.5 to 7.5 times higher than the LC-MS/MS + SPE methodology. This led to an overestimation of the degree of exposure. In addition, the ELISA approach failed to identify α-ZEL, a compound that is 60 times more potent than ZEN. In summary, the use of appropriate sampling and analysis is crucial

TABLE 1.3 Concentrations of ZEN in diets of sows, and of ZEN and its metabolites in the milk of sows as measured by ELISA and LC-MS/MS, combined or not combined with SPE.

	Treatment 1				Treatment 2			
	ZEN	α-ZEL	β-ZEL	ZEN + β-ZEL	ZEN	α-ZEL	β-ZEL	ZEN + β-ZEL
Dietary (μg/kg)	118	N.A.	N.A.	N.A.	312	N.A.	N.A.	N.A.
Levels in milk samples (μg/ml)								
ELISA	–	–	–	0.32	–	–	–	0.60
LC-MS/MS	< LOD	< LOD	< LOD	–	< LOD	< LOD	< LOD	–
LC-MS/MS + SPE	0.12	0.12	0.0	0.12[a]	0.14	0.19	0.0	0.14[a]

<LOD: Below the limit of detection
N.A.: Not applicable
a As calculated

in determining the interventions when suspecting mycotoxin contamination, and the biological matrix must be taken into consideration when selecting a method for mycotoxin detection and quantification.

In recent years, more mycotoxins have been discovered and their chemical properties have been characterised. In general, animals are simultaneously exposed via diet to a mixture of well-described and 'novel' mycotoxins. Chronic exposure will asymptomatically impact animal growth or increase the susceptibility to disease without clear symptoms of mycotoxicosis. To better understand these risks, it is necessary to study how the single and combined mycotoxins impact the functioning of species-specific cell types and organs, considering interactions with microbiota, veterinary drugs and even dietary composition. Conducting this multitude of studies *in vivo*, especially when interactions are considered, would result in an extraordinary number of experimental animals and long-term field trials. Therefore, the use of *in vitro* models, as described in Chapters 13 and 14, will support the assessment of mycotoxin effects in systems mimicking the gastrointestinal and reproductive tracts, respectively.

5 Prevention and remediation

Mycotoxin exposure will remain an issue, and methods to decrease dietary contamination and animal risks may become a routine process in feed manufacturing. As outlined in Chapter 15, the prevention of mycotoxin contamination begins in the field and extends throughout the whole process of transportation, storage and feed manufacture. Feed producers are knowledgeable about many agricultural strategies, but they have not yet consistently included the process of cleaning cereal grains before diet preparation. By removing broken kernels, dust, straw fragments, and other outside components rich in mycotoxins, they have the potential to decrease mycotoxin contamination (see representative images in Figure 1.2). Rejects of cleaned grains for human consumption, e.g. hulls and brans, are common fibre sources used in animal diets and may contribute to mycotoxin exposure. As the global demand for food and feed does not allow for the rejection of a significant amount of feedstuffs due to mycotoxin contamination, these ingredients can still be added to animal diets while respecting inclusion levels that limit mycotoxin exposure.

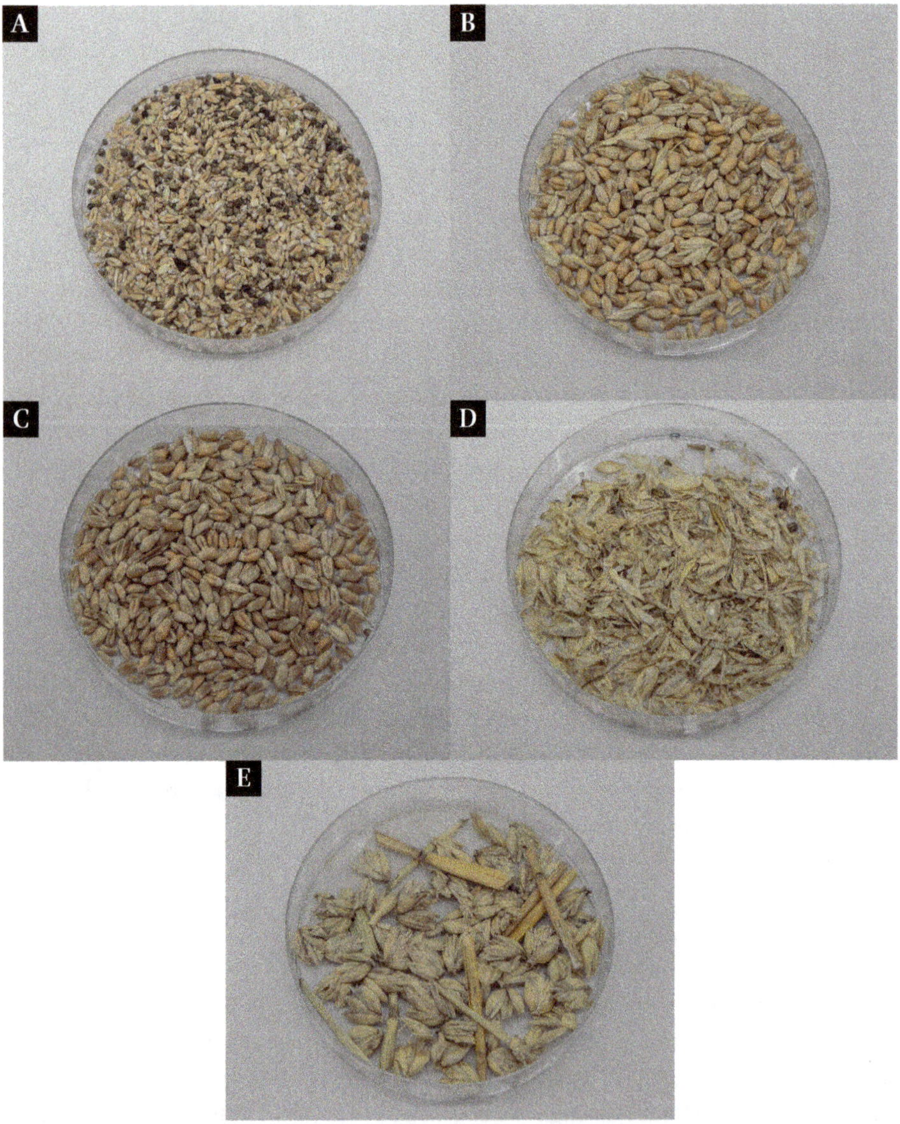

FIGURE 1.2 Samples of DON-contaminated wheat before (A and B) and after (C) mechanical cleaning. Dust (D) and straw pieces (E) were mechanically removed.
PICTURES: ANJA JONGEJAN

Prevention will not stop contamination entirely, and the limited sources of ingredients for dietary production minimises the selection of feedstuffs. However, dietary supplementation with mycotoxin decontaminants and/or additives able to protect the animal during mycotoxin exposure seems to be a promising alternative (Chapter 16). Chapter 17 gives an overview of the toxicological values determined for some mycotoxins in diverse animal species based on the European Food Safety Authority (EFSA) scientific opinions. Meeting strict maximum levels of mycotoxins in animal diets will be challenging in the coming years because climate change is conducive to the proliferation of mycotoxin-producing fungi species (Chapter 18). Consequently, the prevalence of animal diets multi-contaminated with mycotoxins is likely to be far greater. Mycotoxins cannot be fully removed from the final diet, but control measures such as dietary supplementation with feed additives can be applied to minimise their harmful effects.

This book was prepared after many years of direct contact with farmers, nutritionists, veterinarians and researchers involved with the mycotoxin issue. As mycotoxins are a worldwide problem, the book was written by experts from different areas of the globe. The bridge between fundamental and applied research was achieved by keeping each chapter focused on field questions regarding the impact of mycotoxins on the health and welfare of swine, poultry, ruminants (with special attention to cattle), fish, pets and horses, including the negative impact of mycotoxins on the productivity of livestock. Human risks are mentioned when food exposure via animal products may occur, i.e. the transfer of mycotoxins or their metabolites to animal products such as meat, eggs and milk. Moreover, practical solutions are outlined to help the reader to formulate management processes and make pragmatic decisions on the use of feed ingredients.

References

Benthem de Grave, X., Saltzmann, J., Laurain, J., Rodriguez, M.A., Molist, F., Dänicke, S. and Santos, R.R., 2021. Transmission of zearalenone, deoxynivalenol, and their derivatives from sows to piglets during lactation. Toxins 13: 37. https://doi.org/10.3390/toxins13010037.

Bianchini, A. and Bullerman, L.B., 2014. Good food-processing techniques: stability of mycotoxins in processed maize-based foods. Mycotoxin Reduction in Grain Chains, 7. https://doi.org/10.1002/9781118832790.ch7.

Bullerman, L.B., 2003. Spoilage|Fungi in food – an overview. Encyclopedia of Food Sciences and Nutrition. 5511-5522. https://doi.org/10.1016/B0-12-227055-X/01129-9.

Caloni, F., Fossati, P., Anadón, A. and Bertero, A., 2020. Beauvericin: the beauty and the beast. Environmental Toxicology and Pharmacology. 75: 103319. https://doi.org/10.1016/j.etap.2020.103349.

Fujimoto, H. 2011. Yeasts and molds: mycotoxins: classification, occurrence and determination. 792–800. Fuquay, J.W. (ed.) Encyclopedia of Dairy Sciences. 2nd. San Diego, CA, USA. Academic Press.

Fumagalli, F., Ottoboni, M., Pinotti, L. and Cheli, F., 2021. Integrated mycotoxin management system in the feed supply chain: innovative approaches. Toxins 13: 572. https://doi.org/10.3390/toxins13080572.

Groopman, J.D. and Kensler, T.W., 2005. Role of metabolism and viruses in aflatoxin-induced liver cancer. Toxicology and Applied Pharmacology 206: 131-137. https://doi.org/10.1016/j.taap.2004.09.020.

International Feed Industry Federation (IFIF), 2023. Annual report 2022/2023. https://annualreport.ifif.org.

Kossel, A., 1891. Ueber die chemische zusammensetzung der zelle. Du. BoisReymond's Archiv für Anatomie und Physiologie: Physiologische Abteilung 278: 181–186.

Mostrom, M., 2016. Mycotoxins: Toxicology. Encyclopedia of Food and Health. 2016: 43-48. https://doi.org/10.1016/B978-0-12-384947-2.00480-3.

Peraica, M., 2016. Chapter 5 – Mycotoxicosis. Environmental mycology in public health. In: Fungi and Mycotoxins Risk Assessment and Management, Academic Press 2016: 45-49.

Pitt, J.I. and Miller, J.D., 2017. A concise history of mycotoxin research. Journal of Agricultural and Food Chemistry 65: 7021-7033. https://doi.org/10.1021/acs.jafc.6b04494.

Pitt, J.I., 2013. Chapter 30 – Mycotoxins. In: Morrys, G. and Potter, M.E. (eds) Foodborne Infections and Intoxicants (4th ed) Food Science and Technology 2013: 409-418. https://doi.org/10.1016/B978-0-12-416041-5.00030-5.

Santos, R.R. 2021. Risk management systems for prevention and control of contaminants in animal feed. In: Developing Animal Feed Products, Burleigh Dodds, UK. https://doi.org/10.19103/AS.2020.0083.14.

Santos, R.R. and Molist, F., 2020. Effect of different dietary levels of corn naturally contaminated with DON and its derivatives 3+15 Ac-DON and DON-3-glucoside on the performance of broilers. Heliyon 6: e05257. https://doi.org/10.1016/j.heliyon.2020.e05257.

Santos, R.R. and van Eerden, E., 2021. Impaired performance of broiler chickens fed diets naturally contaminated with moderate levels of deoxynivalenol. Toxins 13: 170. https://doi.org/10.3390/toxins13020170.

Statista, 2023. Global pet food sales worldwide from 2010 to 2023. https://www.statista.com/statistics/253953/global-pet-food-sales.

Thirumurugan, D., Cholarajan, A., Raja, S.S.S. and Vijayakumar, R., 2018. An introductory chapter: secondary metabolites. in Secondary Metabolites - Sources and Applications, InTech. https://doi.org/10.5772/intechopen.79766.

van Dongen, P.W.J. and de Groot A., 1995. History of ergot alkaloids from ergotism to ergometrine. European Journal of Obstetrics Gynaecology and Reproductive Biology 60: 109–116. https://doi.org/10.1016/0028-2243(95)02104-z.

CHAPTER 2

Mycotoxins produced in the field

L.A. Rothmann[1], T.V. Masisi[1], B. Flett[2] and B. Janse van Rensburg[2]

[1]University of the Free State, Plant Sciences, Plant Pathology, Bloemfontein, 9301, South Africa
[2]Agricultural Research Institute – Grain Crops – Crop Protection (ARC-GC), Potchefstroom, 2520, South Africa

1 Introduction

Mycotoxins are secondary metabolites that are structurally diverse and produced by a wide range of filamentous fungi (Bullerman and Bianchini, 2014). These toxins may be responsible for defence and survival of the fungi (Thirumurugan et al., 2018). The most common mycotoxins that have been linked to toxic syndromes in livestock include aflatoxins (AFs), alternariol (AOH), beauvericin (BEA), citrinin (CIT), diplodiatoxin, enniatins (ENNs), ergot alkalids (EAs), fumonisins (FUM), ochratoxins, tenuazonic acid (TeA), the trichothecenes deoxynivalenol (DON) and T-2 toxin, and zearaleone (ZEN), which have been implicated in acute and/or chronic outbreaks (Stroka and Gonçalves, 2019).

It is not possible to avoid mycotoxin exposure, because the growth of fungi and subsequent toxin production can occur on several levels of the feed chain: (1) during crop production in the field, (2) during grain storage, (3) during transportation of feeds or crop grains and (4) during feed processing (Mostrom, 2016). In addition, mycotoxins are characterized by their thermal stability, which means that it is impossible to remove them entirely from the animal feed chain (Kuiper-Goodman et al., 1987). The present chapter focuses on the characterization of mycotoxins that play a major role in animal feed contamination.

2 Trichothecenes

There are 200 mycotoxins in the trichothecenes class that are chemically related (McCormick et al., 2011). They are regarded as the largest mycotoxin class with structurally related tricyclic sesquiterpenoids (C3, C4, C7, C8, and C15; Cole et al., 2003; Kumar et al., 2008). Based on these structures, trichothecenes are divided into four distinct groups: type A, B, C and D. Because of their toxic nature and adverse effects on the economy, type A and B mycotoxins have drawn the greatest attention (Polak-Śliwińska and Paszczyk, 2021). The unique distinction between type A and B

 | DOI: 10.1163/9789004712720_3

lies in the residue located at C-8 position (C-8): type B have a keto (carbonyl) group at C-8, whereas type A typically have an esterified hydroxyl group in place of the keto group. Representatives of type A include T-2, HT-2, diacetoxyscirpenol (DAS), and NX-toxin whereas type B are DON and nivalenol (NIV) and their acetylated derivatives (i.e., chemotypes), 3-acetyl-deoxynivalenol (3-ADON) and 15-acetyl-deoxynivalenol (15-ADON), similarly 4-acetylnivalenol (4-ANIV) and fusarenone X, respectively (Boutigny et al., 2011).

2.1 *Causative agents*

Fusarium species are most frequently associated with the production of type A and type B trichothecenes; this fungal genus contains multiple individuals that cause a wide range of plant diseases (Saleh et al., 2015; Yang et al., 2018). Not all *Fusaria* produce trichothecenes, because production is limited to species capable of causing fusarium head blight (FHB) of wheat (Figure 2.1A), or fusarium ear blight/rot (FER) of maize (Figure 2.1B). *Fusarium culmorum, F. cerealis, F. graminearum* species complex (*Fg*SC) and *F. pseudograminearum* are known to produce type B mycotoxins (Beukes et al., 2017), while *F. langsethiae, F. poae,* and *F. sporotrichoides,* and *F. sambucinum* are known to produce type A mycotoxins (Ellner, 2002; Foroud et al., 2019). Prior to 2000, species within the *Fg*SC were believed to represent a single 'worldwide' species; however, due to advances in molecular characterization techniques seven lineages within *Fg*SC were identified by O'Donnell et al. (2000). To date, through the coupling of multi-locus and genealogical concordance of phy-

FIGURE 2.1 (A) Fusarium Head Blight of wheat (Source: Diana Mngomezulu (BSc. Agric Plant Pathology), University of the Free State, (B) Fusarium Ear Rot of maize (Source: Ayesha Shaikh (MSc. Agric Plant pathology), Stellenbosch University).

logenetic species recognition 16 distinct phylogenetic cryptic species of the *Fg*SC have been identified (O'Donnell et al., 2004; Starkey et al., 2007; O'Donnell et al., 2008; Yli-Mattila et al., 2009). The 16 *Fg*SC members have been reported to display differences in the mycotoxin chemotype they produce (Beukes et al., 2017; Hao et al., 2017). Understanding the host-pathogen interaction is thus critical in determining the mycotoxin of concern, as well as the mitigation thereof.

2.2 *Susceptible host plants*

Small grain crops, such as barley, oats, rye and wheat (Bennett and Klich, 2003), as well as maize and sorghum (Chen et al., 2020) are particularly prone to head/ear/panicle rot. All these crops are used in animal feed for their balanced nutritional value, hence animals that consume these infested crops can be subjected to the adverse effects of fungal secondary metabolites. Multiple authors and authorities concur that animals are primarily exposed to DON and NIV through the consumption of cereal grains (EFSA, 2013; Bennett and Klich, 2003). Despite the widespread presence of type B toxins, type A mycotoxin are the most toxic trichothecenes and less frequently found than the other trichothecenes in cereal grain. *F. culmorum* and *Fg*SC members are the predominant causative species of FHB (Foroud et al., 2019), infecting the inflorescence structure of small grain crops. Additionally, FHB associated with maize is known as *Fusarium* Ear Blight/Rot (FER), also caused by members of the *Fg*SC and other non-producing trichothecene species (Mesterházy et al., 2012).

2.3 *Plant disease epidemiology*

Animal feed can be contaminated by *Fusarium* species under conducive environmental conditions during the season and post-harvest, with disastrous consequences (Lahouar et al., 2015). *F. culmorum* and *F. graminearum* are regarded to be the most significant contributors of all the *Fusaria* associated with FHB and FER, and propagules associated with survival are pathogen dependent (Osborne and Stein, 2007). Of all the cereals, oats are the most susceptible to type A producing *Fusaria* (Edwards, 2009). *F. poae, F. langsethiae,* and *F. sporotrichioides* do not cause visible disease symptoms in infected oats. However, grain that appears to be healthy might really contain a lot of mycotoxins, especially T-2 and HT-2 (Kahla et al., 2023). *F. langsethiae* has been identified as the primary generator of T-2 and HT-2 trichothecenes in the cultivation of cereals in Europe (Edwards, 2009; Edwards et al., 2012). Water activity (a_w) of 0.97 and 25°C temperature were found to be the ideal conditions for *F. langsethiae's* colonisation of harvested oats and generation of the HT-2 and T-2 toxins. The majority of *Fusarium* species produce mycotoxins under these typical conditions i.e., *F. culmorum, F. graminearum* and *F. poae* (Imathiu et al., 2013). Warmer temperature (~30 °C), relative humidity (80-98%) and rainfall are essential components for infection by conidia of both *F. graminearum* and *F. culmorum* (Dufault et al., 2006). However, infection at lower temperatures (15 °C) has been observed for ascospores and conidia of *F. graminearum* and *F. culmorum*, respectively (Parry et al., 1995; Doohan et al., 2003; Magan et al., 2006). Mycotoxin production has been reported to have different climatic requirements compared to their fungal producers. DON and NIV production requires high relative humidity (>90%) and lower temperatures, 20 to 25 °C and 20 °C, respectively (Lori et al., 1999; Llorens et al., 2004).

2.4 *Possible conjugation and modification in host plants*

The mycotoxins that fungi make when they infect plants are altered by enzymes found in plants and frequently coupled to more polar compounds, including sugars. These types of toxins are known as masked mycotoxins because they are frequently less dangerous metabolites that are linked to macromolecules or stored in the vacuole of the plant. They pose a serious threat to feed safety due to the difficulty of detection and the possibility that they will revert to their natural forms during the digestion of feed (Berthiller et al., 2015; McCormick et al., 2015). Diverse organisms have exhibited a variety of trichothecene detoxifying methods. A good summary of recognised *in planta* detoxifying techniques is given by Boutigny et al. (2008). One of them is the uridine 5′-diphospho (UDP)-glucuronosyltransferase enzyme, which glycolyses DON at C-3 position to produce DON-3-glucoside (DON-3G) (Poppenberger et al., 2003). Altering the C-3 hydroxyl groups of DON and associated trichothecenes tends to lessen or completely reduce their toxicity. The gene TRI101 encodes for the enzyme acetyltransferase that transforms the hydroxyl group at C-3 (Okubara et al., 2002).

3 Zearalenone

Zearalenone (ZEN), also known as a mycoestrogen, is a phenolic resorcylic acid lactone mycotoxin produced by various *Fusarium* spp. Like trichothecenes (DON and NIV), ZEN is a parent compound that may undergo modifications that can lead to the development of derivatives, α- and β-zearalenol (α- and β-ZEL), which may also occur concurrently in cereals or feed (Liu and Applegate, 2020). ZEN is a non-steroidal oestrogenic mycotoxin mimicking the female hormone oestrogen. Furthermore, it is structurally very similar to 17 β-oestradiol oestrogen and can bind to oestrogen receptors due to this structural resemblance. Consequently, ZEN-contaminated feed is an implied threat to the reproductive systems of livestock (Reddy et al., 2010).

3.1 *Causative agents*

ZEN is mainly produced by *Fusarium* genera, particularly trichothecene producers (Glenn, 2007); these include *F. cerealis, F. culmorum, F. equiseti* and *F. graminearum* Species Complex (*Fg*SC) (Kumar et al., 2008; Beukes et al., 2017). Thus, the species that produce ZEN are the same ones that are able to produce the trichothecenes

DON and NIV. Among the species mentioned, *F. culmorum* and *Fg*SC are the predominant producers of ZEN (Richard, 2007; Yazar and Omurtag, 2008).

3.2 *Susceptible host plants*

ZEN occurs naturally in cereal and grain crops, notably maize, and is among the three top mycotoxins detected in animal feed. Furthermore, ZEN derivatives α- and β-ZEL have also been discovered in maize silage (Schollenberger et al., 2006). To a lesser degree, barley, oats, rice (*Oryza sativa*), sorghum, soybeans, sugar beets and wheat contaminated products have been reported (Bosch and Mirocha, 1992; Zhang et al., 2018).

3.3 *Plant disease epidemiology*

It is essential to note that since ZEN is produced by the same species that produces trichothecenes it is likely to co-occur with trichothecene toxins (Richard, 2007). ZEN is frequently known to co-occur with DON (Grenier and Applegate, 2013). Susceptible host infection occurs during the pre-harvest stage by *Fusarium* spp. Furthermore, this period is also known for the greatest occurrence of toxin accumulation (CCFAC, 2000). Infection during anthesis is when ZEN may build up in the hosts. Bily et al. (2004) argue that ZEN accumulates in crops post-infection. However, these occurrences are indicative that ZEN accumulates in the cereals before harvesting and storage. Additional reports indicate that ZEN may emerge under improper or poor storage conditions which can result in its accumulation (Kuiper-Goodman et al., 1987). This suggests that fungi responsible for ZEN production may have different environmental conditions that favour their development. It should be noted that infected cereal with no detected ZEN may be harvested, but due to conducive storage conditions ZEN accumulation in the feed may be triggered (Zinedine et al., 2007a).

3.4 *Possible conjugation and modification in host plants*

ZEN can be metabolised by plants and fungi by binding to polar residues such as glucose or sulphate. Masked mycotoxins are plant-produced mycotoxin conjugates. ZEN-16G (zearalenone-16-glucoside) and ZEN-14G (zearalenone-14-glucoside) are two examples. A ZEN derivative with a fungal source is ZEN-14S (zearalenone-14-sulfate) (Berthiller et al., 2013; Rychlik et al., 2014). In 2002, a small-scale assessment on samples of wheat grain revealed the first instance of ZEN-14G in naturally infected wheat (Schneweis et al., 2002), where wheat grain had up to 30% more ZEN-14G than ZEN. Wheat grain had up to 30% more ZEN-14G than ZEN. Similarly, the presence of ZEN-14G was discovered in the majority of feed samples, oatmeal samples, etc. (Schneweis et al., 2002). A gene (HvUGT14077) in barley that may metabolise ZEN to ZEN-14G and ZEN-16G was discovered. However, ZEN-16G has not been reported in feed or food (Kovalsky Paris et al., 2014). ZEN-14S was isolated from *F. graminearum* and along with additional *Fusarium* strains. ZEN-14S and also ZEN-14G contaminate cereal, soybean, maize, and their by-products, and consequently animal feed (Kovalsky Paris et al., 2014; Plasencia and Mirocha, 1991).

4 Fumonisins

There are over 15 homologous fumonisins (FUM), which are classified into four distinct groups, A, B, C and P (FA, FB, FC, and FP; Braun and Wink, 2018). Fumonisin B1 (FB_1) comprises the majority of all FUM found in feedstuffs (Gao et al., 2023); it is the primary FUM along with FB_2 and FB_3 and is generated in an approximate 6: 3: 1 ratio (Gil-Serna et al., 2014). FB_3, FB_4, FA_1, and FA_2 are found in trace amounts and do not currently seem to be significant toxins (Haschek and Voss, 2013). This can be explained by the fact that the free hydroxyl group presence and position distinguishes the various FUM, which are polyhydroxyl alkylamines esterified with two carbon acids. While the B series of fumonisins has a free amine, the A series has an acetylated amino group on the amino group (Gil-Serna et al., 2014). Geographically, the quantities of FUM vary frequently between adjacent areas (Haschek and Voss, 2013).

4.1 *Causative agents*

Fusarium verticillioides (previously *F. moniliforme*), *F. proliferatum*, *F. anthophilum*, *F. dlamini*, *F. napiforme*, *F. nygamai*, and *F. oxysporum* produce FUM. Other fungal species, such as *Aspergillus niger* and *Penicillium*, have also been observed to produce FUM (Haschek and Voss, 2013; Braun and Wink, 2018). In comparison to *Aspergillus* and *Penicillium*, *Fusarium* (particularly *F. verticillioides*) is the most common genus for contamination of feed and food (Cao et al., 2013).

4.2 *Susceptible host plants*

Cereals, i.e., barley, maize, millet, oat, rice, rye, and wheat, are among the crops that are most frequently contaminated (Kamle et al., 2019). However, FB_1 is associated with maize and poses a risk to animal health (Haschek and Voss, 2013). Infections of soybean colonised by *F. verticillioides* may occur, although infections are relatively uncommon (0-15%) compared to maize commodities (85%); consequently, soybean products rarely contain FUM (Rodrigues and Naehrer, 2012).

4.3 *Plant disease epidemiology*

Weather conditions, insect damage, and plant characteristics affect maize infection by *F. verticillioides* and FUM accumulation. Warm, dry conditions are ideal for *F. verticillioides* infection of the ear throughout the flowering stage, but warm, moist conditions after silking have also been observed to favour the development of the disease (Munkvold, 2003). Reduced rainfall and maximum temperatures between 30 and 35 °C during flowering have been proven to promote the development of

FIGURE 2.2 Maize infected with *Fusarium verticillioides*.

diseases caused by *F. verticillioides* (Czembor et al., 2019). Variation in accumulation of FUM is associated with a variety of *F. proliferatum* strains and their response to a_w and temperature, where lower levels of FUM were detected at 0.99 a_w and 15 °C and greater accumulation at 25 °C at the same a_w level (Cendoya et al., 2018).

4.4 *Possible conjugation and modification in host plants*

FB_1, FB_2, and FB_3 are the most prevalent, although FB_1 has the widest distribution and greatest toxicity. FUM are easily detectable in their natural form through standard analysis; however, detection in their modified/degraded form is more challenging (Humpf and Voss, 2004). A decline of their biological function, e.g., extrusion and nixtamalisation, may accompany the degradation of natural mycotoxins into modified forms, i.e., hydrolysed fumonisins, reducing the toxicity. The bacterium *Sphingomonas* sp. was discovered to degrade FUM, through the enzymatic activity of aminotransferase and carboxylesterase. Plant microbiomes can metabolise FB_1 toxin into hydrolised fumonisin B_1 (HFB_1) form.

5 Aflatoxins

Aflatoxins (AFs) are fungal secondary metabolites which causes aflatoxicosis in humans and animals, particularly livestock, i.e., cattle, pigs, and poultry (Pleadin et al., 2015). AFs found in mouldy foods and animal feeds can enter the food chain, contaminating human food through the consumption of affected animal products (Shukla et al., 2016). Fungal contamination during crop growth and, more significantly, during post-harvest storage leads to the presence of AFs in both human and animal food and feed. Due to their well-documented toxicity and carcinogenic nature, it is essential to keep aflatoxin exposure from food and feed as low as possible (Early, 2009; van Duijin, 2014).

Several types of AFs are produced; the main AFs include B_1, B_2, G_1, and G_2, along with two other noteworthy types, M_1 and M_2. The M aflatoxins were initially identified in the milk of lactating animals that had ingested aflatoxin-contaminated feed; consequently, they are labelled with an 'M' (Bullerma, 2003). Among the AFs, AFB_1 is the most prevalent in food and is recognised as one of the most toxic and carcinogenic variants (Pleadin et al., 2015).

5.1 *Causative agents*

AFs are secondary metabolites derived from fungal species of the genus *Aspergillus*. AFB_1 is produced by species such as *A. arachidicola, A. bombycis, A. flavus, A. parasiticus, A. nomius, A. minisclerotigenes, A. ochraceoroseus, A. pseudotamarii, and A. rambellii.* Similarly, AFB_2 is produced by many of the same species, including *A. arachidicola, A. flavus, A. minisclerotigenes, A. nomius, and A. parasiticus.* AFG_1 and AFG_2 are synthesised by common soil fungi such as *A. arachidicola, A. bombycis, A. flavus, A. nomius,* and *A. parasiticus*. Additionally, AFM_1 and AFM_2 are derivatives of AFB_1 and AFB_2, produced by *A. flavus* and *A. parasiticus* (Bbosa et al., 2013; Feddern et al., 2013; Popescu et al., 2022: the derivation of these toxins will be discussed in detail in the section on Toxin Conjugation).

5.2 *Susceptible host plants*

AFs are particularly associated with groundnuts, tree nuts (almonds, Brazil nuts, pecans, pistachios and walnuts), crude vegetable oils, figs, spices, and cocoa beans. Among various agricultural products, cottonseed, copra, maize, and rice, are especially prone to aflatoxin contamination. The main types of AFs that are present in oilseeds and vegetable oils include AFB_1, AFG_1, AFB_2, and AFG_2, with AFB_1 and AFG_1 being the most prevalent. Typically, less than 10% of the AFs in seeds, peanuts, and copra are transferred to crude oil during the pressing and extraction process, as AFs are largely protein-bound (van Duijin, 2014; Wang et al., 2022; Molnár et al., 2023).

5.3 *Plant disease epidemiology*

Aflatoxin contamination is most prevalent in Africa and Asia, where climatic conditions favour the growth of aflatoxin-producing fungi both in-field and during storage. The contamination of staple foods, particularly cereals, has attracted significant attention due to its acute and chronic effects on humans and animals. Maize, increasingly used for animal feed due to its lower price, digestibility, palatability, and its ability to improve milk yield and protein content, is especially vulnerable to contamination. Climate change, with its extreme weather patterns, exacerbates this problem for producers every year (Klopfenstein et al., 2013; Heuzé et al., 2017; Kolawole et al., 2022; Popescu et al., 2022; Shabeer et al., 2022).

Aspergillus species, especially *A. flavus*, are the most common contaminants of maize, producing AFB_1, which causes significant economic damage during maize production and storage (Molnár et al., 2023). *Aspergillus* survives in soil and plant residues, and its conidia can be airborne and spread by insects to infect host plants (Payne and Widstrom, 1992). Maize cobs are infected during anthesis through the silk or insect damage and colonisation continues throughout the growing season (Payne and Widstrom, 1992; Varga et al., 2004; Isakeit et al., 2016). Dry periods with high temperatures (≥ 30 °C) and low relative humidity (22 to 24%) promote AFs production by *A. flavus*. Under favourable environmental conditions, such as heat stress and moisture deficit, *A. flavus* outcompetes other fungi and continues to produce toxins even after harvest (Battilani et al., 2008; Molnár et al., 2023). During storage, mould growth and toxin production are influenced by air temperature, humidity and a_w levels in grains. Under favourable environmental conditions, such as heat stress and moisture deficit, *A. flavus* outcompetes other fungi and continues to produce toxins even after harvest (Molnár et al., 2023). Research has shown that *A. flavus* thrives best at 30°C and with lower a_w levels (0.87-0.98 a_w), allowing it to dominate even under conditions that typically inhibit other moulds (Giorni et al., 2016).

6 Ochratoxin A

The chemical structure contributes significantly to the biological ability, toxicity and stability of ochratoxins. Fifteen naturally occurring ochratoxins and derivatives exist, of which ochratoxin A (OTA) and B (OTB) are coupled to phenylalanine, although OTA is the chlorinated derivative of OTB. Ochratoxin C (OTC) is also coupled with phenylalanine, although it is the ethyl ester analogue of OTA (Huff and Hamilton, 1979; El Khoury and Atoui, 2010). In the 1990s OTA received notable attention (i.e., seven and 13 publications on detection methods and nephrotoxicity of OTA, respectively) due to the increased awareness of hazards posed to human and livestock health, which subsequently shaped research, monitoring and legislation (Malir et al., 2016).

6.1 *Causative agents*

OTA is the most prevalent ochratoxin, primarily produced by species within the *Aspergillus* and *Penicillium* genera. OTA was first identified in association with sterilised maize meal inoculated with *A. ochraceus* (van der Merwe et al., 1965), resulting in the affiliated name, ochratoxin (Huff and Hamilton, 1979). Subsequently, OTA was isolated from culturing *Penicillium viridicatum* and detected in whole wheat bread slices inoculated with *P. viridicatum* (van Walbeek et al., 1969). The first report of naturally occurring OTA detection was associated with maize; however, *A. ochraceus* was not successfully isolated from the sample (Shotwell et al., 1969). On another occasion, OTA was detected in wheat naturally infected with *P. viridicatum* (Scott et al., 1970). Subsequently, isolates of *P. viridicatum* were found to be unable to produce OTA, and confirmation of *P. verrucosum* as a new species with the ability to produce OTA rendered this the more likely suspect of earlier reports of OTA production (Pitt, 1987). Special attention is required when investigating causal organisms producing OTA particularly due to the revisions of fungal taxonomy occurring within the *Aspergillus* and *Penicillium* genera, known to co-occur (Malir et al., 2016). *Aspergilli* in the *Circumdati, Flavi* and *Nigri* sections are known to produce OTA, specifically *A. auricomus* (re-identified as *Neopetromyces muricatus*), *A. cretensis, A. flocculosus, A. pseudoelegans, A. roseoglobulosus, A. sclerotiorum, A. sulphureus, A. steynii, A. ochraceus,* and *A. westerdijkiae,* the latter three species known as major causative agents (Frisvad et al., 2004). Previously, *A. alliaceus, A. sulphureus, A. ostianus, A. petrakii* and *A. melleus* were reported to produce OTA (Huff and Hamilton, 1979); however, studies have reported that the latter three species were unable to produce ochratoxins (Frisvad et al., 2004), though trace amounts were detected in authentic strains (Ciegler, 1972; Hesseltine et al., 1972). Although the production of OTA by *A. niger* is rare, it should not be disregarded as high concentrations were detected in Argentinian animal feed (Malir et al., 2016; Dalcero et al., 2002).

6.2 *Susceptible host plants*

An indication of the diversity of OTA susceptible hosts is evident in the European Union (EU) amended regulations of OTA maximum allowable limits in foodstuffs and derivatives, e.g., unprocessed and processed cereal grains (i.e., barley, corn/maize, oats, rye, and wheat), cocoa powder, dried fruit, grape juice and wine, liquorice, olives, pistachios, green and roasted coffee, rice, spices and a variety of seed and soybean (Zinedine et al., 2007b; Duarte et al., 2009; EFSA, 2020). OTA has been detected in maize destined for animal feed, although specific fungi associated with it was not reported (Speijers and van Egmond, 1993; Scudamore and Patel, 2000; Dalcero et al., 2002). Stored winter wheat infected with *P. verrucosum* resulted in the natural contamination of OTA, ranging from 1.2 to 44.9 μg/kg (Limay-Rios et al., 2017). Wheat and rye are common susceptible hosts of *P. verrucosum* (Elmholt, 2003).

6.3 *Plant disease epidemiology*

The research focus of OTA associated with *P. verrucosum* infections has been post-harvest and in storage (Nada et al., 2022); however, Elmholt (2003) states there is a need to investigate the pre-harvest ecology of this pathogen. *P. verrucosum* is suspected to infect and produce OTA during grain development in-field, under cool temperate conditions. Low frequencies of *P. verrucosum* were detected in Danish soils, where conidia were detected in soil of an 18-month field experiment. The survival of conidia in soil suggests the population may establish on soil organic matter and, consequently, splash dispersal could contribute to infection (Elmholt and Hestbjerg, 1999; Elmholt, 2003). Additionally, retained seed is considered an inoculum risk for *P. verrucosum* associated OTA in cropping systems, as contaminated seed increased the pathogen population and likelihood of pre-harvest infections (Elmholt, 2003). In contrast, commodities under storage are susceptible to contamination when conditions are favourable i.e., high moisture with reduced temperatures (~10 °C; Miller, 2008). As a result, mitigation of in-field infections is critical to reducing the likelihood of OTA accumulation during storage (Elmholt, 2003). *Aspergillus ochraceus* and *A. niger* have been isolated from preharvest Algerian wheat indicating a possible risk of in-field OTA accumulation (Riba et al., 2008). However, *Aspergillus* spp. are considered storage fungi, as a result of which signs of compromised grain development and pre-harvest factors are not commonly investigated (Nada et al., 2022).

7 Ergot alkaloids

Ergot alkaloids (EAs), also known as ergolines, are associated with the earliest recordings of mycotoxicosis in humans and animals (IPCS and WHO, 1990). Despite this, in 2007, ergot was declared 'as [an] undesirable substance in animal feed', by the European Food Safety Authority (EFSA), Panel on Contaminants in Food Chain (EFSA, 2012). In 1990, more than 40 EAs had been isolated from *Claviceps sclerotia*, where three classes are proposed according to their derivatives: i) lysergic acid, ii) isolysergic acid and iii) dimethylergolin (IPCS and WHO, 1990). Subsequently, more than 80 EAs have been classified and attributed to fungi outside of *Claviceps* genus and even some species of higher plants (Agriopoulou, 2021). However, in 2014, EAs were regrouped into i) clavines, ii) simple amides of lysergic acid, and iii) ergopeptines (Panaccione et al., 2014). Advances in mycotoxin detection, classification and quantification have resulted in a greater understanding of sources, effects and the impact.

7.1 *Causative agents*

The fungal genus *Claviceps* is known to produce specialised survival structures, i.e., sclerotia, commonly known as ergot, hence the common connection with ergotism (IPCS and WHO, 1990). However, the association between ergot and ergot alkaloid induced mycotoxicosis (i.e., ergotism) should not be misconstrued as only occurring with species within the *Claviceps* genus, within the *Hypocreales* order (EFSA, 2012). EAs can also be produced by related fungi within the *Clavicipitaceae* family, namely, *Balansia, Epichloë, Neotyphodium* and *Periglandula*. Several fungal species within the *Eurotiales* order (EFSA, 2012), e.g., *A. fumigatus* (Panaccione et al., 2014; Reddy et al., 2020) and *P. chennesinum, P. concavorugulosum* and *P. sizovae* (IPCS and WHO, 1990) also have the ability to produce EAs. Additionally, higher plants, *Rivea corymbosa, Ipomoea violacea, I. argyrophylla, I. hildebrandtii, I. tricolor* have also been reported to synthesise EAs (Van Rensburg and Altenkirk, 1974; Kozlovsky and Reshetilova, 1984).

7.2 *Susceptible host plants*

Ergot production in plants occurs from two distinct pathways: i) pathogenic associations between host-pathogen, and ii) endophytic non-pathogenic mutualistic symbiotic relationships. The wide host range, distribution of host species and various production systems, as well as undomesticated host presence in natural environments precipitates the need to understand associated diseases and concomitant mycotoxin contamination under diverse conditions caused by EA producing fungi. Cereal grains and grasses susceptible to *Claviceps* spp. include barley, Kentucky blue grass, oats, pearl millet, rye, sorghum, triticale, wheat and perennial ryegrass.

Tall fescue (*Festuca arundinacea* Scribe = *F. elatior* L.; = *Lolium arundinaceum*) endophytically colonised by the non-pathogenic fungus *Neotyphodium coenophialum* produces ergopeptine alkaloids, e.g., ergovalines. *N. coenophialum* was originally described as *Epichloë typhina*, an *Acremonium* spp., referred to as *Sphacelia typhina* in the conidial state (Lyons et al., 4986). Approximately 90% of pastures in the USA are associated with *Neotyphodium* infected fescue, resulting in significant economic risk and losses associated with raising livestock (EFSA, 2005). Botha et al. (2004) published the first report of fescue foot in South African cattle resulting from both *Neotyphodium* colonised fescue and *C. purpurea* sclerotia contaminated fescue tufts. Episodes of mycotoxicosis associated with *Neotyphodium* toxins in European countries are limited and only reported on certain forage grasses. Additionally, *Neotyphodium lolii* have been reported to colonise *Lolium* host grasses, i.e., perennial ryegrasses, used in the livestock raising pastures of Australia and New Zealand (Aotearoa) (Card et al., 2014; Joost, 1995; Flythe et al., 2016). The contribution of toxicity by EAs to livestock has not been considered as much as that associated with lolitrems, a group of indole-diterpene alkaloids in perennial ryegrasses (Philippe, 2016). Lolitrem B (the most important) and lolitrem A and prasine are mycotoxins of *Epichloë festucae* var. *lolii*, which was previously described as *Neotyphodium lolii*. Other mycotoxins are also related, such as lolicine A and B, lolitriol, lolitrem N, as well as 31-epilolitrem N and 31-epilolitrem F and various aflatrems (Munday-Finch et al., 1998).

7.3 *Plant disease epidemiology*

Susceptible cereal crops and perennial grasses grown for feed and forage purposes may experience infections from EAs producing fungi, i.e., pathogenic and non-pathogenic, under conducive conditions during the season. Notably, both *Claviceps* spp. and *Neotyphodium* spp. are associated with colonising host crops during the growing season. Understanding the epidemiology of fungi leading to ergot contamination in feed and forage crops and grasses is critical to mitigating and managing potential mycotoxicosis in livestock raising.

Ascospores of *Claviceps* spp. released from perithecia of germinated soilborne sclerotia are the primary inoculum source typically associated with the initiation disease in susceptible host plants (Bandyopadhyay et al., 1998; Florea et al., 2017). Immature panicles, housing florets with unfertilised stigmas, are the infection courts where ascospore germination occurs under prolonged periods of high moisture paired with moderate temperatures, i.e., dew, continuous leaf wetness, rainfall events or relative humidity > 80% and 25 to 30 °C, respectively (Thakur, 1991; Bandyopadhyay et al., 1998). Colonisation continues from the stigmatic papillae to the apical region of the ovary, resulting in the ovary being replaced by sphacelia, i.e., hyphal masses. Abundant sticky opaque exuding conidia, from sphacelia, is the first characteristic sign of disease and is commonly referred to as honeydew. Disease progression is marked by the honeydew transforming from opaque to orange droplets. The mass of conidia serve to facilitate pathogen dispersal, within and between fields, through wind, watersplash and contact between infected and healthy plants (IPCS and WHO, 1990). Under favourable conditions, honeydew can drip or wash down the panicle of individual plants increasing the severity within a field. The maturation of sphacelia, observed as miniscule cream hyphal masses, prompts the formation of hardened structures, i.e., sclerotia (namely ergots) between the glumes of diseased panicles. Ergot production signifies the completion of the pathogen lifecycle and serves as a mechanism for survival and dispersal (Bandyopadhyay et al., 1998; Florea et al., 2017).

In contrast, fescue seed infected with endophytic *Neotyphodium* is considered the primary source of infection in mature fescue, where hyphae are found in the intercellular spaces of grass stems and leaf sheaths (Bacon et al., 1977; Botha et al., 2004). Genotypes of fescue, i.e., endophyte-infected and endophyte-free, vary in ergovaline concentrations in straw and seed, as well as their reactions to nitrogen fertilisation (Welty et al., 1994; Joost, 1995). Nitrogen form and application dosage, as well as timing during series of plant maturation (inflorescence to fully ripe stage) can impact toxicosis associated with pastures (see also Chapter 15). There are contradictory reports of the effect of rainfall or water availability under field conditions on the accumulation of ergovaline in pasture grasses (Repussard et al., 2014). The successful development of a novel cultivar, Jesup MaxQ, which does not express EAs, has been reported to reduce negative effects on livestock (Ferguson et al., 2021).

8 Tenuazonic acid and Alternariol

Tenuazonic acid ((5S,8S)-3-acetyl-5-sec-butyltetramic acid, TeA) was initially viewed as an 'unlikely serious mycotoxin problem' (Meronuck et al., 1972). However, for almost 30 years it was stated that '[TeA] probably possesses the highest toxicity of all [*Alternaria* mycotoxins]', which also includes alternariol (AOH) and alternariol monomethyl ether (AME) (Weidenbörner, 2001). In 2011, the EFSA declared *Alternaria* toxins a risk to public health, advising member states to survey natural occurrences of *Alternaria* toxins in food commodities. According to the ESFA (2011) no reports of mycotoxicosis in livestock and companion animals exposed to feed were made and the literature was considered to represent the Scientific Opinion. Although occurrences of *Alternaria*-associated mycotoxins are considered rare, the potential associated risks outweigh this rarity, as a result of which there has been a peak in research and policy interest (ESFA, 2011; Sanchez et al 2016). There has been an increase in detection and sensitivity of quantification methods, as concerns technological advances with a specific focus on TeA (Scott, 2001; Ostry, 2008; Sanchez et al., 2016). An indication of advances in mycotoxin detection and characterisation is evident in the reports of known mycotoxins; in a 2008 review > 30 *Alternaria*-mycotoxins were acknowledged (Ostry, 2008). In 2016, approximately 70 mycotoxins were acknowledged to be produced by *Alternaria* spp. (Sanchez et al., 2016). *Alternaria*-associated mycotoxins were originally classified into three structural classes (Bottalico and Logrieco, 1998), although they are now divided into five classes (EFSA, 2011).

8.1 *Causative agents*

Tenuazonic acid (TeA) was first discovered as a secondary metabolite produced by *Alternaria* in 1957 (Rosett et al., 1957) and is considered the most frequently occurring mycotoxin produced by *Alternaria* spp. Furthermore, *A. alternata* is recognised as the most common *Alternaria* spp. associated with fruits and vegetables and produces the greatest concentrations of TeA (Garganese, 2016). However, TeA has also been reported as products of *Pyricularia oryzae* (syn. *Magnaporthe oryzae*; Umetsu et al., 1972) and *Phoma sorghina* (at present *Epicoccum sorghinum;* Steyn and Rabie, 1976).

The ubiquitous non-host specific fungal genus *Alternaria* is attributed with causing both field, harvest and post-harvest diseases, resulting in the potential development and accumulation of mycotoxins in feedstuffs (Thomma, 2003; Woudenberg et al., 2013). The first and only known species of *Alternaria* at the time*, i.e., A. tenuis* (now *A. alternata;* Thomma 2003), was initially described in 1816. As the genus expanded and taxonomic revisions advanced, 275 *Alternaria* species were recognised through morphological characterisation (Simmons, 2007). The introduction of molecular studies and next-generation sequencing revised the *Alternaria* complex, and delineated nine genera, with 26 sections (= clades), 11 phylogenetic species and one species complex within sect. *Alternaria* (Woudenberg et al., 2013, 2015). Notably, *P. sorghina* (= *E. sorghinum*) is the most frequently isolated species from Brazilian sorghum grain samples, of which strains are capable of producing TeA (Oliveira et al., 2017).

8.2 *Susceptible host plants*

The *Alternaria* sect. *Alternaria* is known to be pathogenic to almost 400 plant species, of which *A. alternata* is responsible for disease in over 100 plant species (Thomma, 2003). TeA, associated with *A. alternata,* has been detected in barley, figs, maize, oats, peanut, rice, rye, sesame seed, sorghum, sunflower, tobacco, tomato, wheat (Meronuck et al., 1972; EFSA, 2011). However, the greatest concentrations of *Alternaria*-associated mycotoxins were detected in the 'legume, nut and oilseed' food group, particularly oats and sunflower seed (EFSA 2011).

8.3 *Plant disease epidemiology*

Despite the high variation within the *Alternaria* spp. the infection pattern in susceptible plant species is considered similar by Thomma (2003). If seedborne, *Alternaria* can initiate infection upon germination, although infection can also occur once the spores are deposited onto infection courts of a host species through wind dispersal. Germ tube initiation is through the stomata, cuticle or wounds and results in small appressoria developing, although this is not always the case as this may be dependent on pathogen virulence. Infection establishment is dependent on host response as well as pathogen enzyme activation and production (Thomma, 2003).

Variation in mycotoxin production as well as resulting concentrations of detected mycotoxins are strongly driven by host-pathogen-environment interactions, at micro-, meso- and macroscale (i.e., biochemical, genetic, morphological and weather parameters; Oliveira et al., 2017). The host source and interaction with the environment should be noted, to correctly attribute the pathogenic causal agent, e.g., *A. alternata* or *P. sorghina,* to the production of TeA, AOH, and AME in order to safely monitor, regulate and manage the occurrence of these mycotoxins.

9 Enniatins

Enniatins (ENNs) are a group of mycotoxins with a cyclic hexadepsipeptide structure, consisting of specific amino acid residues connected by peptide bonds and intra-molecular ester bonds. ENN type A (ENNA and ENNA1) and type B (ENNB, ENNB1, ENNB2, ENNB3, and ENNB4) contain N-methyl-valine or N-methyl-isoleucine, or a combination of these two amino acids (Blais et al., 1992). Currently 29 naturally occurring enniatin analogues have been reported, and seven have been detected in cereals, i.e., enniatins A, A1, B, B1, B2, B3, and B4. The first four listed ENNs are most commonly found in animal feed (Sy-Cordero et al., 2012; Santini et al., 2012). Unlike commonly studied *Fusarium* mycotoxins such as DON, T-2, HT-2, FUM, and ZEN, ENNs have no established limits for their presence in food and feed. However, there are increasing reports of ENNs being found in various feed products and exhibiting toxicity. This raises significant concerns about animal well-being, particularly when combined with other mycotoxins simultaneously.

Beauvericin (BEA), included in the enniatin family, is a minor or emerging mycotoxin. BEA is considered a flexible molecule, i.e., a nucleophile (free electron pairs for bonding are available), liposoluble and able to bind with metallic cations for the translocation within and across membranes. As a result of the structure of BEA, this mycotoxin has toxigenic as well as antibacterial, antifungal, insecticidal and phytotoxic properties (Logrieco et al., 2002), and the negative impact of this mycotoxin has been mostly described *in vitro*.

9.1 *Causative agents*

Fungal genera that produce ENNs include *Fusarium,* as well as *Halosarpheia* and *Verticillium* (Lin et al., 2002; Supothina et al., 2004). The first species of *Fusarium* reported to produce ENNs was *Fusarium orthoceras* var. *enniatinum* (Gauman et al., 1947), although, to date, almost 30 *Fusarium* species have been identified as ENNs producers. In small grain cereals, ENNs production is associated with strains from three common FHB complexes: *Fusarium tricinctum, F. sambucinum*, and *F. sporotrichioides*. In maize and rice, strains from the *F. fujikuroi* complex, along with those from the three aforementioned complexes, contribute to grain contamination. Additionally, ENNs-producing species from four other *Fusarium* complexes (*F. oxysporum, F. incarnatum-equiseti, F. solani* and *F. lateritium*) contaminate various host plants (Stępień and Waśkiewicz, 2013; Villani et al., 2019).

9.2 *Susceptible host plants*

European cereal crops are often contaminated by ENNs, with more than 50% of the samples showing positive results. Geographical latitude and climatic conditions influence concentrations; however, ENNs are found across multiple sites regardless. Burkina Fasso, Cameroon, Egypt, Morocco, Mozambique, Tunisia, and the Shandong province of China, have also reported the presence of ENNs in wheat and maize, although the contamination levels are generally lower than those found in Europe (Gautier et al., 2020). Based on the available data, *F. avenaceum* and *F. tricinctum* were identified as the primary culprits of grain contamination with ENNs. However, there were some variations depending on the specific crop being considered. Specifically, *F. avenaceum* was found to be the main producer of ENNs in triticale, while *F. tricinctum* was associated with the contamination of wheat (durum) and barley (spring). Similarly, the levels of ENNs detected in oat samples from Norway and Sweden were predominantly linked to infections caused by *F. avenaceum* and *F. tricinctum* (Uhlig et al., 2006; Fredlund et al., 2013; Orlando et al., 2019).

9.3 *Plant disease epidemiology*

There is limited available data on the environmental factors that contribute to the production of ENNs in field. Previous research has primarily focused on nutritional factors, particularly carbon and nitrogen sources, to optimise the yield of these cyclic hexadepsipeptides (Madry et al., 1984; Schmidt-Heydt et al., 2008; Lee et al., 2010; Sørensen and Sondergaard, 2014). ENNs can be consistently produced using various carbon sources, and $NaNO_3$ has been identified as the most effective inorganic nitrogen source, although there are variations among different isolates. Sucrose was found to be an optimal carbon source for ENN production, while lactose was less suitable, from *F. oxysporum* and *F. avenaceum* isolates. However, *F. sambucinum* produced the highest yield of ENNs using lactose (Audhya and Russell, 1974; Madry et al., 1984; Lee et al., 2010; Sørensen and Giese, 2013). The impact of temperature and combinations of a_w and temperature on ENNs production has also been investigated. High levels of ENNs were consistently observed across a_w and temperature conditions, where a temperature of 30 °C was found to be optimal for toxin yield in *F. oxysporum*, while *F. avenaceum* produced similar amounts of ENNs under different a_w-temperature combinations. There have been relatively few studies on the influence of abiotic factors on ENNs production; the available data for *F. avenaceum* isolates suggest that culture conditions have a limited effect on the yield of enniatins (Kokkonen et al., 2010; Lee et al., 2010).

10 Phomopsins

Phomopsins (PHOs), a group of cytotoxic and potentially carcinogenic mycotoxins, are known to cause lupinosis, presenting as liver failure and kidney damage (Van Warmelo et al., 1970 as cited in Van Warmelo & Marasas, 1972; EFSA, 2012). The first reports of lupinosis associated with PHOs were attributed to *Diaporthe toxica* infected lupin in sheep from South Africa and Australia, respectively (Van Warmelo & Marasas, 1972; Culvenor et al., 1977). Consequently, research has been concentrated in South Africa and Australia.

Five phomopsins are known, i.e., PHO-A to PHO-E, of which A is considered the most prominent metabolite of the group (EFSA, 2012). Phomopsin C (PHO-C) was identified in 1991 but is still to be fully characterised (Edgar 1991 as cited in ANZFA, 2001). Quantitative data are limited to Australia with a focus on lupin seed and not the stubble, which is the major concern for livestock production (ANZFA, 2001; EFSA, 2012). Consequently, PHOs are considered emerging mycotoxins as they are largely under-reported and understudied (Schryvers et al., 2024). The potential risk for carcinogenicity remains unclear in livestock, although a strong association indicated a relationship between phomopsins and the development of carcinomas in the bile ducts and livers of rats (ANZFA, 2001). Lupinosis sensitivity varies between animals; affected animals include cattle, donkeys, horses, pigs, poultry and sheep

(Van Warmelo & Marasas, 1972; Allen et al., 1979; ANZFA, 2001; EFSA, 2012), though pigs and horses are considered more sensitive to PHOs (EFSA, 2012). To the best of our knowledge there are no studies on the reproductive toxicity of phomopsins (ANZFA, 2001).

There are limited reports of lupinosis, although observations have been made in Australia (Petterson et al., 1985; Wood et al., 1987 & Than et al., 1994 as cited in Battilani et al., 2011), New Zealand (Gardiner, 1965), Poland (Kochman, 1957 as cited in Wood & Sivasithamparam, 1989), South Africa (Van Warmelo & Marasas, 1972), Spain (Soler Rodriguez et al., 1991), Switzerland (Reinhard et al., 2006 as cited in Battilani et al., 2011) and USA (Ostazeski and Wells, 1960). Legislation in Australia and New Zealand has limited PHOs to a maximum of 5 ug/kg in lupin seed and products (ANZFA, 2001). No regulations are currently set in the European Union for PHOs in food destined for human consumption (Schryvers et al., 2024).

10.1 *Causative agents*

Diaporthe toxica (previously *Phomopsis leptostromiformis;* anamorph) is the causative agent of phomopsis stem and pod blight of *Lupinus* spp. (Culvenor et al., 1977, Allen and Hancock 1989 as cited in EFSA, 2012; Van Warmelo & Marasas, 1972). *Diaporthe/Phomopsis* is an endophytic deuteromycete with two varieties, i.e., *P. leptostromiformis* var. *leptostromiformis* and *P. leptostromiformis* var. *occidentalis;* (Punithalingam, 1974 as cited in Gomes et al., 2013) known to cause stem blight of *Lupinus* species.

10.2 *Susceptible host plants*

Lupins are primary hosts for the pathogen (EFSA, 2012). All organs are susceptible to infection (ANZFA, 2001), although livestock exposure is through the high prevalence of PHOs in stubble (Gardiner, 1966 as cited in Wood & Sivasithamparam, 1989; EFSA, 2012). Cultivars of *L. angustifolius* resistant to phomopsis stem and pod blight have been successfully bred in Australia resulting in significant reductions in PHO contamination (Cowling et al., 1987). However, variable *L. albus* genotype responses to *D. toxica* were reported and the delineation of infections of the stem and pods need to be investigated (Cowley et al., 2010).

PHO-A and PHO-C have been isolated once from petioles of clover, *Trifolium subterraneum* (Shivas et al., 1991). Additionally, PHO-A production has been reported on the common agricultural weed, spiny emex, i.e., *Emex australis* (Shivas et al., 1994 as cited in Battilani et al., 2011). Rare reports of possible PHOs produced in chestnut and spoiled mango have been made (ANZFA, 2001). However, these are rare occurrences, and the epidemiological focus is on *D. toxica* infection of lupin.

10.3 *Plant disease epidemiology*

D. toxica is considered a field pathogen infecting *Lupin* spp., however, toxicosis is only known to occur once tissue has senesced (Gardiner, 1966 as cited in Wood & Sivasithamparam, 1989). Conidial infections initiate disease optimally at 25 °C (Kochman, 1957); however, while ascospores can infect plant tissue, their epidemiological significance lies in the long-distance dispersal of the pathogen (Wood & Sivastihamparam, 1989). Mature leaves are more susceptible to infection (Cowley et al., 2012) due to the senescent tissue being present. Rainfall events, > 8 mm, have been considered to increase the risk for phomopsis development triggering PHO contamination. Although, dew and high relative humidity (not reported) are considered sufficient to initiate disease, levels of PHOs seem higher under wetter conditions (Allen et al., 1979). Combinations of rainfall and drying-off events are also considered to promote infection resulting in stem and pod blight leading to PHO contamination (Gardiner, 1966; Cowling and Wood, 1989; EFSA, 2012).

Lesions of leaf and pod blight are considered latent, i.e., infection occurs earlier than when symptoms are observed, resulting in increased risk for grazing on contaminated lupin (Cowley et al., 2012). Pycnidia, asexual fruiting bodies, are a sign that infection has taken place and may act as an indicator for increased risk to grazing (Cowley & Casburn, 2013). Hail damage to *L. albus* near maturity has been associated with pod shattering, resulting in seed remaining in the field along with stubble, increasing the likelihood of PHO intake due to greater volumes of available susceptible tissue (Cowley et al., 2010). *D. toxica* is suspected to be seedborne, which contributes to pathogen dispersal and persistence in the field (Cowley et al., 2010). Additionally, PHOs are retained in the plant material for three to four months while increasing temporal persistence of the pathogen in the field (Allen et al., 1979).

11 Diplodiatoxin, dipmatol, diplonine and chaetoglobosins

Diplodiatoxin contains a β-ketol side chain and the rare β, γ-unsaturated acid unit, and it was detected in *Stenocarpella maydis*-infected maize kernels and diseased stalk material as well as from ears collected during a natural field outbreak (Rogers et al., 2014; Wicklow et al., 2011). *Stenocarpella maydis* was previously known as *Diplodia maydis*, hence the fitting name of dipldiatoxin.

Dipmatol has been characterised as a trihydroxyhexadecanoic acid, and it was isolated from *S. maydis*-infected cultures using chloroform-hexane.

Diplonine is a substituted β-cyclopropylamino acid (Figure 5). This neurotoxin was isolated using methanol extraction from *S. maydis* cultures that had been used to induce diplodiosis in sheep (MRC 2829 and ARC-GCI 6) (Snyman et al., 2011). Rogers et al. (2014) also reported the first recorded detection of diplonine in naturally infected maize ears.

Chaetoglobosins are metabolites that belong to the family of cytochalasins and include chaetoglobosin K and L, and other compounds. Chaetoglobosins K and L are isomers isolated from fermented rice cultures of *S. maydis* (NRRL 53565, 53566 and 53567) using ethyl-acetate (Wicklow et al., 2011). Rogers et al. (2014) isolated additional chaetoglobosins M and O as well as proline and glycine betaines from *S. maydis*-infected maize ears following a field outbreak of *Stenocarpella* ear rot in Illinois, North America, in 2010. These authors also reported the first recorded detection of chaetoglobosins M, O and K in naturally infected maize ears. Chaetoglobosin K had significant antifungal activity when tested against *A. flavus* and *F. verticillioides* (Wicklow et al., 2011). These authors proposed that chaetoglobosins, including chaetoglobosin K and L, are likely to be the key component in the 'unidentified toxin' mixture that is responsible for diplodiosis in cattle and sheep.

11.1 *Causative agents*

Stenocarpella species, including *S. maydis* and *S. macrospora*, are encountered worldwide. The genus was described by Sutton (1980) based primarily on conidiogenesis. The conidiogenic cells of the genus *Stenocarpella* are enteroblastic, phialidic, determinate, discrete (only rarely integrated on one septate conidiophore), cylindrical, colarette and channel minute; the periclinal wall is thickened and formed from the inner cells of the pycnidial wall. The revised typification of the genus *Diplodia* and related genera meant that *Stenocarpella* became the acceptable generic name to use in this context (Sutton, 1980). The genus *Stenocarpella* contains two species: *S. maydis* (Berk.) Sutton (previously named *D. maydis* (Berk.) Sacc) and *S. macrospora* (Earle) Sutton (formerly referred to as *D. macrospora*). No teleomorphic stage of the *S. maydis* species has yet been identified. The nomenclature of the anamorphic stage of this pathogen was a controversial issue for many years. This is clear from the many synonyms of *S. maydis* (Berk.) Sutton which includes *D. maydis* (Berk.) Sacc., *D. zeae* (Schw.) Lev., *Sphaeria maydis* Berk., *S. zeae* Schw., *D. maydicola* Speg., *D. zeae-maydis* Mechtij., *Macrodiplodia zeae* (Schw.) Petrak and Syd., *Hendersonia zeae* (Schw.) Hazsl. and *Phaeostagonosporopsis zeae* (Schw.) Woron. Sutton (1964) suggested the species be accommodated in a distinct genus and later placed both *D. maydis* and *D. macrospora* Earle into the genus *Stenocarpella* after typification of *Diplodia* and related genera (Sutton, 1980).

11.2 *Susceptible host plants*

Maize is considered the only commercial crop hosting *S. maydis* (Flett, 1991; Sutton and Waterston, 1966). In a South African study (Flett and McLaren, 1994), *S. maydis* was reported as the most prevalent pathogen responsible for stalk and ear rots in maize (Figure 1). These infections result in reduced grain quality, yield loss, and plant lodging (Figure 2). *S. macrospora*, on the other hand, is more commonly found in certain areas of KwaZulu-Natal (Flett and Wehner, 1991) and appears to be restricted to lower-altitude maize production regions in South Africa. However, observations in the North West Province of South Africa revealed *S. macrospora* infestations in a Diplodia-infested ear (Dr Belinda Janse van Rensburg, personal communication, 2022). *S. maydis* is widespread in all maize production areas. In KwaZulu-Natal, where both species occur, it is important to determine which *Stenocarpella* spp. is most prevalent in a specific area, which can also vary seasonally.

11.3 *Plant disease epidemiology*

S. maydis overwinters and grows saprophytically on maize stubble. Later, it produces viable pycnidia on infected husks, kernels, and stalks that remain on the land after harvest (Figure 7A and 7B) (Flett and Wehner, 1991; Flett et al., 1992). Pycnidia (Figure 3) are flask-shaped and 150-350 µm in diameter. Conidiophores, which are elongated, straight or curved, are formed from cells of the inner pycnidial wall. Conidia form at the apex of the conidiophores and are cylindrical or ellipsoidal, narrowing at both ends (Photo 4) to a round apex and truncate base (Kellerman et al., 1985). Conidia are released from pycnidia in cirrhi under warm (23 °C - 30 °C) and moist conditions and are wind- or rain-borne (Del Rio and Del Rio, 1991; Shurtleff, 1980). *S. maydis* conidia infect through the leaf attachments onto the maize stalk (Durrel, 1923). Ear rot infections occur by conidia, more frequently invading the ear leaf shank at the base of the ear and less often the ear tips through the silks (Koehler, 1942; White, 1999). After spore germination, the mycelium ramifies from the shank, through the cob, to the grain attachment filling cavities between the embryo, scutellum, coleoptile and plumule (Miller, 1952). Ramification into kernels begins at the embryonic tissues (Vincelli, 1997) and proceeds into the starchy and horny endosperm and pericarp (Bensch, 1995; Miller, 1952). Bird and insect feeding on the ear tips assists the pathogen to enter the ear (White, 1999). Stalk rot develops when *S. maydis* infects maize plants through the crown, mesocotyl, or roots of the maize plant (White, 1999).

The fungus *S. maydis* (Berk.) Sutton produce several secondary metabolites which include diplodiatoxin (Steyn et al., 1972), dipmatol (Ackerman et al., 1995), diplonine (Snyman et al., 2011), and chaetoglobosins K and L (Wicklow et al., 2011). These studies have been done primarily by South African researchers as this was the only region where toxic symptoms on animals were initially observed. The first record of diplodiosis was in South Africa in the region of Mooi River in KwaZulu-Natal province, where an outbreak of 'sickness' in cattle characterised by paralysis following grazing on harvested maize fields was observed (Van der Bijl, 1914). Other animals, other than cattle, suffering from diseases caused by *S. maydis* related mycotoxins included sheep (Theiler, 1927), goats (Kellerman et al., 1985), ducklings (Rabie et al., 1977), broiler chickens, and laying hens (Rabie et al., 1987). No toxic effects have been reported in equidae or pigs (Marasas, 1977; Mitchell, 1919; Theiler, 1927). More recently toxic symptoms have been reported in other countries such as Australia (Darvall, 1964), Argentina (Odriozola et al., 2005) and Brazil (Riet-Correa et al., 2013). The toxicity of these secondary metabolites has not been adequately

FIGURE 2.3 White fungal growth on an entire maize ear caused by S. maydis.

FIGURE 2.4

S. maydis stalk rot infections caused maize stalks to lodge in a field.

FIGURE 2.5 Black pycnidia containing conidia.

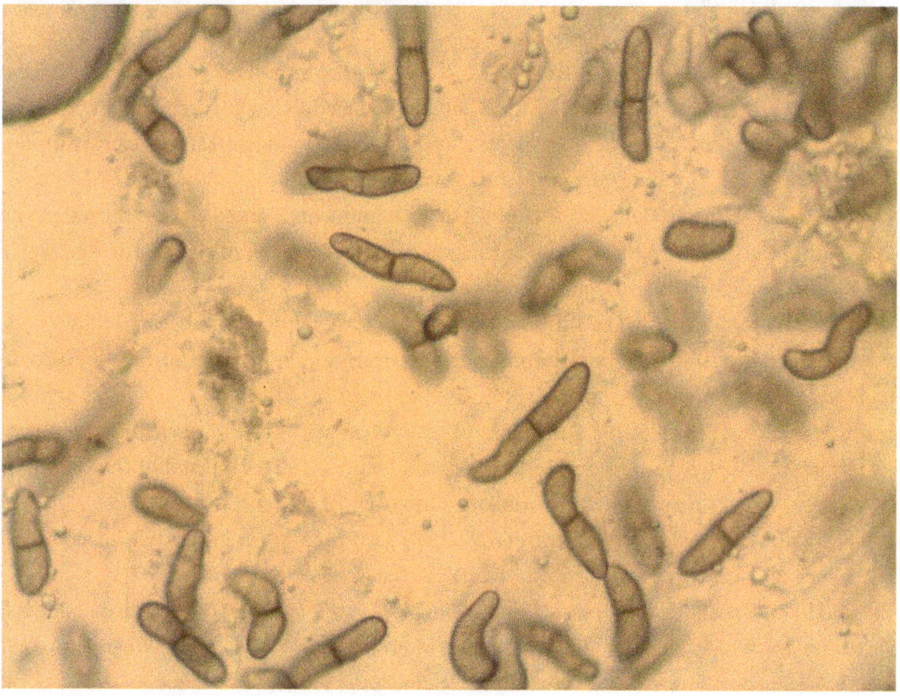

FIGURE 2.6 Cylindrical, two septate conidia of *S. maydis*.

studied worldwide and there is still conjecture as to which metabolite or combinations thereof are responsible for differing symptoms on different animals. Mycotoxins produced by *S. macrospora* are yet to be adequately identified and characterised and their impact on human and animal safety to be determined.

12 Storage mycotoxins

Most storage mycotoxins are produced by *Aspergillus* and *Penicillium* species and are of major importance for ruminants (Chapter 6), pets (Chapter 8), and horses (Chapter 9) because these mycotoxins are produced in silages and spoilages. On silage, fungal growth will be common on surface areas, but the *Penicillium* strains isolated from European silage are micro-aerobic and pH-tolerant and can also invade the inner part of silage bales. Citrinin (CIT), which can be produced by *Penicillium* (*P. citrinum*, *P. verrucosum* and *P. expansum*), *Aspergillus* (*A. carneus*, *A. niveus* and *A. terreus*), and *Monascus* (*M. ruber*) species (Kamle et al., 2022) may also be produced during the storage of pig (Chapter 4) and poultry (Chapter 5) feed. These storage mycotoxins are described within each respective chapter according to their importance in animal production performance and health.

References

Ackerman, L.G.J., Combrink, S., Horak, R.M., Kuhn, M., Learmonth, R.L., Lübben, A., Maharaj, V.J. and Agriopoulou, S., 2021. Ergot alkaloids mycotoxins in cereals and cereal-derived food products: characteristics, toxicity, prevalence, and control strategies. Agronomy 11: 931. https://doi.org/10.3390/agronomy11050931.

Allen, J.G. and Hancock G.R., 1989. Evidence that phomopsins A and B are not the only toxic metabolites produced by *Phomopsis leptostromiformis*. Journal of Applied Toxicology 9: 8389. https:doi.org/10.1002/jat.2550090203.

Allen, J.G., Wood, P.M., Croker, K.P. and Hamblin, J., 1979. Lupinosis-a disease still with us. Accessed from: https://library.dpird.wa.gov.au/cgi/viewcontent.cgi?article=2740&context=journal_agriculture4

Audhya, T.K. and Russell, D.W., 1974. Natural enniatin A, a mixture of optical isomers containing both *erythro*-and *threo-N*-methyl-L-isoleucine residues. Journal of the Chemical Society Perkin 1 7: 743-746. https://doi.org/10.1039/p19740000743.

Australia New Zealand Food Authority (ANZFA) 2001. Phomopsins in Food: A Toxicological Review and Risk Assessment. Australian New Zealand Food Authority.

Bacon, C.W., Porter, J.K., Robbins, J.D. and Luttrell, E.S., 1977. *Epichloë typhina* from toxic tall fescue grasses. Applied and Environmental Microbiology 34: 576. https://doi.org/10.1128/AEM.34.5.576-581.1977.

Bandyopadhyay, R., Frederickson, D.E., Mclaren, N.W., Odvody, G.N. and Ryley, M.J., 1998.

Ergot: a new disease threat to sorghum in the americas and Australia. Plant Disease 82: 356-367. https://doi.org/10.1094/PDIS.1998.82.4.356.

Battilani, P., Barbano, C. and Piva, G., 2008. Aflatoxin B1 contamination in maize related to the aridity index in North Italy. World Mycotoxin Journal, 1: 449-456. https://doi.org/10.3920/WMJ2008.x043.

Bbosa, G.S., Kitya, D. and Ogwal-Okeng, J., 2013. Aflatoxins metabolism, effects on epigenetic mechanisms and their role in carcinogenesis. Scientific Research 5: 1. https://doi.org/10.4236/health.2013.510A1003.

Bennett, J.W. and Klich, M., 2003. Mycotoxins. Clinical Microbiology Reviews 16: 497-516. https://doi.org/10.128/CMR.16.3497-516.2003.

Bensch, M.J., 1995. An evaluation of inoculation techniques inducing *Stenocarpella maydis* ear rot on maize. South African Journal of Plant and Soil 12: 172-174. https://doi.org/10.1080/02571862.1995.10634360.

Berthiller, F., Crews, C., Dall'Asta, C., Saeger, S.D., Haesaert, G., Karlovsky, P., Oswald, I.P., Seefelder, W., Speijers, G. and Stroka, J., 2013. Masked mycotoxins: A review. Molecular Nutrition and Food Research 57: 165-186. https://doi.org/10.1002/mnfr.201100764.

Berthiller, F., Maragos, C.M. and Dall'Asta, C., 2015. Introduction to masked mycotoxins. In: Masked mycotoxins in food: formation, occurrence and toxicological relevance. Royal Society of Chemistry, RSC Publishing; Cambridge, UK. Introduction to masked mycotoxins. 1–13. https://doi.org/10.1039/9781782622574-00001.

Beukes, I., Rose, L.J., Shephard, G.S., Flett, B.C. and Viljoen, A., 2017. Mycotoxigenic *Fusarium* species associated with grain crops in South Africa-A review. South African Journal of Science 113: 1-12. http://dx.doi.org/10.17159/sajs.2017/20160121.

Bily, A.C., Reid, L.M., Savard, M.E., Reddy, R., Blackwell, B.A., Campbell, C.M., Krantis, A., Durst, T., Philogene, B.J.R., Arnason, J.T. and Regnault-Roger, C., 2004. Analysis of *Fusarium graminearum* mycotoxins in different biological matrices by LC/MS. Mycopathologia 157: 117-126. https://doi.org/10.1023/B:MYCO.0000012218.27359.ec.

Blais, L.A., Apsimon, J.W., Blackwell, B.A., Greenhalgh, R. and Miller, J.D., 1992. Isolation and characterization of enniatins from *Fusarium avenaceum* Daom-196490. Canadian Journal of Chemistry 70: 1281-1287. https://doi.org/10.1139/v92-165.

Bosch, U. and Mirocha, C.J., 1992. Toxin production by *Fusarium* species from sugar beets and natural occurrence of zearalenone in beets and beet fibers. Applied environmental Microbiology 58: 3233-3239. https://doi.org/10.1128/aem.58.10.3233-3239.1992.

Botha, C.J., Naudé, T.W., Moroe, M.L. and Rottinghaus, G.E., 2004. Gangrenous ergotism in cattle grazing fescue (*Festuca elatior* L.) in South Africa. Journal of the South African Veterinary Association 75: 45–48. https://doi.org/https://doi.org/10.4102/jsava.v75i1.449.

Bottalico, A. and Logrieco, A., 1998. Toxigenic *Alternaria* species of economic importance. In: Mycotoxins in Agriculture and Food Safety, CRC Press.

Boutigny, A.L., Richard-Forget, F. and Barreau, C., 2008. Natural mechanisms for cereal resistance to the accumulation of *Fusarium* trichothecenes. European Journal of Plant Pathology 121: 411-423. https://doi.org/10.1007/s10658-007-9266-x.

Boutigny, A.L., Ward, T.J., Van Coller, G.J., Flett, B., Lamprecht, S.C., O'Donnell, K. and Viljoen,

A., 2011. Analysis of the *Fusarium graminearum* species complex from wheat, barley and maise in South Africa provides evidence of species-specific differences in host preference. Fungal Genetics and Biology 48: 914-920. https://doi.org/10.1016/j.fgb.2011.05.005.

Braun, M.S. and Wink, M., 2018. Exposure, occurrence, and chemistry of fumonisins and their cryptic derivatives. Comprehensive Reviews in Food Science and Food Safety 17: 769-791. https://doi.org/10.1111/1541-4337.12334.

Bullerman, L.B. 2003. 'Mycotoxins |Classification', in B. Cabarello (ed). Encyclopedia of Food Sciences and Nutrition. Academic Press, 4080-4089.

Bullerman, L.B. and Bianchini, A., 2014. Good food-processing techniques: stability of mycotoxins in processed maize-based foods. In: Leslie, J.F. and Logrieco, A.F. (eds) Mycotoxin Reduction in Grain Chains 7. https://doi.org/10.1002/9781118832790.ch7.

Cao, A., Santiago, R., Ramos, A.J., Marín, S., Reid, L.M. and Butrón, A., 2013. Environmental factors related to fungal infection and fumonisin accumulation during the development and drying of white maize kernels. International Journal of Food Microbiology 164: 15-22. https://doi.org/10.1016/j.ijfoodmicro.2013.03.012.

Card, S.D., Philip Rolston, M., Lloyd-West, C and Hume, D.E., 2014. Novel perennial ryegrass-*Neotyphodium* endophyte associations: relationships between seed weight, seedling vigour and endophyte presence. Symbiosis 62: 51-62. https://doi.org/10.1007/s13199-014-0271-5.

CCFAC, 2000. Codex Committee on Food Additives and Contaminants. Posting date. Joint FAO/WHO Expert Committee on Food Additives: Position paper on zearalenone. Publication CCFAC 00/19. Codex Alimentarius Commission, Rome, Italy.

Cendoya, E., Chiotta, M.L., Zachetti, V., Chulze, S.N. and Ramirez, M.L., 2018. Fumonisins and fumonisin-producing *Fusarium* occurrence in wheat and wheat by products: a review. Journal of Cereal Science 80: 158-166. https://doi.org/10.1016/j.jcs.2018.02.010.

Chen, X., Wu, Q., Gao, Y., Zhang, J., Wang, Y., Zhang, R., Zhou, Y., Xiao, M., Xu, W. and Huang, R., 2020.The role of deep roots in sorghum yield production under drought conditions. Agronomy, 10: 611. http://dx.doi.org/10.3390/agronomy10040611.

Ciegler, A., 1972. Bioproduction of ochratoxin A and penicillic acid by members of the *Aspergillus ochraceus* group. Canadian Journal of Microbiology 18: 72-100. https://doi.org/10.1139/m72-100.

Cole, R., Schweikert, M. and Jarvis B., 2003. Handbook of fungal secondary metabolites. Academic Press; San Diego, CA, USA.

Cowley, R.B., Ash, G.J., Harper, J.D.I. and Luckett, D.J., 2010. Evidence that *Diaporthe toxica* infection of *Lupinus albus* is an emerging concern for the Australian lupin industry. Australasian Plant Pathology 39: 146-153. https://doi.org/10.1071/AP09067.

Cowley, R.B., Luckett, D.J., Harper, J.D. and Ash, G.J., 2012. Development of a reliable and rapid detached leaf assay to detect resistance to the fungal disease phomopsis leaf blight, caused by *Diaporthe toxica*, in *Lupinus albus*. Canadian Journal of Plant Pathology 34: 401-409. https://doi.org/10.1080/07060661.2012.705327.

Cowling, W.A., Hamblin, J., Wood, P.M. and Gladstones, J.S., 1987. Resistance to phomopsis stem blight in *Lupinus angustifolius* L. 1. Crop Science 27: 648-652. https://doi.org/10.2135/cropsci1987.0011183X002700040007x.

Culvenor, C.C.J., Beck, A.B., Clarke, M., Cockrum, P.A., Edgar, J.A., Frahn, J.L., Jago, M.V., Lanigan, G.W., Payne, A.L., Peterson, J.E., Petterson, D.S., Smith, L.W. and White, RR., 1977. Isolation of toxic metabolites of *Phomopsis leptostromiformis* responsible for lupinosis. Australian Journal of Biological Sciences 30: 269-277. https://doi.org/10.1071/bi9770269.

Czembor, E., Waskiewicz, A., Piechota, U., Puchta, M., Czembor, J.H. and Stepien, L., 2019. Differences in ear rot resistance and *Fusarium verticillioides*-produced fumonisin contamination between Polish currently and historically used maize inbred lines. Frontiers in Microbiology 10: 449. https://doi.org/10.3389/fmicb.2019.00449.

Dalcero, A., Magnoli, C., Halla, C., Chiacchiera, S.M., Palacio, G. and Rosa, C.A.R., 2002. Detection of ochratoxin A in animal feeds and capacity to produce this mycotoxin by *Aspergillus* section *Nigri* in Argentina. Food Additives and Contaminants 19: 1065-1072. https://doi.org/10.1080/02652030210151895.

Darvall, P.M., 1964. Mouldy corn cobs, a danger to cows. Queensland Agricultural Journal 90: 692-693.

Del Rio, M.L. and Del Rio, L., 1991. Dispersion of *Stenocarpella maydis* (Berk.) Sutton in maize crops. CEIBA 32: 133-140.

Doohan, F.M., Brennan, J. and Cooke, B.M., 2003. Influence of climatic factors on *Fusarium* pathogenic to cereals. European Journal of Plant Pathology 109: 755-768. https://doi.org/10.1023/A:1026090626994.

Duarte, S.C., Pena, A. and Lino, C.M., 2009. Ochratoxin A non-conventional exposure sources – a review. Microchemical Journal 93: 115-120. https://10.1016/j.microc.2009.06.002.

Dufault, N.S., De Wolf, E.D., Lipps, P.E. and Madden, L.V., 2006. Role of temperature and moisture in the production and maturation of *Gibberella zeae* perithecia. Plant Disease 90: 637-644. https://doi.org/10.1094/PD-90-0637.

Durrell, L.W., 1923. Dry rot of corn. Iowa Agricultural Experimental Station Research Bulletin 77: 31.

Early, R., 2009. 'Pathogen control in primary production: crop foods', in C.W Blackburn and J. Peter (eds). *Foodborne Pathogens.* Academic Press, 205-279.

Edgar, J.A., 1991. Phomopsins: antimicrotubule mycotoxins. In: Handbook of Natural Toxins, (Eds. R.F. Keeler and A.T. Tu), Marcel Dekker Inc, 371-395.

Edwards, S.G., 2009. *Fusarium* mycotoxin content of UK organic and conventional oats. Food Additives and Contaminants Part A 26: 1063-1069. https://doi.org/10.1080/02652030902788953.

Edwards, S.G., Imathiu, S.M., Ray, R.V., Back, M. and Hare, M.C., 2012. Molecular studies to identify the Fusarium species responsible for HT-2 and T-2 mycotoxins in UK oats. International Journal of Food Microbiology 156: 168-175. https://doi.org/10.1016/j.ijfoodmicro.2012.03.020.

EFSA CONTAM Panel (EFSA Panel on Contaminants in the Food Chain), 2013. Scientific Opinion on risks for animal and public health related to the presence of nivalenol in food and feed. The EFSA Journal 11: 3262. https://doi.org/10.2903/j.efsa.2013.3262.

EFSA CONTAM Panel (EFSA Panel on Contaminants in the Food Chain), Schrenk, D., Bodin, L., Chipman, J.K., del Mazo, J., Grasl-Kraupp, B., Hogstrand, C., Hoogenboom, L.,Leblanc, J.C., Nebbia, C.S., Nielsen, E., Ntzani, E., Petersen, A., Sand, S., Schwerdtle, T., Vleminckx, C., Wallace, H., Alexander, J., Dall'Asta, C., Mally, A., Metzler, M., Binaglia, M., Horváth, Z., Steinkellner, H. and Bignami, M., 2020. Scientific Opinion on the risk assessment of ochratoxin A in food. The EFSA Journal 18: 6113. https://doi.org/10.2903/j.efsa.2020.6113.

EFSA on Contaminants in the Food Chain (CONTAM), 2011. Scientific Opinion on the risks for animal and public health related to the presence of *Alternaria* toxins in feed and food. The EFSA Journal 9: 2407. https://doi.org/10.2903/j.efsa.2011.2407.

EFSA Panel on Contaminants in the Food Chain (CONTAM), 2012. Scientific Opinion on Ergot alkaloids in food and feed. The EFSA Journal 10: 2798. https://doi.org/10.2903/j.efsa.2012.2798.

EFSA Panel on Contaminants in the Food Chain, 2012. Scientific Opinion on the risks for animal and public health related to the presence of phomopsins in feed and food. EFSA journal 10: 2567. https://doi.org/10.2903/j.efsa.2012.2567.

EFSA, 2005. Opinion of the scientific panel on contaminants in food chain on a request from the commission related to ergot as undesirable substance in animal feed. The EFSA Journal 225: 1-27. https://doi.org/10.2903/j.efsa.2005.225.

El Khoury, A. and Atoui, A., 2010. Ochratoxin A: general overview and actual molecular status. Toxins 2: 461-493. https://doi.org/10.3390/toxins2040461.

Ellner, F.M., 2002. Mycotoxins in potato tubers infected by *Fusarium sambucinum*. Mycotoxin Research 18: 57–61. https://doi.org/10.1007/BF02946697.

Elmholt, S. and Hestbjerg, H., 1999. Field ecology of the ochratoxin A-producing *Penicillium verrucosum*: survival and resource colonisation in soil. Mycopathologia 147: 67-81. https://doi.org/10.1023/a:1007003106343.

Elmholt, S., 2003. Ecology of the Ochratoxin A Producing *Penicillium verrucosum:* occurrence in field soil and grain with special attention to farming system and on-farm drying practices. Biological Agriculture and Horticulture 20: 311-337. https://doi.org/10.1080/01448765.2003.9754976.

Feddern, V., Dors, G.C., Tavernari, F.D.C., Mazzuco, H., Cunha, A., Krabbe, E.L. and Scheuermann, G.N., 2013. Aflatoxins importance on animal nutrition. Aflatoxins-Recent Advances and Future Prospects 171-195.

Ferguson, T.D., Vanzant, E.S. and McLeod, K.R., 2021. Endophyte infected tall fescue: plant symbiosis to animal toxicosis. Frontiers in Veterinary Science 8: 1548. https://doi.org/10.3389/fvets.2021.774287.

Flett, B.C. and McLaren, N.W., 1994. Optimum disease potential for evaluating resistance to *Stenocarpella maydis* ear rot in corn hybrids. Plant Disease 78: 587-589.

Flett, B.C. and Wehner, F.C., 1991. Incidence of *Stenocarpella* and *Fusarium* cob rots in monoculture maize under different tillage systems. Journal of Phytopathology 133: 327-333.

Flett, B.C., 1991. Crop plants as hosts and non-hosts of *Stenocarpella maydis*. Phytophylactica 23: 237 – 238.

Flett, B.C., Wehner, F.C. and Smith, M.F., 1992. Relationship between maize stubble placement in soil and survival of *Stenocarpella maydis* (*Diplodia maydis*). Journal of Phytopathology 134: 33-38.

Florea, S., Panaccione, D.G. and Schardl, C.L., 2017. Ergot Alkaloids of the Family *Clavicipitaceae*. Phytopathology 107: 504-518. https://doi.org/10.1094/PHYTO-12-16-0435-RVW.

Flythe, M.D., Ohkubo, T., Bradford, B., Wang, Y., Coufal-Majewski, S., Stanford, K., McAllister, T., Blakley, B., McKinnon, J. and Vieira Chaves, A., 2016. Impacts of cereal ergot in food animal production. 3: 15. https://doi.org/10.3389/fvets.2016.00015.

Foroud, N.A., Baines, D., Gagkaeva, T.Y., Thakor, N., Badea, A., Steiner, B., Bürstmayr, M. and Bürstmayr, H., 2019. Trichothecenes in cereal grains–An update. Toxins 11: 634. https://doi.org/10.3390/toxins11110634.

Fredlund, E., Gidlund, A., Sulyok, M., Börjesson, T., Krska, R., Olsen, M. and Lindblad, M., 2013. Deoxynivalenol and other selected *Fusarium* toxins in Swedish oats — Occurrence and correlation to specific *Fusarium* species. International Journal of Food Microbiology 167: 276-283. https://doi.org/10.1016/j.ijfoodmicro.2013.06.026.

Frisvad, J.C., Frank, J.M., Houbraken, J.A.M.P., Kuijpers, A.F.A. and Samson, R.A., 2004. New Ochratoxin A producing species of *Aspergillus* section *Circumdati*. Studies in Mycology 50: 23–43.

Gao, Z., Luo, K., Zhu, Q., Peng, J., Liu, C., Wang, X., Li, S. and Zhang, H., 2023. The natural occurrence, toxicity mechanisms and management strategies of Fumonisin B1: A review. Environmental Pollution 320: 121065. https://doi.org/10.1016/j.envpol.2023.121065.

Gardiner, M.R., 1965. The pathology of lupinosis of sheep: Gross-and histo-pathology. Pathologia veterinaria 2: 417-445.

Gardiner, M.R., 1966. Abstract. Fungus-induced toxicity in lupinosis. British Veterinary Journal 122: 508-516.

Garganese, F., Schena, L., Siciliano, I., Prigigallo, M.I., Spadaro, D., De Grassi, A., Ippolito, A. and Sanzani, S.M., 2016. Characterization of citrus-associated *Alternaria* species in Mediterranean areas. PLoS ONE 11: e0163255. https://doi.org/10.1371/journal.pone.0163255.

Gäumann, E., Roth, S., Ettlinger, L., Plattner, P.A. and Nager, U., 1947. Enniatin, ein neues, gegen Mykobakterien wirksames Antibiotikum. Experientia 3: 202–203.

Gautier, C., Pinson-Gadais, L. and Richard-Forget, F., 2020. *Fusarium* Mycotoxins Enniatins: An Updated Review of Their Occurrence, the Producing *Fusarium* Species, and the Abiotic Determinants of Their Accumulation in Crop Harvests. Journal of Agriculture and Food Chemistry 68: 4788-4798. https://doi.org/10.1021/acs.jafc.0c00411.

Gil-Serna, J., Vázquez, C., González-Jaén, M.T. and Patiño, B., 2014. Mycotoxins | Toxicology. In: Batt CA, Tortorello M Lou (eds) Encyclopedia of food microbiology, second ed. Elsevier, Oxford 887–892.

Giorni, P., Bertuzzi, T. and Battilani, P., 2016. Aflatoxin in maize, a multifaceted answer of *Aspergillus flavus* governed by weather, host-plant and competitor fungi. Journal of Cereal Science 70: 256-262. https://doi.org/10.1016/j.jcs.2016.07.004.

Glenn, A.E., 2007. Mycotoxigenic *Fusarium* species in animal feed. Animal Feed Science and Technology 137: 213-240. https://doi.org/10.1016/j.anifeedsci.2007.06.003.

Gomes, R.R., Glienke, C., Videira, S.I.R., Lombard, L., Groenewald, J.Z. and Crous, P.W., 2013. Diaporthe: a genus of endophytic, saprobic and plant pathogenic fungi. Persoonia-Molecular Phylogeny and Evolution of Fungi 31: 1-41. 10.https://doi.org/3767/003158513X666844.

Grenier B., Applegate T.J. Modulation of Intestinal function following mycotoxin ingestion: meta-analysis of published experiments in animals. Toxins. 2013. 5:396-430. https://doi.org/10.3390/toxins5020396.

Hao, J.J., Xie, S.N., Sun, J.,Yang, G.Q., Liu, J. Z., Xu F., Ru, Y.Y. and Song Y.L., 2017. Analysis of *Fusarium graminearum* Species Complex from Wheat–Maise Rotation Regions in Henan (China). Plant Sciences 101: 720-725. https://doi.org/10.0194/PDIS-06-16-0912-RE.

Haschek, W.M. and Voss, K.A., 2013. Chapter 39 – Mycotoxins. In: Haschek, W.M., Rousseaux, C.G. and Wallig, M.A. (eds) Haschek and Rousseaux's Handbook of Toxicologic Pathology (Third Edition), Academic Press.

Hesseltine, C.W., Vandegraft, E.E., Fennell, D.I., Smith, M.I. and Shotwell, O.l., 1972. *Aspergilli* as ochratoxin producers. Mycologia 64: 539-550.

Heuzé V., Tran G., Lebas F., 2017. Maize grain. Feedipedia, a programme by INRAE, CIRAD, AFZ and FAO. https://feedipedia.org/node/556 Date Accessed 19 July 2024.

Huff, W.E. and Hamilton, P.B., 1979. Mycotoxins – their biosynthesis in Fungi: Ochratoxins – metabolites of combined pathways. Journal of Food Protection 42: 815.820. https://doi.org/10.4315/0362-028X-42.10.815.

Humpf, H.U. and Voss, K.A., 2004. Effects of thermal food processing on the chemical structure and toxicity of fumonisin mycotoxins. Molecular Nutrition and Food Research 48: 255-269. https://doi.org/10.1002/mnfr.200400033.

Imathiu, S.M., Edwards, S.G., Ray, R.V. and Back, M.A., 2013. *Fusarium langsethiae*–a HT-2 and T-2 toxins producer that needs more attention. Journal of Phytopathology 161: 1-10. https://doi.org/10.1111/jph.12036.

International Programme on Chemical Safety (IPCS) and World Health Organization (WHO), 1990. Selected mycotoxins: ochratoxins, trichothecenes, ergot / published under the joint sponsorship of the United Nations Environment Programme, the International Labour Organisation and the World Health Organization. World Health Organization. https://iris.who.int/handle/10665/39552.

Isakeit, T., Isakeit, T., Murray, S. and Pekar, J. 2016, December. Influence of corn GMO traits for insect control on aflatoxin and fumonisin contamination in Texas. In *Phytopathology* (Vol. 106, No. 12, pp. 46-46). 3340 Pilot Knob Road, St Paul, MN 55121 USA: Amer Phytopathological Soc.

Joost, R.E., 1995. *Acremonium* in fescue and ryegrass: boon or bane? A review. Journal of Animal Science 73: 881–888. https://doi.org/10.2527/1995.733881X.

Kahla, A., Verheecke-Vaessen, C., Delpino-Deelias, M., Gutierrez-Pozo, M., Medina, A., Magan, N. and Doohan, F., 2023. Acclimatisation of *Fusarium langsethiae, F. poae* and *F. sporotrichioides* to elevated CO2: Impact on fungal growth and mycotoxin production on oat-based media. International Journal of Food Microbiology 394: 110176. https://doi.org/10.1016/j.ijfoodmicro.2023.110176.

Kamle, M., Mahato, D.K., Devi, S., Lee, K.E., Kang, S.G. and Kumar, P., 2019. Fumonisins: Impact on agriculture, food, and human health and their management strategies. Toxins 11: 328. https://doi.org/10.3390/toxins11060328.

Kamle, M., Mahato, D.K., Gupta, A., Pandhi, S., Sharma, N., Sharma, B., Mishra, S., Arora, S., Selvakumar, R., Saurabh, V., Dhakane-Lad, J., Kumar, M., Barua, S., Kumar, A., Gamlath, S. and Kumar, P., 2022. Citrinin mycotoxin contamination in food and feed: impact on agriculture, human health, and detection and management strategies. Toxins 14: 85. https://doi.org/10.3390/toxins14020085.

Kellerman, T.S., Rabie, C.J., van der Westhuizen, G.C., Kriek, N.P. and Prozesky, L., 1985. Induction of diplodiosis, a neuromycotoxicosis, in domestic ruminants with cultures of indigenous and exotic isolates of *Diplodia maydis*. Onderstepoort Journal of Veterinary Research 52: 35-42.

Klopfenstein, T.J., Erickson, G.E. and Berger, L.L. 2013. Maize is a critically important source of food, feed, energy and forage in the USA. Field Crops Research 153: 5-11.

Kochman, J., 1957. Abstract. Studies on the patchiness of Lupin stems caused by *Phomopsis leptostromiformis* (Kühn) Bubak. Acta Agrobotanica 6: 117-143. https://www.cabidigitallibrary.org/doi/full/10.5555/19581101866.

Koehler, B., 1942. Natural mode of entrance of fungi into corn ears and some symptoms that indicate infection. Journal of Agricultural Research 64: 421-442.

Kokkonen, M., Ojala, L., Parikka, P. and Jestoi, M., 2010. Mycotoxin production of selected *Fusarium* species at different culture conditions. International Journal of Food and Microbiology 143: 17-25. https://doi.org/10.1016/j.ijfoodmicro.2010.07.015.

Kolawole, O., Siri-Anusornsak, W., Petchkongkaw, A., Meneely, J. and Elliott, C., 2022. The efficacy of additives for the mitigation of aflatoxins in animal feed: a systematic review and network meta-analysis. Toxins 14: 707. https://doi.org/10.3390/toxins14100707.

Kovalsky Paris, M.P., Schweiger, W., Hametner, C., Stückler, R., Muehlbauer, G.J., Varga, E., Krska, R., Berthiller, F. and Adam, G., 2014. Zearalenone-16-O-glucoside: a new masked mycotoxin. Journal of Agricultural and Food Chemistry 62: 1181-1189. https://doi.org/10.1021/jf405627d.

Kozlovsky, A.G. and Reshetilova, T.A., 1984. Regulation of the biosynthesis of ergot alkaloids by *Penicillium sizovae*. Folia Microbiologica 29: 301-305. https://D978-0-08-023179-2.50015-0.

Kuiper-Goodman, T., Scott, P. and Watanabe, H., 1987. Risk assessment of the mycotoxin zearalenone. Regulatory Toxicology and Pharmacology 7: 253-306. https://doi.org/10.1016/0273-2300(87)90037-7.

Kumar, V., Basu, M.S. and Rajendran, T.P., 2008. Mycotoxin research and mycoflora in some commercially important agricultural commodities. Crop protection 27: 891-905. https://doi.org/10.1016/j.cropro.2007.12.011.

Lahouar, A., Crespo-Sempere, A., Marín, S., Saïd, S. and Sanchis, V., 2015. Toxigenic molds in Tunisian and Egyptian sorghum for human consumption. Journal of Stored Products Research 63: 57-62. https://doi.org/10.1016/j.jspr.2015.07.001.

Lee, S.H., Lee, J.K., Nam, Y.J., Lee, S.H., Ryu, J.G. and Lee, T., 2010. Population structure of

Fusarium graminearum from maise and rice in 2009 in Korea. The Plant Pathology Journal 26: 321-327. https://doi.org/10.5423/PPJ.2010.26.4.321.

Limay-Rios, V., Miller, J.D. and Schaafsma, A.W., 2017. Occurrence of *Penicillium verrucosum*, ochratoxin A, ochratoxin B and citrinin in on-farm stored winter wheat from the Canadian Great Lakes Region. PLoS One. 12: e0181239. https://doi.org/10.1371/journal.pone.0181239.

Lin, Y., Wang, J., Wu, X., Zhou, S., Vrijmoed, L.L.P. and Jones, E.B.G., 2002. A novel compound enniatin G from the mangrove fungus *Halosarpheia* sp. (strain #732) from the South China sea. Australian Journal of Chemistry 55: 225–227. https://doi.org/10.1071/CH01164.

Liu, J. and Applegate, T., 2020. Zearalenone (ZEN) in livestock and poultry: dose, toxicokinetics, toxicity and estrogenicity. Toxins 12: 377. https://doi.org/10.3390/toxins12006377.

Llorens, A., Mateo, R., Hinojo, M.J., Valle-Algarra, F.M., and Jimenez, M., 2004. Influence of environmental factors on the biosynthesis of type B trichothecenes by isolates of *Fusarium* spp. from Spanish crops. International Journal of Food Microbiology 94: 43–54. https://doi.org/10.1016/j.ijfoodmicro.2003.12.017.

Logrieco, A., Rizzo, A., Ferracane, R. and Ritieni, A., 2002. Occurrence of beauvericin and enniatins in wheat affected by *Fusarium avenaceum* heat blight. Applied and Environmental Microbiology 68: 82-85. https://doi.org/10.1128/AEM.68.1.82-85.2002.

Lori, G., Salerno, M.I., Wolcan, S., Gimenez, J., and Basil, G., 1999. *Fusarium* species from a forest nursery soil in Western Patagonia and reduction of their population by soil solarisation. *Journal of Plant Disease and Protection* 106: 363-371. http://www.jstor.org/stable/43216175.

Lyons, P.C., Platrner, R.D. and Bacon, C.W., 1986. Occurrence of peptide and clavine ergot alkaloids in tall fescue grass. Science 232: 487-489. https://doi.org/10.1126/science.3008328.

Madry, N., Zocher, R., Grodzki, K. and Kleinkauf, H., 1984. Selective synthesis of depsipeptides by the immobilized multienzyme enniatin synthetase. Applied Microbiology and Biotechnology 20: 83–86. https://doi.org/10.1007/BF00252582.

Magan, J., Hope, R. and Aldred, D., 2006. Ecophysiology of *Fusarium culmorum* and mycotoxin production. Advances in Food Mycology 571: 123-136. https://doi.org/10.1007/0-387-28391-9_7.

Malir, F., Ostry, V., Pfohl-Leszkowicz, A., Malir, J., Toman, J., 2016. Ochratoxin A: 50 years of research. Toxins 8: 191. https://doi.org/10.3390/toxins8070191.

Marasas, W.F.O., 1977. The genus *Diplodia*. In: Wyllie, T.D. and Moorehouse, L.G. (eds.) Mycotoxic fungi, mycotoxins, mycotoxicoses. An encyclopedic handbook, Vol. I. Marcel Decker, Inc., New York, pp. 119-128.

McCormick, S.P., Kato, T., Maragos, C.M., Busman, M., Lattanzio, V.M., Galaverna, G., Dall-Asta, C., Crich, D., Price, N.P. and Kurtzman, C.P., 2015. Anomericity of T-2 toxin-glucoside: Masked mycotoxin in cereal crops. Journal of agricultural and food chemistry 63: 731-738. https://doi.org/10.1021/jf504737f.

McCormick, S.P., Stanley, A.M., Stover, N.A. and Alexander, N.J., 2011. Trichothecenes: from simple to complex mycotoxins. Toxins 3: 802-814. https://doi.org/10.3390/toxins3070802.

Meronuck, R.A., Steele, J.A., Mirocha, C.J. and Christensen, C.M., 1972. Tenuazonic Acid, a toxin produced by *Alternaria alternata*. Applied Microbiology 23: 613–617. https://doi.org/10.1128/am.23.3.613-617.1972.

Mesterházy Á., Lemmens M. and Reid L.M., 2012. Breeding for resistance to ear rots caused by *Fusarium* spp. in maise—A review. Plant Breeding 131: 1-19. https://doi.org/10.1111/j.1439-0523.2011.01936.x.

Miller, J.D., 2008. Mycotoxins in small grains and maize: old problems, new challenges. Food Additives and Contaminants 25: 219-230. https://doi.org/10.1080/02652030701744520.

Miller, J.H., 1952. The presence of internal mycelium in corn grains in relation to external symptoms of corn ear rot. Phytopathology 4: 286.

Mitchell, D.T., 1919. Poisoning of cattle by *Diplodia*-infected maize. South African Journal of Science 16: 446-452. https://hdl.handle.net/10520/AJA00382353_2404.

Molnár, K., Rácz, C., Dövényi-Nagy, T., Bakó, K., Pusztahelyi, T., Kovács, S., Adácsi, C., Pócsi, I. and Dobos, A., 2023. The effect of environmental factors on mould counts and AFB_1 toxin production by *Aspergillus flavus* in maize. Toxins 15: 227. https://doi.org/10.3390/toxins15030227.

Mostrom, M., 2016. Mycotoxins: Toxicology. Encyclopedia of Food and Health, North Dakota State University, Fargo, ND, USA, 43-48.

Munday-Finch, S.C., Wilkins, A.L. and Miles, C.O., 1998. Isolation of Lolicine A, Lolicine B, Lolitriol, and Lolitrem N from Lolium perenne Infected with *Neotyphodium lolii* and evidence for the natural occurrence of 31-Epilolitrem N and 31-Epilolitrem F. Journal of Agricultural and Food Chemistry 46: 590-598. https://doi.org/10.1021/jf9706787.

Munkvold, G.P., 2003. Cultural and genetic approaches to managing mycotoxins in maize. Annual Review of Phytopathology 41: 99-116. https://doi.org/10.1146/annurev.phyto.41.052002.095510.

Nada, S., Nikola, T., Bozidar, U., Ilija, D. and Andreja, R., 2022. Prevention and practical strategies to control mycotoxins in the wheat and maize chain. Food Control 136: 108855. https://doi.org/10.1016/j.foodcont.2022.108855.

O'Donnell, K., Ward, T.J., Aberra, D., Kistler, H.C., Aoki, T., Orwig, N., Kimura, M., Bjørnstad, Å. and Klemsdal, S.S., 2008. Multilocus genotyping and molecular phylogenetics resolve a novel head blight pathogen within the *Fusarium graminearum* species complex from Ethiopia. Fungal Genetics and Biology 45: 1514-1522. https://doi.org/10.1016/j.fgb.2008.09.002.

O'Donnell, K., Ward, T.J., Geiser, D.M., Kistler, H.C. and Aoki, T., 2004. Genealogical concordance between mating type locus and seven other nuclear genes supports formal recognition of nine phylogenetically distinct species within the *Fusarium graminearum* clade. Fungal Genetics and Biology 41: 600-623. https://doi.org/10.1016/j.fgb.2004.03.003.

O'Donnell, K., Kistler, H.C., Tacke, B.K. and Casper, H.H., 2000. Gene genealogies reveal global phylogeographic structure and reproductive isolation among 55 lineages of *Fusarium graminearum*, the fungus causing wheat scab. Proceedings of the National Academy of Sciences 97: 7905-7910. https://doi.org/10.1073/pnas.130193297.

Odriozola, E., Odeon, A., Canton, G., Clemente, G. and Escande, A., 2005. *Diplodia maydis*: a cause of death of cattle in Argentina. New Zealand Veterinary Journal 53: 160e161. https://doi.org/10.1080/00480169.2005.36496.

Okubara, P., Blechl, A., McCormick, S., Alexander, N., Dill-Macky, R. and Hohn, T., 2002. Engineering deoxynivalenol metabolism in wheat through the expression of a fungal trichothecene acetyltransferase gene. Theoretical and Applied Genetics 106: 74-83. https://doi.org/10.1007/s00122-002-1066-2.

Oliveira, R.C., Gonçalves, S.S., Oliveira, M.S., Dilkin, P., Mallmann, C.A., Freitas, R.S., Bianchi, P. and Correa, B., 2017. Natural occurrence of tenuazonic acid and *Phoma sorghina* in Brazilian sorghum grains at different maturity stages. Food Chemistry 230: 491-496. https://doi.org/10.1016/j.foodchem.2017.03.079.

Orlando, B., Grignon, G., Vitry, C., Kashefifard, K. and Valade, R., 2019. *Fusarium* species and enniatin mycotoxins in wheat, durum wheat, triticale and barley harvested in France. Mycotoxin Research 35: 369-380. https://doi.org/10.1007/s12550-019-00363-x.

Osborne, L.E. and Stein, J.M., 2007. Epidemiology of Fusarium head blight on small-grain cereals. International journal of food microbiology 119: 103-108. https://doi.org/10.1016/j.ijfoodmicro.2007.07.032.

Ostazeski, S.A. and Wells, H.D., 1960. A phomopsis stem blight of yellow lupine (*Lupinus luteus* L.). The Plant Disease Reporter 44: 66-67.

Ostry, V., 2008. *Alternaria* mycotoxins: an overview of chemical characterization, producers, toxicity, analysis and occurrence in foodstuffs. World Mycotoxin Journal 1: 175-188. https://doi.org/10.3920/WMJ2008.x013.

Panaccione, D.G., Beaulieu, W.T. and Cook, D., 2014. Bioactive alkaloids in vertically transmitted fungal endophytes. Functional Ecology 28: 299–314. https://doi.org/10.1111/1365-2435.12076.

Parry, D.W., Jenkinson, P. and McLeod, L., 1995. Fusarium ear blight (scab) in small grain cereals–a review. Plant Pathology 44: 207-238. https://doi.org/10.1111/j.1365-3059.1995.tb02773.x.

Payne, G.A. and Widstrom, N.W. 1992. Aflatoxin in maize. Critical Reviews in Plant Sciences 10: 423-440.

Petterson, D.S., Peterson, J.E., Smith, L.W., Wood, P.M. and Culvenor, C.C.J., 1985. Bioassay of the contamination of lupin seed by the mycotoxin phomopsin. Australian Journal of Experimental Agriculture 25: 434-439. https://doi.org/10.1071/EA9850434.

Philippe, G., 2016. Lolitrem B and indole diterpene alkaloids produced by endophytic fungi of the genus *Epichloë* and their toxic effects in livestock. Toxins 8: 47. https://doi.org/10.3390/toxins8020047.

Pitt, J.I., 1987. *Penicillium viridicatum, Penicillium verrucosum,* and production of ochratoxin A. Applied Environmental Microbiology 53: 266-269. https://doi.org/10.1128/aem.53.2.266-269.1987.

Plasencia, J. and Mirocha, C.J., 1991. Isolation and characterization of zearalenone sulfate produced by *Fusarium* spp. Applied and Environmental Microbiology 57: 146-150. https://doi.org/10.1128/aem.57.1.146-150.1991.

Pleadin, J., Kovačević, D. and Perković, I. 2015. Impact of casing damaging on aflatoxin B1 concentration during the ripening of dry-fermented meat sausages. Journal of Immunoassay and Immunochemistry 36: 655-666. https://doi.org/10.1080/15321819.2015.1032306.

Polak-Śliwińska, M. and Paszczyk, B., 2021. Trichothecenes in food and feed, relevance to human and animal health and methods of detection: a systematic review. Molecules 26: 454. https://doi.org/10.3390/molecules26020454.

Popescu, R.G., Rădulescu, A.L., Georgescu, S.E. and Dinischiotu, A. 2022. Aflatoxins in feed: types, metabolism, health consequences in swine and mitigation strategies. Toxins 14: 853. https://doi.org/10.3390/toxins14120853.

Poppenberger, B., Berthiller, F., Lucyshyn, D., Sieberer, T., Schuhmacher, R., Krska, R., Kuchler, K., Glossl, J., Luschnig, C. and Adam, G., 2003. Detoxification of the *Fusarium* mycotoxin deoxynivalenol by a UDP-glucosyltransferase from *Arabidopsis thaliana*. Journal of Biological Chemistry 278: 47905–47914. https://doi.org/10.1074/jbc.M307552200.

Punithalingam, E., 1974. Studies on Sphaeropsidales in culture. II. Mycological Papers 136: 63.

Rabie, C.J., Du, J.J. and Hayes, J.P., 1987. Toxicity of *Diplodia maydis* to broilers, ducklings, and laying chicken hens. Poultry Science 66: 1123-1128. https://doi.org/10.3382/ps.0661123.

Rabie, C.J., Van Rensburg, S.J., Kriek, N.P.J. and Lübben, A., 1977. Toxicity of *Diplodia maydis* to laboratory animals. Applied and Environmental Microbiology 34: 111-114. https://doi.org/10.1128/aem.34.2.111-114.1977.

Reddy, K.R.N., Salleh, B., Saad, B., Abbas, H.K., Abel, C.A. and Shier, W.T., 2010. An overview of mycotoxin contamination in foods and its implications for human health. Toxin reviews 29: 3-26. https://doi.org/10.3109/15569541003598553.

Reddy, P., Hemsworth, J., Guthridge, K.M., Vinh, A., Vassiliadis, S., Ezernieks, V., Spangenberg, G.C. and Rochfort, S.J., 2020. Ergot alkaloid mycotoxins: physiological effects, metabolism and distribution of the residual toxin in mice. Scientific Reports 10: 9714. https://doi.org/10.1038/S41598-020-66358-2.

Repussard, C., Zbib, N., Tardieu, D. and Guerre, P., 2014. Endophyte infection of tall fescue and the impact of climatic factors on ergovaline concentrations in field crops cultivated in southern France. Journal of Agricultural and Food Chemistry 62: 9609–9614. https://doi.org/10.1021/jf503015m.

Riba, A., Mokrane, S., Mathieu, F., Lebrihi, A. and Sabaou, N., 2008. Mycoflora and ochratoxin A producing strains of *Aspergillus* in Algerian wheat. International Journal of Food Microbiology 122: 85-92. https://doi.org/10.1016/j.ijfoodmicro.2007.11.057.

Richard, J.L., 2007. Some major mycotoxins and their mycotoxicoses—An overview. International journal of food microbiology 119: 3-10. https://doi.org/10.1016/j.ijfoodmicro.2007.07.019.

Riet-Correa, F., Rivero, R., Odriozola, E., Adrien, M.D., Medeiros, R.M.T. and Schild, A.L., 2013. Mycotoxicoses of ruminants and horses. Journal of Veterinary Diagnostic Investigation 25: 692-708. https://doi.org/10.1177/1040638713504572.

Rodrigues, I. and Naehrer, K., 2012. Prevalence of mycotoxins in feedstuffs and feed surveyed worldwide in 2009 and 2010. Phytopathologia Mediterranea 51: 175-192. https://doi.org/10.14601/Phytopathol_Mediterr-9693.

Rogers, K.D., Cannistra, J.C., Gloer, J.B. and Wicklow, D.T., 2014. Diplodiatoxin, chaetoglobosins, and diplonine associated with a field outbreak of Stenocarpella ear rot in Illinois. Mycotoxin Research 30: 61-70. https://doi.org/10.1007/s12550-014-0188-0.

Rosett, T., Sankhala, R.H., Stickings, C.E., Taylor, M.E. and Thomas, R., 1957. Studies in the biochemistry of micro-organisms. 103. Metabolites of *Alternaria tenuis auct*; culture filtrate products. Biochemistry Journal 67: 390-400. https://doi.org/10.1042/bj0670390.

Rychlik, M., Humpf, H.U., Marko, D., Dänicke, S., Mally, A., Berthiller, F., Klaffke, H. and Lorenz, N., 2014. Proposal of a comprehensive definition of modified and other forms of mycotoxins including "masked" mycotoxins. Mycotoxin Research 30: 197-205. https://doi.org/10.1007/s12550-014-0203-5.

Saleh, A.A., El_Komy, M.H., Eranthodi, A., Hamoud, A.S. and Molan, Y.Y., 2015. Variation in a molecular marker for resistance of Saudi date palm germplasm to *Fusarium oxysporum f. sp. albedinis* the causal agent of Bayoud disease. European journal of plant pathology 143: 507-514. https://doi.org/10.1007/s10658-015-0703-y.

Sanchez, P.L., Venema, D., Mol, H., Spanjer, M., de Stoppelaar, J., Pfeiffer, E. and de Nijs, M., 2016. *Alternaria* toxins and conjugates in selected foods in the Netherlands. Food Control 69: 153-159. https://doi.org/10.1016/j.foodcont.2016.04.001.

Santini, A., Meca, G., Uhlig, S. and Ritieni, A., 2012. Fusaproliferin, beauvericin and enniatins: occurrence in food – a review. World Mycotoxin Journal 5: 71-81. https://doi.org/10.3920/WMJ2011.1331.

Schloß, S., Koch, M., Rohn, S. and Maul, R., 2015. Development of a SIDA-LC-MS/MS method for the determination of phomopsin A in legumes. Journal of Agricultural and Food Chemistry 63: 10543-10549. https://doi.org/10.1021/acs.jafc.5b04792.

Schmidt-Heydt, M., Magan, N. and Geisen, R., 2008. Stress induction of mycotoxin biosynthesis genes by abiotic factors. FEMS Microbiology Letters 284: 142–149. https://doi.org/10.1111/j.1574-6968.2008.01182.x.

Schneweis, I., Meyer, K., Engelhardt, G. and Bauer, J., 2002. Occurrence of zearalenone-4-β-D-glucopyranoside in wheat. Journal of Agricultural and Food Chemistry 50: 1736-1738. https://doi.org/10.1021/jf010802t.

Schollenberger, M., Müller, H.M., Rüfle, M., Suchy, S., Planck, S. and Drochner, W., 2006. Natural occurrence of 16 *Fusarium* toxins in grains and feedstuffs of plant origin from Germany. Mycopathologia 161: 43-52. https://doi.org/10.1007/s11046-005-0199-7.

Schryvers, S., Jacxsens, L., Croubels, S., Vonck, S., Miserez, B., Van De Steene, J., Necchi Rohers, G. and Eeckhout, M., 2024. Quinolizidine alkaloids and phomopsin A in animal feed containing lupins: co-occurrence and carry-over into veal products. Food Additives and Contaminants: Part A 1-15. https://doi.org/10.6084/m9.figshare.26008246.v1.

Scott, P.M., 2001. Analysis of agricultural commodities and foods for *Alternaria* mycotoxins. Journal of AOAC International 84: 1809-1817. https://doi.org/10.1093/jaoac/84.6.1809.

Scott, P.M., van Walbeek, W., Harwig, J. and Fennell, D.I., 1970. Occurrence of a mycotoxin, Ochratoxin A, in wheat and isolation of Ochratoxin A and citrinin producing strains of *Penicillium viridicatum*. Canadian Journal in Plant Science 50: 583–585. https://doi.org/10.4141/cjps70-108.

Scudamore, K.A. and Patel, S., 2000. Survey for aflatoxins, ochratoxin A, zearalenone and fumonisins in maize. Food Additives and Contaminants 17: 407–416.

Shabeer, S., Asad, S., Jamal, A. and Ali, A., 2022. Aflatoxin contamination, its impact and management strategies: an updated review. Toxins 14: 307. https://doi.org/10.3390/toxins14050307.

Shivas, R., Allen, J. and Williamson, P., 1991. Infraspecific variation demonstrated in Phomopsis leptostromiformis using cultural and biochemical techniques. Mycological Research 95: 320-323. https://doi.org/10.1016/S0953-7562(09)81241-0.

Shivas, R.G., Allen, J.G., Edgar, J.A., Cockrum, P.A., Gallagher, P.F., Ellis, Z. and Harvey, M., 1994. In: S.M. Colegate and P.R. Dorling (Eds.), Plant-associated toxins. CAB International, Wallingford, UK, pp. 161-166.

Shotwell, O.L., Hesseltine, C.W. and Goulden, M.L., 1969 Ochratoxin A: Occurrence as natural contaminant of a corn sample. Applied Microbiology 17: 765–766. https://doi.org/10.1128/am.17.5.765-766.1969.

Shukla, S., Kim, D.H., Chung, S.H., and Kim, M. 2016. 'Occurrence of aflatoxins in fermented food products' in J. Frias et al. (eds). Fermented Foods in Health and Disease Prevention. Academic Press, pp 653-674.

Shurtleff, M.C., 1980. Compendium of corn diseases. 2nd Edition. American Phytopathological Society, St. Paul, MN, USA.

Simmons, E.G., 2007. *Alternaria*: an identification manual, CBS Fungal Biodiversity Center, Utrecht, The Netherlands.

Snyman, L.d., Kellerman, T.S., Vleggar, R., Flett, B.C., Basson, K.M. and Schultz, R.A., 2011. Diplonine, a neurotoxin isolated from cultures of the fungus *Stenocarpella maydis* (Berk.) Sacc. that induces diplodiosis. Journal of Agricultural and Food Chemistry 59: 9039-9044. https://doi.org/10.1021/jf202735e.

Soler Rodriguez, F., Miguez Santiyan, M.P., Pedrera Zamorano, J.D., and Roncero Cordero, V. 1991. An outbreak of lupinosis in sheep. Veterinary and Human Toxicology 33: 492-494.

Sørensen, J.L. and Sondergaard, T.E., 2014. The effects of different yeast extracts on secondary metabolite production in *Fusarium*. International Journal of Food Microbiology 170: 55-60. https://doi.org/10.1016/j.ijfoodmicro.2013.10.024.

Sørensen, J.L., Giese, H., 2013. Influence of carbohydrates on secondary metabolism in *Fusarium avenaceum*. Toxins 5: 1655–1663. https://doi.org/10.3390/toxins5091655.

Speijers, G.J.A. and Van Egmond, H.P., 1993. Worldwide ochratoxin A levels in food and feeds. In: Creppy, E.E., Castegnaro, M., Dirheimer, G. (eds.) Human ochratoxicosis and its pathologies. Montrouge: Colloque INSERM/John Libbey Eurotext Ltd. p 85-100.

Starkey, D.E., Ward, T.J., Aoki, T., Gale, L.R., Kistler, H.C., Geiser, D.M., Suga, H., Tóth, B., Varga, J. and O'Donnell, K., 2007. Global molecular surveillance reveals novel Fusarium head blight species and trichothecene toxin diversity. Fungal Genetics and Biology 44: 1191-1204. https://doi.org/10.1016/j.fgb.2007.03.001.

Stępień, L. and Waśkiewicz, A., 2013. Sequence divergence of the enniatin synthase gene in relation to production of beauvericin and enniatins in *Fusarium* species. Toxins 5: 537-555. https://doi.org/10.3390/toxins5030537.

Steyn, P.S. and Rabie, C.J., 1976. Characterization of magnesium and calcium tenuazonate from Phoma sorghina. Phytochemistry 15, 1977–1979. https://doi.org/10.1016/S0031-9422(00)88860-3.

Stroka, J. and Gonçalves, C., 2019. Mycotoxins in food and feed: an overview. Encyclopedia of Food Chemistry 1: 401-419. https://doi.org/10.1016/B978-0-08-100596-5.21801-5.

Supothina, S., Isaka, M., Kirtikara, K., Tanticharoen, M., Thebtaranonth, Y., 2004. Enniatin production by the entomopathogenic fungus *Verticillium hemipterigenum* BCC 1449. Journal of Antibiotics 57: 732–738. https://doi.org/10.7164/antibiotics.57.732.

Sutton, B.C. and Waterston, J.M., 1966. *Diplodia maydis*. CMI Descriptions of 525 pathogenic fungi and bacteria. Set 9. No. 84. CAB International, Willingford, UK.

Sutton, B.C., 1964. Coelomycetes III. Annellolacinia Gen. Nov., Aristastoma, Phaeocytostroma, Seimatosporium, etc. Commonwealth Mycological Institute.

Sutton, B.C., 1980. The Coelomycetes: *fungi imperfecti* with pycnidia, acervuli, and stromata. Commonwealth Mycological Institute, Kew, Surrey, England.

Sy-Cordero, A.A., Pearce, C.J. and Oberlies, N.H., 2012. Revisiting the enniatins: a review of their isolation, biosynthesis, structure determination, and biological activities. Journal of Antibiotics 65: 541-549. https://doi.org/10.1038/ja.2012.71.

Thakur, R.P., 1991. Influence of temperature and wetness duration on infection of pearl millet by *Claviceps fusiformis*. Phytopathology 81: 835. https://doi.org/10.1094/PHYTO-81-835.

Theiler, A., 1927. Die diplodiosis der rinder und scafe in Süd-Afrika. Deutschen Tierärztlichen Wochenschrift 35: 395-399.

Thirumurugan, D., Cholarajan, A., Raja, S.S. and Vijayakumar, R., 2018. An introductory chapter: secondary metabolites. In: Vijayakumar, R. and Raja, S.S. (eds.) Secondary Metabolites-Sources and Applications. IntechOpen, London, UK. https://doi.org/10.5772/INTECHOPEN.79766.

Thomma, B.P.H.J., 2003. *Alternaria spp.*: from general saprophyte to specific parasite. Molecular Plant Pathology 4: 225-236. https://doi.org/10.1046/j.1364-3703.2003.00173.x.

Uhlig, S., Torp, M. and Heier, B.T., 2006. Beauvericin and enniatins A, A1, B and B1 in Norwegian grain: a survey. Food Chemistry 94: 193-201. https://doi.org/10.1016/j.foodchem.2004.11.004.

Umetsu, N., Kaji, J. and Tamari, K., 1972. Isolation of tenuazonic acid from blast-diseased rice plants. Agricultural and Biological Chemistry 37: 451–452. https://doi.org/1080/00021369.1973.10860692.

Van der Bijl, P.A., 1914. Preliminary investigation on the deterioration of maize infected with *Diplodia zeae* (Schw.) Lev. Transactions of the Royal Society of South Africa 4: 231-239.

Van der Merwe, K.J., Steyn, P.S., Fourie, L., Scott, D.B. and Theron, J.J., 1965. Ochratoxin A, a toxic metabolite produced by *Aspergillus ochraceus* Wilh. Nature 205: 1112-1113. https://doi.org/10.1038/2051112a0.

Van Duijin, G. 2014. 'Safety of Food and Beverages: Oils and Fats' in Y. Motarjemi (ed). Encyclopedia of Food Safety. Academic Press, pp 315-323.

Van Rensburg, S.J. and Altenkirk, B., 1974. *Claviceps purpurea*-ergotism. In: Mycotoxins, Elsevier, New York.

van Walbeek, W., Scott, P.M., Harwig, J. and Lawrence, J.W., 1969. *Penicillium viridicatum*

Westling: A new source of Ochratoxin A. Canadian Journal of Microbiology 15: 1281–1285. https://doi.org/10.1139/m69-232.

Van Warmelo, K.T. and Marasas, W.F.O., 1972. *Phomopsis leptostromiformis*: the causal fungus of lupinosis, a mycotoxicosis, in sheep. Mycologia 64: 316-324. https://doi.org/10.2307/3757835.

Varga, J., Tóth, B., Mesterházy, Á., Téren, J. and Fazekas, B., 2004. Mycotoxigenic fungi and mycotoxins in foods and feeds in Hungary. An overview on toxigenic fungi and mycotoxins in Europe 123-139.

Villani, A., Proctor, R.H., Kim, H.S., Brown, D.W., Logrieco, A.F., Amatulli, M.T., Moretti, A. and Susca, A., 2019. Variation in secondary metabolite production potential in the *Fusarium incarnatum-equiseti* species complex revealed by comparative analysis of 13 genomes. BMC Genomics 20: 314. https://doi.org/10.1186/s12864-019-5567-7.

Vincelli, P. 1997. Ear rot of corn caused by *Stenocarpella maydis* (= *Diplodia maydis*). Univ. of Kentucky Coop. Ext. Service Rep. PPA-43. Univ. of Kentucky College of Agric., Lexington, KY.

Wang, Y., Zhou, Y., Qin, Y. and Wang, L. 2022. Effect of environmental factors on the aflatoxin production by *Aspergillus flavus* during storage in upland rice seed using response surface methodology. LWT 169: 113977. https://doi.org/10.1016/j.lwt.2022.113977.

Weidenbörner, M., 2001. T. In: Encyclopedia of Food Mycotoxins. Springer, Berlin, Heidelberg. https://doi.org/10.1007/978-3-662-04464-3_19.

Welty, R.E., Craig, A.M. and Azevedo, M.D., 1994. Variability of ergovaline in seeds and straw and endophyte infection in seeds among endophyte-infected genotypes of tall fescue. Plant Disease 78: 845-849. https://doi.org/10.1094/PD-78-0845.

White, D., 1999. Compendium of corn diseases. 3rd ed. APS Press, St. Paul, MN.

Wicklow, D.T., Rogers, K.D., Dowd, P.F. and Gloer, J.B., 2011. Bioactive metabolites from *Stenocarpella maydis*, a stalk and ear rot pathogen of maize. Fungal Biology 115: 133-142. https://doi.org/10.1016/j.funbio.2010.11.003.

Wood, P.M., Petterson, D.S., Hancock, G.R. and Brown, G.A., 1987. Distribution of seed infected with Phomopsis leptostromiformis and of phomopsin A within a lupin crop. Australian Journal of Experimental Agriculture 27: 77-79. https://doi.org/10.1071/EA9870077.

Wood, P. McR. and Sivasithamparam, K., 1989. *Diaporthe woodii* (anamorph *Phomopsis leptostromiformis*)—A toxigenic fungus infecting cultivated lupins. Mycopathologia 105: 79-86. https://doi.org/10.1007/BF00444029.

Woudenberg, J.H.C., Groenewald, J.Z., Binder, M. and Crous, P.W., 2013. *Alternaria* redefined. Studies in Mycology 75: 171-212. https://doi.org/10.3114/sim0015.

Woudenberg, J.H.C., Seidl, M.F., Groenewald, J.Z., de Vries, M., Stielow, J.B., Thomma, B.P.H.J. and Crous, P.W., 2015. *Alternaria* section *Alternaria*: Species, formae speciales or pathotypes? Studies in Mycology 82: 1-21. https://doi.org/10.1016/j.simyco.2015.07.001.

Yang, J.W., Nam, S.S., Lee, H.U., Choi, K.H., Hwang, S.G. and Paul, N.C., 2018. *Fusarium* root rot caused by *Fusarium solani* on sweet potato (*Ipomoea batatas*) in South Korea. Canadian Journal of Plant Pathology 40: 90-95. https://doi.org/10.1080/07060661.2017.1394914.

Yazar, S. and Omurtag, G.Z. 2008. Fumonisins, trichothecenes and zearalenone in cereals. International Journal of Molecular Sciences 9: 2062-2090. https://doi.org/10.3390/ijms9112062.

Yli-Mattila, T., Gagkaeva, T., Ward, T.J., Aoki, T., Kistler, H.C. and O'Donnell, K., 2009. A novel Asian clade within the *Fusarium graminearum* species complex includes a newly discovered cereal head blight pathogen from the Russian Far East. Mycologia 101: 841-852. https://doi.org/10.3852/08-217.

Zhang, G.L., Feng, Y.L., Song, J.L. and Zhou, X.S. 2018. Zearalenone: A mycotoxin with different toxic effect in domestic and laboratory animals' granulosa cells. Frontiers in Genetics 9: 667. https://doi.org/10.3389/fgene.2018.00667.

Zinedine A., Brera C., Elakhdari S., Catano C., Debegnac F.R., Angelini S., De Santis B., Faid M., Benlemlih M., Minardi V. and Miraglia M., 2017b. Natural occurrence of mycotoxins in cereals and spices commercialized in Morocco. Food Control 17: 868–874. https://doi.org/10.1016/j.foodcont.2005.06.001.

Zinedine, A., Soriano, J.M., Moltó, J.C. and Mañes, J., 2007a. Review on the toxicity, occurrence, metabolism, detoxification, regulations and intake of zearalenone: an oestrogenic mycotoxin. Food Chemistry and Toxicology 45: 1-18. https://doi.org/10.1016/j.fct.2006.07.030.

CHAPTER 3

The mycotoxin interactions in the frame of the chemical exposome: from the toxicodynamics to the pharmacokinetic outcomes

*I. Pavlov[1], D. Payros[2,3], I.P. Oswald[3] and I. Alassane-Kpembi[1]**

[1]Department of Veterinary Biomedicine, Faculty of Veterinary Medicine, Université de Montréal, 3200 Rue Sicotte, Saint-Hyacinthe, Québec, J2S 2M2, Canada; [2]IRSD, INSERM, Université de Toulouse, INRAE, ENVT, UPS, Toulouse, France; [3]Toxalim, Research Centre in Food Toxicology, Université de Toulouse, INRAE, ENVT, INP PURPAN, UPS, Toulouse France
**correspondence: imourana.alassane-kpembi@umontreal.ca*

1 Introduction

As shown in Chapter 2, most fungi can produce several mycotoxins simultaneously. Several fungi can contaminate commodities at the same time or in quick succession. As a result, co-contamination of feed and feedstuffs by several mycotoxins is the rule, not the exception. A global survey conducted on 74,821 samples of feed and feed raw materials collected from 100 countries indicated that 64% of all samples contained at least two mycotoxins (Tolosa et al., 2021). Other reports consider 30% to 100% of food and feed samples on a global level to be co-contaminated (Marin et al., 2013).

The reality of the exposure to a cocktail of mycotoxins is also substantiated by a wealth of data from biomonitoring studies performed to evaluate human exposure. In the European Food Consumption Validation (EFCOVAL) project, which included 600 men and women from Belgium, the Czech Republic, France, the Netherlands, and Norway, only 5% of the population tested positive for only one mycotoxin, while some blood samples were observed to contain up to 12 mycotoxins (De Ruyck et al., 2020). Likewise, up to five mycotoxins have been reported in biological fluids from severe cases of co-exposure in other biomonitoring studies conducted in southern Europe and southern Africa (Solfrizzo et al., 2014; Warth et al., 2012). An equally important animal biomonitoring study in 2012 also indicated the simultaneous presence of at least three mycotoxins in 85% of pig urine samples collected in Swedish slaughterhouses (Gambacorta et al., 2019). Unfortunately, animal biomonitoring is not yet performed on a routine basis as it is in humans, and only a few studies are available.

 | DOI: 10.1163/9789004712720_4

The toxicity of mycotoxins, like other feed contaminants, has most often been studied individually. Although this approach may be useful for establishing the toxicological profile of an individual element and determining health-based guidance values for regulatory purposes, it does not allow for the assessment of the risks associated with a cocktail of contaminants. Compounds present at low concentrations, which on their own do not produce significant effects, can become more harmful due to toxicodynamic interactions and elicit substantial mixture responses (Thrupp et al., 2018). Therefore, an increasing number of studies are being devoted to the exploration of the toxicodynamic interactions of the mycotoxin mixtures humans and animals may be exposed to.

Co-exposure to mycotoxins reproduces on a smaller scale a more complex reality. The exposome concept has recently been coined to figure out the whole chemical exposure, among other environmental factors, relevant for adverse health and welfare effects (Marín et al., 2018). It is described as the totality of internal organism exposure with regard to exogenous chemicals, their biotransformation products, and endogenous chemicals that are produced or altered in response to external stressors and thereby may pose threats to health (Huhn et al., 2021). Research is therefore needed into the way in which mycotoxins interact with other elements in the exposome to jointly modify the overall risk of disease.

Not only does the exposome directly affect health outcomes, but it is becoming clear that it also modulates and interacts with drug-metabolising enzymes, xenobiotic-sensing receptors, and drug transporters, giving rise to changes in drug action and/or adverse effects during treatment (Pristner and Warth, 2020). Assuming that the presence of mycotoxins in food or feed may affect medication in the body, some studies have gone deeper into toxicokinetic interactions, i.e. the effect of mycotoxin exposure on the transport and metabolism of concurrently administered drugs. The importance of such studies is directly linked to the pharmacotherapy outcomes for drug-treated humans or animals, or the public health outcomes for consumers of drug-treated animals for food.

The present chapter summarises our current knowledge of the toxicodynamic and pharmacokinetic interactions in which mycotoxins may be involved with other xenobiotics, because of their contemporaneous effects on human or animal organisms.

2 Toxicodynamic interactions of mycotoxins

2.1 *Theoretical considerations*

Owing to the importance and complexity of the toxicodynamic interactions of chemical mixtures, several regulatory bodies have codified the general principles of their analysis (Figure 3.1) (Wilbur, 2004). Testing for a toxicodynamic interaction in mixture toxicity requires comparing the measured effect for the mixture to a

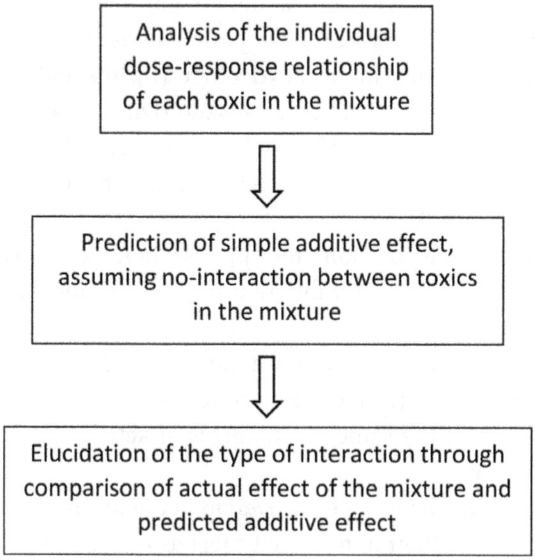

FIGURE 3.1 Stepwise analysis of the toxicological interaction in a mixture.

predicted no interaction effect, the so-called additive effect. This prediction of no interaction is based on the measured toxicities of each compound in the mixture. A stronger-than-predicted effect indicates synergism, whereas a lower-than-predicted effect is defined as antagonism.

Different methodological approaches have been used to define a reference additive effect in mycotoxin interaction studies (Alassane-Kpembi et al., 2017).

Some mycotoxin combination studies considered the additive effect to equal the arithmetic sum of the sizes of the effects for individual mycotoxins (Gayathri et al., 2015). However, there is now a large consensus in the biomedical fields that the combined effect of a mixture does not simply equal the sum of the single effects measured for individual compounds when tested separately (Boedeker and Backhaus, 2010).

Many other mycotoxin combination studies used factorial design experiments to investigate the interactions (Lei et al., 2013). The general assumption behind the factorial design is that when testing the effects of a mixture of non-interacting compounds in different combination patterns, the effect of any compound can be predicted by subtracting the mean of the groups not containing that compound from the mean of the other groups (Groten et al., 2001). Even though interaction is detected by such an approach, the type of interaction regarding additivity, synergism or antagonism is not clearly indicated and must be inferred indirectly (Bhat and Ahangar, 2007).

Besides the arithmetic definition of interaction and the factorial design approach, more advanced theoretical biology model-based approaches are gaining increased attention in the field of mycotoxicology. The Bliss independence criterion also known as Response Addition, the Loewe additivity model also called the Concentration or Dose Addition, and the Median Effect Principle of the Mass action law are the main theoretical biology models that have been applied to mycotoxin mixtures (Kifer et al., 2020).

The Bliss independence criterion only applies when the assessed mycotoxins in the mixture exert their toxicity via different modes and possibly different sites of action (Szabó-Fodor et al., 2019).

The pivotal assumption of the Loewe additivity model and its algebraic extension, the interaction index, is that the mycotoxins under investigation act on the same biological sites, by the same mechanisms of action and differ only in their potency (Oh et al., 2017).

Because it is unlikely that there is no overlap in the mechanisms of action of the compounds assessed for their joint toxicity, just as it is unlikely that their biological targets are strictly the same, a number of mycotoxin studies favour the Median-Effect Equation of the Mass action Law and its extension, the combination index-isobologram method (Aupanun et al., 2019; Kolf-Clauw et al., 2013; Vejdovszky et al., 2017ab). This last approach is independent of the mechanisms of action and just considers both the potency (i.e. EC_{50}) and the shape of the dose-effect curves for each mycotoxin and their mixture (Chou, 2006).

2.2 *Interactions between mycotoxins*

The combined effect of DON and other co-occurring type B trichothecene mycotoxins on intestinal health was assessed applying the combination index-isobologram method (Figure 3.2) (Alassane-Kpembi et al., 2013; 2015; 2017; Cheat et al., 2016). Using proliferating human and porcine intestinal cells, as well as fully differentiated porcine jejunal explants and intestinal loops, it was observed that low doses of trichothecene mycotoxins, clearly below the current No-Observed Adverse Effect Levels (NOAEL), synergistically impair the renewing of epithelial cells and increase the expression of pro-inflammatory genes in the intestine, with magnitude of synergy ranging from 2 to 10. These synergies were mainly observed at low concentrations. At higher concentrations, combinations of mycotoxins showed additive or antagonistic effects.

Considering the frequent co-occurrence of type B trichothecene mycotoxins in diet, and the concentrations of toxins to which consumers and animals are exposed, these results highlight the need for a re-evaluation of the risk associated with DON and its congeners in human and animal, taking into account combined exposure patterns.

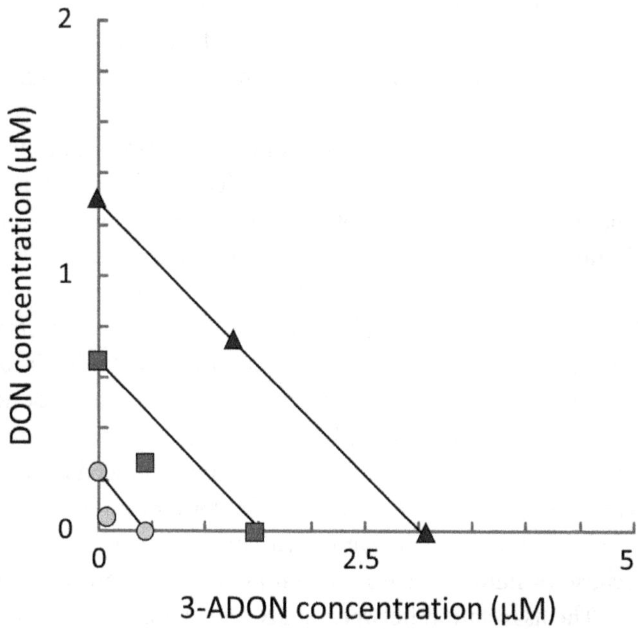

FIGURE 3.2 Isobologram illustrating the combined cytotoxicity of trichothecenes DON and 3-ADON at concentrations eliciting 10% (circles), 30% (squares) or 50% (triangles) inhibition of the viability of intestinal Caco-2 cells. In this type of graph, the additive combined effect follows the diagonal line between the effective concentrations of each single toxin. If the measured combined effect of two toxins is above or below the diagonal line, it indicates an antagonistic or a synergistic effect of the combination respectively.
NOTE. The figure was created from (Alassane-Kpembi et al., 2013).

Synergistic interactions induced by low doses of other *Fusarium* mycotoxins have also been reported for hepatotoxicity and alteration of the expression of biotransformation enzymes, using a human hepatocyte line (Smith et al., 2017).

Besides the combined toxicity of the only regulated mycotoxins, combination patterns involving the emerging mycotoxins are also under scrutiny. For example, synergistic oestrogenic activity has been identified for ZEN and the emerging mycotoxin AOH, using a human cell line (Vejdovszky et al., 2017a). Conversely, the ten most represented emerging mycotoxins in pig feed demonstrated no interaction with DON for cytotoxicity toward the porcine enterocyte line IPEC-J2 (Khoshal et al., 2019). The combined toxicities of all the tested mixtures were not significantly different from that of DON alone. It thus appears that the co-occurrence of the tested emerging mycotoxins with DON does not exacerbate the toxicity of this mycotoxin alone in pig enterocytes.

2.3 *Interactions of mycotoxins with other feed contaminants*

Many contaminants such as pesticides, heavy metals, toxins, etc. may be present in feed, but the toxicity of their mixtures with mycotoxins is only beginning to be analysed.

Due to their broad use, pesticides frequently contaminate feed and can therefore interact with mycotoxins. Most studies, however, are dedicated to human exposure. One study from Spain indicated that 62% of samples of a red wine contained AFB_2 and the fungicide metalaxyl (Pérez-Ortega et al., 2012). *Fusarium* and *Alternaria* toxins, too, have been detected in mixtures with fungicides, insecticides, or herbicides in feed (Mol et al., 2014). A few studies have investigated the effects of these mixtures. For example, the mixture of ZEN and the insecticide dichlorodiphenyltrichloroethane (DTT) has been shown to increase testosterone production by Leydig cells, which suggests that ZEN-DTT co-exposure may exacerbate adverse reproductive effects (Eze et al., 2018). Likewise, the patulin-induced hepatotoxicity in mice is significantly enhanced in the presence of chlorpyrifos (Lu et al., 2019).

Mycotoxins can also be found in association with heavy metals or other chemical contaminants because of natural environmental pollution or industrial practices (Kovač et al., 2021). The association between DON and cadmium showed that the type of interaction between these two contaminants varies from antagonism to synergy, depending on the investigated organ (Le et al., 2018). At the intestinal level, the effects induced by the co-exposure on the barrier function are similar to the effects of each contaminant taken individually (Luo et al., 2019). Likewise, the combined endocrine disruptor activity of bisphenol A, a food contaminant that arises from its use in plastics, and the mycoestrogens ZEN and AOH appeared moderately antagonistic to almost additive, when tested in human endometrial cells (Aichinger et al., 2020).

A largely overlooked pattern of co-occurrence of food and feed contaminants is the simultaneous exposure to fungal and bacterial toxins, although some non-fungal microbial metabolites, e.g. chloramphenicol, nonactin and geldanamycin, have been reported along with mycotoxin contamination in staple products in Africa (Adetunji et al., 2014). One of the rare mixtures of fungal and bacterial toxins that has been assessed so far is the bacterial toxin cereulide and DON that were found to co-contaminate a frequently consumed maize porridge in West and Central Africa (Abia et al., 2017). As with DON, which is also called 'vomitoxin' due to its characteristic symptom of acute intoxication, the main symptoms of acute cereulide poisoning include nausea and emesis (Ehling-Schulz et al., 2004). Both toxins appeared to synergise in inducing cytotoxicity and altering the barrier function and the immune response of the human intestine (Beisl et al., 2020; 2021). Attention needs to be paid to whether and to what extent the observed synergy at sub-emetic doses of cereulide and DON affects the acute reference dose established for DON based on its emetic activity.

Another study investigated the combined effects of DON and colibactin, a genotoxin that is produced by some *Escherichia coli* strains and thought to be associated with the onset of colorectal cancers in an increasing number of human beings from developed countries (Payros et al., 2017). A significant increase in DNA damage was observed in intestinal epithelial cells of rats colonised by *E. coli* strains producing colibactin and co-exposed to DON, compared to animals colonised with *E. coli* strains unable to produce colibactin, or animals exposed only to DON. This highlighting of the exacerbation of the genotoxicity of colibactin by DON raises questions about the synergism between the mycotoxins contaminating diet and gut microbiota regarding intestinal carcinogenesis. More data is required to answer these questions and deepen our understanding of the consequences of co-exposure to fungal and bacterial toxins in humans and in agriculturally important animals.

2.4 *Interactions of mycotoxins with feed constituents*

Toxicodynamic interactions can arise between mycotoxins and some plant constituents. The soya isoflavone genistein has been investigated for combined oestrogenic activity with ZEN or AOH (Vejdovszky et al., 2017b). Depending on the tested concentration ranges, combinations of the phytoestrogen with either ZEN or AOH showed synergism and antagonism, synergy being the main interaction for the combinations of ZEN with genistein. Conversely, the hop polyphenols xanthohumol and 8-prenyl-naringenin that co-occur in beer with ZEN antagonised the oestrogenic effects of the *Fusarium* toxin and its reductive metabolite α-ZEL, at concentrations that can be reached by moderate beer consumption (Aichinger *et al.*, 2018). Also, it recently emerged that the aetiology of the Balkan endemic nephropathy that was first associated with OTA exposure, is more likely to be linked to chronic exposure to the plant toxin aristolochic acid (Marin et al., 2018). However, a study in rodents has also shown that OTA exacerbates the formation of DNA adducts caused by aristolochic acid in the kidneys, suggesting that both compounds may interact in promoting the development of cancer (Stiborová et al., 2015).

The issue of interaction between mycotoxins and certain plant constituents is also acute in the context of the constantly growing need for livestock feed production. Besides contaminating mycotoxins, livestock feed also contains considerable amounts of plant metabolites, such as phytoestrogens, that can act as endocrine disruptors, causing reproductive dysfunctions and even resulting in complete infertility (Mostrom and Evans, 2011). Although cancer development is not expected to be an issue in farm animals, the chronic oxidative stress caused by these toxins and their interaction with other compounds may result in a negative impact on health and welfare.

3 Toxicokinetic interactions of mycotoxins in xenobiotic metabolism and transport

Xenobiotics, including drugs, are processed in the body and eliminated via various chemical reactions. These reactions can be split into three phases (I-III), which actually tell us nothing about their order of occurrence. Phase I metabolising enzymes are mainly represented by the cytochrome p450 (CYP) superfamily and are responsible for oxidation, reduction, and hydrolysis reactions. The primary role of Phase II enzymes, such as UDP-glucuronosyltransferases, sulfotransferases, N-acetyltransferases, glutathione S-transferase (GST) and methyltransferases, is to turn the intermediate metabolites of phase I into more polar end products ready for excretion. The Phase III of biotransformation reactions describes the reactions and proteins involved in the transport of the xenobiotic metabolites across different membranes. The main proteins involved in the transport of xenobiotics are P-glycoprotein or multidrug resistance protein 1 (P-gp, MDR1), multidrug resistance associated protein 2 (MRP2) and breast cancer resistance protein (BCRP) (Almazroo et al., 2017). These proteins are located in the apical membrane of epithelial cells (e.g. in the intestinal wall and blood-brain barrier), where they can actively extrude a variety of structurally diverse drugs and drug metabolites.

The biotransformation of xenobiotics is under the upstream control of three main regulators, the so-called xenosensors, which are the aryl hydrocarbon receptor (AhR), constitutive androstane receptor (CAR) and pregnane X receptor (PXR) (Hakkola et al., 2018).

Several publications have established the interactions of mycotoxins with all the stages of the metabolism of xenobiotics. Especially in drugs, this could have pharmacotherapy outcomes for the patient, whether human or animal. It could also translate into public health outcomes for the consumers of food animals.

3.1 *Interaction of mycotoxins with xenosensing*

Xenosensing is a phenomenon whereby a compound interacts with specific transcription factors resulting in the activation of a battery of genes that mediate the metabolism and transport of xenobiotics, thereby enhancing their elimination by the organism (Plant and Aouabdi, 2009). One the most deeply characterised mechanisms for xenosensing activation by mycotoxins is the interaction between ZEN and the human pregnane X receptor (hPXR) (Ding et al., 2006). ZEN binds and causes conformational changes in hPXR that favour the release of the corepressor protein N-CoR, along with recruitment of the coactivator proteins Steroid Receptor Coactivator-1, Glucocorticoid Receptor-Interacting Protein 1 and PPAR-Binding protein.

Because the activation of xenosensors represents the molecular basis of an important class of drug interactions, ZEN is under suspicion of being a significant source of food/feed drug interactions (Ding et al., 2006; Tolson and Wang, 2010).

Indeed, ZEN also activates CAR and AhR both in human and in animal (Ayed-Boussema et al., 2011; Chang et al., 2017; Shi et al., 2018).

Besides ZEN, several other mycotoxins, including AFB_1, PAT, FB_1, OTA and the *Alternaria* toxins AOH and AME can activate either PXR and CAR, or AhR (Ayed-Boussema et al., 2012; Hassen et al., 2009; Mary et al., 2015; Schreck et al., 2012).

3.2 *Interactions of mycotoxins with the metabolism of xenobiotics*

Rodents and farm animals exposed to mycotoxins show increased or decreased activity of their biotransformation enzymes (Galtier et al., 2008). Feeding pigs or rabbits with AFB_1 or T-2 toxin contaminated feed decreases their liver levels of the CYP isoenzymes 1A, 2A, 2B and 3A (Guerre et al., 2000; Meissonnier *et al.*, 2008). By contrast, sub-chronic exposure of rodents to the structurally related type B trichothecene mycotoxins DON and NIV increases hepatic levels not only for the CYP isoenzyme 2B, but also the phase II GST isoenzymes α and π (Gouze et al., 2006; Yabe et al., 1993). As far as FB_1 is concerned, its administration to rats selectively increases their hepatic expression of CYP isoforms 1A1 and 4A1 and decreases the expression of CYP1A2 and 2C11 (Martinez-Larrañaga et al., 1996; Spotti et al., 2000).

The implications of such effects of mycotoxins are better understood with pharmacokinetic investigations using prototypical drugs that rely on specific biotransformation enzymes to be metabolised. Typical examples are midazolam (MDZ) and enrofloxacin (EFX).

MDZ is an important anaesthetic drug that is mainly metabolised in human liver by CYP3A4, 3A5, and 3A7, resulting in the production of hydroxylated metabolites (Patki et al., 2003). It is used as a model substrate for the CYP3A biotransformation pathway. The emerging mycotoxin ENNB1, which is also metabolised by CYP3A, inhibits MDZ biotransformation by human liver microsomes by decreasing its hydroxylation rate (Ivanova et al., 2019). Likewise, exposure to the CYP3A inducer ZEN significantly modifies the absolute oral bioavailability of MDZ in pigs (Schelstraete et al., 2019). This suggests that some dosage adjustment may be needed in case of organism exposure to ENNB1 or ZEN in order to maintain MDZ concentrations within its therapeutic and safety windows.

EFX is a broad-spectrum fluoroquinolone antibiotic that is frequently used in poultry medicine. In contrast to mammal species, the biotransformation of EFX in its phase I metabolite ciprofloxacin is very limited in poultry (Devreese et al., 2014). Sub-chronic exposure of broiler chicken to AFB_1 decreases the oral absorption of EFX (Kalpana et al., 2015). Moreover, compared to control animals, EFX residues in liver, kidney and fat persist for a longer time in AFB_1-exposed broiler chickens, and ciprofloxacin is detectable in tissues of the only mycotoxin-exposed chickens (Kalpana et al., 2012). This suggests that a scaling of the withdrawal period defined for food animals following administration of EFX may be needed in case of poultry exposure to AFB_1.

3.3 *Interactions of mycotoxins with the transport of xenobiotics*

Mycotoxins interfere with the transport of xenobiotics including drugs via their effects on efflux proteins. They exert contrasting effects on the expression of efflux proteins, especially MDR1 and BCRP proteins that confer cellular resistance to anticancer drugs due to acquisition of a multidrug resistance phenotype (Breedveld et al., 2006).

Induction of MDR1, MRP1, and MRP2 proteins is observed following exposure to ZEN, while the *Aspergillus fumigatus* mycotoxin fumitremorgin C reverses the multidrug resistance profile of cancer cells by drastically inhibiting BCRP (González-Lobato et al., 2010; Prouillac et al., 2009). Pre-treatment with fumitremorgin C has been shown to increase the cytotoxic activities of topotecan in ovarian and small cell lung cancer and mitoxantrone in castration-resistant prostate cancer (Scholar, 2007; Teo et al., 2019). However, the neurotoxic effects of fumitremorgin C, characterised by tremours and tetanic convulsions, preclude use *in vivo* despite its great potential as a chemosensitiser (Ahmed-Belkacem et al., 2006). Instead, it served as a basis for the development of Ko143, a non-toxic tetracyclic analogue with highly potent and specific inhibitory activity on BCRP (Allen et al., 2002).

In farm animals, a decreased or complete loss of MDR1 expression is observed in the ileum of pigs fed a diet contaminated with 1 or 3 mg/kg DON, respectively (Van der Heyden et al., 2009), while feeding approx. 18 mg/kg FUM negatively impacts the gene expression of MDR1 in the jejunum and significantly decreases the oral absorption of the MDR1 substrate drug EFX (Antonissen et al., 2017;). These interactions between mycotoxin exposure and drug regimen are of the utmost importance since they can determine pharmacotherapy outcomes.

4 Conclusions

The simultaneous exposure of organisms to mycotoxins and other types of xenobiotics can result in multifaceted interactions. Toxicodynamic interactions for the combined effects of mycotoxins and other contaminants eliciting similar toxicity endpoints have attracted by far the most attention, though preliminary data highlight the neglected potential for pharmacokinetic interaction of mycotoxins with drugs.

The studies addressing the effects of co-exposure to a cocktail of mycotoxins or mixture of mycotoxins and other food/feed contaminants, or food/feed constituents have showed that their combined toxicity cannot always be predicted based on their individual effects. Regardless of the conceptual definition of no interaction favoured in those studies, the type of interaction in terms of additivity, synergy or antagonism varies accordingly with the mixtures analysed, and even with the dose ranges tested.

Demonstrations, or even allegations, of synergism have always heightened concerns about health risks. However, the implications of the additive interactions between mycotoxins whose exposure is close to their tolerable dose intake (TDI) and other food contaminants should not be overlooked. In fact, the threshold dose for the toxic effects may be exceeded in case of exposure to the mixtures.

Many mycotoxins interfere with xenosensing. Hence, they bear a significant potential for pharmacokinetic interaction with concurrently administrated drugs. The mycotoxins modulate the expression of important classes of drug metabolising enzymes and transport proteins, which means they could modify the bioavailability, metabolite profile and elimination kinetics of interacting drugs. Owing to such interactions, some dosage adjustment may be needed to maintain drug concentrations within therapeutic and safety windows in case of organism exposure to mycotoxins. Hence, more attention should be paid to the impact of mycotoxin exposure on the fate in the organism of pharmaceuticals with a narrow therapeutic window or low therapeutic index. From the public health standpoint, a scaling of the withdrawal periods following the usage of veterinary drugs may be required for food animal products. Both the toxicodynamic and pharmacokinetic interaction potentials should be considered for the characterisation of the risk associated with mycotoxins.

References

Abia, W.A., Warth, B., Ezekiel, C.N., Sarkanj, B., Turner, P.C., Marko, D., Krska, R. and Sulyok, M. 2017. Uncommon toxic microbial metabolite patterns in traditionally home-processed maize dish (fufu) consumed in rural Cameroon. Food and Chemical Toxicology 107: 10-19. https://doi.org/10.1016/j.fct.2017.06.011.

Adetunji, M., Atanda, O., Ezekiel, C.N., Sulyok, M., Warth, B., Beltran, E., Krska, R., Obadina, O., Bakare, A. and Chilaka, C.A., 2014. Fungal and bacterial metabolites of stored maize (Zea mays, L.) from five agro-ecological zones of Nigeria. Mycotoxin Research 30: 89-102. https://doi.org/10.1007/s12550-014-0194-2.

Ahmed-Belkacem, A., Pozza, A., Macalou, S., Pérez-Victoria, J., Boumendjel, A. and Di Pietro, A., 2006. Inhibitors of cancer cell multidrug resistance mediated by breast cancer resistance protein (BCRP/ABCG2). Anticancer Drugs 17: 239-243. https://doi.org/10.1097/00001813-200603000-00001.

Aichinger, G., Beisl, J. and Marko, D., 2018. The hop polyphenols xanthohumol and 8-prenyl-naringenin antagonize the estrogenic effects of *Fusarium* mycotoxins in human endometrial cancer cells. Frontiers in Nutrition 5: 85. https://doi.org/10.3389/fnut.2018.00085.

Aichinger, G., Pantazi, F. and Marko, D., 2020. Combinatory estrogenic effects of bisphenol A in mixtures with alternariol and zearalenone in human endometrial cells. Toxicology Letters 319: 242-249. https://doi.org/10.1016/j.toxlet.2019.10.025.

Alassane-Kpembi, I., Kolf-Clauw, M., Gauthier, T., Abrami, R., Abiola, F.A., Oswald, I.P. and

Puel, O., 2013. New insights into mycotoxin mixtures: the toxicity of low doses of Type B trichothecenes on intestinal epithelial cells is synergistic. Toxicology and Applied Pharmacology 272: 191-198. https://doi.org/10.1016/j.taap.2013.05.023.

Alassane-Kpembi, I., Puel, O. and Oswald, I.P., 2015. Toxicological interactions between the mycotoxins deoxynivalenol, nivalenol and their acetylated derivatives in intestinal epithelial cells. Archives of Toxicology 89: 1337-46. https://doi.org/10.1007/s00204-014-1309-4.

Alassane-Kpembi, I., Puel, O., Pinton, P., Cossalter, A.M., Chou, T.C. and Oswald, I.P., 2017. Co-exposure to low doses of the food contaminants deoxynivalenol and nivalenol has a synergistic inflammatory effect on intestinal explants. Archives of Toxicology 91: 2677-2687. https://doi.org/10.1007/s00204-016-1902-9.

Alassane-Kpembi, I., Schatzmayr, G., Taranu, I., Marin, D., Puel, O. and Oswald, I.P., 2017. Mycotoxins co-contamination: Methodological aspects and biological relevance of combined toxicity studies. Critical Reviews in Food Science and Nutrition 57: 3489-3507. https://doi.org/10.1080/10408698.2016.1140632.

Allen, J.D., van Loevezijn, A., Lakhai, J.M., van der Valk, M., van Tellingen, O., Reid, G., Schellens, J.H.M., Koomen, G.J. and Schinkel, A.H., 2002. Potent and specific inhibition of the breast cancer resistance protein multidrug transporter in vitro and in mouse intestine by a novel analogue of fumitremorgin C. Molecular Cancer Therapeutics 1: 417-425.

Almazroo, O.A., Miah, M.K. and Venkataramanan, R. 2017. Drug metabolism in the liver. Clinical Liver Disease 21: 1-20. https://doi.org/10.1016/j.cld.2016.08.001.

Antonissen, G., Devreese, M., De Baere, S., Martel, A., Van Immerseel, F. and Croubels, S., 2017. Impact of *Fusarium* mycotoxins on hepatic and intestinal mRNA expression of cytochrome P450 enzymes and drug transporters, and on the pharmacokinetics of oral enrofloxacin in broiler chickens. Food and Chemical Toxicology 101: 75-83. https://doi.org/10.1016/j.fct.2017.01.006.

Aupanun, S., Phuektes, P., Poapolathep, S., Alassane-Kpembi, I., Oswald, I.P. and Poapolathep, A. 2019. Individual and combined cytotoxicity of major trichothecenes type B, deoxynivalenol, nivalenol, and fusarenon-X on Jurkat human T cells. Toxicon 160: 29-37. https://doi.org/10.1016/j.toxicon.2019.02.006.

Ayed-Boussema, I., Pascussi, J.M., Maurel, P., Bacha, H. and Hassen, W., 2011. Zearalenone activates pregnane X receptor, constitutive androstane receptor and aryl hydrocarbon receptor and corresponding phase I target genes mRNA in primary cultures of human hepatocytes. Environmental Toxicology Pharmacology 31: 79-87. https://doi.org/10.1016/j.etap.2010.09.008.

Ayed-Boussema, I., Pascussi, J.M., Rjiba, K., Maurel, P., Bacha, H. and Hassen, W., 2012. The mycotoxin, patulin, increases the expression of PXR and AhR and their target cytochrome P450s in primary cultured human hepatocytes. Drug and Chemical Toxicology 35: 241-250. https://doi.org/10.3109/01480545.2011.592194.

Beisl, J., Pahlke, G., Abeln, H., Ehling-Schulz, M., Del Favero, G., Varga, E., Warth, B., Sulyok, M., Abia, W., Ezekiel C.N. and Marko, D., 2020. Combinatory effects of cereulide and deoxynivalenol on *in vitro* cell viability and inflammation of human Caco-2 cells. Archives of Toxicology 94: 833-844. https://doi.org/10.1007/s00204-020-02658-w.

Beisl, J., Varga, E., Braun, D., Warth, B., Ehling-Schulz, M., Del Favero, G. and Marko, D., 2021. Assessing mixture effects of cereulide and deoxynivalenol on intestinal barrier integrity and uptake in differentiated human Caco-2 cells. Toxins 13: 189. https://doi.org/10.3390/toxins13030189.

Bhat, A.S. and Ahangar, A.A., 2007. Methods for detecting chemical-chemical interaction in toxicology. Toxicology Mechanisms and Methods 17: 441-450. https://doi.org/10.1080/153710601177654.

Boedeker, W. and Backhaus, T., 2010. The scientific assessment of combined effects of risk factors: different approaches in experimental biosciences and epidemiology. European Journal of Epidemiology 25: 539-546. https://doi.org/10.1007/s10654-0109464-2.

Breedveld, P., Beijnen, J.H. and Schellens, J.H.M., 2006. Use of P-glycoprotein and BCRP inhibitors to improve oral bioavailability and CNS penetration of anticancer drugs. Trends in Pharmacological Sciences 27: 17-24. https://doi.org/10.1016/j.tips.2005.11.009.

Chang, S., Su, Y., Sun, Y., Meng, X., Shi, B. and Shan, A., 2018. Response of the nuclear receptors PXR and CAR and their target gene mRNA expression in female piglets exposed to zearalenone. Toxicon. 151: 111-118. https://doi.org/10.1016/j.toxicon.2018.06.081.

Cheat, S., Pinton, P., Cossalter, A.M., Cognie, J., Vilariño, M., Callu, P., Raymond-Letron, I., Oswald, I.P. and Kolf-Clauw, M. 2016. The mycotoxins deoxynivalenol and nivalenol show *in vivo* synergism on jejunum enterocytes apoptosis. Food Chemistry and Toxicology 87: 45-54. https://doi.org/10.1016/j.fct.2015019.

Chou, T.C. 2006. Theoretical basis, experimental design, and computerized simulation of synergism and antagonism in drug combination studies. Pharmacological Reviews 58: 621-681. https://doi.org/10.1124/pr.58.3.10.

De Ruyck, K., Huybrechts, I., Yang, S.; Arcella, D.; Claeys, L., Abbeddou, S., De Keyzer, W., De Vries, J., Ocke, M., Ruprich, J., De Boevre, M. and De Saeger, S., 2020. Mycotoxin exposure assessments in a multi-center European validation study by 24-hour dietary recall and biological fluid sampling. Environment International 137: 105539. https://doi.org/10.1016/j.envint.2020.105539.

Devreese, M., Antonissen, G., De Baere, S., De Backer, P. and Croubels, S., 2014. Effect of administration route and dose escalation on plasma and intestinal concentrations of enrofloxacin and ciprofloxacin in broiler chickens. BMC Veterinary Research 10: 1-9. https://doi.org/10.1186/s12917-014-0289-1.

Ding, X., Lichti, K. and Staudinger, J.L., 2006. The mycoestrogen zearalenone induces CYP3A through activation of the pregnane X receptor. Toxicological Sciences 91: 448-455. https://doi.org/10.1093/toxsci/kfj163.

Ehling-Schulz, M., Fricker, M. and Scherer, S., 2004. *Bacillus cereus*, the causative agent of an emetic type of food-borne illness. Molecular Nutrition and Food Research 48: 479-487. https://doi.org/10.1002/mnfr.200400055.

Eze, U.A., Huntriss, J., Routledge, M.N. and Gong, Y.Y., 2018. Toxicological effects of regulated mycotoxins and persistent organochloride pesticides: *In vitro* cytotoxic assessment of single and defined mixtures on MA-10 murine Leydig cell line. Toxicology *In Vitro* 48: 93-103. https://doi.org/10.1016.j.tiv.2017.12.019.

Galtier, P.M., Meissonnier, G.M., Laffitte, J., Oswald, I.P. and Loiseau, N., 2008. Molecular interaction between mycotoxins and liver enzymes involved in drug metabolism in rodents and farm animals. Krmiva 50: 205-213. https://hrcak.srce.hr.28963.

Gambacorta, L., Olsen, M. and Solfrizzo, M., 2019. Pig urinary concentration of mycotoxins and metabolites reflects regional differences, mycotoxin intake and feed contaminations. Toxins 11: 378. https://doi.org/10.3390/tosinx11070378.

Gayathri, L., Dhivya, R., Dhanasekaran, D., Periasamy, V.S., Alshatwi,A.A. and Akbarsha, M.A. 2015. Hepatotoxic effect of ochratoxin A and citrinin, alone and in combination, and protective effect of vitamin E: *In vitro* study in HepG2 cell. Food and Chemical Toxicology 83: 151-163. https://doi.org/10.1016/j.fct.2015.06.009.

González-Lobato, L., Real, R., Prieto, J.G., Alvarez, A.I. and Merino, G., 2010. Differential inhibition of murine Bcrp1/Abcg2 and human BCRP/ABCG2 by the mycotoxin fumitremorgin C. European Journal of Pharmacology 644: 41-48. https://doi.org/10.1016/j.ejphar.2010.07.016.

Gouze, M.E., Laffitte, J., Rouimi, P., Loiseau, N., Oswald, I.P. and Galtier, P., 2006. Effect of various doses of deoxynivalenol on liver xenobiotic metabolizing enzymes in mice. Food and Chemical Toxicology 44: 476-483. https://doi.org/10.1016/j.fct.2005.08.020.

Groten, J.P., Feron, V.J. and Sühnel, J., 2001. Toxicology of simple and complex mixtures. Trends in Pharmacologica Sciences 22: 316-22. https://doi.org/10.1016/s0165-6147(00)01720-x.

Guerre, P., Eeckhoutte, C., Burgat, V. and Galtier, P., 2000. The effects of T-2 toxin exposure on liver drug metabolizing enzymes in rabbit. Food Additives and Contaminants 17: 1019-1026. https://doi.org/10.1080/02652030050207819.

Hakkola, J., nasconi, C., Coeke, S., Richert, L., Anderson, T.B. and Pelkonen, O., 2018. Cytochrome P450 induction and xeno-sensing receptors pregnane X receptor, constitutive androstane receptor, aryl hydrocarbon receptor and peroxisome proliferator-activated receptor α at the crossroads of toxicokinetics and toxicodynamics. Basic Clinical Pharmacology and Toxicology 123: 42-50. https://doi.org/10.1111/bcpt.13004.

Hassen, W., Ayed, I., Pascussi, J.M., Maurel, P. and Bacha, H., 2009. Effects of the mycotoxins aflatoxin B_1, zearalenone and ochratoxine A on nuclear receptors PXR, CAR and AhR expression and corresponding CYP 450 in primary cultured human hepatocytes. Toxicology Letters 189: S74. https://doi.org/10.1016/j.toxlet.2009.06.222.

Huhn, S., Escher, B.I., Krauss, M., Scholz, S., Hackermüller, J. and Altenburger R., 2021. Unravelling the chemical exposome in cohort studies: routes explored and steps to become comprehensive. Environmental Sciences Europe 33: 17. https://doi.org/10.1186/s12302-020-00444-0.

Ivanova, L., Denisov, I.G., Grinkova, Y.V., Sligar, S.G. and Faeste, C.K., 2019. Biotransformation of the mycotoxin Enniatin B_1 by CYP P450 3A4 and potential for drug-drug interactions. Metabolites 9: 158. https://doi.org/10.3390/metabo9080158.

Kalpana, S., Aggarwal, M., Srinivasa Rao, G. and Malik, J.K., 2012. Effects of aflatoxin B_1 on tissue residues of enrofloxacin and its metabolite ciprofloxacin in broiler chickens. Environmental Toxicology and Pharmacology 33: 121-126. https://doi.org/10.1016/j.etap.2011.11.005.

Kalpana, S., Srinivasa Rao, G. and Malik, J.K., 2015. Impact of aflatoxin B_1 on the pharmacokinetic disposition of enrofloxacin in broiler chickens. Environmental Toxicology and Pharmacology 40: 645-649. https://doi.org/10.1016/j.etap.2015.08.023.

Khoshal, A.K., Novak, B., Martin, P.G.P., Jenkins, T., Neves, M., Schatzmayr, G., Oswald, I.P. and Pinton, P., 2019. Co-occurrence of DON and emerging mycotoxins in worldwide finished pig feed and their combined toxicity in intestinal cells. Toxins 11: 727. https://doi.org/10.3390/toxins11120727.

Kifer, D., Jakšić, D. and Šegvić Klarić, M., 2020. Assessing the effect of mycotoxin combinations: which mathematical model is (the most) appropriate? Toxins 12: 153. https://doi.og/10.3390/toxins12030153.

Kolf-Clauw, M., Sassahara, M., Lucioli, J., Rubira-Gerez, J., Alassane-Kpembi, I., Lyazhri, F., Borin, C. and Oswald, I.P. 2013. The emerging mycotoxin, enniatin B_1, down-modulates the gastrointestinal toxicity of T-2 toxin *in vitro* on intestinal epithelial cells and ex *vivo* on intestinal explants. Archives of Toxicology 87: 2233-2241. https://doi.org/10.1007/s00204-013-1067-8.

Kovač, M., Bulaic, MJakovljevic, J., Nevistic, A., Rot, T., Kovač, T., Sarkanj, I.D. and Sarkanj, B. 2021. Mycotoxins, pesticide residues, and heavy metals analysis of Croatian cereals. Microorganisms 9: 216. https://doi.org.10.3390/microoragnisms9020216.

Le, T.H., Alassane-Kpembi, I., Oswald, I.P. and Pinton, P., 2018. Analysis of the interactions between environmental and food contaminants, cadmium and deoxynivalenol, in different target organs. Science of the Total Environment. 622-623: 841-848. https://doi.org/10.1016/j.scitotenv.2017.12.014.

Lei, M., Zhang, N. and Qi, D., 2013. *In vitro* investigation of individual and combined cytotoxic effects of aflatoxin B_1 and other selected mycotoxins on the cell line porcine kidney 15. Experimental and Toxicologic Pathology 65: 1149-1157. https://doi.org/10.1016/j.tp.2013.05.007.

Lu, S., Liu, S., Cui, J., Liu, X., Zhao, C., Fan, L., Yin, S. and Hu., H. 2019. Combination of Patulin and Chlorpyrifos synergistically induces hepatotoxicity via inhibition of catalase activity and generation of reactive oxygen species. Journal of Agricultural and Food Chemistry 67: 11474-11480. https://doi.org/10.1021/acs.jafc.9b04814.

Luo, S., Terciolo, C., Bracarense, A.P.F.L., Payros, D., Pinton, P. and Oswald, I.P. 2019. *In vitro* and *in vivo* effects of a mycotoxin, deoxynivalenol, and a trace metal, cadmium, alone or in a mixture on the intestinal barrier. Environment International 132: 105082. https://doi.org/10.1016/j.envint.2019.105082.

Marin, D.E., Pistol, G.C., Gras, M., Palade, M. and Taranu, I. 2018. A comparison between the effects of ochratoxin A and aristolochic acid on the inflammation and oxidative stress in the liver and kidney of weanling piglets. Naunyn Schmiedeberg's Archives of Pharmacology 391: 1147-1156. https://doi.org/10.1007/s00210-018-1538-9.

Marín, S., Cano-Sancho, G., Sanchis, V. and Ramos, A.J., 2018. The role of mycotoxins in the human exposome: Application of mycotoxin biomarkers in exposome-health studies. Food and Chemical Toxicology 121: 504-518. https://doi.org/10.1016/j.fct.2018.09.039.

Marin, S., Ramos, A.J., Cano-Sancho, G. and Sanchis, V. 2013. Mycotoxins: occurrence, toxicology, and exposure assessment. Food Chem Toxicol 60: 218-237. https://doi.org/10.1016/j.fct.2013.07.047.

Martinez-Larrañaga, M.R., Anadón, A., Diaz, M.J., Fernandez, R., Sevil, B., Fernandez-Cruz, M.L., Fernandez, M.C., Martinez, M.A. and Anton, R., 1996. Induction of cytochrome P4501A1 and P4504A1 activities and peroxisomal proliferation by fumonisin B1. Toxicology and Applied Pharmacology 141: 185-194. https://doi.org/10.1006/taap.1996.0275.

Mary, V.S., Valdehita, A., Navas, J.M. and Rubinstein, H.R., 2015. Effects of aflatoxin B_1, fumonisin B_1 and their mixture on the aryl hydrocarbon receptor and cytochrome P450 1A induction. Food and Chemical Toxicology 75: 104-111. https://doi.org/10.1016/j.fct.2014.10.030.

Meissonnier, G.M., Laffitte, J., Raymond, I., Benoit, E., Cossalter, A.M., Pinton, P., Bertin, G., Oswald, I.P. and Galtier, P., 2008. Subclinical doses of T-2 toxin impair acquired immune response and liver cytochrome P450 in pigs. Toxicology 247: 46-54. https://doi.org/10.1016/j.tox.2008.02.003.

Mol, J.G.J., de Rijk, T.C., van Egmond, H.J. and de Jong, J. 2014. Occurence of mycotoxins and pesticides in straw and hay used as animal feed. (RIKILT report; No. 2014.006). RIKILT Wageningen UR. https://edepot.wur.nl/313921.

Mostrom, M. and Evans, T.J., 2011. Chapter 52 - Phytoestrogens, in reproductive and developmental toxicology, R.C. Gupta, Editor. Academic Press: San Diego. 707-722.

Oh, S.Y., Cedergreen, N., Yiannikouris, A., Swamy, H.V.L.N. and Karrow, N.A., 2017. Assessing interactions of binary mixtures of Penicillium mycotoxins (PMs) by using a bovine macrophage cell line (BoMacs). Toxicology and Applied Pharmacology 318: 33-40. https://doi.org/10.1016/j.taap.2017.0015.

Patki, K.C., Von Moltke, L.L. and Greenblatt, D.J., 2003. *In vitro* metabolism of midazolam, triazolam, nifedipine, and testosterone by human liver microsomes and recombinant cytochromes p450: role of cyp3a4 and cyp3a5. Drug Metabolismo and Disposition 31: 938-944. https://doi.org/10.1124/dmd.31.7.938.

Payros, D., Dobrinndt, U., Martin, P., Secher, T., Bracarense, A.P.F.L., Boury, M., Laffitte, J., Pinton, P., Oswald, E. and Oswald, I.P., 2017. The food contaminant deoxynivalenol exacerbates the genotoxicity of gut microbiota. mBio 8: e00007-17. https://doi.org/10.1128/mBio.00007-17.

Pérez-Ortega, P., Gilbert-López, B., Garcia-Reyes, J.F., Ramos-Matos, N. and Molina-Díaz, A., 2012. Generic sample treatment method for simultaneous determination of multiclass pesticides and mycotoxins in wines by liquid chromatography-mass spectrometry. Journal of Chromatography A 1249: 32-40. https://doi.org/10.1016/j.chroma.2012.06.020.

Plant, N. and Aouabdi, S., 2009. Nuclear receptors: the controlling force in drug metabolism of the liver? Xenobiotica 39: 597-605. https://doi.org/10.1080/00498250903098218.

Pristner, M. and Warth, B., 2020. Drug-exposome interactions: The next frontier in precision medicine. Trends in Pharmacological Sciences 41: 994-1005. https://doi.org/10.1016/j.tips.2020.09.012.

Prouillac, C., Videmann, B., Mazallon, M. and Lecoeur, S., 2009. Induction of cells differentiation and ABC transporters expression by a myco-estrogen, zearalenone, in human choriocarcinoma cell line (BeWo). Toxicology 263: 100-107. https://doi.org/10.1016/j.tox.2009.06.023.

Schelstraete, W., Devreese, M. and Croubels, S. 2019. Impact of subacute exposure to T-2 toxin and zearalenone on the pharmacokinetics of midazolam as CYP3A probe drug in a porcine animal model: a pilot study. Frontiers in Pharmacology 10: 399. https://doi.org/10.10.3389/fphar.2019.00399.

Scholar, E., in xPharm: The Comprehensive Pharmacology Reference, S.J. Enna and D.B. Bylund, Editors. 2007, Elsevier: New York. p. 1-6.

Schreck, I., Deigendesch, U., Burkhardt, B., Marko, D. and Weiss, C., 2012. The Alternaria mycotoxins alternariol and alternariol methyl ether induce cytochrome P450 1A1 and apoptosis in murine hepatoma cells dependent on the aryl hydrocarbon receptor. Archives of Toxicology 86: 625-632. https://doi.org/10.1007/s00204-011-0781-3.

Shi, B., Su, Y., Chang, S., Sun, Y., Meng, X. and Shan, A., 2017. Vitamin C protects piglet liver against zearalenone-induced oxidative stress by modulating expression of nuclear receptors PXR and CAR and their target genes. Food and Function 8: 3675-3687. https://doi.org/10.1039/c7fo01301a.

Smith, M.C., Hymery, N., Troadec, S., Pwtowski, A., Coton, E. and Madec, S., 2017. et al., Hepatotoxicity of fusariotoxins, alone and in combination, towards the HepaRG human hepatocyte cell line. Food and Chemical Toxicology 109: 439-451. https://doi.org/10.1016/j.fct.2017.09.022.

Solfrizzo, M., Gambacorta, L. and Visconti, A., 2014. Assessment of multi-mycotoxin exposure in southern Italy by urinary multi-biomarker determination. Toxins 6: 523-538. https://doi.org/10.3390.toxins6020523.

Spotti, M., Maas, R.F., de Nijs, C.M. and Fink-Gremmels, J., 2000. Effect of fumonisin B($_1$) on rat hepatic P450 system. Environmental Toxicology and Pharmacology 8: 197-204. https://doi.org/10.1016/s1382-6689(00)00040-5.

Stiborová, M., Bárta, F., Levová, K., Hodek, P., Frei, E., Arlt, V.M.d Schmeiser, H.H., 2015. The influence of ochratoxin A on DNA adduct formation by the carcinogen aristolochic acid in rats. Archives of Toxicology 89: 2141-2158. https://doi.org/10.1007/s00204-014-1360-1.

Szabó-Fodor, J., Szabó, A., Kócsó, D., Marosi, K., Bóta, B, Kachlek, M., Mézes, M., Balogh, K., Kövér, G., Nagy, I. and Glávits, R., Kovács, M., 2019. Interaction between the three frequently co-occurring *Fusarium* mycotoxins in rats. Journal of Animal Physiology and Animal Nutrition 103: 370-382. https://doi.org/10.1111/jpn.13013.

Teo, M.Y., Rathkopf, D.E. and Kantoff, P., 2019. Treatment of Advanced Prostate Cancer. Annual Review of Medicine 70: 479-499. https://doi.org/10.1146/annurev-med-051517-011947.

Thrupp, T.J., Runnalls, T.J., Scholze, M., Kugathas, S., Kortenkamp, A. and Sumpter J.P., 2018. The consequences of exposure to mixtures of chemicals: Something from 'nothing' and 'a lot from a little' when fish are exposed to steroid hormones. Science of the Total Environment 619-620: 1482-1492. https://doi.org/10.1016/j.scitotenv.2017.11.081.

Tolosa, J., Rodríguez-Carrasco, Y., Ruiz, M.J. and Vila-Donat, P., 2021. Multi-mycotoxin occurrence in feed, metabolism and carry-over to animal-derived food products: A review. Food and Chemical Toxicology 158: 112661. https://doi.org/10.1016/j.fct.2021.112661.

Tolson, A.H. and Wang, H., 2010. Regulation of drug-metabolizing enzymes by xenobiotic receptors: PXR and CAR. Advanced Drug Delivery Reviews 62: 1238-1249. https://doi.org/10.1016/j.addr.2010.08.006.

Van der Heyden, S., Gossens, J., Vandenbroucke, V., Vercauteren, G., Chiers, K., Pasmans, F., Haesebrouck, F., De Backers, P., Croubels, S. and Ducatelle, R., 2009. Reduced expression of intestinal p-glycoprotein following ingestion of deoxynivalenol (DON) contaminatad feed in pigs. Journal of Comparative Pathology, 141: 272. https://doi.org/10.1016/j.jcpa.2009.07.035.

Vejdovszky, K., Hahn, K., Braun, D., Warth, B. and Marko, D. 2017a. Synergistic estrogenic effects of *Fusarium* and *Alternaria* mycotoxins *in vitro*. Archives of Toxicology 91: 1447-1460. https://doi.org/10.1007/s00204-016-1795-7.

Vejdovszky, K., Schmidt, V., Warth, B. and Marko, D. 2017b. Combinatory estrogenic effects between the isoflavone genistein and the mycotoxins zearalenone and alternariol *in vitro*. Molecular Nutrition and Food Research 61: 1600526. https://doi.org/10.1002/mnfr.201600526.

Warth, B., Parich, A., Atehnkeng, J., Bandyopadhyay, R., Schuhmacher, R., Sulyok, M. and Krska, R., 2012. Quantitation of mycotoxins in food and feed from Burkina Faso and Mozambique using a modern LC-MS/MS multitoxin method. Journal of Agricultural and Food Chemistry 60: 9352-9363. https://doi.org/10.1021/jf302003n.

Wilbur, S.B., 2004. Guidance manual for the assessment of joint toxic action of chemical mixtures. United States, Agency for Toxic Substances and Disease Registry, Division of Toxicology. Syracuse Research Corporation. URL : https://stacks.cdc.gov/view/cdc/6935.

Yabe, T., Hashimoto, H., Sekijima, M., Degawa, M., Hashimoto, Y., Tashiroto, F.and Yeno, Y., 1993. Effects of nivalenol on hepatic drug-metabolizing activity in rats. Food and Chemical Toxicology 31: 573-581. https://doi.org/10.1016/0278-6915(93)90207-f.

CHAPTER 4

Effects of mycotoxins in pigs

A.P.F.R.L. Bracarense[1], M. Souza[1] and R.R. Santos[2]*

[1]Laboratory of Animal Pathology, Universidade Estadual de Londrina, Londrina, PR, Brazil; [2]Schothorst Feed Research, Lelystad, the Netherlands
**correspondence: anapaula@uel.br*

1 Introduction

Apart from occasional acute mycotoxicosis outbreaks, dietary mycotoxin contamination is one of the main causes of impaired production performance, subclinical diseases, and susceptibility to infectious diseases in swine. A worldwide 10-year survey, covering over 74,000 samples of feed and feed raw materials (maize, wheat, and soybean), detected contamination in 88% of the samples and 64% were multi-contaminated with two or more mycotoxins (Gruber-Dorninger et al., 2019). These authors demonstrated that DON and FUM were the most common contaminants found in all feedstuffs (64% and 60%, respectively), followed by ZEN (45%), AFB_1 (23%), T-2 toxin (19%) and OTA (15%). Other mycotoxins such as EAs were detected in cereals in Europe, North America, and Asia (Poapolathep et al., 2021). The occurrence of mycotoxins shows distinct regional tendencies. Climatic conditions and agricultural practices are major influencing factors (Gruber-Dorninger et al., 2019). However, most of the contaminated feedstuffs have a common characteristic: in general, the mycotoxin levels comply with regulated maximum guidance values established for feedstuffs intended for the production of animal feed and, in most cases, co-contamination occurs.

Pigs are more sensitive to mycotoxin toxicity because of a high oral bioavailability of several mycotoxins (Goyarts and Dänicke, 2006; Devreese et al., 2014; Sun et al., 2022) and a low ability to detoxify these contaminants into less toxic products, e.g. ZEN degrades to a greater extent into α-ZEL rather than into β-ZEL (Zöllner et al., 2002; Benthem de Grave et al., 2021). Although pigs are very sensitive to mycotoxin exposure, acute symptoms of mycotoxicosis are uncommon and occasional. Some mycotoxins such as FB_1 have a very low bioavailability in pigs, being approximately 4% (Prelusky et al., 1996; Fodor et al., 2008). However, this mycotoxin is frequently present in pig diets and the negative impact after chronic exposure should not be denied. The absorbed fraction has a long biological half-life due to enterohepatic recirculation and there is the possibility of accumulation in tissues such as liver and kidneys (Prelusky et al., 1994, 1996; Pierron et al., 2016a).

 | DOI: 10.1163/9789004712720_5

A meta-analytical study analysed results from nearly 100 published experiments (*in vitro, ex vivo* and *in vivo*) with special attention given to the doses used. The authors suggested three categories of exposure levels: realistic doses (representative of field conditions), occasional doses (unfavourable weather conditions) and unrealistic doses (unlikely to occur in nature) (Grenier and Applegate, 2013). In this chapter, data involving realistic doses were reported and discussed. Therefore, the main goal of this chapter is to discuss the effects of mycotoxins on pig's health, production performance, intestinal integrity and function, and nutrient utilization, after dietary exposure to realistic doses of mycotoxins.

2 Occurrence of mycotoxins in pig diets

Raw materials used to prepare pig diets may vary according to geographic area, being based on either maize-soybean meal or wheat-soybean when formulating feed. These ingredients are known to have frequent multi-contamination with mycotoxins. Other feedstuffs like barley can also be added to the diets at levels often reaching 40%, as well as complementary ingredients such as brans, hulls, distillers' dried grains with solubles (DDGS), rootlets, middlings, sugar beet pulp, and oils at different inclusion levels. All of these ingredients d can, to some degree, be a source of mycotoxin contamination.

Reduction of costs in pig production can be achieved by formulation of diets with nutrient-dense industrial co-products, such as the by-product of maize ethanol fermentation (for beer production or fuel ethanol), the DDGS. The fermentation process does not degrade the mycotoxins present in the maize, and the DON and FUM concentrations in the DDGS are usually greater by a factor of between two and five when compared to the grain of origin (Bothast et al., 1992; Hanschmann et al., 2007). This increase in concentration is a result of a three-fold decrease in the dry mass of DDGS when compared to the original grain, and the mycotoxins are not contained in the distilled ethanol, but concentrated in the by-product (Wu and Munkvold, 2008). Maize DDGS contaminated with ZEN will usually have twice the concentration of ZEN than that in the original maize. However, its conversion into α- and β-ZEL may be induced by brewing yeasts (Matsuura and Yoshizawa, 1985). In maize DDGS, no increase was observed in the concentration of DON-3G, which can be hydrolysed in the fermentation process (Schaafsma et al., 2009). This probably contributed to the increased DON contamination. Molds can grow in these by-products during inappropriate storage and transport, resulting in the risk of the of further mycotoxin production. Increased inclusion of maize DDGS in pig diets, e.g. an increase from 5% to 20%, has been associated with significant losses for swine producers resulting from mycotoxin contamination (Wu and Munkvold, 2008). A meta-analytical study showed that every 1% increase in the inclusion level of maize DGGS for growing-finishing pigs may result in a decrease in average daily gain (ADG) of 0.10% and in average daily feed intake (ADFI) of -0.09%. The study also showed that this negative effect can be more prominent in nursery piglets (Jang et al., 2021). These authors did not consider mycotoxin contamination in their study, but AFs and *Fusarium* mycotoxins are often detected in maize DDGS. Studies showed a 3.2-factor increase in AFs levels (from 204 to 652 μg/kg) and a 1.7-factor increase in FUM levels (from 15 to 25.3 mg/kg) in maize DDGS (Wu and Munkvold, 2008). DDGS produced from maize with an average concentration of 1.80 mg/kg DON presented final DON levels ranging from 2.9 to 6.97 mg/kg (Schaafsma et al., 2009). As described in a global survey, maize DDGS can be highly contaminated with several mycotoxins, reaching extreme high levels for DON (84.86 mg/kg; average of 1.49 mg/kg), FUM (26.83 mg/kg; average of 814 μg/kg), ZEN (2.9 mg/kg; average of 102 μg/kg), and AFB_1 (340 μg/kg; average of 11 μg/kg) (Gruber-Dorninger et al., 2019). Wheat DDGS can also be a source of ENNs, DON, and OTA contamination (Mortensen et al., 2014). Total ENNs levels in wheat was measured in Norway as up to 7.4 mg/kg (Uhlig et al., 2006), this could be of concern if a by-product of this feedstuff entered the pig feed chain.

Although barley energy content is lower than that of maize, this feedstuff remains a common ingredient in pig diet at inclusion levels close to 40% or even 60% (Kim et al., 2014). Contamination of barley with DON, T-2 toxin, FUM, ZEN, AFB_1, and OTA was described (Gruber-Dorninger et al., 2019). Moreover, in France, ENNs were detected in barley, especially in spring barley, at levels up to 1.32 mg/kg (Orlando et al., 2019) and up to 5.66 mg/kg in Norway (Uhlig et al., 2006). Barley rootlets are the main malting by-product and are recognized as a potential feed ingredient for pig diet due to their rich protein and fibre content (Neylon et al., 2020; Van Baelen et al., 2024). However, as for DDGS, mycotoxin levels in this barley by-product are higher than in the barley of origin due to the amount of husks and outer layers that are accumulated. The presence of FB_1 (100% of samples; 104 - 2043 μg/kg) and AFB_1 (18% of samples; 19 - 45 μg/kg) was reported in barley rootlets, as well as other mycotoxins such as DON, ZEN, and OTA (Mastanjevic et al., 2019). There is a significant possibility of wheat by-products from brewing and from milling (wheat middlings) having high levels of contamination with mycotoxins. For instance, in studies performed in Croatia and in China, high levels of *Fusarium* toxins were observed in wheat germ/rootlets (5 to 12.6 mg/kg DON and 1.7 mg/kg NIV) and in wheat middlings (1.02 mg/kg to 12.63 mg/kg DON and 57 μg/kg to 1.20 mg/kg ZEN) (Ma et al., 2018; Mastanjevic et al., 2019). In both cases, all samples were contaminated with at least two mycotoxins.

Sugar beet pulp, another major by-product of the sugar industry, is often added to pig diets at an inclusion level close to 5%. Although there had been previous reports of the presence of ZEN in 31 of 75 sugar beet samples at levels ranging from 13 to 47 μg/kg (Bosch and Mirocha, 1992), sugar beet pulp was not considered a source of mycotoxin contamination because it was believed that toxins should be removed during diffusion (Huijbregts et al., 2013). However, an analysis of 40 samples of sugar beet pulp showed the presence of OTA in 2.5% of the samples (up to 15 µg/kg), MPA in 12.5% of the samples (up to 1.5 mg/kg), and ZEN in 7.5% of the samples (up to 6.9 mg/kg) (Lopes et al., 2023). In a study evaluating the exposure of sows to ZEN during lactation, the diet was prepared with naturally contaminated sugar beet pulp (Benthem de Grave et al., 2021). In this previous study, the dietary inclusion level of sugar beet pulp was 5% and this ingredient contained 4.6 mg/kg ZEN, 190 µg/kg DON, 156 µg/kg DON-3G, 52 µg/kg 3-ADON + 15-ADON, 10 µg/kg AOH, 10 µg/kg AME, 244 µg/kg BEA, 5.6 µg/kg ENNA, 16.3 µg/kg ENNB, 12 µg/kg ENNB1 and 12 µg/kg Roquefortin C.

Oil products can be contaminated with various mycotoxins, but at very low levels due to their removal during extraction and at significantly lower levels than in crude oil due to degradation during the refining process, e.g. AFs levels in crude rice bran oil have decreased by a factor of 30 (956 µg/l to 28 µg/l) after refining (Guerre, 2016).

Pigs will be unequivocally exposed to mycotoxins via diet. A survey performed in Spain, based on 228 feed samples and focused on pig diets, found that 4.4% of the feed samples were contaminated as follows: with DON, 0.9%; with T-2-toxin, 50%; with FB_1, 30%; with FB_2; with ZEN, 7% ; were contaminated with OTA, none; with AFB_1, 3% (Arroyo-Manzanares et al., 2019). These authors also tested the diets for the presence of another 12 mycotoxins and found that all feed samples were contaminated with at least one mycotoxin (ENNB) and that 40% of them contained at least five mycotoxins. The average levels of these mycotoxins were usually low (from 0.33 µg AFG_1/kg diet to 403 µg FB_1/kg diet), but could reach 741 µg ZEN/kg diet (Arroyo-Manzanares et al., 2019). In another study, in Italy, 100 pig dietary samples were evaluated for the presence of 12 mycotoxins (Nualkaw et al., 2020). These authors observed that co-contamination was the norm, as 91% of the feed samples were contaminated with ZEN, 85% with FB_1, 77% with FB_2, 43% with DON and 34% were contaminated with AFB_1. However, the mycotoxin levels were often below 170 µg/kg diet. Analysis of pig diets in the USA to identify only ZEN and type-B trichothecene mycotoxins, showed that 77% of the diets contained multiple contaminants, 94% were contaminated with DON, 33% with 3-ADON, 57% with 15-ADON, 1% with NIV and 47% with ZEN (Pack et al., 2021). Analysis of 100 pig feed samples collected in Thailand showed that more than 49% were contaminated with EAs – mostly ergocryptine (40%; 0.25 to 101 µg/kg), ergosine (37%; 0.53 to 9.7 µg/kg) and ergotamine (37%; 0.27 to 13.5 µg/kg) (Poapolathep et al., 2021). As in the aforementioned surveys, the mycotoxins levels were generally below national advisory limits. In Bulgaria, more than 90% of pig feed samples were reported to be contaminated with CIT (55 µg/kg). At slaughter pigs were presenting altered kidneys (enlarged or pale) (Talmaciu et al., 2008). It was also reported that 45% of Belgian pig feed was contaminated with CIT (Meerpoel et al., 2018).

3 Main mycotoxins affecting pigs

3.1 *Deoxynivalenol*

After ingestion, DON is rapidly absorbed by pigs, where maximum serum levels are reached around 1.5 hours after ingestion with an elimination half-life of five to six hours. Depending on the exposure, acute or chronic, 54% and 89% of DON orally administered via feed, respectively, can be recovered in the serum from pigs. DON is rapidly excreted and after 24 h no remarkable concentration is found in serum (Goyarts and Dänicke, 2006).

DON is known as a vomitoxin as vomiting is one of the acute symptoms evidenced in pigs. Exposure to high doses of DON can also lead to diarrhoea and feed refusal. Decreased FI can be associated with the anorectic effect of DON due to an increased serotonin turnover (Reddy et al. 2018), the release of satiety hormones from enteroendocrine cells (Peng et al. 2017) and a pro-inflammatory status in the intestine (Alassane-Kpembi et al. 2017). These acute symptoms are only observed when the pigs are exposed to extremely high levels of this mycotoxin (2.8 mg/kg DON) (EFSA, 2017). Another *Fusarium* mycotoxin, NIV, is considered more harmful to intestine integrity than DON (Cheat et al., 2015) However, studies using realistic doses of this mycotoxin are scarce. In a study, pigs were fed synthetic NIV over three weeks and no vomiting symptoms were observed even after exposure to 5 mg/kg NIV. However, erosions, mucosa thickening and haemorrhage in the stomach and small intestine were observed starting at an exposure to 2.5 mg/kg NIV (Pettersson and Hedman, 1997).

3.2 *T-2 toxin*

The recovery of T-2 toxins after oral administration to pigs (0.1 and 0.4 mg/kg BW) is low (Schelstraete et al., 2020). Toxicokinetic studies in pigs are lacking, probably due to the fact that this mycotoxin has a rapid metabolization and clearance after ingestion. Symptoms of acute toxicity include vomiting, stomach ulcers, bloody diarrhoea, as well as severe dermatitis because dermal exposure to this mycotoxin also causes local inflammation (Meneely et al., 2023). Exposure of growing pigs to 0.5 mg/kg T-2 over three weeks resulted in decreased feed intake and impaired liver function (Rafai et al., 1995a). When the exposure level was increased to 2 mg/kg T-2, immunosuppression was reported (Rafai et al., 1995b). Topical exposure to 15 mg/kg T-2 (3% of total body surface area; 4 mg T-2 per cm^2) caused an abrupt rise in rectal temperature after 24 hours, a dog-sitting position after two to three days that disappeared two days later, fatigue and anorexia with persistent fever after one week, and skin lesions that persisted for two weeks (Pang et al., 1987).

3.3 *Fumonisins*

In pigs, ingested FUM is poorly bioavailable (3%-5%), rapidly distributed to liver and kidney (distribution half-life of 6 minutes), biotransformed mainly into $pHFB_1$

and HFB_1 and excreted mostly via the faecal route (Schertz et al., 2018). FB_1 is mainly excreted in the faeces and urine. Pigs fed a total of 360 mg of FB_1 over 10 days excreted 247.5 mg (69% of the FB_1) in this period (Fodor et al., 2008). The metabolism and excretion of FUM occurs primarily in the liver and kidneys. However, the enterohepatic circulation contributes to extended time of FUM within the organism. Although the deleterious effect of FUM even with a low oral bioavailability is unclear, it was also suggested that there is a higher bioavailability and/or greater conversion to active metabolites at a low dose of FB_1, accumulation in tissues, and increased uptake of FB_1 derivatives that are converted back into FB_1 (Shier, 2000).

After exposure to FUM, sphingolipid metabolism inhibits ceramide synthase, an enzyme playing a central role in sphingolipid biosynthesis. Changes in the sphinganine:sphingosine (Sa/So) relationship are associated with the toxicity of FUM (Pierron et al., 2016a). Pigs and horses (see Chapter 9) are the species that are most sensitive to FUM. Specifically, outbreaks of fatal pulmonary oedema in pigs have been reported and connected with FUM exposure since the 1980s (Harrison et al., 1990). FUM also affects other organs such as intestine, heart, lung, liver, kidney, and lymphoid tissue. Haemorrhage and inflammatory infiltrate were reported in the heart, degenerative cell changes in the kidney and liver, while in the lungs and lymphoid organs (spleen and lymph nodes) alveolar oedema and follicular depletion were observed, respectively (Terciolo et al., 2019). Acute toxicity is reported in pigs receiving high levels of FUM (92 mg/kg) in the diet. Weaned piglets (16-24 kg BW) fed maize contaminated with *Fusarium moniliforme* for seven days presented severe dyspnoea caused by pulmonary oedema and hydrothorax. These symptoms were also observed in other piglets fed the contaminated maize for 14 or 28 days (Harrison et al., 1990). These authors estimated that the FUM dietary levels were 105 to 155 mg/kg, which are unexpected for a natural dietary contamination. Furthermore, the contaminated maize was deficient in nutrients, which may have increased the susceptibility of the animals. Pulmonary oedema is characterized by wet lungs and the presence of fluid in the airways and thoracic cavity (Haschek et al., 2001). Exposure of weaned pigs (20 kg BW) to FB_1 over three days (15 mg/kg BW FUM; daily) or five weeks (2 mg/kg BW FUM; daily) did not result in pulmonary oedema (Gumprecht et al., 2001).

3.4 *Zearalenone*

ZEN is a mycoestrogen with a structure similar to 17β-oestradiol, which enables this mycotoxin to bind oestrogen receptors consequently affecting the reproductive ability of animals (Greenman et al, 1979). After ingestion, 61%-85% of this mycotoxin is absorbed and rapidly metabolised in the intestine and in the liver. Less than 2% can be detected in the blood (Schelstraete et al., 2020). Among the various metabolites formed, isomer α-ZEL has a 60- to 90-fold higher affinity for the oestrogen receptor and isomer β-ZEL a 2.5- to 5-fold lower affinity. Hence, the pathway leading to the production of α-ZEL can be considered as an enhancement of toxicity, whereas the pathway leading to the production of β-ZEL can be considered as a detoxification process (Malekinejad et al., 2006). Variation in the rate and extent of α- or β-ZEL production accounts for significant differences in susceptibility to ZEN toxicity between species. Pigs are particularly susceptible as they predominantly convert ZEN to α-ZEL, whereas poultry and cattle predominantly metabolise ZEN to β-ZEL (Devreese et al., 2013; see also Chapters 5 and 6).

Pigs are the most sensitive species to ZEN exposure. This is evidenced by fertility and reproductive disorders induced by this mycotoxin due to the dominant biotransformation of ZEN into α-ZEL. α-ZEL has 60-90 times the potency of ZEN. Piglets and gilts are more sensitive to ZEN than sows. Acute exposure is characterized by swelling and reddening of the vulva, increase in the uterus weight, prolonged cycle, and sub- or infertility (EFSA, 2017). Exposure of piglets, gilts, and sows to ZEN is often underestimated because feed samples are usually only analysed when a case of mycotoxicosis is suspected. However, edible enrichment materials such as straws and other fibres, from sources such as hay, may contain a considerable amount of mycotoxins including ZEN. Reproductive dysfunction was reported on a farm, where sows were fed diets mixed with hay pellets contaminated with ZEN (1 mg/kg) and ZEN-14S (0.5 mg/kg), resulting in piglets being born weak, presenting swollen external genitals and a mortality rate of 30% during lactation (Hennig-Pauka et al., 2018). A two-year study performed on 49 Swedish pig farms reported the presence of mycotoxins in straw used for bedding, where ZEN (up to 1.8 mg/kg) was present in 54% of the samples (Nordkvist and Häggblom, 2014). In this study, 86% of the straw samples were contaminated with DON with an average concentration of 0.9 mg/kg, but where the majority of the samples had high levels of DON, occasionally reaching 11.20 mg/kg. T-2/HT-2 were detected in 22% of the samples reaching levels as high as 0.3 mg/kg.

3.5 *Aflatoxins*

Aflatoxin is a non-immunogenic, low molecular weight and lipophilic compound, rapidly absorbed by passive diffusion through mucous membranes and reaching the liver by the portal system. Metabolization occurs in two steps. The first one is mediated by the enzyme CYP, responsible for hydrolysis, reduction and oxidation reactions. During the second step, the main event is conjugation, when DNA adducts are formed (pro-carcinogen) (Gregorio et al., 2015). After oral challenge (5 µg/kg) the highest plasmatic concentration was observed 3 hours after challenge, with AFB_1 being detected for a period of 10 days. More than 80% of aflatoxins were absorbed in the upper part of the gastrointestinal system. Tissue distribution is highest in the liver, followed by bile, kidney, lung, spleen, fat, brain and muscles (Lüthy et al., 1980). More than 20 aflatoxin molecules are known, but AFB_1 is the molecule with the highest toxicity. Aflatoxin M_1 (AFM_1) is a product of its biotransformation, is present in the milk of several species, including pigs and is also considered to be a grade 1 carcinogen (IARC, 2002). The liver is the earliest target for AFB_1 toxicity in both humans and animals. An AFB_1-diet (0.32 mg/kg, for 28 days) induced histological changes in piglets' livers characterized by inflammatory infiltrate in the portal system, fibrosis, hydropic degeneration of hepatocytes and vacuolation of periportal parenchyma cells. This AFB_1-diet induced hepatic gene overexpression of genes involved in inflammation and immune responses. Liver oxidative stress was also induced by the mycotoxin challenge (Taranu et al., 2020). Other effects reported were a linear increase in the serum albumim:globulin ratio and in the hepatic enzymes AST (aspartate aminotransferase), ALP (alkaline phosphatase) and GGT (gamma glutamyl transferase). The increase in hepatic enzymes is a consequence of the toxic effects on the liver, where AF is predominantly metabolized via CYP (Wang et al., 2020). Interestingly, AFB_1 (0.32 mg/kg for 30 days) also induced toxic effects on mesenteric lymph nodes of piglets by reducing the antioxidant status by 20.5%, increasing lipid peroxidation levels and exerting a pro-inflammatory effect (Marin et al., 2020). Acute aflatoxicosis is reported in young animals (2-5 months) fed unrealistic levels of AFB_1 (3.4 mg/kg). The clinical signs observed were fever, weight loss, tachycardia, tachypnoea, lethargy, muscle tremours, muscle weakness, diarrhoea, and death within 48 hours. The gross findings were generalized jaundice, ascites, hydropericardium, petechial haemorrhages in the mesentery, subcutaneous oedema and an enlarged, friable, yellow-orange liver with reddish diffuse multifocal areas. Microscopic findings observed were swelling of hepatocytes, hepatocellular fatty degeneration, centrilobular hepatocellular necrosis, bile duct proliferation and cholestasis (Olinda et al., 2016). The liver is the most affected organ in acute aflatoxicosis cases since it is the main site of mycotoxin metabolization.

3.6 *Ochratoxin A*

OTA is initially absorbed in the stomach and subsequently in the small intestine at the proximal jejunum. It can be metabolized in the kidney, the liver (via CYP) and in the intestines. Hydrolysis, hydroxylation, lactone-opening, and conjugation are the major metabolic pathways (Wu et al., 2011). OTA is hydrolysed to OTα (a less toxic metabolite) by carboxypeptidase A and chymotrypsin, as well as by microorganisms from the digestive tract. Distribution in tissues follows the subsequent pattern kidney > liver > muscle > adipose tissue. The strong affinity bond between OTA and plasma proteins delays its elimination. This increases the half-life period, which is up to 3-5 days in pigs after oral administration. The inhibition of protein synthesis is the primary toxic effect of OTA (Marin et al., 2009). Acute OTA toxicity was described in weaned piglets fed 1-2 mg/kg of OTA for six days. The clinical signs were fever, emesis, depression, reduction in FI and loss of body weight (BW), followed by diarrhoea, polyuria, polydipsia, dehydration and death. Urine changes were observed in relation to the control group, such as a reduction in the specific gravity, increased concentrations of protein, presence of granular casts and necrotic renal tubular epithelial cells in the sediment. Gross findings were dehydration, thin and atonic intestine walls, pale tan to yellowish liver, oedematous and slightly congested mesentery and peripheral lymph nodes and pale tan kidneys. Microscopically the most severe changes were found in the kidneys with necrosis of epithelial cells of the proximal convoluted tubules and occasionally with the presence of casts. In the intestine, congestion, mucosal lesions with necrosis foci and loss of surface epithelium were observed. In the liver, the main finding was fatty change. The lymph nodes showed slight oedema and congestion (Szczech et al., 1973). In pigs OTA is well known to induce the mycotoxic porcine nephropathy (Heussner and Bingle, 2015). Swine is one of the most sensitive species to OTA, with an oral LD_{50} of 1 mg/kg BW. Pigs' sensitivity is a consequence of OTA tissue accumulation, due to a long serum half-life (≈ 30 times longer than in poultry), sustained by the high affinity of mycotoxin for proteins and by enterohepatic circulation and biliary excretion. These might be responsible for toxin accumulation and prolonged elimination from the body (Duarte et al., 2011).

3.7 *Citrinin*

There is a scarcity of studies on this nephrotoxic mycotoxin, which may have an additive effect when present with OTA (EFSA, 2012). Oral bioavailability of CIT ranges from 37 to 44% in pigs, with rapid absorption (0.9 h after ingestion) and slow elimination (around 20 hours) (Meerpoel et al., 2020a). Studies focusing on dietary exposure to CIT were performed at non-realistic doses (20 to 100 mg/kg BW) resulting in kidney failure at the highest tested concentration, and in impaired growth performance at the lowest tested one (Friis et al., 1969). Dietary exposure to 20 µg CIT/kg BW does not result in toxicity in pigs (Sándor et al., 1991), but no information is available regarding chronic exposure to diets naturally contaminated with this mycotoxin.

3.8 *Ergot alkaloids*

Data related to the absorption, distribution, metabolism, and excretion (ADME) of EAs are limited because of the variety of these mycotoxins and most of the metabolites produced are still unknown (EFSA, 2024). Piglets and growing-finishing pigs exposed to EAs (4.66 mg/kg feed) presented an increase in the weight of the heart and spleen relative to BW (Mainka et al., 2005). When EAs were present at a dietary concentration of 7 mg/kg for sows, a reduction in milk production or agalactia was observed due to the substantial decrease in the release of prolactin caused by these mycotoxins (Kopinski et al., 2007). Although sorghum ergot (*Claviceps africana*) has been considered less toxic than rye ergot (*C. purpurea*), because it mainly produces dihydroergosine, sorghum ergot is also able to decrease serum prolactin and causes failure to lactate after farrowing (Blaney et al., 2000). Based on the sum of the EAs, the EFSA Panel considered 0.6 mg/kg as a Reference Point in complete feed for adverse health effects in pigs (EFSA, 2024). For more details, see Chapter 17.

3.9 *Other mycotoxins*

Emerging, or more appropriately described as neglected, modified and masked mycotoxins are commonly present in animal diets. These mycotoxins are not routinely included in feed analysis and due to the paucity of toxicological data, no regulatory levels are available.

In contrast to poultry (Chapter 5), in pigs, ENNB1 exhibits a high oral bioavailability of 91% and the plasma peak is already observed after 15 minutes after oral administration (Devreese et al., 2014). This suggests that adverse effects can be expected in pigs. ENNs are known by their ionophoric characteristics and for their ability to inhibit membrane-bound enzymes including those involved in efflux pumps (Ivanova et al., 2017). Although ENNs are often present in pig diets, the absence of reports demonstrating acute toxicity indicates that chronic exposure causes asymptomatic effects, resulting in impaired production performance. Other *Fusarium* mycotoxins will be present in diets contaminated with ENNs, and these interactions should be investigated.

Most pig-related studies including BEA and AOH were performed *in vitro* (Schoevers et al., 2016, 2020, 2021; see Chapter 14) or combined with other mycotoxins such as the study combining BEA, ENNs and DON (Novak et al., 2021). Based on the *in vitro* studies, gilts are more sensitive to BEA than sows (Schoevers et al., 2021), and both BEA (Schoevers et al., 2016) and AOH (Schoevers et al., 2020) are detrimental to reproductive cells and embryo development.

Masked mycotoxins such as DON-3G can cause biological toxicity (Berthiller et al., 2013). DON-3G can be hydrolysed in the intestinal tract of pigs resulting in increased DON exposure (Nagl et al., 2014). Diets naturally contaminated with DON will often contain 3-ADON, 15-ADON, and DON-3G. Both 3- and 15-ADON are rapidly deacetylated in the upper intestinal tract, resulting in increased exposure to free DON (Ericsen et al., 2003; Broekaert et al., 2015). The levels of DON-3G will usually be 4 to 10 times lower than the DON level, whereas much lower levels of 3- and 15-ADON are commonly detected in the final diet. Based on their common presence in DON-contaminated diets, it is recommended that the possible additional exposure to DON after ingestion of 3-ADON, 15-ADON, and DON-3G be calculated because of the additional harm caused by these mycotoxins after deacetylation and hydrolysis in the intestine.

4 Effects of mycotoxins on nutrient digestibility, and growth and reproduction performance

Production performance and nutrient digestibility are closely associated. This highlights the importance of understanding how mycotoxins modulate the activity of enzymes and transporters involved in nutrient digestion and uptake, and the consequences for nutrient digestibility and for metabolizable energy. DON is known to induce changes in intestinal mucosa. However, little is known about the impact of these alterations on nutrient digestibility. Growing pigs, fed a diet naturally contaminated with 4 mg/kg DON, presented an increase in the apparent ileal digestibility (AID) of leucine, valine, aspartic acid, glycine, and serine as well as phosphorus and calcium. However, no effect was observed in dry matter (DM) and energy AID and the apparent total tract digestibility (ATTD) of dry matter and energy decreased (Bouchard et al., 2019). These authors also observed that DON was associated with increased faecal concentrations of acetate, butyrate, valerate and total volatile fatty acid. They suggested that the decrease in the ATTD of energy could explain the negative effect of DON on pig growth (Bouchard et al., 2019). Feeding newly weaned pigs with DON (3.2 mg/kg) did not affect DM, gross energy, nitrogen, and ether extract (EE) (Holanda and Kim, 2020). However, pigs fed a multi-contaminated diet with 180 µg/kg AFB_1, 1 mg/kg DON, and 9 or 14 mg/kg FB_1 presented a reduction in AID of crude protein (CP), gross energy, and EE (Kim et al., 2019).

The effect of dietary FB_1 on nutrient digestibility in pigs was evaluated after exposure to one of 5, 10 or 15 mg/kg FB_1 over six months. The feeding trial was divided into three physiological phases weanling (starter), peri-pubertal (grower) and pubertal (finisher). A significant decrease in the apparent digestibility of EE was detected in pigs fed the contaminated diets during the weaning phase. At the peri-pubertal phase, the digestibility of EE and CP reduced as the levels of FB_1 increased (Gbore and Egbunike, 2007).

Diets contaminated with ZEN (2 mg/kg) over 21 days did not affect nutrient digestibility in nursery gilts (Hauschild et al., 2007). However, decreases in apparent digestibility of CP, gross energy, and apparent metabolic rate of gross energy were observed in nursery gilts exposed to a lower ZEN concentration (1 mg/kg) for 24 days (Jiang et al., 2010). Synthetic ZEN was added to the diets in both trials. The initial BW of the gilts was 12 kg, each replicate consisted of one gilt and each treatment was repeated 4-5 times. Energy, CP, and mineral levels were similar in both diets, and the only difference between these studies related to the ingredients used to prepare them. The diets used in the study from Hauschild et al. (2007) were prepared with maize (59.82%) and soybean meal (31.52%). The diet in the study from Jiang et al. (2010) consisted of maize (53.0%), soybean meal (24.76%), wheat powder (6.50%) and wheat middlings (5.0%). This may have increased the risk of mycotoxin multi-contamination in the latter diets.

The reduction in the growth rate of challenged pigs is explained by the combined effects of reduction in ADFI and the efficiency of protein deposition. Several factors, including dose, duration of exposure, age, and sex, can modulate the negative effect of mycotoxins on animal performance. Most of the studies focus on the effects of mycotoxins in growing pigs and, unfortunately, there is a scarcity of studies based on nursery pigs. A meta-analytical study of growing pigs reported an average reduction of 26% in ADFI, an increase of 25% in feed conversion ratio (FCR) and a decrease of 23% in ADG in animals fed a DON-contaminated diet.

However, a high level mycotoxin exposure is needed to reach these values. For each additional 1 mg/kg of mycotoxins in the diets, a reduction of 0.28% in the ADG was reported for DON and a reduction of 0.17% for FUM (Andretta et al. 2012). No differences in ADFI and ADG were observed in growing pigs fed 1, 2 or even 5 mg/kg DON for five weeks (Sayyari et al. 2018). Significant decreases in ADFI, ADG, and BW were reported in finishing pigs fed diets contaminated with 3 to 5 mg/kg DON. The duration of the exposure varied from six to ten weeks. Reductions in ADFI (9 to 32%) and ADG (27 to 60%) resulted in lower final BW (Dänicke et al., 2004; Serviento et al., 2018; Wellington et al., 2020). As shown in the laboratory, younger animals have a reduced capacity for or quantity of hepatic enzymes to detoxify mycotoxins and to manage the stress caused by them. Similarly, younger pigs were more affected by mycotoxins than were growing-finishing pigs (Andretta et al. 2012). Besides the lower capacity to cope with mycotoxin exposure, stress during weaning may increase the susceptibility of piglets to mycotoxins. The younger the piglet during weaning, the lower the activity of antioxidant enzymes (Yu et al., 2021). Moreover, it was shown that 4-weeks-old piglets have a higher absorption rate of ZEN-14-glucoside (ZEN-14G) (94%) than 8-weeks-old ones (61%) (Catteuw et al., 2020). The influence of sex was described in growing pigs' performance. Exposure to DON reduced ADFI by 3% in females and 20% in males, while ADG decreased by 2% in females and 34% in males (Andretta et al., 2012). These authors also observed a higher sensitivity of male pigs to AFs exposure than females, where ADFI decreased by 24% in females and 30% in males, and ADG decreased by 23% in females and 27% in males.

When growing pigs were fed a diet contaminated with T-2 toxin at a concentration of 0.5 mg/kg during 3 weeks ADFI decreased by 10% but no effect on ADG was observed (Rafai et al., 1995a). A reduced body weight gain (BWG) was observed when piglets were exposed to 1 mg/kg T-2 toxin for 8 weeks (Weaver et al., 1978). However, dietary exposure of nursery piglets to realistic doses of T-2 toxin (83 μg/kg) for 18 days resulted in decreased BWG (Verbrugghe et al., 2012).

Dietary exposure to 5 mg/kg FB_1 modulates several appetite-related brain-gut peptides (including neuropeptide Y (NPY), peptide YY (PYY), ghrelin and obestatin) in growing pigs (Liu et al., 2021). However, growth performance trials in growing and nursery pigs showed that much higher levels of this mycotoxin are needed to impair growth performance. Weaned male pigs (7 kg BW) fed diets containing one of 0.2, 5, 10, or 15 mg/kg FB_1 over 24 weeks only presented a significant decrease in ADFI and impaired FCR if the diet contained 15 mg/kg FB_1. ADG decreased after exposure to 10 mg/kg FB_1 (Gbore, 2009). Furthermore, this author showed that pigs fed a diet containing 10 mg/kg FB_1 presented a delay in the puberty and those fed 5 mg/kg FB_1 were heavier when starting puberty. No effect on growth performance was observed in nursery pigs exposed to a diet naturally contaminated with 17.35 mg/kg FB_1 and 4.51 mg/kg FB_2. A negative impact was first observed when the piglets were a fed a diet with 25.21 mg/kg FB_1 and 7.46 mg/kg FB_2 (Rao et al., 2020).

Exposure of gilts to a diet artificially contaminated with ZEN (0.2, 0.4 and 0.8 mg/kg for 28 days) reduced the ADG linearly, increased ADFI linearly and impaired FCR (Wang et al., 2012). Feeding nursery gilts a diet artificially contaminated with ZEN at 5-fold higher level (1 mg/kg) than the EU recommended maximum level (Chapter 17) over 35 days did not affect growth performance (Liu et al., 2020). Suckling piglets exposed to ZEN via colostrum (0.08 ng/ml ZEN and 0.22 ng/ml α-ZEL) and milk (0.12 ng/ml ZEN and 0.16 ng/ml α-ZEL) presented a numerical increase in creep feed intake without affecting growth (Benthem de Grave et al., 2021). The serum levels of the intestinal hormone glucagon-like peptide 1 (GLP1) were significantly decreased together with an increase in the calprotectin serum levels in these piglets. GLP1 plays a role in satiety but is also linked to inflammatory processes.

Barrows (± 35 days) fed a diet highly contaminated with AFB_1 (250 μg/kg) for 10 weeks showed a reduced ADFI but no effect on the ADG was observed (Rustemeyer et al., 2011). Weaning piglets fed an AFB_1 diet (320 μg/kg) for 30 days exhibited a decreased BW and diarrhoea (Grosu et al., 2019). Exposure to AFs (0.25 and 0.3 mg/kg; 88% AFB_1) for 35 days decreases pigs' ADG and ADFI linearly (Wang et al., 2020). Aflatoxin-challenged growing pigs' showed 16% lower FI, and consequently 22% reduction in weight gain. Each additional 1 mg/kg of aflatoxin in the diets causes a reduction of 3.9% in pig weight gain (Andretta et al., 2012). Nursery pigs fed a diet naturally contaminated with AFs below the EU recommended levels (20 μg/kg) for 35 days showed no significant effect on growth performance, serum IgG, lipid peroxidation or on TNF-α levels (Sun et al., 2015). Male piglets are more sensitive to AFs than female ones, probably because of a higher clearance of AFs through urine in female pigs (Lee et al., 2017).

It is assumed that each 1 mg/kg of OTA in pigs' diets results in an average reduction of 12% in ADG. This can be a result of a decrease in ADFI or of a direct toxic effect of OTA on animals' metabolism (OTA is a potent protein synthesis inhibitor), or both (Battacone et al., 2010). Growing pigs fed an OTA diet (250 μg/kg; 119 days) showed reduced final BW, ADG, and impaired FCR (≈ -7%), while ADFI was not affected. The lower growth was probably attributable to reduced functionality of the kidneys and the liver, both organs showed high OTA concentrations (Malagutti et al., 2005). Nursery pigs fed an OTA-contaminated diet (400 μg/kg; 42 days) showed a reduced final BW, ADG and ADFI, and an increased FCR (Gan et al., 2017). Growth performance was not affected in weaned pigs fed an OTA diet (0.18 μg/kg; 43 days) below the European Commission recommend levels (Bernardini et al., 2014).

Growing-finishing pigs were fed diets containing contaminated rye, resulting in dietary levels of EAs of 0.05, 0.60, and 4.66 mg/kg. The authors only observed a decrease in ADFI and ADG when the pigs were exposed to the highest tested dosage, but carcass quality was not affected and no symptoms of intoxication were present (Mainka et al., 2005). Feeding growing pigs an ergot-contaminated barley diet containing 1.62 mg/kg EAs resulted in reduced ADFI, but this effect disappeared in the finisher period (Holden and Zimmerman, 1998).

Studies focusing on co-contamination of piglets' diets, e.g., 1 mg/kg DON + 260 µg/kg ZEN, reported a significant decrease in the BWG and ADFI, whereas the mono-contaminated diet (DON or ZEN) did not affect growth performance (Jia et al., 2020). Growing-finishing pigs fed a diet naturally contaminated with 0.5 mg/kg DON and 0.8 mg/kg FUM over 16 weeks had an impaired growth performance leading to reduced hot carcass weight, carcass yield, longissimus muscle area and a reduced percentage of carcass fat-free lean (Yang et al., 2020). When nursery piglets were exposed for 14 days to a diet artificially contaminated with 2.6 mg/kg BEA, 1.35 mg/kg ENNB + ENNB1 together or with 2.5 mg/kg DON alone, growth performance was not affected. DON on its own did induce a decrease in ADFI. However, a combination of all these mycotoxins (3.6 mg/kg BEA, 1.8 mg/kg ENNB + ENNB1, and 2.0 mg/kg DON) resulted in decreased weight gain (Novak et al., 2021).

After 56 days of exposure of gilts to diets naturally contaminated with ZEN (up to 145 µg/kg), DON (up to 533 µg/kg), and AFB_1 (up to 10 µg/kg), they presented a decreased ADG and impaired reproductive function (Zhuo et al., 2022). The analysis of different studies may lead to controversial results regarding growth performance. These inconsistencies are probably related to the method of dietary contamination, i.e. single *versus* multi-contamination and artificial *versus* natural, where natural contamination includes the presence of multiple mycotoxins and their metabolites or synergistic effects between co-occurring mycotoxins. These findings emphasize that co-contamination should be considered in the evaluation of the toxicity.

5 Effects of mycotoxins on the gastrointestinal health of pigs and susceptibility to diseases

The gastrointestinal tract (GIT) is the first system to come into contact with DON-contaminated feed. DON is moderately absorbed in the initial portions of the GIT. Thus, a substantial portion of non-absorbed toxin remains within the lumen of the GIT, resulting in a gut epithelium exposure to a very high proportion of the toxin ingested. Available data in pigs fed contaminated diets describe atrophy and fusion of villi, decreased villi height and cell proliferation, a reduced number of goblet cells and lymphocytes (Bracarense et al., 2012). In addition, an inflammatory milieu is established as evidenced by an up-regulation of several cytokines. The expression of cell protein junctions (intestinal tight junctions) is decreased resulting in changes to cell permeability and a predisposition to pathogen translocation (Pierron et al., 2016a). In practice, outbreaks of infectious diseases combined with dietary contamination with DON may increase the use of antibiotics and the period required to treat the affected animals. Piglets experiencing *Streptococcus suis* or other infections and being fed diets contaminated with DON at levels ranging from

0.8 to 1.1 mg/kg presented a higher incidence of meningitis and joint inflammation, exhibited disease symptoms over an increased period (3.5 vs. 6.4 days of sickness; increase in 80%) and required increased use of antibiotics (1.6 vs. 2.1 doses; increase in 31%) (Figure 4.1; personal data of the authors).

Dietary exposure of nursery pigs to 15 and 83 µg/kg T-2 toxin over 23 days *in vitro* decreased the *Salmonella* Typhimurium population in the caeca tended to decrease it in the jejunum, ileum and colon, but increased bacterial translocation (Verbrugghe et al., 2012). It is important to mention that the authors exposed the pigs to T-2 toxin only and that the diet was not contaminated with any other mycotoxin – a very rare condition.

Ingestion of diets contaminated with DON and ZEN induces transient modifications in the composition and functions of faecal microbiota in finishing pigs. A decrease in the relative abundances of the families *Ruminococcaceae, Streptococcaceae* and *Veillonellaceae* and an increase of *Erysipelotrichaceae* were reported, but these changes were not affecting the microbiota populations in the long-term (Le Sciellour et al., 2020).

Exposure to FUM induces changes in intestinal tissue, mainly the jejunum. The histological changes are reduced villi height, villi fusion and atrophy, apical necrosis of villi and flattening of enterocytes. As with exposure to DON, enterocytes exposed to FUM showed a reduced expression of tight junctions (Basso et al., 2013) leading to increased permeability and this may have facilitated bacterial translocation

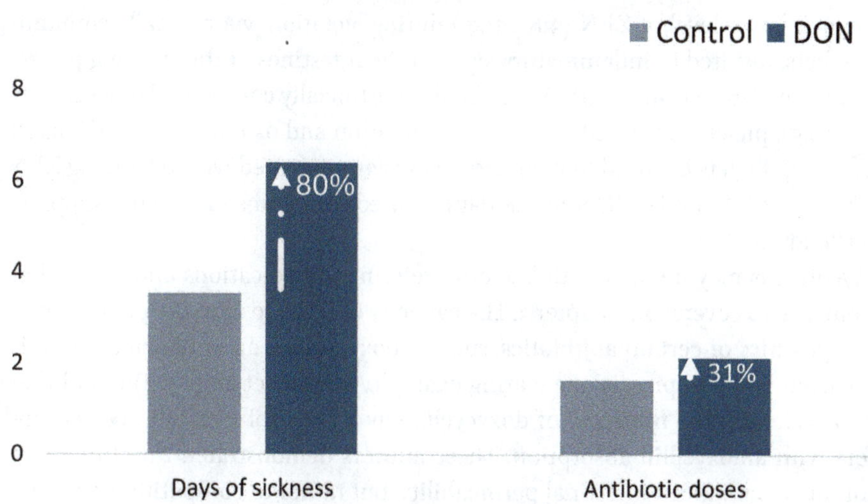

FIGURE 4.1 Average days of sickness and of administered antibiotic doses in piglets fed diets naturally contaminated with DON at marginal levels (below 0.15 mg/kg) or moderately contaminated with DON (between 0.8 and 1.1 mg/kg).
SOURCE: the authors.

across the intestine. In fact, nursery pigs, challenged with realistic doses of FB_1 (5 to 8 mg/kg) and pathogenic strains of *Escherichia coli,* presented translocation of bacteria to mesenteric lymph nodes and dissemination to the lungs, liver and spleen, highlighting the potential for sepsis (Oswald et al., 2003).

The reproductive tract experiences the main effects of ZEN. Some studies have also reported its ability to induce toxicity and generate adverse effects in the GIT. Gilts exposed to a ZEN-contaminated diet (1 mg/kg for 35 days) showed shortening, retardation and branching dysfunction of intestinal villi (Zhang et al., 2021). An *ex vivo* study using pigs' jejunal samples demonstrated an increased number of goblet cells and infiltrated lymphocytes in the lamina propria in intestinal explants exposed to ZEN (Soler et al., 2020). Similar results were described in the intestines of pigs fed ZEN (40 µg/kg BW, six weeks) contaminated diets (Przybylska-Gornowicz et al., 2015). Gilts fed for six weeks with mono or co-contaminated diets (1 mg/kg ZEN and 0.3 mg/kg DON) showed increased numbers of inflammatory cells in the intestinal lamina propria (Lewczuk et al., 2016). Pigs exposed to ZEN exhibited deceleration of gastrointestinal motility (an effect that can contribute to increased accumulation of ZEN into the intestine), inhibition of cell proliferation, apoptosis, decreased faecal bacterial counts in the colon and increased proliferative activity of lymphocytes in their intestines (Gajęcka et al., 2016). Exposure of sows to 2.8 mg/kg ZEN during gestation induced intestinal oxidative stress and inflammation, characterized by the increased expression of interleukins-1α and 6, as well as of TNF-α. The suckling piglets of these sows had a reduced intestinal population of *Bifidobacteria* spp. and *Lactobacillus*. Furthermore, while *Enterobacteriaceae* populations decreased in sows, they increased in the piglets (Liu et al., 2018). Exposure of sows to a much lower level of ZEN (318 µg/kg) during lactation, via naturally contaminated diets, resulted in inflammatory signs in the intestines of the suckling piglets (Benthem de Grave et al., 2021). When fed diets artificially contaminated with ZEN (290 µg/kg), piglets presented signs of inflammation and oxidative stress (Bulgaru et al., 2023). Piglets exposed to diets artificially contaminated with 0.8 mg/kg ZEN combined with 8 mg/kg DON for 28 days showed symptoms of immunosuppression (Reddy et al., 2018).

Mycotoxins may interact with bacteria, veterinary medications and with other mycotoxins, as covered in Chapter 3. The capacity of DON to alter the pharmacokinetic dynamics of certain antibiotics, such as doxycycline, an antibacterial that is often given orally to pigs, is one glaring example. Mendel et al. (2022) found that DON accelerates the transport of doxycycline by a factor of five, but does not interfere with amoxycillin absorption. These authors demonstrated that this effect is not due to enhanced intestinal permeability, but rather to competition between DON and the antibiotic for an efflux transporter. They also demonstrated that it will probably interfere with the pharmacodynamic and safety of the antibiotic.

Besides inducing intestinal inflammation (immunostimulatory effect) and increasing susceptibility to infectious diseases, exposure to mycotoxins may also decrease vaccine efficacy (immunosuppressive effect) (Pierron et al., 2016b). Weaned piglets fed with a diet contaminated with 50 μg/kg OTA over 30 days had a decreased expression of inflammatory cytokines in the colon (IL-6, IL-8, IL-12, IL-17A, and IL-18) and in the kidneys (IL-17A and IL-10), due to an inhibition of the synthesis of proteins which act as effectors in inflammation (Marin et al., 2017). OTA-induced immunomodulation has been reported in farm animals. Impairment of the immune system function renders animals more susceptible to several infectious and parasitic diseases, thus increasing the severity of the infection and an enhancement of the toxic effects of OTA, a consequence of both humoral and cell-mediated immune response suppression (Duarte et al., 2011). Exposure to a diet highly contaminated with AFB_1 (385 μg/kg) impaired the cell-mediated response to vaccines in pigs (Meissonnier et al., 2008). Weaned piglets vaccinated against the porcine reproductive and respiratory syndrome virus (PRRSV) were fed diets contaminated with DON (2 mg/kg) over four weeks. When these pigs were exposed to PRRSV, 20% of the animals were unable to present an efficient PRRSV-specific antibody response (Rückner et al., 2022). Impaired vaccination efficacy was also related to exposure to FB_1, T-2 toxin, AF, and OTA (Pierron et al., 2016b).

6 Mycotoxin carry-over in swine food products

Worldwide, pork meat and pork meat products are considered important sources of proteins for human consumption. However, they can also be a source of chronic exposure to mycotoxins for humans. Most *Fusarium* mycotoxins are found at negligible levels in pork meat. Although DON is poorly detoxified and highly distributed, the carry-over rates indicate a limited risk to the consumer of the edible tissues of food animals (Tolosa et al., 2021). The carry-over of DON + DOM-1 decreases between bile and back fat, following this order: bile > kidney > liver > serum > muscle > back fat. Only negligible amounts of DON were found in the back fat of pigs fed a diet highly contaminated with DON (6.68 mg/kg) over 12 weeks (Goyarts et al., 2007).

The carry-over of FUM from feed to pig's meat is not considered toxicologically relevant. FB_1, FB_2 and the main metabolites may be found in limited amounts in tissues (EFSA, 2018). Pigs fed a diet contaminated with 3 mg/kg of radiolabelled FB_1 showed an accumulation of 160 and 65 ng/g in the liver and kidneys, respectively. and no accumulation in muscle and adipose tissue (Prelusky et al., 1996).

Pigs fed a diet containing 250 μg/kg of ZEN for 90 days showed levels of ZEN and α-ZEL below 1.24 μg/kg in the liver, and levels of ZEN and α-ZEL below 0.56 μg/kg in muscles and kidneys. (Dänicke and Winkler, 2015). ZEN was detected in pork sausages, pork liver and pork meat with a mean concentration of 0.62 ± 0.14 μg/kg, and a maximum concentration of 4.31 μg/kg in pork meat (Pleadin et al., 2015a). The presence of α-ZEL in 27% of meat-based infant food (maximum 30.5 μg/kg) was also reported (Meucci et al., 2011). Besides the detection of low concentrations of ZEN and its metabolites in pork products, the risk of exposure for humans is considered lower compared to the ingestion of food based contaminated grains (Tolosa et al., 2021).

The transfer of AFB_1 to meat products is considered negligible, but this contamination cannot be discounted. In a survey of Croatian meat products, AFB_1 was detected at a maximum level 1.3 µg/kg in only 5% of the samples. Contamination with AFB_1 was observed in 11% of the liver sausages and pâté (up to 1.7 µg/kg AFB_1) probably because this mycotoxin is mostly accumulated in the liver (Pleadin et al., 2015b).

The accumulation of OTA in the tissues of animals can pose a threat to food safety by disseminating mycotoxins throughout the food chain. Several studies revealed different levels of OTA in the meat of slaughtered pigs, while OTA levels were particularly high in the blood serum and kidneys of pigs. Pork products made from pig blood or organs such as the kidney or liver have often been found to be contaminated with OTA (Marin et al., 2017; Vlachou et al., 2022). Finishing pigs fed a diet contaminated with OTA (300 µg/kg) over 30 days, resulted in an accumulation of mycotoxin residues in muscle and adipose tissue (5 µg/kg) (Pleadin et al., 2013). In Croatia, OTA was detected in fermented sausages and hams at levels up to 5.10 µg/kg and 9.95 µg/kg, respectively. These levels were far higher than the maximum level of 1.0 µg/kg stipulated for pork products (Pleadin et al., 2015b).

Exposure to pork products contaminated with CIT is of low concern for humans. The carry-over of CIT in edible tissues from growing pigs fed a diet contaminated with CIT (1 mg/kg) over 3 weeks was determined as 0.1% for muscle (0.6 µg/kg), 0.5% in skin-fat (5 µg/kg), 1.5% in kidneys (1.5 µg/kg), and 2% in liver (2.3 µg/kg) (Meerpoel et al., 2020b).

7 Conclusions

Pigs may be exposed to a combination of mycotoxins during gestation – through the placenta, during the lactation period (via colostrum and milk) and, after birth, through feed. Acute symptoms are uncommon, while chronic exposure is occurring on a regular basis. While diagnosing mycotoxicosis may not be straightforward, it is possible to observe losses in production performance. Despite the fact that mycotoxins have a diverse array of target organs, exposure to them will lead to intestinal damage and inflammation, increasing susceptibility to bacterial translocation and infection. The immune system will not always be stimulated by mycotoxins, but may also experience suppression, thereby reducing the effectiveness of vaccination. The sensitivity of pigs will depend on their age and sex (nursery pigs are more sensitive than growing-finishing ones), male pigs tend to be more sensitive than female ones (e.g., AFB_1 and DON exposure). Multi-mycotoxin exposure will determine the level of harm. Pragmatic application of the maximum recommended levels of mycotoxins in pig diets must consider age, sex, and the animal's category, i.e., growth for meat production or for reproduction. Although two different levels of ZEN are recommended for piglets, gilts, and sows, such a categorised recommendation should also be applied to other mycotoxins because nursery piglets are often more sensitive to mycotoxin exposure, especially when experiencing stress during weaning. In addition to the parental mycotoxin levels that are regularly measured in the feeds, it is important to consider the potential contribution of masked mycotoxins that will be present in the diet alongside the parental one.

References

Alassane-Kpembi, I., Puel, O., Pinton, P., Cossalter, A.M., Chou, T.C. and Oswald, I.P., 2017. Co-exposure to low doses of the food contaminants deoxynivalenol and nivalenol has a synergistic inflammatory effect on intestinal explants. Archives of Toxicology 91: 2677-2687. https://doi.org/10.1007/s00204-016-1902-9.

Andretta, I., Kipper, M., Lehnen, C.R., Hauschild, L., Vale, M.M. and Lovatto, P.A., 2012. Meta-analytical study of productive and nutritional interactions of mycotoxins in growing pigs. Animal 6: 1476-1482. https://doi.org/10.1017/S1751731111002278.

Arroyo-Manzanares, N., Rodríguez-Estévez, V., Arenas-Fernández, P., García-Campaña, A.M. and Gámiz-Gracia, L., 2019. Occurrence of mycotoxins in swine feeding from Spain. Toxins 11: 342. https://doi.org/10.3390/toxins11060342.

Basso, K., Gomes, F. and Bracarense, A.P., 2013. Deoxynivanelol and fumonisin, alone or in combination, induce changes on intestinal junction complexes and in E-cadherin expression. Toxins 5: 2341-2352. https://doi.org/10.3390/toxins5122341.

Battacone, G., Nudda, A. and Pulina, G., 2010. Effects of Ochratoxin A on Livestock Production. Toxins 2: 1796-1824. https://doi.org/10.3390/toxins2071796.

Benthem de Grave, X., Winkler, J., Laurain, J., Rodriguez, M., Molist, F., Dänicke, S. and Santos, R.R., 2021. Transmission of zearalenone, deoxynivalenol, and their derivatives from sows to piglets during lactation. Toxins 13: 37. https://doi.org/10.3390/toxins13010037.

Bernardini, C., Grilli, E., Duvigneau, J.C., Zannoni, A., Tugnoli, B., Gentilini, F., Bertuzzi, T., Spinozzi, S., Camborata, C., Bacci, M.L., Piva, A. and Forni, M., 2014. Cellular stress marker alteration and inflammatory response in pigs fed with an ochratoxin contaminated diet. Research in Veterinary Science 97: 244-250. https://doi.org/10.1016/j.rvsc.2014.07.018.

Berthiller, F., Crews, C., Dall'Asta, C., Saeger, S. D., Haesaert, G., Karlovsky, P., Oswald, I. P., Seefelder, W., Speijers, G. and Stroka, J., 2013. Masked mycotoxins: a review. Molecular Nutrition and Food Research 57: 165–186. https://doi.org/10.1002/mnfr.201100764.

Blaney, B.J., Maryam, R., Murray, S.A. and Ryley, M.J., 2003. Alkaloids of the sorghum ergot pathogen (*Claviceps africana*): assay methods for grain and feed and variation between sclerotia/sphacelia. Australian Journal of Agricultural Research 54: 167–175. http://dx.doi.org/10.1071/AR02095.

Bosch, U. and Mirocha, C.J., 1992. Toxin production by *Fusarium* species from sugar beets and natural occurrence of zearalenone in beets and beet fibers. Applied Environmental Microbiology 58: 3233–3239. https://doi.org/10.1128/aem.58.10.3233-3239.1992.

Bothast, R.J., Bennett, G.A., Vancauwenberge, J.E. and Richard, J.L., 1992. Fate of fumonisin B1 in naturally contaminated corn during ethanol fermentation. Applied Environmental Microbiology 58: 233–236. https://doi.org/10.1128/aem.58.1.233-236.1992.

Bouchard, M.J., Chorfi, Y., Létourneau-Montminy, M.P. and Guay, F., 2019. Effects of deoxynivalenol and sodium meta-bisulphite on nutrient digestibility in growing pigs. Archives of Animal Nutrition 73:360-373. https://doi.org/10.1080/1745039X.2019.1641369.

Bracarense, A.P., Lucioli, J., Grenier, B., Drociunas, P. G., Moll, W.D., Schatzmayr, G. and Oswald, I.P., 2012. Chronic ingestion of deoxynivalenol and fumonisin, alone or in interaction, induces morphological and immunological changes in the intestine of piglets. British Journal of Nutrition 107: 1776-1786. https://doi.org/10.1017/S0007114511004946.

Broekaert, N., Devreese, M., De Mill, T., Fraeyman, S., Antonissen, G., De Baere, S., De Backer, P., Vermeulen, A. and Croubels, S., 2015. Oral bioavailability, hydrolysis, and comparative toxicokinetics of 3-acetyldeoxynivalenol and 15-acetyldeoxynivalenol in broiler chickens and pigs. Journal of Agriculture and Food Chemistry 63: 8734-8742. https://doi.org/10.1021/acs.jafc.5b03270.

Bulgaru, V.C., Pertea, A.M., Grosu, I.A., Anghel, A.C., Pistol, G.C., Marin, D.E., Dinischiotu, A. and Taranu, I., 2023. Effects and underlying mechanisms of zearalenone mycotoxin at concentrations close to the EC Recommendation on the colon of piglets after weaning. Agriculture 13: 1372. https://doi.org/10.3390/agriculture13071372.

Catteuw, A., Devreese, M., De Baere, S., Antonissen, G., Huybrechts, B., Ivanova, L., Uhlig, S., Martens, A., De Saeger, S., De Boevre, M. and Croubels, S., 2020. Toxicokinetic studies in piglets reveal age-related differences in systemic exposure to zearalenone, zearalenone-14-glucoside, and zearalenone-14-sulfate. Journal of Agricultural and Food Chemistry 68: 7757-7764. https://doi.org/10.1021/acs.jafc.0c01943.

Cheat, S., Gerez, J.R., Cognie, J., Alassane-Kpembi, I., Bracarense, A.P.F., Raymond-Letron, I., Oswald, I.P. and Kolf-Claw, M., 2015. Nivalenol has a greater impact than deoxynivalenol on pig jejunum mucosa in vitro on explants and in vivo on intestinal loops. Toxins 7: 1945-1961. https://doi.org/10.3390/toxins7061945.

Dänicke, S. and Winkler, J., 2015. Invited review: Diagnosis of zearalenone (ZEN) exposure of farm animals and transfer of its residues into edible tissues (carry over). Food and Chemical Toxicology 84: 225-249. https://doi.org/10.1016/j.fct.2015.08.009.

Dänicke, S., Valenta, H., Klobasa, F., Döll, S., Ganter, M. and Flachowsky, G., 2004. Effects of graded levels of *Fusarium* toxin contaminated wheat in diets for fattening pigs on growth performance, nutrient digestibility, deoxynivalenol balance and clinical serum characteristics. Archives of Animal Nutrition 58: 1-17. https://doi.org/10.1080/00039420 31000161045.

Devreese, M., Broekaert, N., De Mil, T., Fraeyman, S., De Backer, P. and Croubels, S., 2014. Pilot toxicokinetic study and absolute oral bioavailability of the *Fusarium* mycotoxin enniatin B1 in pigs. Food and Chemical Toxicology 63: 161-165. https://doi.org/10.1016/j.fct.2013.11.005.

Devreese, M., De Backer, P. and Croubels, S., 2013. Overview of the most important mycotoxins for the pig and poultry husbandry. Vlaams Diergeneeskundig Tijdschrift 82: 171-180. http://hdl.handle.net/1854/LU-4125525.

Duarte, S. C., Lino, C. M. and Pena, A., 2011. Ochratoxin A in feed of food-producing animals: An undesirable mycotoxin with health and performance effects. Veterinary Microbiology 154: 1-13. https://doi.org/1310.1016/j.vetmic.2011.05.006.

EFSA CONTAM Panel (EFSA Panel on Contaminants in the Food Chain), Schrenk, D., Bignami, M., Bodin, L., Chipman, J.K., del Mazo, J., Grasl-Kraupp, B., Hogstrand, C., Leblanc, J.C., Nebbia, C.S., Nielsen, E., Ntzani, E., Petersen, A., Sand, S., Schwerdtle, T., Vleminckx, C.,Wallace, H., Gropp, J., Mulder, P., Oswald, I.P., Woutersen, R., Ruiz, J.A.G., Rovesti, E. and Hoogenboom,L.R., 2024. Risks for animal health related to the presence of ergot alkaloids in feed. EFSA Journal 22: 8496. https://doi.org/10.2903/j.efsa.2024.8496.

EFSA CONTAM Panel (EFSA Panel on Contaminants in the Food Chain). Knutsen, H.K., Alexander, J., Barregard, L., Bignami, M., Bruschweiler, B., Ceccatelli, S., Cottrill, B., Dinovi, M., Grasl-Kraupp, B., Hogstrand, C., Hoogenboom, L.R., Nebbia, C.S., Oswald, I.P., Petersen, A., Rose, M., Roudot, A.C., Schwerdtle, T., Vleminckx, C., Vollmer, G., Wallace, H., De Saeger, S., Eriksen, G.S., Farmer, P., Fremy, J.M., Gong, Y.Y., Meyer, K., Naegeli, H., Parent-Massin, D., Rietjens, I., van Egmond, H., Altieri, A., Eskola, M., Gergelova, P., Ramos Bordajandi, L., Benkova, B., Dorr, B., Gkrillas, A., Gustavsson, N., van Manen, M. and Edler, L., 2017. Scientific Opinion on the risks to human and animal health related to the presence of deoxynivalenol and its acetylated and modified forms in food and feed. EFSA Journal 15: 4718. https://doi.org/10.2903/j.efsa.2017.4718.

EFSA Panel on Contaminants in the Food Chain (CONTAM). Alexander, J., Benford,D., Boobis, A., Ceccatelli, S., Cottrill, B., Cravedi, J.P., Di Domenico, A., Doerge, D., Dogliotti, E., Edler, L., Farmer, P., Filipič, M., Fink-Gremmels, J., Fürst, P., Guérin, T., Knutsen, H.K., Machala, M., Mutti, A., Rose, M., Schlatter, J., and van Leeuwen, R., 2012. Scientific Opinion on the risks for public and animal health related to the presence of citrinin in food and feed. EFSA Journal 10 :2605. https://doi.org/10.2903/j.efsa.2012.2605.

EFSA Panel on Contaminants in the Food Chain (CONTAM). Knutsen, H.K., Alexander, J., Barregård, L., Bignami, M., Brüschweiler, B., Ceccatelli, S., Cottrill, B., Dinovi, M., Edler, L., Grasl-Kraupp, B., Hogstrand, C., Hoogenboom, L.R., Nebbia, C.S., Petersen, A., Rose, M., Roudot, A.C., Schwerdtle, T., Vleminckx, C., Vollmer, G., Wallace, H., Dall'Asta, C., Eriksen, G.S., Taranu, I., Altieri, A., Roldán-Torres, R., Oswald, I.P., 2018. Risks for animal health related to the presence of fumonisins, their modified forms and hidden forms in feed. EFSA Journal 16: e05242. https://doi.org/10.2903/j.efsa.2018.5242.

Eriksen, G.S., Pettersson, H. and Lindberg, J.E., 2003. Absorption, metabolism and excretion of 3-acetyl DON in pigs. Archives in Animal Nutrition 57: 335–345. https://doi.org/10.1080/00039420310001607699.

Fodor, J., Balogh, K., Weber, M., Miklós, M., Kametler, L., Pósa, R., Mamet, R., Bauer, J., Horn, P., Kovács, F. and Kovács, M., 2008. Absorption, distribution and elimination of fumonisin B1 metabolites in weaned piglets. Food Additives and Contaminants: Part

A: Chemistry, Analysis, Control, Exposure and Risk Assessment 25: 88-96. https://doi.org/10.1080/02652030701546180.

Friis, P., Hasselager, E. and Krogh, P., 1969. Isolation of citrinin and oxalic acid from Penicillium viridicatum Westling and their nephrotoxicity in rats and pigs. Acta Pathologica et Microbiologica Scandinavica. 77: 559-560.

Gajęcka, M., Zielonka, Ł. and Gajęcki, M., 2016. Activity of zearalenone in the porcine intestinal tract. Molecules 22:18. https://doi.org/10.3390/molecules22010018.

Gan, F., Hou, L., Zhou, Y., Liu, Y., Huang, D., Chen, X. and Huang, K., 2017. Effects of ochratoxin A on ER stress, MAPK signaling pathway and autophagy of kidney and spleen in pigs. Environmental Toxicology 32: 2277-2286. https://doi.org/10.1002/tox.22443.

Gbore, F. A. and Egbunike, G. N., 2007. Influence of dietary fumonisin B1 on nutrient utilization by growing pigs. Livestock Research for Rural Development 19: article #93. http://www.lrrd.org/lrrd19/7/gbor19093.htm.

Gbore, F.A., 2009. Growth performance and puberty attainment in growing pigs fed dietary fumonisin B1. Journal of Animal Physiology and Animal Nutrition 93: 761-767. https://doi.org/10.1111/j.1439-0396.2008.00866.x.

Goyarts, T. and Dänicke, S., 2006. Bioavailability of the *Fusarium* toxin deoxynivalenol (DON) from naturally contaminated wheat for the pig. Toxicology Letters 163: 171-182. https://doi.org/10.1016/j.toxlet.2005.10.007.

Goyarts, T., Dänicke, S., Valenta, H. and Ueberschär, K.H., 2007. Carry-over of *Fusarium* toxins (deoxynivalenol and zearalenone) from naturally contaminated wheat to pigs. Food Additives & Contaminants 24: 369-380. https://doi.org/10.1080/02652030600988038.

Greenman, D.L., Mehta, R.G. and Wittliff, J.L., 1979. Nuclear interaction of *Fusarium* mycotoxins with estradiol binding sites in the mouse uterus. Journal of Toxicology and Environmental Health 5: 593-598. https://doi.org/10.1080/15287397909529771.

Gregorio, M.C.D., Bordin, K., Souto, P.C.M.D.C., Corassin, C.H. and Oliveira, C.A.F., 2015. Comparative biotransformation of aflatoxin B1 in swine, domestic fowls, and humans. Toxin Reviews 34: 142-150. https://doi.org/10.3109/15569543.2015.1091979.

Grenier, B. and Applegate, T.J., 2013. Modulation of intestinal functions following mycotoxin ingestion: meta-analysis of published experiments in animals. Toxins 5: 396-430. https://doi.org/10.3390/toxins5020396.

Grosu, I. A., Pistol, G. C., Taranu, I. and Marin, D. E., 2019. The impact of dietary grape seed meal on healthy and aflatoxin B1 afflicted microbiota of pigs after weaning. Toxins 11: 25. https://doi.org/10.3390/toxins11010025.

Gruber-Dorninger, C., Jenkins, T. and Schatzmayr, G., 2019. Global mycotoxin occurrence in feed: a ten-year survey. Toxins 11: 375. https://doi.org/10.3390/toxins11070375.

Guerre, P., 2016. Worldwide mycotoxins exposure in pig and poultry feed formulations. 8: 350. https://doi.org/10.3390/toxins8120350.

Gumprecht, L.A., Smith, G.W., Constable, P.C. and Haschek, W.M., 2001. Species and organ specificity of fumonisin-induced endothelial alterations: potential role in porcine pulmonary edema. Comparative Study 160: 71-79. https://doi.org/10.1016/s0300-483x(00)00444-3.

Hanschmann, G., Krieg, D. and Saechsische, D., 2007. Behavior of *Fusarium* toxins during ethanol production from polluted grain. Sachsische Landesanstalt Landwirtschaft 62: 290–296. https://doi.org/10.1007/BF02959272.

Harrison, L.R., Colvin, B.M., Greene, J.T., Newman, L.E. and Cole Jr, J.R., 1990. Pulmonary edema and hydrothorax in swine produced by fumonisin B1, a toxic metabolite of *Fusarium moniliforme*. Journal of Veterinary Diagnostic Investigation 2: 217-221. https://doi.org/10.1177/104063879000200312.

Haschek, W.M., Gumprecht, L.A., Smith, G., Tumbleson, M.E. and Constable, P.D., 2001. Fumonisin toxicosis in swine: an overview of porcine pulmonary edema and current perspectives. Environmental Health Perspectives 109: 251-257. https://doi.org/10.1289/ehp.01109s2251.

Hauschild, L., Lovatto, P., Cheila, R., Lehnen, D'avila Verardi, A., Guarez, G. and Mallmann, C., 2007. Digestibilidade e metabolismo de dietas de suínos contendo zearalenone com adição de organoaluminossilicato. Pesquisa Agropecuária Brasileira 42: 219-224. https://doi.org/10.1590/S0100-204X2007000200010.

Hennig-Pauka, I., Koch, FJ., Schaumberger, S., Woechtl, B., Novak, J., Sulyok, M. and Nagl, V., 2018. Current challenges in the diagnosis of zearalenone toxicosis as illustrated by a field case of hyperestrogenism in suckling piglets. Porcine Health Management 4: 18. https://doi.org/10.1186/s40813-018-0095-4.

Heussner, A. H. and Bingle, L. E. H., 2015. Comparative ochratoxin toxicity: a review of the available data. Toxins 7: 4253-4282. https://doi.org/10.3390/toxins7104253.

Holanda, D.M. and Kim, S.W., 2020. Efficacy of mycotoxin detoxifiers on health and growth of newly-weaned pigs under chronic dietary challenge of deoxynivalenol. Toxins 12: 311. https://doi.org/10.3390/toxins12050311.

Holden, P. and Zimmerman, D. 1998. Effect of ergot-contaminated barley on growing-finishing pig performance. Asl-R1482, report of the Iowa State University. https://www.extension.iastate.edu/Pages/ansci/swinereports/asl-1482.pdf .

Huijbregts, T., Legrand, G., Hoffmann, C., Olsson, R. and Olsson, A., 2013. Long-term storage of sugar beet in North-West Europe. COBRI https://www.nordicbeet.nu/wp-content/uploads/2016/04/COBRI-storage-report-2013-final-131004.pdf.

IARC, I.A.F.R.O.C., 2002. Some traditional herbal medicines, some mycotoxins, naphthalene and styrene. Lyon: IARC: International Agency for Research on Cancer. 82: 171-274.

Ivanova, L., Uhlig, S., Devreese, M., Croubels, S. and Faeste, C.K., 2017. Biotransformation of the mycotoxin enniatin B1 in pigs: a comparative *in vitro* and *in vivo* approach. Food and Chemical Toxicology 105: 506-517. https://doi.org/10.1016/j.fct.2017.04.041.

Jang, J.C., Zeng, Z., Urriola, P.E. and Shurson, G.C., 2021. Effects of feeding corn distillers dried grains with solubles diets without or with supplemental enzymes on growth performance of pigs: a meta-analysis. Translational Animal Science 5: txab029. https://doi.org/10.1093/tas/txab029.

Jia, R., Liu, W., Zhao, L., Cao, L. and Shen, Z., 2020. Low doses of individual and combined deoxynivalenol and zearalenone in naturally moldy diets impair intestinal functions via inducing inflammation and disrupting epithelial barrier in the intestine of piglets. Toxicology Letters 333: 159-169. https://doi.org/10.1016/j.toxlet.2020.07.032.

Jiang, S.Z., Yang, Z.B., Yang, W.R., Yao, B.Q., Zhao, H., Liu, F.X., Chen, C.C. and Chi, F., 2010. Effects of feeding purified zearalenone contaminated diets with or without clay enterosorbent on growth, nutrient availability, and genital organs in post-weaning female pigs. Asian-Australasian Journal of Animal Sciences 23: 74-81. https://doi.org/10.5713/ajas.2010.90242.

Kim, B.G., Wulf, D.M., Maddock, R.J., Peters, D.N., Pedersen, C., Liu, Y. and Stein, H.H., 2014. Effects of dietary barley on growth performance, carcass traits and pork quality of finishing pigs. Revista Colombiana de Ciencias Pecuarias 27: 102-113.

Kim, S.W., Holanda, D.M., Gao, X., Park, I. and Yiannikouris, A., 2019. Efficacy of a yeast cell wall extract to mitigate the effect of naturally co-occurring mycotoxins contaminating feed ingredients fed to young pigs: impact on gut health, microbiome, and growth. Toxins 11: 633. https://doi.org/10.3390/toxins11110633.

Kopinski, J.S., Blaney, B.J., Downing, J.A., McVeigh, J.F. and Murray, S.A., 2007. Feeding sorghum ergot (*Claviceps africana*) to sows before farrowing inhibits milk production. Australian Veterinary Journal 85: 169–176. https://doi.org/10.1111/j.1751-0813.2007.00139.x.

Le Sciellour, M., Zemb, O., Serviento, A.M. and Renaudeau, D., 2020. Transient effect of single or repeated acute deoxynivalenol and zearalenone dietary challenge on fecal microbiota composition in female finishing pigs. Animal 14: 2277-2287. https://doi.org/10.1017/S1751731120001299.

Lee, H.S., Lindahl, J., Nguyen-Viet, H., Khong, N.V., Nghia, V.B., Xuan, H.N. and Grace, D., 2017. An investigation into aflatoxin M1 in slaughtered fattening pigs and awareness of aflatoxins in Vietnam. BMC Veterinary Research 13: 363. https://doi.org/10.1186/s12917-017-1297-8.

Lewczuk, B., Przybylska-Gornowicz, B., Gajęcka, M., Targońska, K., Ziółkowska, N., Prusik, M. and Gajęcki, M., 2016. Histological structure of duodenum in gilts receiving low doses of zearalenone and deoxynivalenol in feed. Experimental and Toxicologic Pathology 68:157-166. https://doi.org/10.1016/j.etp.2015.11.008.

Liu, M., Zhu, D., Guo, T., Zhang, Y., Shi, B., Shan, A., Chen, Z., 2018. Toxicity of zearalenone on the intestines of pregnant sows and their offspring and alleviation with modified halloysite nanotubes. Journal of the Science of Food and Agriculture 98: 698-706. https://doi.org/10.1002/jsfa.8517.

Liu, Q., Li, F., Huang, L., Chen, W., Li, Z. and Wang, C., 2021. FumDSB can reduce the toxic effects of fumonisin b_1 by regulating several brain-gut peptides in both the hypothalamus and jejunum of growing pigs. Toxins. 13: 874. https://doi.org/10.3390/toxins13120874.

Liu, X., Xu, C., Yang, Z., Yang, W., Huang, L., Wang, S., Liu, F., Liu, M., Wang, Y. and Jiang, S., 2020. Effects of dietary zearalenone exposure on the growth performance, small intestine disaccharidase, and antioxidant activities of weaned gilts. Animals 10: 2157. https://doi.org/10.3390/ani10112157.

Lopes, P., Sobral, M.M.C., Lopes, G.R., Martins, Z.E., Passos, C.P., Petronilho, S. and Ferreira,

I.M.P.L.V.O., 2023. Mycotoxins' prevalence in food industry by-products: a systematic review. Toxins 15: 249. https://doi.org/10.3390/toxins15040249.

Lüthy, J., Zweifel, U. and Schlatter, C., 1980. Metabolism and tissue distribution of [14C] aflatoxin B1 in pigs. Food and Cosmetics Toxicology 18: 253-256. https://doi.org/10.1016/0015-6264(80)90103-0.

Ma, R., Zhang, L., Liu, M., Su, Y.T., Xie, W.M., Zhang, N.Y., Dai, J.F., Wang, Y., Rajput, S.A., Qi, D.S., Karrow, N.A. and Sun, L.H., 2018. Individual and combined occurrence of mycotoxins in feed ingredients and complete feeds in China. Toxins 10: 113. https://doi.org/10.3390/toxins10030113.

Mainka, S., Dänicke, S., Böhme, H., Ueberschär, K.H., Polten, S. and Hüther, L., 2005. The influence of ergot-contaminated feed on growth and slaughtering performance, nutrient digestibility and carry over of ergot alkaloids in growing-finishing pigs. Archives of Animal Nutrition 59: 377-395. http://dx.doi.org/10.1080/17450390500352970.

Malagutti, L., Zannotti, M., Scampini, A. and Sciaraffia, F., 2005. Effects of Ochratoxin A on heavy pig production. Animal Research 54: 179-184. https://doi.org/10.1051/animres:2005019.

Malekinejad, H., Colenbrander, B. And Fink-Gremmels, J., 2006. Hydroxysteroid dehydrogenases in bovine and porcine granulosa cells convert zearalenone into its hydroxylated metabolites α-zearalenol and β-zearalenol. Veterinary Research Communications 30: 445-453. https://doi.org/10.1007/s11259-006-3325-1.

Marin, D. E., Bulgaru, C. V., Anghel, C. A., Pistol, G. C., Dore, M. I., Palade, M. L. and Taranu, I., 2020. Grape seed waste counteracts aflatoxin B1 toxicity in piglet mesenteric lymph nodes. Toxins 12: 800. https://doi.org/10.3390/toxins12120800.

Marin, D. E., Garanu, I., Tabuc, C. and Burgehelea, M., 2009. Ochratoxin: nature, origin, metabolism and toxic effects in pigs. Archiva Zootechnica 12:5-17.

Marin, D. E., Pistol, G. C., Gras, M. A., Palade, M. L. and Taranu, I., 2017. Comparative effect of ochratoxin A on inflammation and oxidative stress parameters in gut and kidney of piglets. Regulatory Toxicology and Pharmacology 89: 224-231. https://doi.org/10.1016/j.yrtph.2017.07.031.

Mastanjevic, K., Lukinac, J., Jukic, M., Sarkanj, B., Krstanovic, V. and Mastanjevic, K., 2019. Multi-(myco)toxins in malting and brewing by-products. Toxins 11: 30. https://doi.org/10.3390/toxins11010030.

Matsuura, Y. and Yoshizawa, T., 1985. Conversion of zearalenone, an estrogenic mycotoxin, by brewing microorganisms. Journal of the Food Hygienic Society of Japan 26: 24–28. https://doi.org/10.3358/shokueishi.26.24.

Meerpoel, C., Vidal, A., di Mavungu, J.D., Huybrechts, B., Tangni, Devreese, M., Croubels, S. and De Saeger, S., 2018. Development and validation of an LC–MS/MS method for the simultaneous determination of citrinin and ochratoxin a in a variety of feed and foodstuffs. Journal of Chromatography A 1580: 100–109. https://doi.org/10.1016/j.chroma.2018.10.039.

Meerpoel, C., Vidal, A., Huybrechts, B., Tangni, E.K., De Saeger, S., Croubles, S. and Devreese, M., 2020a. Comprehensive toxicokinetic analysis reveals major interspecies differences in absorption, distribution and elimination of citrinin in pigs and broiler chickens. Food Chemistry and Toxicology 141: 111365. https://doi.org/10.1016/j.fct.2020.111365.

Meerpoel, C., Vidal, A., Tangni, E.K., Huybrechts, B., Couck, L., De Rycke, R., De Bels, L., De Saeger, S., Van den Broeck, W., Devreese, M. and Croubels, S., 2020b. A study of carry-over and histopathological effects after chronic dietary intake of citrinin in pigs, broiler chickens and laying hens. Toxins 12: 719. https://doi.org/10.3390/toxins12110719.

Meissonnier, G.M., Pinton, P., Laffitte, J., Cossalter, A.M., Gong, Y.Y., Wild, C.P., Bertin, G., Galtier, P. and Oswald, I.P., 2008. Immunotoxicity of aflatoxin B1: impairment of the cell-mediated response to vaccine antigen and modulation of cytokine expression. Toxicology and Applied Pharmacology 231: 142-149. https://doi.org/10.1016/j.taap.2008.04.004.

Mendel, M., Karlik, W., Latek, U., Chtopecka, M., Nowacka-Kozak, E., Pietruszka, K. and Jedziniak, P., 2022. Does Deoxynivalenol affect amoxicillin and doxycycline absorption in the gastrointestinal tract? ex vivo study on swine jejunum mucosa explants. Toxins 14: 743. https://doi.org/10.3390/toxins14110743.

Meneely, J., Greer, B., Kolawole, O. and Elliott, C., 2023. T-2 and HT-2 toxins: toxicity, occurrence and analysis: a review. Toxins 15: 481. https://doi.org/10.3390/toxins15080481.

Meucci, V., Soldani, G., Razzuoli, E., Saggese, G. and Massart, F., 2011. Mycoestrogen pollution of Italian infant food. The Journal of Pediatrics 159: 278-283. https://doi.org/10.1016/j.jpeds.2011.01.028.

Mortensen, A., Granby, K., Eriksen, F.D., Cederberg, T.L., Friis-Wandall, S., Simonsen, Y., Broesbøl-Jensen, B. and Bonnichsen, R., 2014. Levels and risk assessment of chemical contaminants in byproducts for animal feed in Denmark. Journal of Environmental Science and Health Part B 49: 797–810. https://doi.org/10.1080/03601234.2014.938546.

Nagl, V., Woechtl, B., Schwartz-Zimmermann, H. E., Hennig-Pauka, I., Moll, W. D., Adam, G. and Berthiller, F., 2014. Metabolism of the masked mycotoxin deoxynivalenol-3-glucoside in pigs. Toxicology Letters 229: 190–197. https://doi.org/10.1016/j.toxlet.2014.06.032.

Neylon, E., Arendt, E.K., Lynch, K.M., Zannini, E., Bazzoli, P., Monin, T. and Sahin, A.W., 2020. Rootlets, a malting by-product with great potential. Fermentation 6: 117. https://doi.org/10.3390/fermentation6040117.

Nordkvist, E. and Häggblom, P., 2014. *Fusarium* mycotoxin contamination of cereals and bedding straw at Swedish pig farms. Animal Feed Science and Technology 198: 231-237. https://doi.org/10.1016/j.anifeedsci.2014.10.002.

Novak, B., Hasuda, A.M., Ghanbari, M., Maruo, V.M., Bracarense, A.P.F.R.L., Neves, M., Emsenhuber, C., Wein, S., Oswald, I.P., Pinton, P. and Schatzmayr, D., 2021. Effects of *Fusarium* metabolites beauvericin and enniatins alone or in mixture with deoxynivalenol on weaning piglets. Food and Chemical Toxicology 158: 112719. https://doi.org/10.1016/j.fct.2021.112719.

Nualkaw, K., Poapolathep, S., Zhang, Z., Zhang, Q., Giorgi, M., Li, P., Logrieco, A.F. and Poapolathep A., 2020. Simultaneous determination of multiple mycotoxins in swine, poultry and dairy feeds using ultra high performance liquid chromatography-tandem mass spectrometry. Toxins 12: 253. https://doi.org/10.3390/toxins12040253.

Olinda, R.G., Lima, J.M., De Lucena, R.B., Do Vale, A. M., Batista, J.S., De Barros, C.S.L., Riet-Correa, F. and Dantas, A. F. M., 2016. Acute aflatoxicosis in swines in Northeastern Brazil. Acta Scientiae Veterinariae 44: 6. https://doi.org/10.22456/1679-9216.84459.

Orlando, B., Grignon, G., Vitry, C., Kashefifard, K. and Valade, R., 2019. *Fusarium* species and enniatin mycotoxins in wheat, durum wheat, triticale and barley harvested in France. Mycotoxin Research 35: 369-380. https://doi.org/10.1007/s12550-019-00363-x.

Oswald, I.P., Desautels, C., Laffitte, J., Fournout, S., Peres, S.Y., Odin, M., Le Bars, P., Le Bars, J. and Fairbrother, J.M., 2003. Mycotoxin fumonisin B1 increases intestinal colonization by pathogenic *Escherichia coli* in pigs. Applied and Environmental Microbiology 69: 5870-5874. https://doi.org/10.1128/AEM.69.10.5870-5874.2003.

Pack, E.D., Weiland, S., Musser, R. And Schmale, D.G., 2121. Survey of zearalenone and type-B trichothecene mycotoxins in swine feed in the USA. Mycotoxin Research 37: 297-313. https://doi.org/10.1007/s12550-021-00442-y.

Pang, V.F., Swanson, S.P., Beasley, V.R., Buck, W.B. and Haschek, W.M., 1987. The toxicity of T-2 toxin in swine following topical application:I. Clinical signs, pathology, and residue concentrations. Fundamental and Applied Toxicology 9: 41-49. https://doi.org/10.1016/0272-0590(87)90152-7.

Peng, Z., Chen, L., Xiao, J., Zhou, X., Nüssler, A.K., Liu, L., Liu, J. and Yang, W., 2017. Review of mechanisms of deoxynivalenol-induced anorexia: The role of gut microbiota. Journal of Applied Toxicology 37: 1021-1029. https://doi.org/10.1002/jat.3475.

Petterson, H. and Hedman, R., 1997. Toxicity and metabolism of nivalenol in farm animals. Cereal Research Communications 25: 423-427. https://www.jstor.org/stable/23785398.

Pierron, A., Alassane-Kpembi, I. and Oswald, I.P., 2016a. Impact of two mycotoxins deoxynivalenol and fumonisin on pig intestinal health. Porcine Health Management 2: 21. https://doi.org/10.1186/s40813-016-0041-2.

Pierron, A., Alassane-Kpembi, I. and Oswald, I.P., 2016b. Impact of mycotoxin on immune response and consequences for pig health. Animal Nutrition 2: 63-68. https://doi.org/10.1016/j.aninu.2016.03.001.

Pleadin, J., Mihaljević, Ž., Barbir, T., Vulić, A., Kmetič, I., Zadravec, M., Brumen, V. and Mitak, M., 2015a. Natural incidence of zearalenone in Croatian pig feed, urine and meat in 2014. Food Additives and Contaminants: Part B Surveillance 8: 277-283. https://doi.org/10.1080/19393210.2015.1089946.

Pleadin, J., Perši, N., Kovačević, D., Vahčić, N., Scortichini, G. and Milone, S., 2013. Ochratoxin A in traditional dry-cured meat products produced from sub-chronic-exposed pigs. Food Additives and Contaminants: Part A: Chemistry, Analysis, Control, Exposure and Risk Assessment 30:1827-1836. https://doi.org/10.1080/19440049.2013.825817.

Pleadin, J., Staver, M.M., Vahcic, N., Kovacevic, D., Milone, S., Saftic, L. And Scortichini, G., 2015b. Survey of aflatoxin B1 and ochratoxin A occurrence in traditional meat products coming from Croatian households and markets. Food Control 52: 71-77. https://doi.org/10.1016/j.foodcont.2014.12.027.

Poapolathep, S., Klangkaew, N., Zhang, Z., Giorgi, M., Logrieco, A.F. and Poapolathep, A., 2021. Simultaneous determination of ergot alkaloids in swine and dairy feeds using ultra high-performance liquid chromatography-tandem mass spectrometry. Toxins 2021 13: 724. https://doi.org/10.3390/toxins13100724.

Prelusky, D.B., Trenholm, H.L. and Savard, M.E., 1994. Pharmacokinetic fate of 14C-labelled fumonisin B1 in swine. Natural Toxins. 2: 73–80. https://doi.org/10.1002/nt.2620020205.

Prelusky, D.B., Trenholm, H.L., Rotter, B.A., Miller, J.D., Savard, M.E., Yeung, J.M. and Scott, P.M., 1996. Biological fate of fumonisin B1 in food-producing animals. Advances in Experimental Medicine and Biology 392: 265-278. https://doi.org/10.1007/978-1-4899-1379-1_23.

Przybylska-Gornowicz, B., Tarasiuk, M., Lewczuk, B., Prusik, M., Ziółkowska, N., Zielonka, Ł., Gajęcki, M. and Gajęcka, M., 2015. The effects of low doses of two *Fusarium* toxins, zearalenone and deoxynivalenol, on the pig jejunum. A light and electron microscopic study. Toxins 7: 4684-4705. https://doi.org/10.3390/toxins7114684.

Rafai, P., Bata, A., Ványi, A., Papp, Z., Brydl, E., Jakab, L., Tuboly, S. and Túry, E., 1995a. Effect of various levels of T-2 toxin on the clinical status, performance and metabolism of growing pigs. Veterinary Records 136: 485-489. https://doi.org/10.1136/vr.136.19.485.

Rafai, P., Tuboly, S., Bata, A., Tilly, P., Ványi, A., Papp, Z., Jakab, L. and Túry, E., 1995b. Effect of various levels of T-2 toxin in the immune system of growing pigs. Veterinary Records 136: 511-514. https://doi.org/10.1136/vr.136.20.511.

Rao, Z.X., Tokach, M.D., Woodworth, J.C., DeRouchey, J.M., Goodband, R.D., Calderón, H.I. and Dritz, S.S., 2020. Effects of fumonisin-contaminated corn on growth performance of 9 to 28 kg nursery pigs. Toxins 12: 604. https://doi.org/10.3390/toxins12090604.

Reddy, K.E., Song, J., Lee, H.J., Kim, M., Kim, D.W., Jung, H.J., Kim, B., Lee, Y., Yu, D., Kim, D.W., Oh, Y.K. and Lee, S.D., 2018. Effects of high levels of deoxynivalenol and zearalenone on growth performance, and hematological and immunological parameters in pigs. Toxins 10: 114. https://doi.org/10.3390/toxins10030114.

Rückner, A., Plagge, L., Heenemann, K., Harzer, M., Thaa, B., Winkler, J. And Dänicke, S., 2022. The mycotoxin deoxynivalenol (DON) can deteriorate vaccination efcacy against porcine reproductive and respiratory syndrome virus (PRRSV) at subtoxic levels. Porcine Health Management 8: 13. https://doi.org/10.1186/s40813-022-00254-1.

Rustemeyer, S. M., Lamberson, W. R., Ledoux, D. R., Wells, K., Austin, K. J. and Cammack, K. M., 2011. Effects of dietary aflatoxin on the hepatic expression of apoptosis genes in growing barrows. Journal of Animal Science 89: 916-925. https://doi.org/10.2527/jas.2010-3473.

Sándor, G., Busch, A., Watzke, H., Reek, J. and Ványi, A., 1991. Subacute toxicity testing of ochratoxin-A and citrini in swine. Acta Veterinaria Hungarica 39: 149-160.

Sayyari, A., Fæste, C.K., Hansen, U., Uhlig, S., Framstad, T., Schatzmayr, D. and Sivertsen, T., 2018. Effects and biotransformation of the mycotoxin deoxynivalenol in growing pigs fed with naturally contaminated pelleted grains with and without the addition of *Coriobacteriaceum* DSM 11798. Food Additives and Contaminants: Part A: Chemistry, Analysis, Control, Exposure and Risk Assessment 35: 1394-1409. https://doi.org/10.1080/19440049.2018.1461254.

Schaafsma, A.W., Limay-Rios, V., Paul, D.E. and Miller, J.D., 2009. Mycotoxins in fuel ethanol co-products derived from maize: a mass balance for deoxynivalenol. Journal of Food Science and Agriculture 89: 1574-1580. https://doi.org/10.1002/jsfa.3626.

Scheltraete, W., Devreese, M. and Croubels, S., 2020. Comparative toxicokinetics of *Fusarium* mycotoxins in pigs and humans. Food and Chemical Toxicology 137: 111140. https://doi.org/10.1016/j.fct.2020.111140.

Schertz, H., Kluess, J., Frahm, J., Schatzmayr, D., Dohnal, I., Bichl, G., Schwartz-Zimmermann, H., Breves, G. and Dänicke, S., 2018. Oral and intravenous fumonisin exposure in pigs—a single-dose treatment experiment evaluating toxicokinetics and detoxification. Toxins 10: 150. https://doi.org/10.3390/toxins10040150.

Schoevers, E.J., Santos, R.R. and Roelen, B.A.J., 2020. Alternariol disturbs oocyte maturation and preimplantation development. Mycotoxin Research 36: 93-101. https://doi.org/10.1007/s12550-019-00372-w.

Schoevers, E.J., Santos, R.R. and Roelen, B.A.J., 2021. Susceptibility of oocytes from gilts and sows to beauvericin and deoxynivalenol and its relationship with oxidative stress. Toxins 13: 260. https://doi.org/10.3390/toxins13040260.

Schoevers, E.J., Santos, R.R., Fink-Gremmels, J. and Roelen, B.A.J., 2016. Toxicity of beauvericin on porcine oocyte maturation and preimplantation embryo development. Reproductive Toxicology 65: 159-169. https://doi.org/10.1016/j.reprotox.2016.07.017.

Serviento, A.M., Brossard, L. and Renaudeau, D., 2018. An acute challenge with a deoxynivalenol-contaminated diet has short- and long-term effects on performance and feeding behavior in finishing pigs. Journal of Animal Science 96: 5209-5221. https://doi.org/10.1093/jas/sky378.

Shier, W.T., 2000. The fumonisin paradox: A review of research on oral bioavailability of fumonisin B1, a mycotoxin produced by *Fusarium moniliforme*. Journal of Toxicology: Toxin Reviews 19: 161–187. https://doi.org/10.1081/TXR-100100319.

Soler, L., Stella, A., Seva, J., Pallarés, F.J., Lahjouji, T., Burlet-Schiltz, O. and Oswald, I.P., 2020. Proteome changes induced by a short, non-cytotoxic exposure to the mycoestrogen zearalenone in the pig intestine. Journal of Proteomics 224: 103842. https://doi.org/10.1016/j.jprot.2020.103842.

Sun, Y., Jiang, J., Mu, P, Lin, R., Wen, J. and Deng, Y., 2022. Toxicokinetics and metabolism of deoxynivalenol in animals and humans. Archives in Toxicology 96: 2639-2654. https://doi.org/10.1007/s.00204-022-03337-8.

Sun, Y., Park, I., Guo, J., Weaver, A.C. and Kim, S.W., 2015. Impacts of low level aflatoxin in feed and the use of modified yeast cell wall extract on growth and health of nursery pigs. Animal Nutrition 1: 177-183. https://doi.org/10.1016/j.aninu.2015.08.012.

Sun, Y., Park, I., Guo, J., Weaver, A.C. and Kim, S.W., 2015. Impacts of low level aflatoxin in feed and the use of modified yeast cell wall extract on growth and health of nursery pigs. Animal Nutrition 1: 177-183. https://doi.org/10.1016/j.aninu.2015.08.012.

Szczech, G. M., Carlton, W. W., Tuite, J. and Caldwell, R., 1973. Ochratoxin A toxicosis in swine. Veterinary Pathology 10: 347-364. https://doi.org/10.1177/030098587301000408.

Talmaciu, E., Sandu, I. and Banu, T., 2008. Researches regarding the fungal contamination and the presence of citrinin in feed. Fungi and Mycotoxins 2: 212-217.

Taranu, I., Hermenean, A., Bulgaru, C., Pistol, G.C., Ciceu, A., Grosu, I. A. and Marin, D.E., 2020. Diet containing grape seed meal by-product counteracts AFB_1 toxicity in liver of pig after weaning. Ecotoxicology and Environmental Safety 203: 110899. https://doi.org/10.1016/j.ecoenv.2020.110899.

Terciolo, C., Bracarense, A.P., Souto, P.C.M.C., Cossalter, A.M., Dopavogui, L., Loiseau, N., Oliveira, C.A.F., Pinton, P. and Oswald, I.P., 2019. Fumonisins at doses below EU regulatory limits induce histological alterations in piglets. Toxins 11: 548. https://doi.org/10.3390/toxins11090548.

Tolosa, J., Rodríguez-Carrasco, Y., Ruiz, M.J. and Vila-Donat, P., 2021. Multi-mycotoxin occurrence in feed, metabolism and carry-over to animal-derived food products: A review. Food and Chemical Toxicology 158: 112661. https://doi.org/10.1016/j.fct.2021.112661.

Uhlig, S., Torp, M. and Heier, B.T., 2006. Beauvericin and enniatins A, A1, B and B1 in Norwegian grain: a survey. Food Chemistry 94: 193-201. https://doi.org/10.1016/j.foodchem.2004.11.004.

Van Baelen, C., Montagne, L., Ferchaud, S., Prunier, A. and Lebret, B., 2024. Feeding strategy in organic pig farming as a lever to improve various quality dimensions of pork. Animal 18: 101190. https://doi.org/10.1016/j.animal.2024.101190.

Verbrugghe, E., Vandenbroucke, V., Dhaenens, M., Shearer, N., Goossens, J., De Saeger, S., Eeckhout, M., D'Herde, K., Thompson, A., Deforce, D., Boyen, F., Leyman, B., Van Parys, A., De Backer, P., Haesebrouck, F., Croubels, S. and Pasmans, F., 2012. T-2 toxin induced *Salmonella* Typhimurium intoxication results in decreased *Salmonella* numbers in the cecum contents of pigs, despite marked effects on *Salmonella*-host cell interactions. Veterinary Research 43: 22. https://doi.org/10.1186/1297-9716-43-22.

Vlachou, M., Pexara, A., Solomakos, N. and Govaris, A., 2022. Ochratoxin A in slaughtered pigs and pork products. Toxins. 14: 67. https://doi.org/10.3390/toxins14020067.

Wang, D., Lindemann, M. D. and Estienne, M. J., 2020. Effect of folic acid supplementation and dietary protein level on growth performance, serum chemistry and immune response in weanling piglets fed differing concentrations of aflatoxin. Toxins. 12: 651. https://doi.org/10.3390/toxins12100651.

Wang, J. P., Chi, F. and Kim, I. H., 2012. Effects of montmorillonite clay on growth performance, nutrient digestibility, vulva size, faecal microflora, and oxidative stress in weaning gilts challenged with zearalenone. Animal Feed Science and Technology 178: 158-166. https://doi.org/10.1016/j.anifeedsci.2012.09.004.

Weaver, G.A., Kurtz, H.J., Baresv, F.Y., Chi, M.S., Mirocha, C. J., Behrens, J.C. and Robison, T. S., 1978. Acute and chronic toxicity of T-2 mycotoxin in swine. Veterinary Records 103: 531-535.

Wellington, M.O., Bosompem, M.A., Petracek, R., Nagl, V. And Columbus, D.A., 2020. Effect of long-term feeding of graded levels of deoxynivalenol (DON) on growth performance, nutrient utilization, and organ health in finishing pigs and DON content in biological samples. Journal of Animal Science 98:skaa378. https://doi.org/10.1093/jas/skaa378.

Wu, F. and Munkvold, G.P., 2008. Mycotoxins in ethanol co-products: modeling economic impacts on the livestock industry and management strategies. Journal of Agriculture and Food Chemistry 56: 3900-3911. https://doi.org/10.1021/jf072697e.

Wu, Q., Dohnal, V., Huang, L., Kuča, K., Wang, X., Chen, G. and Yuan, Z., 2011. Metabolic pathways of ochratoxin A. Current Drug Metabolism 12: 1-10. https://doi.org/10.2174/138920011794520026.

Yang, Z., Urriola, P.E., Hilbrands, A., Johnston, L.J. and Shurson, G.C., 2020. Effects of feeding high-protein corn distillers dried grains and a mycotoxin mitigation additive on growth performance, carcass characteristics, and pork fat quality of growing-finishing pigs. Translational Animal Science 4: txaa051. https://doi.org/10.1093/tas/txaa051.

Yu, L., Li, H., Peng, Z., Ge, Y., Liu, J., Wang, T., Wang, H. and Dong. L., 2021. Early weaning affects liver antioxidant function in piglets. Animals 11: 2679. https://doi.org/10.3390/ani11092679.

Zhang, Q., Huang, L., Leng, B., Li, Y., Jiao, N., Jiang, S., Yang, W. and Yuan, X., 2021. Zearalenone affect the intestinal villi associated with the distribution and the expression of ghrelin and proliferating cell nuclear antigen in weaned gilts. Toxins 13: 736. https://doi.org/10.3390/toxins13100736.

Zhuo, Y., Yang, P., Hua, L., Zhu, L., Zhu, X., Han, X., Pang, X., Xu, S., Jiang, X., Lin, Y., Che, L., Fang, Z., Feng, B., Wang, J., Li, J., Wu, D., Huang, J. and Jin, C., 2022. Effects of chronic exposure to diets containing moldy corn or moldy wheat bran on growth performance, ovarian follicular pool, and oxidative status of gilts. Toxins, 14: 413. https://doi.org/10.3390/toxins14060413.

Zöllner, P., Jodlbauer, J., Kleinova, M., Kahlbacher, H., Kuhn, T., Hochsteiner, W. and Lindner, W., 2002. Concentration levels of zearalenone and its metabolites in urine, muscle tissue, and liver samples of pigs fed with mycotoxin-contaminated oats. Journal of Agriculture and Food Chemistry 50: 2494-2501. https://doi.org/10.1021/jf0113631.

CHAPTER 5

Effects of mycotoxins in poultry

G. Antonissen

Department of Pathobiology, Pharmacology and Zoological Medicine, Ghent University, Faculty of Veterinary Medicine, Merelbeke, Belgium
Gunther.Antonissen@UGent.be

1 Introduction

The poultry and feed industries share a common challenge for the future: to ensure the production of safe and nutritious animal protein in an environmentally sustainable manner. The provision of high-quality, safe and nutritious feed is essential to improve animal health, welfare and productivity. The presence of mycotoxins in feed poses a serious threat to animal welfare and presents significant research hurdles in feed toxicology.

The impact of climate change on the global distribution of mycotoxigenic fungi and their toxins adds complexity to this challenge. While acute mycotoxicosis from high doses is currently rare, ingestion of low to moderate levels of mycotoxins is more common, and though this often does not result in overt poisoning, such contamination levels can still adversely affect gut health, immune function and susceptibility to infectious diseases.

In addition, the co-occurrence of mycotoxins is common, amplifying their toxic effects beyond those of individual mycotoxins. Research efforts have traditionally focused on mycotoxins such as AFs, OTA, DON, T-2 toxin, FUM and ZEN. However, there is still a lack of toxicological data on commonly encountered modified mycotoxins and emerging variants.

Given that mycotoxins can affect gut health and the innate and adaptive immune systems, their potential impact on future poultry production is significant. However, clinical manifestations of mycotoxicosis are likely to be less common.

2 Occurrence of mycotoxins in poultry diets

As for pigs, poultry diets are usually prepared with maize-soybean meal or wheat-soybean based diets and, depending on the geographical area, some other ingredients such as sunflower seed meal, rye, barley, DDGS, as well as fibre sources such as oat hulls and wheat bran, all of which are sources of mycotoxin contamination. As a result, poultry feed is certain to be contaminated more often with a variety of mycotoxins.

 | DOI: 10.1163/9789004712720_6

Regardless of the geographical area, mycotoxins will be present in poultry feed. A survey in Sub-Saharan Africa reported that, depending on the mycotoxins, 19 to 100% of the poultry diets were contaminated with DON, T-2 toxin, FUM, ZEN, AFs, and OTA. In this survey, the authors observed that from the samples contaminated with AFs, 15 to 62% had AFB_1 at levels above the EU limits (Ochieng et al., 2021). In the Philippines, in more than 1700 poultry diets evaluated over 10 years, the AFs levels in 44.5 to 94% samples were above 20 µg/kg, sometimes reaching 792 µg/kg (Arim et al., 2003). When the presence of mycotoxins was analysed in Spanish poultry feed, it was reported that 71% of the samples had DON and 66% had ZEN, whereas AFB_1 and OTA were present in 5 and 13 % of the samples, respectively (Muñoz-Solano and González-Peñas, 2023). It is worth noting that the levels of AFB_1 contamination in all the feed samples were below 7 µg/kg, which is much lower than the EU limit. Analysis of poultry feeds collected from 49 farms in Argentina showed the presence of several fungal species including *Fusarium, Aspergillus* and *Penicillium* in 69.6, 43.5, and 45.6% of the samples, respectively (Greco et al., 2014). These authors also showed that all samples were contaminated with FUM at concentrations ranging from 0.2 to 6 mg/kg diet, whereas AFs (4.2 to 37.7 µg/kg), DON (220 to 318 µg/kg), T-2 toxin (50 to 107 µg/kg), ZEN (50 to 96 µg/kg), and OTA (5 to 13.2 µg/kg) were present in at least 80% of the samples. In all the aforementioned studies, diets were usually contaminated with at least two mycotoxins.

3 Main mycotoxins affecting poultry

The severity of the impact of mycotoxins on poultry health depends on several factors, including the type of mycotoxin, the level and duration of exposure, whether exposure is to a single mycotoxin or a combination, as well as factors related to the animals themselves (such as species, breed, age, immune status and general health) and management practices (such as biosecurity measures, environmental conditions such as temperature and feed composition). Specific mycotoxins can affect different organs or tissues, including the liver, kidneys, gastrointestinal and respiratory tracts, reproductive and urinary systems, brain and skin (Devegowda and Murthy, 2005). Acute disease episodes typically result from high-dose ingestion of mycotoxins and are characterised by mortality and well-defined clinical symptoms (Figure 5.1).

3.1 *Trichothecenes*

DON and T-2 toxin are rapidly absorbed and characterised by a low oral bioavailability of 19.3% and 20.9% for DON in broiler chickens and turkey poults, respectively (Osselaere et al., 2013a; Devreese et al., 2015a) and 17.1% for T-2 toxin (Sun et al., 2015). DON and T-2 toxin are de-epoxidised by intestinal commensal bacteria, and both toxins are additionally metabolised by phase II biotransformation in the

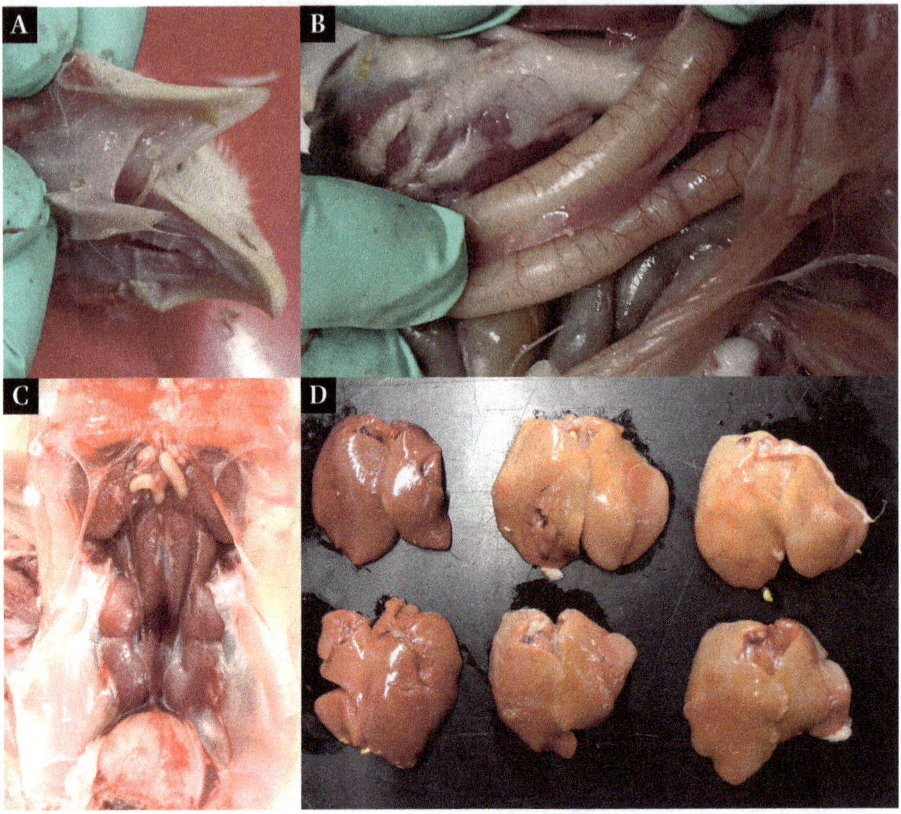

FIGURE 5.1 Illustrative photographs of poultry post-mortem analysis reveal lesions in the oral cavity (A) and congested intestines (B) following exposure to a combination of DON and FUM toxins. Enlarged kidneys are observed after exposure to OTA (C). Furthermore, damaged livers exhibit a range of colours, progressing from normal red to nearly light yellow, corresponding to varying levels of exposure to AFs. The liver becomes paler as the AF level increases (D).

intestinal mucosa and liver. While, glucuronidation of DON has been identified as a major biotransformation pathway in mammals (Maul et al., 2012), DON-3α-glucuronide is only a minor phase II metabolite in broilers and turkey poults. It seems that poultry biotransform DON more extensively to the sulphate conjugate, such as DON-3α-sulphate (DON-3S) (Devreese et al., 2015a). In poultry, DON metabolites are excreted via the bile and urine and only small amounts of unmodified DON are found in the excreta.

Trichothecenes exert their effects by binding non-covalently to the peptidyl transferase site of ribosomes, thereby inhibiting protein synthesis. Among them, T-2 toxin and DAS are prominent type A trichothecenes, while DON and NIV are well-known type B trichothecenes. DON is particularly widespread worldwide. Although DON-contaminated grain is sometimes diverted to poultry feed due to its relatively low toxicity in poultry compared to pigs – which show feed refusal and vomiting in response (Hoerr et al., 2020) – trichothecene mycotoxins still cause chemical irritation in the upper digestive tract, leading to necrosis of the oral mucosa (known as black tongue) and erosion of the gizzard.

Gizzard erosion as a result of DON exposure has been documented in chickens fed high levels of dietary contamination (49-83 mg/kg diet) (Gjevre et al., 2013). However, even lower concentrations, such as 3 mg/kg feed of NIV-contaminated diets, have resulted in gizzard erosions in chickens (Hedman et al., 1995). The more potent type A trichothecenes, T-2 toxin and DAS, have been shown to induce oral lesions in chicks at low levels of 0.4 and 1 mg/kg, respectively, and higher levels in the diet (Wyatt et al., 1972; Richard, 2007). It is noteworthy that oral lesions heal rapidly after removal of the toxin. The differential diagnosis should include other causes of gizzard erosions such as adenoviruses, biogenic amines, tannins or non-steroidal anti-inflammatory drugs (NSAIDs), as well as other causes of oral lesions such as poxviruses, candidiasis, protozoa, trichomoniasis, overdosage of organic acids or liquid methionine due to technical problems with injectors, or excessive levels of copper sulphate in feed or drinking water.

In addition to mucosal necrosis in the upper GIT, trichothecenes can cause immunosuppression by depleting lymphoid and haematopoietic tissues, necrosis and haemorrhage in the liver and gallbladder, and necrosis of the feather epithelium leading to feather deformities (Awad et al., 2013; Hoerr, 2020).

Research has shown that trichothecene mycotoxins can cause more severe toxic effects in turkeys than in chickens. Although, the toxicokinetic behaviour of DON in broiler chickens and turkey poults is comparable (low absolute oral bioavailability, rapid absorption and elimination, extensive biotransformation to DON-3S), relative differences in DON-3S/DON ratios exist between both species, with less extensive biotransformation in turkeys, which might explain the hypothesised difference in sensitivity of both poultry species to DON (Devreese et al., 2015a).

Studies with broiler chickens have shown that DON causes intestinal damage, such as decreased villus heights and histological damage, at concentrations as low as 1.9 mg/kg feed, which are considered adverse for animal health. For turkeys, some studies reported adverse effects on zootechnical parameters and intestinal morphology at concentrations as low as 1.7 mg/kg feed. In contrast, no such intestinal adverse effects were identified for ducks and laying hens, with Reference Points confirmed at 7 mg/kg and 5 mg/kg feed, respectively. The EFSA CONTAM Panel derived a Reference Point of 0.6 mg/kg feed for both chickens and turkeys, applying an uncertainty factor to the lowest observed adverse effect levels (EFSA, 2023a).

3.2 *Fumonisins*

FUM are characterised by a very low oral bioavailability. In laying hens, only 0.71% (F) is absorbed after oral administration of 2 mg [14C]FB_1/kg BW (Vudathala et al., 1994). In turkeys and ducks, a low F was also demonstrated after dosing 100 mg FB_1/kg BW, namely 2-2.3% and 3.2%, respectively (Tardieu et al., 2008; Tardieu et al.,

2009). Evidence for a very low bioavailability (i.e. plasma levels around 2.5 ng/mL (LOQ)) after oral dosing has been provided also for the metabolite hydrolysed FB_1 (Antonissen et al., 2020). Kinetic studies indicate a rapid elimination of FUM, particularly in chickens and ducks with very limited tendency to tissue accumulation.

As discussed in the previous chapters, FUM have a striking structural similarity to the long chain backbones of sphingolipids. Consequently, they competitively inhibit sphinganine N-acyl transferase (ceramide synthase), resulting in a disruption of ceramide and sphingolipid metabolism (Merrill, 2001). The primary biological effects of this enzyme blockade include inhibition of sphingolipid synthesis and accumulation of free sphinganine (Sa), accompanied by a lesser increase in sphingosine (So) levels in tissues, serum and urine. The increase in the Sa/So ratio in tissues and body fluids following exposure to FUM serves as an appropriate biomarker of exposure in both humans and animals (Qiu and Liu, 2001). Elevated Sa and Sa/So levels have been observed even at low feed concentrations (2 mg FB_1/kg feed) in different poultry species (EFSA, 2018).

In contrast to poultry, clinical outbreaks associated with FUM have been reported in horses (equine leukoencephalomalacia, ELEM; see Chapter 9) and pigs (porcine pulmonary oedema, PPE; see Chapter 4), with these species considered to be the most susceptible to the effects of FUM. However, such lesions are not observed in poultry. Overall, poultry are generally considered to be quite resistant to the adverse effects of FUM, although there are species-specific differences; laying hens and broilers are less sensitive to FUM than turkeys and ducks (Voss et al., 2007). FUM adversely affect the liver, intestines and immune system of poultry, thereby reducing performance. Based on the effects of FUM on intestines in chickens, in particular a decrease in crypt depth with a LOAEL of 2.5 mg/kg feed, the EFSA CONTAM Panel concluded that the Reference Point for adverse animal health effects is only 1 mg/kg feed, while the previously derived NOAEL was 20 mg/kg feed (EFSA, 2022).

3.3 *Zearalenone*

In different poultry species (broiler chickens, laying hens, and turkey poults) ZEN is characterised by a low oral bioavailability (F% = 6.87–10.28%) (Devreese et al., 2015b). After intestinal absorption, ZEN is mainly metabolised by the liver to various metabolites, including the two isomers α- and β-ZEL, where compared to ZEN, β-ZEL has a 2.5- to 5-fold lower affinity for the oestrogen receptor, whereas α-ZEL has a 60- to 92-fold higher binding affinity. Variation in the rate and extent of α- or β-ZEL production accounts for significant differences in susceptibility to ZEN toxicity between species. As shown in the previous chapter, pigs are particularly susceptible as they predominantly convert ZEN to α-ZEL, whereas poultry predominantly metabolise ZEN to β-ZEL. Although, the β-ZEL isomer is mainly formed in broiler chickens (α-ZEL/β-ZELPO = 0.153) and laying hens (αZEL/β-ZELPO = 0.027), this is significantly different in turkey poults (α-ZEL/β-ZELPO = 0.749), which supports the hypothesis of increased sensitivity of turkey poults (Devreese et al., 2015b).

ZEN has been found in the diets of broiler breeders, resulting in reduced egg production associated with ascites and cystic inflammation of the oviduct. In males, reduced comb and testicular weight and reduced fertility may be observed. In addition, in laying hens, ZEN can have adverse effects on both egg production and egg quality (Hoerr, 2020). Unfortunately, this author did not report the ZEN dietary levels, age of the birds, breed, and the duration of the exposure.

In general, poultry demonstrate significant resistance to ZEN, as evidenced by studies conducted on chickens and turkeys. In male and female broiler chickens, lymphocyte count was the only endpoint significantly affected by diets fortified with pure ZEN. The studies determined a LOAEL of 30 mg ZEN/kg BW per day and a NOAEL of 7.5 mg ZEN/kg BW per day, corresponding to dietary exposure levels of 200 mg/kg and 50 mg/kg, respectively. For turkeys, based on vent swelling in both females and males, a LOAEL of 19 mg ZEN/kg BW per day and a NOAEL of 91 mg ZEN/kg BW per day were identified, with dietary ZEN concentrations of 200 mg/kg and 100 mg/kg, respectively (EFSA, 2017).

3.4 *Aflatoxins*

Due to the lipophilic character and low molecular weight, AFs are rapidly absorbed from the GIT (Bbosa et al., 2013). In broiler chickens, AFG_1 was characterised by a low oral bioavailability of 14.4% (Rejeb et al., 2020). Furthermore, after oral administration of 14C-labelled AFB1 to laying hens, 71% of the dose was recovered within 7 days post-administration. In the same study, only 28% of the administered dose of AFB_1 was recovered during the first 24 hours (Sawhney et al., 1973).

AFB_1 undergoes various biotransformation processes in the liver. The phase I metabolisation primarily includes O-de-alkylation, keto-reduction, hydroxylation and epoxidation. The majority of phase I reactions take place in the microsomal compartment of the hepatocyte. The metabolic activation of AFs (via 8,9-epoxidation) leads to their binding to DNA and RNA, mainly in the liver. As a result, AFs exert significant biological effects, including carcinogenicity, immunosuppression, mutagenicity and teratogenicity (Devreese et al., 2013). AFB_1 has been classified by the IARC as a Group 1 compound due to its potent carcinogenic properties, including in humans (IARC, 1993). The discovery of AFs followed an outbreak of an unknown disease in turkeys in England in 1960, which was eventually linked to toxic peanut meal imported from Brazil (Richard, 2008).

A negative impact in growth, feed conversion, egg production, processing yield, pigmentation, fertility and hatchability is observed after exposure to AFB_1 at levels from 0.4 to 1 mg/kg diet (Monson et al., 2015). Exposure to low levels of AFs in poultry feed can affect growth, feed conversion, egg production, processing yield, pigmentation, fertility and hatchability (Monson et al., 2015). Poultry species are particularly susceptible to AFs, with even minimal contamination of feed with AFB_1

resulting in significant adverse health effects. Susceptibility varies between poultry species, with turkey chicks, ducklings and goslings being most susceptible, while quail, chickens and guinea fowl are more resistant (Arafa et al., 1981; Monson et al., 2015). The CYP enzymes activate the metabolic conversion of AFB_1 into its active product, AFBO (AFB_1-8,9-*exo*-epoxide). The susceptibility of turkeys to AFB_1 compared to other poultry species may be partially explained by the fact that turkeys produce 1.8 and 3.5 more AFBO than quails and chickens, respectively (Lozano and Diaz, 2006). Furthermore, these authors reported that the male birds, especially among ducks and turkeys, produce more AFBO than the female of the species. The sex-related regulation of CYP activity was previously described in birds (Pampori and Shapiro, 1993).

Aflatoxicosis in poultry is characterised by liver enlargement due to steatosis and hyperplasia of the biliary tree, resulting in a pale, enlarged, yellow and friable liver and an enlarged gallbladder filled with bile. The differential diagnosis should include other causes of fatty liver such as nutritional (e.g. excessive dietary energy) and hormonal (e.g. oestrogen-induced fatty liver syndrome) as well as viral inclusion body hepatitis. In addition, birds may exhibit petechial haemorrhages and bruising in multiple organs due to increased capillary fragility and altered clotting factor synthesis as a result of aflatoxin exposure (Hoerr, 2020).

AFs also affect skin pigmentation by inhibiting intestinal absorption of carotenoid pigments and sequestering pigments in the liver. Notably, aflatoxin toxicity is influenced by dietary factors; low dietary fat, protein, riboflavin and vitamin D3 levels combined with high tannic acid levels exacerbate aflatoxicosis. Increasing CP levels can mitigate the effects of low-dose aflatoxin exposure (Hoerr, 2020).

Chronic ingestion of AFs over a prolonged period (0.5-1.5 mg/kg diet, 42 weeks) has been shown to affect renal function in laying hens, resulting in urinary losses of calcium, phosphate and sodium (Martinez-de-Anda et al., 2010).

Besides producing AFs, *A. flavus* is also responsible for the production of CPA. The BW of young broiler chickens was significantly decreased after exposure to very high levels of CPA (34 mg/kg) (Gentles et al., 1999). Eggshell quality was impaired when laying hens were fed diets containing CPA (2.5 or 5 mg/kg BW), also resulting in mortality after seven days' exposure to the highest tested concentration (Suksupath, 1993). Unfortunately, there is a lack of studies indicating the impact of this mycotoxin on poultry production and healthy at realistic dietary levels.

3.5 *Ochratoxin A*

OTA demonstrates almost complete overall absorption in different poultry species incl. broiler chickens, laying hens, ducks and turkey poults, exhibiting extensive binding to serum proteins and widespread distribution across various tissues, particularly the kidneys and liver. Specifically, turkeys show very high plasma protein binding and a lower volume of distribution (Vd), whereas ducks exhibit lower plasma protein binding and a higher Vd. In chickens, OTA is metabolised through hydroxylation and dechlorination processes. Additionally, OTA clearance is slower in turkeys compared to other poultry species. Elimination of OTA primarily occurs via bile and excreta (Devreese et al., 2018; EFSA, 2023b).

OTA, the most abundant and highly toxic of the ochratoxins, exhibits several mechanisms to disrupt cellular physiology. Its primary effect is to inhibit the phenylalanine tRNA complex, thereby disrupting phenylalanine metabolism. In addition, OTA stimulates lipid peroxidation, leading to cellular damage. Molecularly, it targets histone acetyltransferases, which are critical in regulating many cellular processes such as DNA damage repair and mitosis (Devreese et al, 2013, 2018).

OTA is characterised by enterohepatic recirculation being reabsorbed in the intestine (Roth et al., 1988). Hence, exposure to OTA can result in nephrotoxic, hepatotoxic, teratogenic and immunotoxic effects. Lesions may manifest as renal and hepatic discolouration, visceral gout and enteritis. Histologically, renal lesions include oedema, degeneration of proximal and distal tubules, and interstitial fibrosis (Hoerr, 2020).

In hens, 0.1 mg OTA/kg feed, the lowest dose tested, given for 21 days, significantly reduced egg mass production and deteriorated feed to egg ratio. Therefore, the EFSA CONTAM Panel considers the concentration of 0.1 mg OTA/kg feed for chickens as the LOAEL, suggesting a Reference Point of 0.03 mg OTA/kg feed (EFSA, 2023b; for more details, see Chapter 17).

3.6 *Citrinin*

As with OTA, citrinin (CIT) is characterised by complete overall absorption and extensive binding to serum proteins (Meerpoel et al., 2020a).

Dietary exposure of broiler chickens and laying hens to 3 and 3.5 mg/kg CIT, respectively, did not result in macroscopic abnormalities, but mitochondrial kidneys were swollen and degenerated, probably due to oxidative stress (Meerpoel et al., 2020a). Only extremely high dietary levels of CIT impaired growth performance of growing chickens, where 220 mg/kg diet exposure for three weeks decreased BW in 8%, and 440 mg/kg diet was needed to impair feed efficiency (Nelson et al., 1981). Dietary intake of 150 mg/kg CIT resulted in increased water intake and decreased FI in chickens (Kirby et al., 1987). These tested concentrations, however, are not considered realistic for a diet naturally contaminated with mycotoxins.

When diets are naturally contaminated, disease outbreak due to mycotoxicosis may occur. A broiler chicken farm recorded a mortality rate of 26.4% after dietary exposure of 16,000 birds to 2 mg/kg CIT, where high mortality increased 14 days post-hatching, reaching a peak almost two weeks later. Besides mortality, the other birds had a decrease in feed and water intake. Post-mortem analysis revealed swollen liver, pancreas and kidneys, and intestinal congestion, indicating a combination of CIT exposure and inclusion body hepatitis (Mariappan et al., 2018).

3.7 *Ergot alkaloids*

The effects of ergot alkaloids (EAs) result from their activity as ligands for a wide array of adrenergic, serotonergic and dopaminergic receptors (Klotz, 2015). Birds that consume feed contaminated with EAs exhibit several adverse effects, including impaired performance parameters, poor feathering, respiratory difficulties, diarrhoea, increased mortality. In addition, as seen in other species, birds are prone to gangrene, especially in the legs, beaks, and crests. Lower prolactin levels are also noted, impacting both incubation and behaviour. The decrease in prolactin results in low circulating gonadotropins, leading to noticeable ovarian regression (EFSA, 2024).

Based on the sum of the EAs, the EFSA Panel considered the following as Reference Points in complete feed for adverse animal health effects: for chickens for fattening and hens 2.1 and 3.7 mg/kg, respectively, and for ducks 0.2 mg/kg (EFSA, 2024). For more details, see Chapter 17. Unfortunately, there is very limited information available on the toxicokinetics of EAs.

3.8 *Emerging mycotoxins*

The category of 'emerging' mycotoxins is not routinely detected or regulated and has received considerably less scientific attention than the more well-known mycotoxins such as AFs, FUM, trichothecenes, ZEN and OTA.

Frequently mentioned in this category are EAs, ENNs, BEA, moniliformin and *Alternaria* toxins (Jestoi, 2008; Gruber-Dorninger et al., 2017). Advances in liquid chromatography coupled with mass spectrometric methods have facilitated the analysis of an almost exhaustive list of mycotoxins in various food and feed matrices (Sulyok et al., 2007; Streit et al., 2013). These comprehensive screenings have revealed a high prevalence of so-called 'emerging' mycotoxins. Fraeyman et al. (2016) showed that both ENNB1 and ENNB exhibit poor absorption after oral administration in broiler chickens, with absolute oral bioavailabilities of 5% and 11% for ENNs B1 and B, respectively, resulting in limited systemic exposure to these mycotoxins. Therefore, systemic adverse effects due to the consumption of contaminated feed are likely to be limited. Nevertheless, concentrations of ENNs in the GIT can be significant and *in vitro* studies have demonstrated the cytotoxicity of ENNs in various cell types. ENNB was more cytotoxic than NIV, DON, and ZEN in Caco-2 cells (EFSA, 2014). Furthermore, incubation with BEA, ENNA, ENNA1, and ENNB1 resulted in a decrease in IPEC-J2 cell viability, while ENNB showed significantly lower cytotoxicity towards this porcine intestinal cell line (Fraeyman et al., 2018).

3.9 *Masked or modified mycotoxins*

In some cases, the clinical symptoms resulting from the consumption of mycotoxin-contaminated feed exceed what would be expected based on the level of feed contamination alone. This anomaly has led to the identification of masked or modified mycotoxins, so called because of their ability to evade detection by standard analytical methods (Rychlik et al., 2014). At the level of food and feed, mycotoxins can be biologically modified by conjugation by plants and fungi, or chemically modified by either thermal or non-thermal means, such as during food and feed processing. These modified mycotoxins pose potential risks to animal and human health as they can be hydrolysed to their free toxin forms in the digestive tract of animals, potentially leading to unexpectedly high levels of toxicity.

As the issue of modified toxins continues to emerge, it is not surprising that there is a lack of comprehensive toxicological data for most of these compounds (Berthiller et al., 2012; Freire and Sant'Ana, 2018). Broekaert et al. (2015) showed that in broiler chickens the absorbed fractions of 3-ADON (18.2%) and especially 15-ADON (42.2%) are higher than or equal to that of DON (10.6%). Furthermore, 3-ADON, which is reported to be less toxic than DON, is completely hydrolysed to DON before systemic absorption, whereas 15-ADON, which is reported to be more toxic, undergoes incomplete hydrolysis. This creates a 'worst-case scenario' for broilers, whereby each mole of 15-ADON can be as toxic as four moles of DON, and whereby 3-ADON, despite being less toxic, is considered to be as toxic as DON due to complete pre-systemic hydrolysis.

Furthermore, the hydrolysis of modified mycotoxins may differ between animal species; for example, broiler chickens do not hydrolyse DON-3G to DON, unlike pigs where complete presystemic hydrolysis occurs (Broekaert et al., 2017).

4 Cumulative health risk assessment of co-occurring mycotoxins

Mycotoxigenic fungi typically have the ability to produce multiple mycotoxins, and feed raw materials often harbour different fungal species simultaneously. In particular, compound poultry feeds are prone to multiple mycotoxin contamination due to their composition from several raw materials (Kovalsky et al., 2016). Although the co-occurrence of mycotoxins is common, existing legislation (EC, 2006) does not address the risks associated with co-contamination.

For many years, toxicological studies on mycotoxins have mainly focused on single-molecule studies, which provided insights into the individual and system-dependent potency of different mycotoxins. Animal responses to exposure to multiple mycotoxins can range from additive effects, where the response reflects that of each individual toxin, to synergistic effects, where the combined response exceeds the sum of the individual responses, and less commonly antagonistic effects, where the combined response is less than predicted (Grenier and Oswald, 2011; Streit et al., 2012). However, there is a paucity of *in vivo* data reporting the effects of mycotoxin co-occurrence (Grenier and Oswald, 2011).

For example, Verma et al. (2007) observed an additive interaction between AFs and OTA on egg production and feed efficiency in laying hens. Tessari et al. (2006) demonstrated a synergistic decrease in antibody titres against Newcastle disease following simultaneous exposure to AFs and FUM. More recently, Kolawole et al. (2020) found that mixtures of low to moderate levels of ZEN, DON and FUM and ZEN, DON and DAS had a significant interaction effect on feed efficiency in broilers. Young broiler chickens (14-day-old) were more sensitive to DON than older ones, especially when exposed to diets naturally contaminated with a mixture of *Fusarium* mycotoxins such as DON, 3-ADON and 15-ADON, resulting in liver damage, liver cysts, and ascites (Santos et al., 2021).

5 Effects of mycotoxins on growth and reproduction performance

Mycotoxins can negatively impact zootechnical parameters in a poultry farm, and have been associated with increased mortality, reduced FI, reduced BWG, resulting in a higher FCR. Egg production and quality are also adversely affected; hens exposed to mycotoxins often lay fewer eggs, and the eggs they do produce may have compromised shell integrity and internal quality, which can affect both marketability and hatchability.

In terms of reproductive performance, mycotoxins can increase embryonic mortality and negatively influence hatchability rates and day-old chick quality. These reproductive issues can stem from the direct toxic effects of mycotoxins on developing embryos or from the overall weakened condition of the breeding stock. An example of a realistic outcome also comes from recent data of Kolawole et al. (2020). In his long-term evaluation (18 successive trials) of a commercial farm, it was shown that natural contamination with levels below the EU recommendations for mycotoxins has a significant impact of 2.5 points in FCR, confirming the synergistic effects of the mycotoxins (FUM, ZEN, DON and DAS) and the decreased broiler performance in commercial conditions. Furthermore, Wang et al. (2021) found that residual levels of fumonisins and DON negatively affected hatchability, with hatchability rates decreasing by up to 25% in affected populations.

6 Effects of mycotoxins on the gastrointestinal health of poultry and susceptibility to diseases

At present, clinical mycotoxicosis resulting from high doses is uncommon. However, even the ingestion of low to moderate levels of these toxins can lead to a variety of metabolic, physiological and immunological disorders, with the GIT being a major target organ. The intestinal epithelium, a single layer of cells lining the intestinal lumen, acts as a selective barrier. It facilitates the absorption of nutrients and water from the gut lumen into the bloodstream, while acting as the largest and most important barrier to the passage of harmful substances in the GIT, including foreign antigens, microorganisms and their toxins (Alassane-Kpembi et al., 2019).

Mycotoxins compromise the intestinal barrier and induce leaky gut by modulating intestinal epithelial integrity and epithelial cell renewal and repair (see Figure 5.2). Various *in vitro* and *ex vivo* studies measuring transepithelial electrical resistance (TEER) and intestinal mRNA expression of genes encoding tight junction proteins have shown that mycotoxins such as DON and FUM can increase intestinal epithelial permeability (Osselaere et al., 2013b; Antonissen et al., 2014; see also Chapter 13). Mycotoxins can also negatively affect the viability and proliferation of intestinal epithelial cells (Antonissen et al., 2014; Antonissen et al., 2015a).

In addition to affecting intestinal epithelial cells, mycotoxins also affect other components of the intestinal barrier, including the mucus layer, secretion of antimicrobial peptides, chemokines, cytokines and oxidative stress. For example, feeding broiler chickens diets contaminated with DON (4.3-4.6 mg/kg feed) and/or FUM (22.9-25.4 mg/kg feed) at levels close to the EU maximum guidance levels was observed to affect the duodenal mucus layer by reducing the expression of the intestinal mucin 2 (MUC2) gene and altering the composition of duodenal mucin monosaccharides, such as N-acetyl-neuraminic acid, N-acetyl-galactosamine and galactose (Antonissen et al., 2015b). Several mycotoxins can also modulate the production of inflammatory cytokines in the gut. For example, exposure to DON induces a pro-inflammatory response characterised by increased expression of interleukin (IL)-1α, IL-1β, IL-8 and TNF-α (Alassane-Kpembi et al., 2019; Paraskeuas et al., 2021).

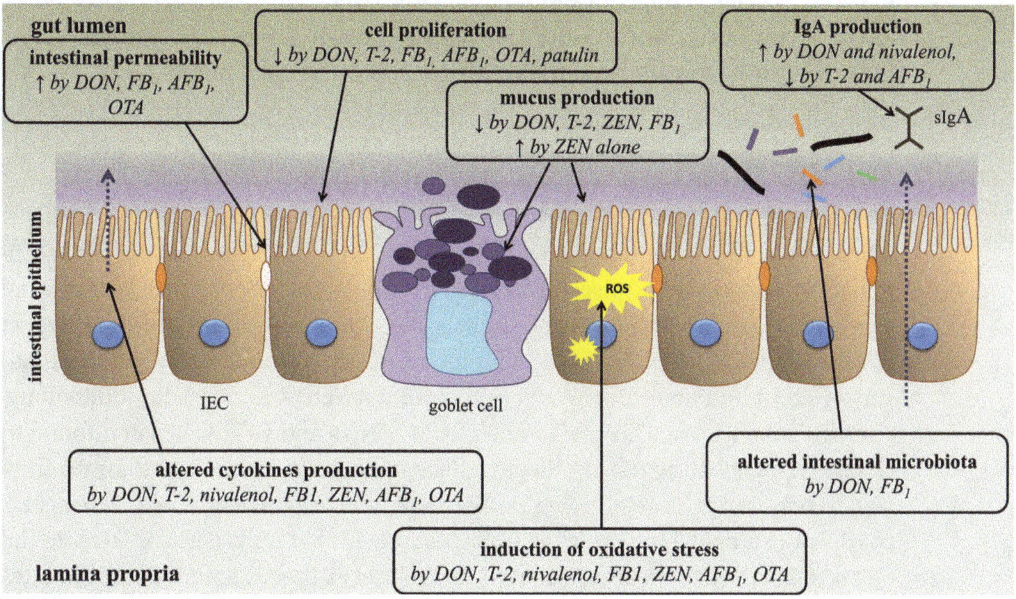

FIGURE 5.2 Impact of mycotoxins on intestinal (©Ghent University – Gunther Antonissen)

The adverse effects of mycotoxins on gastrointestinal physiology and the immune system can increase the susceptibility of animals to common infectious diseases in animal production. These diseases include coccidiosis, *Clostridium perfringens* necrotic enteritis, campylobacteriosis, colibacillosis, salmonellosis and ascaridiosis. In broilers, a delayed cell-mediated immune response to coccidiosis was observed in chickens fed a diet naturally contaminated with *Fusarium* mycotoxins (Girgis et al., 2008, 2010). In addition, a greater incidence and severity of *Eimeria* (*E.*) *maxima* and *E. tenella* lesions and a higher oocyst yield were observed in chickens fed a diet contaminated with DON and/or FUM. Feeding a diet contaminated with DON (3-4 mg/kg feed) or FUM (18.6 mg FB_1+FB_2/kg feed) predisposes broiler chickens to the development of *Clostridium perfringens*-induced necrotic enteritis (Antonissen et al., 2014; Antonissen et al., 2015a). These effects are consistent with the negative effects observed on selected components of the chicken host intestinal barrier, such as villus height, tight junctions, mucus, oxidative stress and microbiota homeostasis. Furthermore, feeding pigeons a DON-contaminated diet (approximately 3.5 mg/kg) resulted in an increased percentage of pigeons shedding *Salmonella* compared to birds fed a control diet, 87% versus 74%, respectively (Antonissen et al., 2016). DON has also been shown to affect the *Campylobacter jejuni* infection profile in broilers. Co-exposure to 5 mg/kg DON in poultry feed and *C. jejuni* resulted in increased *C. jejuni* colonisation in the gut as well as increased gut permeability (Ruhnau et al., 2020).

Furthermore, reduced immune responses, characterised by lower antibody titres to vaccines such as Infectious Bronchitis Virus (IBV) and Newcastle Disease Virus (NDV), have been linked to the immunosuppressive effects of T-2 toxin, AFB_1, FUM, and CPA (Kamalavenkatesh et al., 2005; Tessari et al., 2006).

7 Mycotoxin carry-over in poultry food products

Mycotoxins have the potential to be excreted in urine and faeces and can also accumulate in eggs, meat and visceral organs. While several studies suggest that chicken eggs and meat play a minor role in human dietary mycotoxin exposure (Meerpoel et al., 2020b; Tangni et al., 2020), chronic exposure to these low levels could adversely affect human health (Wang et al., 2018; Tardieu et al., 2021). Residues of mycotoxins such as AFs, OTA, CIT, ZEN, FUM, ENNs and BEA have been found in chicken eggs, albeit with very low transfer rates reported to be less than 1% from feed to eggs (Sypecka et al., 2004; Tardieu et al., 2021; Fakhri et al., 2021; Wang et al., 2021). For example, Prelusky et al. (1989) estimated that only 0.31% of DON in the diet is transferred to the eggs. Similarly, Sypecka et al. (2004) reported transfer rates of less than 0.3% for ZEN and its metabolites from contaminated feed to eggs, and for AFB_1 the transfer rate was about 0.003% (Adegbeye et al., 2020).

8 Conclusions

In summary, the presence of mycotoxins in poultry feed poses a significant threat to animal health and represents a major research challenge in the field of nutritional toxicology. At present, acute mycotoxicosis resulting from high doses is rare. However, ingestion of low to moderate levels of mycotoxins is more common, though typically does not result in overt intoxication. Nevertheless, these low to moderate levels of contamination can affect gut health and immune function. Consequently, mycotoxin-induced changes in digestive physiology, together with direct effects on microbial metabolism, can disrupt the intestinal microbial balance. This dysregulation can lead to increased inflammatory signalling, increased epithelial permeability and increased susceptibility to infectious diseases in animal.

References

Adegbeye, M.J., Reddy, P.R.K., Chilaka, C.A., Balogun, O.B., Elghandour, M.M., Rivas-Caceres, R.R. and Salem, A.Z., 2020. Mycotoxin toxicity and residue in animal products: Prevalence, consumer exposure and reduction strategies–A review. Toxicon 177: 96-108. https://doi.org/10.1016/j.toxicon.2020.01.007.

Alassane-Kpembi, I., Pinton, P. and Oswald, I. P., 2019. Effects of mycotoxins on the intestine. Toxins 11: 159. https://doi.org/10.3390/toxins11030159.

Antonissen, G., Croubels, S., Pasmans, F., Ducatelle, R., Eeckhaut, V., Devreese, M., Verlinden, M., Haesebrouck, F., Eeckhout, M., De Saeger, S., Antlinger, B., Novak, B., Martel, A. and Van Immerseel, F., 2015a. Fumonisins affect the intestinal microbial homeostasis in broiler chickens, predisposing to necrotic enteritis. Veterinary Research 46: 98. https://doi.org/10.1186/s13567-015-0234-8.

Antonissen, G., Haesendonck, R., Devreese, M., Broekaert, N., Verbrugghe, E., De Saeger, S., Audenaert, K., Haesebrouck, F., Pasmans, F., Ducatelle, R., Croubels, S. and Martel, A., 2016. The impact of deoxynivalenol on pigeon health: occurrence in feed, toxicokinetics and interaction with salmonellosis. Plos one 11: e0168205. https://doi.org/10.1371/journal.pone.0168205.

Antonissen, G., Van Immerseel, F., Pasmans, F., Ducatelle, R., Haesebrouck, F., Timbermont, L., Verlinden, M., Janssens, G.P.J, Eeckhaut, V., Eeckhout, M., De Saeger, S., Hessenberger, S., Martel, A. and Croubels, S., 2014. The mycotoxin deoxynivalenol predisposes for the development of *Clostridium perfringens*-Induced Necrotic Enteritis in Broiler Chickens. PLoS ONE 9: e108775. https://doi.org/10.1371/journal.pone.0108775.

Antonissen, G., Van Immerseel, F., Pasmans, F., Ducatelle, R., Janssens, G. P., De Baere, S., Mountzouris, K.C., Su, S., Wong, E.A., De Meulenaer, B., Verlinden, M., Devreese, M., Haesebrouck, F., Novak, B., Dohnal, I., Martel, A. and Croubels, S., 2015b. Mycotoxins deoxynivalenol and fumonisins alter the extrinsic component of intestinal barrier in broiler chickens. Journal of Agricultural and Food Chemistry 63: 10846-10855. https://doi.org/10.1021/acs.jafc.5b04119.

Antonissen, G., De Baere, S., Novak, B., Schatzmayr, D., den Hollander, D., Devreese, M., and Croubels, S., 2020. Toxicokinetics of hydrolyzed fumonisin B1 after single oral or intravenous bolus to broiler chickens fed a control or a fumonisins-contaminated diet. Toxins 12: 413. https://doi.org/10.3390/toxins12060413.

Arafa, A.S., Bloomer, R.J., Wilson, H.R., Simpson, C.F. and Harms, R.H., 1981. Susceptibility of various poultry species to dietary aflatoxin. British Poultry Science 22: 431-436. https://doi.org/10.1080/00071688108447906.

Arim, R.H., 2003. Mycotoxin contamination of food and feeds in the Philippines. Mycotoxins 2003: 167-173.

Awad, W., Ghareeb, K., Böhm, J. and Zentek, J., 2013. The toxicological impacts of the *Fusarium* mycotoxin, deoxynivalenol, in poultry flocks with special reference to immunotoxicity. Toxins 5: 912-925. https://doi.org/10.3390/toxins5050912.

Bbosa, G.S., Kitya D., Lubega, A., Ogwal-Okeng, J., Anokbonggo, W.W. and Kyegombe, D.B., 2013. Review of the biological and health effects of aflatoxins on body organs and body systems. In: Razzaghi-Abyaneh, M. Aflatoxins-recent advances and future prospects 12: 239-265. https://doi.org/10.5772/51201.

Berthiller, F., Crews, C., Dall'Asta, C., Saeger, S.D., Haesaert, G., Karlovsky, P., Oswald, I.P., Seefelder, W., Speijers, G. and Stroka, J., 2012. Masked mycotoxins: a review. Molecular Nutrition and Food Research 57: 165-186. https://doi.org/10.1002/mnfr.201100764.

Broekaert, N., Devreese, M., De Mil, T., Fraeyman, S., Antonissen, G., De Baere, S., De Backer, P., Vermeulen, A. and Croubels, S., 2015. Oral bioavailability, hydrolysis, and comparative toxicokinetics of 3-acetyldeoxynivalenol and 15-acetyldeoxynivalenol in broiler chickens and pigs. Journal of agricultural and food chemistry 63: 8734-8742. https://doi.org/10.1021/acs.jafc.5b03270.

Broekaert, N., Devreese, M., van Bergen, T., Schauvliege, S., De Boevre, M., De Saeger, S., Vanhaecke, L., Berthiller, F., Michlmayr, H., Malachová, A., Adam, G., Vermeulen, A. and Croubels, S., 2017. In vivo contribution of deoxynivalenol-3-β-d-glucoside to deoxynivalenol exposure in broiler chickens and pigs: Oral bioavailability, hydrolysis and toxicokinetics. Archives of Toxicology 91: 699-712. https://doi.org/10.1007/s00204-016-1710-2.

Devegowda, G., and Murthy, T.N.K., 2005. Mycotoxins: Their effects in poultry and some practical solutions. The Mycotoxin Blue Book, 25-56.

Devreese, M., Antonissen, G., Broekaert, N., De Baere, S., Vanhaecke, L., De Backer, P. and Croubels, S., 2015b. Comparative toxicokinetics, absolute oral bioavailability, and biotransformation of zearalenone in different poultry species. Journal of Agricultural and Food Chemistry 63: 5092–5098. https://doi.org/10.1021/acs.jafc.5b01608.

Devreese, M., Antonissen, G., Broekaert, N., De Mil, T., De Baere, S., Vanhaecke, L., De Backer, P. and Croubels, S., 2015a. Toxicokinetic study and oral bioavailability of deoxynivalenol in turkey poults, and comparative biotransformation between broilers and turkeys. World Mycotoxin Journal 8: 533–539. https://doi.org/10.3920/WMJ2014.1843.

Devreese, M., Croubels, S., De Baere, S., Gehring, R. and Antonissen, G., 2018. Comparative toxicokinetics and plasma protein binding of ochratoxin A in four avian species. Journal of agricultural and food chemistry 66: 2129-2135. https://doi.org/10.1021/acs.jafc.7b06048.

Devreese, M., De Backer, P. and Croubels, S., 2013. Overview of the most important mycotoxins for the pig and poultry husbandry. Vlaams Diergeneeskundig Tijdschrift 82: 171-180. http://hdl.handle.net/1854/LU-4125525.

EFSA CONTAM Panel (EFSA Panel on Contaminants in the Food Chain), 2014. Scientific Opinion on the risks to human and animal health related to the presence of beauvericin and enniatins in food and feed. EFSA Journal 12: 3802, 174 pp. https://doi.org/10.2903/j.efsa.2014.38022.

EFSA CONTAM Panel (EFSA Panel on Contaminants in the Food Chain), Schrenk, D., Bignami, M., Bodin, L., Chipman, J.K., del Mazo, J., Grasl-Kraupp, B., Hogstrand, C., Leblanc, J.C., Nielsen, E., Ntzani, E., Petersen, A., Sand, S., Schwerdtle, T., Vleminckx, C., Wallace, H., Dänicke, S., Nebbia, C.S., Oswald, I.P., Rovesti, E., Steinkellner, H. and Hoogenboom, L.R., 2022. Scientific Opinion on the assessment of information as regards the toxicity of fumonisins for pigs, poultry and horses. EFSA Journal 20: 7534, 16 pp. https://doi.org/10.2903/j.efsa.2022.7534.

EFSA CONTAM Panel (EFSA Panel on Contaminants in the Food Chain), Schrenk, D., Bignami, M., Bodin, L., Chipman, J.K., delMazo, J., Grasl-Kraupp, B., Hogstrand, C., Leblanc, J.C., Nielsen, E., Ntzani, E., Petersen, A., Sand, S., Schwerdtle, T., Vleminckx, C., Wallace, H., Dänicke, S., Nebbia, C.S., Oswald, I.P., Rovesti, E., Steinkellner, H. and Hoogenboom, R.L., 2023a. Scientific Opinion on the assessment of information as regards the toxicity of deoxynivalenol for horses and poultry. EFSA Journal 21: 7806, 30 pp. https://doi.org/10.2903/j.efsa.2023.7806.

EFSA CONTAM Panel (EFSA Panel on Contaminants in the Food Chain), Schrenk, D., Bignami, M., Bodin, L., Chipman, J.K., del Mazo, Grasl-Kraupp, B., Hogstrand, C., Hoogenboom, L.R., Leblanc, J.C., Nebbia, C.S., Nielsen, E., Ntzani, E., Sand, S., Schwerdtle, T., Vleminckx, C., Wallace, H., Gropp, J., Antonissen, G., Rychen, G., Gómez Ruiz, J.A., Innocenti, M.L., Rovesti, E. and Petersen, A., 2023b. Risks for animal health related to the presence of ochratoxin A (OTA) in feed. EFSA Journal 21: 1–89. https://doi.org/10.2903/j.efsa.2023.8375.

EFSA CONTAM Panel (EFSA Panel on Contaminants in the Food Chain), Schrenk, D., Bignami, M., Bodin, L., Chipman, J.K., del Mazo,/ Grasl-Kraupp, B., Hogstrand, C., Leblanc, J.C., Nebbia, C.S., Nielsen, E., Ntzani, E., Petersen, A., Sand, S., Schwerdtle, T., Vleminckx, C., Wallace, H., Gropp, J., Mulder, P., Oswald, I.P., Woutersen, R., Gómez Ruiz, J.A., Rovesti, E. and Hoogenboom, L.R., 2024. Risks for animal health related to the presence of ergot alkaloids in feed. EFSA Journal 22: e8496. https://doi.org/10.2903/j.efsa.2024.8496.

EFSA Panel on Contaminants in the Food Chain (CONTAM), Knutsen, H.K., Alexander, J., Barregård, L., Bignami, M., Brüschweiler, B., Ceccatelli, S., Cottrill, B., Dinovi, M., Edler, L., Grasl-Kraupp, B., Hogstrand, C., Hoogenboom, L.R., Nebbia, C.S., Petersen, A., Rose, M., Roudot, A.C., Schwerdtle, T., Vleminckx, C., Vollmer, G., Wallace, H., Dall'Asta, C., Dänicke, S., Eriksen, G.S., Altieri, A., Roldán-Torres, R. and Oswald, I.P., 2017. Risks for animal health related to the presence of zearalenone and its modified forms in feed. EFSA Journal 15: e04851. https://doi.org/10.2903/j.efsa.2017.4851.

EFSA Panel on Contaminants in the Food Chain (CONTAM), Knutsen, H.K., Alexander, J., Barregård, L., Bignami, M., Brüschweiler, B., Ceccatelli, S., Cottril, B., Dinovi, M., Edler, L., Grasl-Kraupp, B., Hogstrand, C., Hoogenboom, C.S., Nebbia, C.S., Petersen, A., Rose, M., Roudot, A.C., Schwerdtle, T., Vleminckx, C., Wallace, H., Dall'Asta, C., Eriksen, G.S., Taranu, I., Altieri, A., Roldán-Torres, R. and Oswald, I.P., 2018. Scientific opinion on the risks for animal health related to the presence of fumonisins, their modified forms and hidden forms in feed. EFSA Journal 16: 5242, 144 pp. https://doi.org/10.2903/j.efsa.2018.5242

EFSA CONTAM Panel (EFSA Panel on Contaminants in the Food Chain), 2014. Scientific Opinion on the risks to human and animal health related to the presence of beauvericin and enniatins in food and feed. EFSA Journal 12: 3802, 174 pp. https://10.2903/j.efsa.2014.3802.

Fakhri, Y., Sarafraz, M., Nematollahi, A., Ranaei, V., Soleimani-Ahmadi, M., Thai, V.N. and Mousavi Khaneghah, A., 2021. A global systematic review and meta-analysis of concentration and prevalence of mycotoxins in birds' egg. Environmental Science and Pollution Research 28: 59542-59550. https://doi.org/10.1007/s11356-021-16136-y.

Fraeyman, S., Devreese, M., Antonissen, G., De Baere, S., Rychlik, M. and Croubels, S., 2016. Comparative oral bioavailability, toxicokinetics, and biotransformation of enniatin B1 and enniatin B in broiler chickens. Journal of Agricultural and Food Chemistry 64: 7259-7264. https://doi.org/10.1021/acs.jafc.6b02913.

Fraeyman, S., Meyer, E., Devreese, M., Antonissen, G., Demeyere, K., Haesebrouck, F. and Croubels, S., 2018. Comparative in vitro cytotoxicity of the emerging *Fusarium* mycotoxins beauvericin and enniatins to porcine intestinal epithelial cells. Food and Chemical Toxicology 121: 566-572. https://doi.org/10.1016/j.fct.2018.09.053.

Freire, L. and Sant'Ana, A.S., 2018. Modified mycotoxins: an updated review on their formation, detection, occurrence, and toxic effects. Food and Chemical Toxicology 111: 189-205. https://doi.org/10.1016/j.fct.2017.11.021.

Gentles, A., Smith, E.E., Kubena, L.F., Duffus, E., Johnson, P., Thompson, J., Harvey, R.B. and Edrington, T.S., 1999. Toxicological evaluations of cyclopiazonic acid and ochratoxin A in broilers. Poultry Science 78: 1380-1384. https://doi.org/10.1093/ps/78.10.1380.

Girgis, G.N., Barta, J.R., Girish, C.K., Karrow, N.A., Boermans, H.J. and Smith, T.K., 2010. Effects of feed-borne *Fusarium* mycotoxins and an organic mycotoxin adsorbent on immune cell dynamics in the jejunum of chickens infected with *Eimeria maxima*. Veterinary Immunology and Immunopathology, 138: 218-223. https://doi.org/10.1016/j.vetimm.2010.07.018.

Girgis, G.N., Sharif, S., Barta, J.R., Boermans, H.J. and Smith, T.K., 2008. Immunomodulatory effects of feed-borne *Fusarium* mycotoxins in chickens infected with coccidia. Experimental Biology and Medicine 233: 1411-1420. https://doi.org/10.3181/0805-RM-173.

Gjevre, A.G., Kaldhusdal, M. and Eriksen, G.S., 2013. Gizzard erosion and ulceration syndrome in chickens and turkeys: a review of causal or predisposing factors. Avian Pathology 42: 297-303. https://doi.org/10.1080/03079457.2013.817665.

Greco, M.V., Franchi, M.L., Golba, S.L.R., Pardo, A.G. and Pose, G.N., 204. Mycotoxins and mycotoxigenic fungi in poultry feed for food-producing animals. Scientific World Journal 2014: 9682215. https://doi.org/10.1155/2014/968215.

Grenier, B.O.I.P. and Oswald, I., 2011. Mycotoxin co-contamination of food and feed: meta-analysis of publications describing toxicological interactions. World Mycotoxin Journal 4: 285-313. https://doi.org/10.3920/WMJ2011.1281.

Gruber-Dorninger, C., Novak, B., Nagl, V. and Berthiller, F., 2017. Emerging mycotoxins: Beyond traditionally determined food contaminants. Journal of Agricultural and Food Chemistry 65: 7052-7070. https://doi.org/10.1021/acs.jafc.6b03413.

Hedman, R., Pettersson, H., Engström, B., Elwinger, K. and Fossum, O., 1995. Effects of feeding nivalenol-contaminated diets to male broiler chickens. Poultry Science 74: 620-625. https://doi.org/10.3382/ps.0740620.

Hoerr, F.J., 2020. Mycotoxicosis. In: Diseases of Poultry, Swayne, D.E. (ed.) 31: 1330-1348. https://doi.org/10.1002/9781119371199.ch31.

IARC, 1993. Monographs on the evaluation of the carcinogenic risk of chemicals to humans: some naturally occurring substances. Food items and constituents, heterocyclic aromatic amines and mycotoxins. Lyon, France. IARC Monographs on the Evaluation of Carcinogic Risks to Humans 56.

Jestoi, M., 2008. Emerging *Fusarium*-mycotoxins fusaproliferin, beauvericin, enniatins, and moniliformin—A review. Critical Reviews in Food Science and Nutrition 48: 21-49. https://doi.org/10.1080/10408390601062021.

Kamalavenkatesh, P., Vairamuthu, S., Balachandran, C., Manohar, B.M. and Raj, G.D., 2005. Immunopathological effect of the mycotoxins cyclopiazonic acid and T-2 toxin on broiler chicken. Mycopathologia 159: 273–279. https://doi.org/10.1007/s11046-004-7321-0.

Kirby, L.K., Nelson, MT.S., Halley, J.T. and Beasley, J.N., 1987. Citrinin toxicity in young chicks. Poultry Science 66: 966-968. https://doi.org/10.3382/ps.0660966.

Klotz, J.L., 2015. Activities and effects of ergot alkaloids on livestock physiology and production. Toxins 7: 2801-2821. https://doi.org/10.3390/toxins7082801.

Kolawole, O., Graham, A., Donaldson, C., Owens, B., Abia, W. A., Meneely, J., Alcorn, M.J., Connolly, L. and Elliott, C.T., 2020. Low doses of mycotoxin mixtures below EU regulatory limits can negatively affect the performance of broiler chickens: A longitudinal study. Toxins 12: 433. https://doi.org/10.3390/toxins12070433.

Kovalsky, P., Kos, G., Nährer, K., Schwab, C., Jenkins, T., Schatzmayr, G., Sulyok, M. and Krska, R., 2016. Co-occurrence of regulated, masked and emerging mycotoxins and secondary metabolites in finished feed and maize—An extensive survey. Toxins 8: 363. https://doi.org/10.3390/toxins8120363.

Lozano, M.C. and Diaz, G.J., 2006. Microsoma land cytosolic biotransformation of aflatoxin B1 in four poultry species. British Poutry Science 47: 734-741. https://doi.org/10.1080/00071660601084390.

Mariappan, A.K., Munusamy, P., Latheef, S.K., Singh, S.D. and Dhama, K., 2018. Hepatonephropathology associated with inclusion body hepatitis complicated with citrinin mycotoxicosis in a broiler farm. Veterinary World 11: 112-117. https://doi.org/10.14202/vetworld.2018.112-117.

Martínez-de-Anda, A., Valdivia, A.G., Jaramillo-Juárez, F., Reyes, J.L., Ortiz, R., Quezada, T., de Luna, M.C. and Rodríguez, M.L., 2010. Effects of aflatoxin chronic intoxication in renal funtion of laying hens. Poultry Science 89: 1622-1628. https://doi.org/10.3382/ps.2010-00763.

Maul, R., Warth, B., Kant, J.S., Schebb, N.H., Krska, R., Koch, M. and Sulyok, M., 2012. Investigation of the hepatic glucuronidation pattern of the *Fusarium* mycotoxin deoxynivalenol in various species. Chemical Research in Toxicology 25: 2715-2717. https://doi.org/10.1021/tx300348x.

Meerpoel, C., Vidal, A., Huybrechts, B., Tangni, E.K., De Saeger, S., Croubels, S. and Devreese, M., 2020a. Comprehensive toxicokinetic analysis reveals major interspecies differences in absorption, distribution and elimination of citrinin in pigs and broiler chickens. Food and Chemical Toxicology 141: 111365. https://doi.org/10.1016/j.fct.2020.111365.

Meerpoel, C., Vidal A., Tangni E.K., Huybrechts B., Couck L., De Rycke R., De Bels L., De Saeger S., Broeck W.V.D., Devreese M. and Croubels, S., 2020b. A Study of carry-over and histopathological effects after chronic dietary intake of citrinin in pigs, broiler chickens and laying hens. Toxins 12: 719. https://doi.org/10.3390/toxins12110719.

Merrill, A.H., Sullards, M.C., Wang, E., Voss, K.A. and Riley, R.T., 2001. Sphingolipid metabolism: Roles in signal transduction and disruption by fumonisins. Environmental Health Perspectives 109: 283-289. https://doi.org/10.1289/ehp.01109s2283.

Monson, M.S., Coulombe, R.A. and Reed, K.M., 2015. Aflatoxicosis: lessons from toxicity and responses to aflatoxin B1 in poultry. Agriculture 5: 742-777. https://doi.org/10.3390/agriculture5030742.

Muñoz Solano, B. and González Peñas, E., 2023. Cooccurrence of mycotoxins in feed for cattle, pigs, poultry, and sheep in Navarra, a region of Northern Spain. Toxins 15: 172. https://doi.org/10.3390/toxins15030172.

Nelson, T.S., Beasley, J.N., Kirby, L.K., Johnson, Z.B., Ballam, G.C. and Campbell, M.M., 1981. Citrinin toxicity in growing chicks 1. Poultry Science 60: 2165-2166. https://doi.org/10.3382/ps.0602165.

Ochieng, P.E., Scippo, M.L., Kemboi, D.C., Croubels, S., Okoth, S., Kang'ethe, E.K., Doupovec, B., Gathumbi, J.K., Lindahl, J.F. and Antonissen, G., 2021. Mycotoxins in poultry feed and feed ingredients from Sub-Saharan Africa and their impact on the production of broiler and layer chickens: a review. Toxins 13: 633. https://doi.org/10.3390/toxins13090633.

Osselaere, A., Devreese, M., Goossens, J., Vandenbroucke, V., De Baere, S., De Backer, P., and Croubels, S., 2013a. Toxicokinetic study and absolute oral bioavailability of deoxynivalenol, T-2 toxin and zearalenone in broiler chickens. Food and Chemical Toxicology 51: 350-355. https://doi.org/10.1016/j.fct.2012.10.006.

Osselaere A., Santos R., Hautekiet V., De Backer P., Chiers K., Ducatelle R. and Croubels S., 2013b. Deoxynivalenol impairs hepatic and intestinal gene expression of selected oxidative stress, tight junction and inflammation proteins in broiler chickens, but addition of an adsorbing agent shifts the effects to the distal parts of the small intestine. PloS ONE 8: e69014. https://doi.org/10.1371/journal.pone.0069014.

Pampori, N.A. and Shapiro, B.H., 1993. Sexual dismorphism in avian hepatic monooxygenases. Biochemical Pharmacology 46: 885-890. https://doi.org/10.1016/0006-2952(93)90498-l.

Paraskeuas, V., Griela, E., Bouziotis, D., Fegeros, K., Antonissen, G. and Mountzouris, K.C., 2021. Effects of deoxynivalenol and fumonisins on broiler gut cytoprotective capacity. Toxins 13: 729. https://doi.org/10.3390/toxins13100729.

Prelusky, D.B., Hamilton, R.M. and Trenholm, H.L., 1989. Transmission of residues to eggs following long-term administration of 14C-labelled deoxynivalenol to laying hens. Poultry Science 68: 744–748. https://doi.org/10.3382/ps.0680744.

Qiu, M.F. and Liu, X.M., 2001. Determination of sphinganine, sphingosine and Sa/So ratio in urine of humans exposed to dietary fumonisin B1. Food Additives and Contaminants 18: 263-269. https://doi.org/10.1080/02652030117470.

Rejeb, R., De Baere, S., Devreese, M., Ducatelle, R., Croubels, S., Ayed, M.H., Ghorbal, A. and Antonissen, G., 2020. Calcination improves the *in vivo* efficacy of a montmorillonite clay to bind aflatoxin G1 in broiler chickens: a toxicokinetic approach. Toxins 12: 660. https://doi.org/10.3390/toxins12100660.

Richard, J.L., 2007. Some major mycotoxins and their mycotoxicoses—An overview. International Journal of Food Microbiology 119: 3-10. https://doi.org/10.1016/j.ijfoodmicro.2007.07.019.

Richard, J.L., 2008. Discovery of aflatoxins and significant historical features. Toxin Reviews 27: 171-201. https://doi.org/10.1080/15569540802462040.

Roth, A., Chakor, A., Creepy, E.E., Kane, A., Roschenthaler, R. and Dirheimer, G. 1988. Evidence for an enterohepatic circulation of ochratoxin A in mice. Toxicology 48: 293–308. https://doi.org/10.1016/0300-483x(88)90110-2.

Ruhnau, D., Hess, C., Grenier, B., Doupovec, B., Schatzmayr, D., Hess, M. and Awad, W.A., 2020. The mycotoxin Deoxynivalenol (DON) promotes *Campylobacter jejuni* multiplication in the intestine of broiler chickens with consequences on bacterial translocation and gut integrity. Frontiers in Veterinary Science 7: 573894. https://doi.org/10.3389/fvets.2020.573894.

Rychlik, M., Humpf, H. U., Marko, D., Dänicke, S., Mally, A., Berthiller, F., Klaffke, H. and Lorenz, N., 2014. Proposal of a comprehensive definition of modified and other forms of mycotoxins including "masked" mycotoxins. Mycotoxin Research 30: 197-205. https://doi.org/10.1007/s12550-014-0203-5.

Santos, R.R., Ooosterveer-van der Doelen, M.A.M., Tersteeg-Zijderveld, M.H.G., Molist, F., Mézes, M. and Gehring, R., 2021. Susceptibility of broiler chickens to deoxynivalenol exposure via artificial or natural dietary contamination. Animals 11: 989. https://doi.org/10.3390/ani11040989.

Sawhney, D.S., Vadehra, D.V. and Baker, R.C., 1973. The metabolism of 14 C aflatoxins in laying hens. Poultry Science 52: 1302-1309. https://doi.org/10.3382/ps.0521302.

Streit, E., Schatzmayr, G., Tassis, P., Tzika, E., Marin, D., Taranu, I., Tabuc, C., Nicolau, A., Aprodu, I., Puel, O. and Oswald, I.P., 2012. Current situation of mycotoxin contamination and co-occurrence in animal feed—Focus on Europe. Toxins 4: 788-809. https://doi.org/10.3390/toxins4100788.

Streit, E., Schwab, C., Sulyok, M., Naehrer, K., Krska, R. and Schatzmayr, G., 2013. Multi-mycotoxin screening reveals the occurrence of 139 different secondary metabolites in feed and feed ingredients. Toxins 5: 504-523. https://doi.org/10.3390/toxins5030504.

Suksupath, S., 1993. Cyclopiazonic acid toxicity in laying hens. PhD thesis. The University of Sydney, Camden, Australia.

Sulyok, M., Krska, R. and Schuhmacher, R., 2007. A liquid chromatography/tandem mass spectrometric multi-mycotoxin method for the quantification of 87 analytes and its application to semi-quantitative screening of moldy food samples. Analytical and Bioanalytical Chemistry 389: 1505-1523. https://doi.org/10.1007/s00216-007-1542-2.

Sun, Y.X., Yao, X., Shi, S.N., Zhang, G.J., Xu, L.X., Liu, Y.J., and Fang, B. H., 2015. Toxicokinetics of T-2 toxin and its major metabolites in broiler chickens after intravenous and oral administration. Journal of Veterinary Pharmacology and Therapeutics 38: 80-85. https://doi.org/10.1111/jvp.12142.

Sypecka, Z., Kelly, M. and Brereton, P., 2004. Deoxynivalenol and zearalenone residues in eggs of laying hens fed with a naturally contaminated diet: Effects on egg production and estimation of transmission rates from feed to eggs. Journal of Agricultural and Food Chemistry 52: 5463-5471. https://doi.org/10.1021/jf040039d.

Tangni E.K., Van Pamel, E., Huybrechts B., Delezie, E., van Hoeck, E. and Daesseleire, E., 2020. Carry-over of some *Fusarium* mycotoxins in tissues and eggs of chickens fed experimentally mycotoxin-contaminated diets. Food and Chemical Toxicology 145: 111715. https://doi.org/10.1016/j.fct.2020.111715.

Tardieu, D., Bailly, J.D., Skiba, F., Grosjean, F. and Guerre, P., 2008. Toxicokinetics of fumonisin B1 in turkey poults and tissue persistence after exposure to a diet containing the maximum European tolerance for fumonisins in avian feeds. Food and Chemical Toxicology 46: 3213-3218. https://doi.org/10.1016/j.fct.2008.07.013.

Tardieu, D., Bailly, J.D., Benlashehr, I., Auby, A., Jouglar, J.Y. and Guerre, P., 2009. Tissue persistence of fumonisin B1 in ducks and after exposure to a diet containing the maximum European tolerance for fumonisins in avian feeds. Chemico-Biological Interactions 182: 239-244. https://doi.org/10.1016/j.cbi.2009.06.009.

Tardieu, D., Travel, A., Le Bourhis, C., Metayer, J.P., Mika, A., Cleva, D., Boissieu, C. and Guerre, P., 2021. Fumonisins and zearalenone fed at low levels can persist several days in the liver of turkeys and broiler chickens after exposure to the contaminated diet was stopped. Food Chemistry and Toxicology 148: 111968. https://doi.org/10.1016/j.fct.2021.111968.

Tessari, E.N.C., Oliveira, C.A.F.D., Cardoso, A.L.S.P., Ledoux, D.R. and Rottinghaus, G.E., 2006. Effects of aflatoxin B1 and fumonisin B1 on body weight, antibody titres and histology of broiler chicks. British Poultry Science 47: 357-364. https://doi.org/10.1080/00071660600756071.

Verma, J., Johri, T.S. and Swain, B.K., 2007. Effect of aflatoxin, ochratoxin and their combination on protein and energy utilisation in white leghorn laying hens. Journal of the Science of Food and Agriculture 87: 760-764. https://doi.org/10.1002/jsfa.2655.

Voss, K.A., Smith, G.W. and Haschek, W.M., 2007. Fumonisins: toxicokinetics, mechanism of action and toxicity. Animal Feed Science and Technology 137: 299–325. https://doi.org/10.1016/j.anifeedsci.2007.06.007.

Vudathala, D., Prelusky, D.M.A., Trenholm, H. and Miller, J., 1994. Pharmacokinetic fate and pathological effects of 14C-fumonisin B1 in laying hens. Natural Toxins 2: 81-88 . https://doi.org/10.1002/nt.2620020206.

Wang, L., Zhang, Q., Yan, Z., Tan, Y., Zhu, R., Yu, D., Yang, H. and Wu, A., 2018. Occurrence and quantitative risk assessment of twelve mycotoxins in eggs and chicken tissues in China. Toxins 10: 477. https://doi.org/10.3390/toxins10110477.

Wang, Y., Quan, H., Li, X., Li, Q., Haque, M. A., Shi, Q., Fu, Q. and He, C., 2021. Contamination with fumonisin B and deoxynivalenol is a threat to egg safety and contributes to Gizzard ulcerations of newborn chickens. Frontiers in Microbiology 12: 676671. https://doi.org/10.3389/fmicb.676671.

Wyatt, R.D., Weeks, B.A., Hamilton, P.B. and Burmeister, H.R., 1972. Severe oral lesions in chickens caused by ingestion of dietary fusariotoxin T-2. Applied Microbiology 24: 251-257. https://doi.org/10.1128/am.24.2.251-257.1972.

CHAPTER 6

Effects of mycotoxins in cattle

A. Gallo

Department of Animal Science, Food and Nutrition (DIANA), Facoltà di Scienze Agrarie, Alimentari e Ambientali, Università Cattolica del Sacro Cuore, Piacenza, Italy
antonio.gallo@unicatt.it

1 Introduction

The nutritional needs of ruminants are satisfied through the use of various raw materials that can vary widely at farm level, such as concentrates and forages. In brief, ruminant diets include cereals, protein feeds, their by-products, as well as hay and silages. These forages can be contaminated by several exometabolites of mycotoxigenic fungi that increase and diversify the risk of mycotoxin exposure in ruminants compared to swine and poultry, which have less varied diets (Fink-Gremmels, 2008; Gallo et al., 2015). Therefore, the variability of the raw materials used to feed livestock is also reflected in the variability with which these matrices can be contaminated by mycotoxins. The various raw materials may be contaminated not only by a specific mycotoxin but by several mycotoxins at the same time, a phenomenon known as co-exposure (Pestka, 2007; Gallo et al., 2021a). The broad distribution of *Fusarium* spp. allows for the widespread presence of mycotoxins produced by these fungi in both temperate and warm environments and consequently found in feedstuffs for livestock use. Besides this, evidence suggests the greatest exposure for ruminants to many other mycotoxins is due to forage contamination. These toxic compounds can be produced by different species of *Alternaria spp., Aspergillus flavus* and *A. fumigatus, Penicillium roqueforti* and *P. paneum, Monascus ruber* and other mycotoxigenic fungi, as previously reviewed by Gallo et al. (2015).

Both maize silage and wheat silage represent the main source of DON contamination in cattle diets. Specifically, the average concentrations were 859 µg DON/kg for maize silage and around 621 µg DON/kg for wheat silage (Reisinger et al., 2019). A significant proportion of maize silage samples (67.7 %) were also found to be contaminated by ZEN (Reisinger et al., 2019), although another trial conducted reported that the level of ZEN contamination was 17.8 % of maize silage and 21.2 % of maize grain samples (Weaver et al., 1986a). As far as FUM are concerned, the highest contamination occurs in silage samples taken from the top layer and from the sides of the silage. These are the areas that present greater air infiltration, less fermentation and consequently may present different mycotoxin contamination

 | DOI: 10.1163/9789004712720_7

values than the other points of the silage, explaining the high variability of mycotoxins ranging from 120 to 1840 µg/kg DM pre-fermentation and 340 to 2490 µg/kg DM post-fermentation (Pereyra et al., 2008a). Generally, the incidence and concentration of AFs are low in well-preserved silages compared with other mycotoxins (Richard et al., 2009; Alonso et al., 2013) and were present in 33% of 7,049 livestock feed samples at an average concentration of 63 µg/kg as reported in a survey conducted by Rodrigues and Naehrer (2012). High concentrations of AFs can occur in poorly made silages or in silages made with diseased maize plants (Pereyra et al., 2008a). Values up to 51 µg/kg of DM of AFB_1 was detected in maize silage stored in bunkers in France (Richard et al., 2009).

The term mycotoxicosis indicates the syndromes found in animals following ingestion, skin contact with or inhalation of mycotoxins. In particular, these secondary metabolites may present a wide range of different effects, including teratogenic, mutagenic and carcinogenic, in the animals involved. Effects due to mycotoxin exposure differ depending on the type of toxin, exposure dose, target organs, sex and age. Mycotoxins can cause acute or chronic health effects. Acute manifestations, which are generally rare at farm level, involve clear signs of disease that can be traced back to mycotoxins and in some cases even result in death. Chronic manifestations, on the other hand, cause a range of effects that are not clearly attributable to mycotoxin exposure, such as food refusal and reduced intake, decreased productivity, reduced fertility and immunosuppression (Phillips et al., 1990; Mostrom and Jacobsen, 2011). Not only acute but also, and especially, chronic manifestations can cause major economic losses: indeed, with chronic manifestations, intervention may be too late (as the symptoms are not clearly traceable to mycotoxin exposure) and this may further promote negative effects on animal organisms (EFSA, 2011). Severe economic losses are due to ambiguous clinical changes in animal growth, reduction in the amount of food ingested or refusal of food, altered nutrient absorption, effects on endocrine metabolism and impact on immune status (Zain, 2011). The aim of this chapter is to present the main mycotoxigenic fungi that can affect the health and production performance of ruminants, with particular emphasis on cattle.

2 Rumen function

Mycotoxins are responsible for causing different effects on both the rumen microflora and the entire digestive process. Fink-Gremmels (2008) reported that ruminants are less sensitive than monogastrics to mycotoxin exposure because the rumen acts as an active defence against mycotoxins. More precisely, the rumen is able to bind these secondary metabolites to components present in the rumen (such as fibres, walls of yeast or bacteria) and ruminal bacteria and protozoa assist in the deactivation/degradation of these substances (Gallo and Masoero, 2009; Spanghero et al., 2023).

The rumen microbiota is considered a defence against mycotoxins as it allows different substances to be bio-converted. However, some authors (Upadhaya et al., 2010) stated that the metabolization of these substances differs according to species, age, sex and breed, as all these factors are able to act at the level of the composition of ruminal microbes. It was also reported that ruminants might be affected by the action of mycotoxins at the ruminal level, which may damage the ruminal microbiota and consequently interfere with and increase the number of other mycotoxins escaping microbial degradation (Fink-Gremmels, 2008). Mycotoxins in fact possess antimicrobial, antiprotozoal and antifungal activity and at the same time are responsible for alterations in microbial fermentation with consequent alterations in the nutrients that the organism is able to utilize, thus influencing the composition of the milk (Seeling et al., 2006). It has been reported, for example, that AFB_1, administered at doses of 0.2 to 0.8 mg/kg, in dairy cows causes a reduction in ruminal motility (Cook et al., 1986), while inhibition of the multiplication of ruminal microorganisms has been reported to occur with fusaric acid (May et al., 2000). Also, others (Colombatto et al., 2003; Morgavi et al., 2013) reported that the presence of mycotoxins adversely affects the cellulolysis of the rumen and also protein synthesis.

Fusaric acid inhibits *Ruminococcus albus* and *Methanobrevibacter ruminantium*, which are important ruminal microorganisms that degrade cellulose in the rumen and neutralize ruminal hydrogen (May et al., 2000). It can therefore be inferred that, despite being able to act as a primary defence, the rumen is affected by the activity of these secondary metabolites and also causes imbalances at a physiological level. It has been reported, for example, that AFB_1, DON, gliotoxin and PAT cause a reduction in rumen dry matter and also a reduction in the digestibility of neutral detergent fibre (NDF) (Gallo et al., 2015). The reduced digestibility of fibre in the ruminal compartment induces the utilization of highly fermentable carbohydrates, which leads to a drop in pH also at the ruminal level (Guo et al., 2020). Another consequence of mycotoxins at the ruminal level, is the reduction of protein synthesis in the rumen, which promotes the onset of a negative protein balance (Rodrigues et al., 2009; Rodrigues, 2014).

3 Gut health

It was demonstrated in monogastrics that mycotoxins possess the ability to alter intestinal permeability by causing imbalances in the intestinal barrier (Grenier and Applegate, 2013; Antonissen et al., 2014). Indeed, they are able to interact with the intestinal mucosa as well as the mucus (Przybylska-Gornowicz et al., 2018) and gut microbiota (Yang et al., 2017). Mycotoxins also affect the enteric nervous system (Makowska et al., 2018). In particular, following mycotoxin exposure, the enteric nervous system undergoes both adaptative and reparative processes, with the possibility of triggering changes in chemical coding in neurons and nerve fibres in the stomach and duodenum.

The detrimental effects of mycotoxins on the function of the ruminant intestinal system has been frequently neglected. However, a rumen that is not functioning optimally may also allow mycotoxins to interact with the intestine, which may be susceptible to DON, NIV, FB_1, and ENNB (Reisinger et al., 2019). Beef cattle that were naturally exposed to acute intoxication with AFs exhibited intestinal congestion (Elgioushy et al., 2020). Despite the fact that the authors only reported the concentrations of AFs (5 to 20 μg/kg) in the feed pellet, it is possible that other mycotoxins were also present in the diet. Juvenile goats (2–3 months old) were subjected to intestinal congestion and hemorrhagy as a result of T-2 toxin (10–20 mg/kg feed), while the rumen exhibited no significant changes (Nayakwadi et al., 2020). In addition, post-mortem analysis revealed intestinal inflammation in dairy cows that were naturally contaminated with *Fusarium* mycotoxins (DON and ZEN) for six months on a Polish farm. The cows were exhibiting non-specific disease symptoms, and the analysis of their blood revealed the presence of DON (18 to 24 ng/ml) and ZEN (6 to 14 ng/ml) (Marczuk et al., 2012).

4 Subclinical, pro-inflammatory, and immunosuppressive effects of mycotoxins

Mycotoxin exposure frequently occurs at levels that result in subclinical symptoms that are not clearly attributed to mycotoxicosis, making diagnosis more challenging. Feeding mycotoxin-contaminated feed causes a drop in performance due to metabolic problems such as ketosis, placental retention, dislocated abomasum, mastitis, metritis, lameness, and elevated somatic cell counts.

Mycotoxins, particularly AFs and trichothecenes, cause reduced resistance to environmental and microbial stressors and are therefore responsible for increased susceptibility to disease (CAST, 2003), and alter the immune system, specifically the number of macrophages, lymphocytes and erythrocytes (Upadhaya et al., 2010). Grass silage, contaminated at high levels with DON, caused the toxic syndrome in cattle, characterized by an increase in visible inflammatory reactions such as mastitis and laminitis (Speijers and Speijers, 2004), while other authors showed that animals with lameness were characterized by longer intervals between parturition and first insemination and conception (Pirestani and Toghyani, 2010); however, this could be attributable to the presence of other mycotoxins in the feed, such as ZEN.

Vaccination efficacy is affected by the cow exposure to mycotoxins. Cows fed a DON-contaminated diet with 5.66 mg/kg DM and 8.31 mg/kg DM had a different antibody response when vaccinated against bovine viral diarrhoea virus (BVDV), where the lowest DON dosage increased antibody response, and the highest DON dosage decreased antibody response (Dänicke et al., 2018). Failure in the acquired immunity was reported in lambs exposed to 2 mg/kg AFs during 37 days (Fernandez et al., 1997).

5 Oxidative stress

Oxidative stress is the term used to describe the excessive cell production of reactive oxygen species (ROS) and reactive nitrogen species (RNS), which are not promptly counteracted by the antioxidant systems responsible for maintaining redox homeostasis (Overton et al., 2017). In particular, oxidative stress has the capacity to activate the innate immune response. Mycotoxins also possess mechanisms to cause immune and oxidative stress responses; this can be exerted at three different levels:

- on the microbiota of the GIT and thus consequently on the epithelial cells of the GIT and on the immune cells of the intestine;
- directly in the GIT on both epithelium and immune cells;
- on organs and tissues following the transfer of metabolites into the bloodstream.

Both DON and ZEN act on mammary cells that are more sensitive to oxidative stress and apoptosis. The increased sensitivity is due to the fact that these mammary cells are extremely metabolically active, especially during lactation (Fu et al., 2019).

The various trials conducted in this area have led to the inference that the antioxidant system can be influenced by the level of mycotoxins and the duration of exposure to these substances. After exposure to mycotoxins, there is a severe oxidative condition that initially favours the activation of endogenous antioxidant mechanisms and increases enzyme activity (Sulzberger et al., 2016), whereas prolonged exposure to mycotoxins causes a decrease in antioxidant defences, allowing the suppression of antioxidant enzymes (Hou et al., 2013).

6 Main *Fusarium* mycotoxins affecting ruminants

6.1 *Deoxynivalenol*

Once ingested, DON is converted into de-epoxy-DON (called DOM-1 or de-DON) and its respective conjugated forms within the body. DON is rapidly distributed in the tissues. In cows and sheep, it has a half-life of 4 hours, disappearing 24 hours after ingestion (Prelusky et al., 1984, 1986). After being efficiently absorbed in the GIT, DON is excreted in the urine as a glucuronidated form, and in general the excretion of DON and related metabolites occurs mainly in the faeces and minimally in the urine (Gallo et al. 2022; see also Chpater 12). Although most ingested DON is converted to its less toxic form (DOM-1) in the rumen, Dänicke et al. (2002) observed a reduction in both the efficiency of microbial synthesis (by 24%) and the flow of total protein to the duodenum (by 25%) following the administration of *Fusarium* contaminated grains, while Seeling et al. (2006) noted that exposure to DON causes an alteration in the profile of short-chain fatty acids (SCFA) possibly caused by a change in the composition of the microbial communities.

DON causes gastrointestinal disorders, soft stools, immunosuppression and a general decrease in performance due to both reduced FI (EFSA, 2004a; Pestka et al., 2004) and the increased demand for energy by the immune system in order to function optimally (Fink-Gremmels, 2008). Ruminants absorb only 7-10% of ingested DON (Kiessling et al., 1984; Gallo et al., 2022) because the ruminal epithelium acts as a barrier (Seeling et al., 2006) and before absorption most DON undergoes deacetylation and deepoxidation by protozoa and ruminant bacteria (Kiessling et al., 1984).

Gallo et al. (2020) conducted an experiment to study the effects following a medium period of exposure (21 days) to 733 µg/kg DON and observed that the DON group showed less milk production (36.4 kg/day/cow) than the control group (37.7 kg/day/cow). Mycotoxin exposure resulted in negative effects on milk parameters, as the treated group exhibited worse curd firmness than the control. Finally, the DON group also presented a down-regulation of genes related to immune and inflammatory functions, suggesting a DON immunosuppressive effect.

6.2 *T-2 toxin*

Cattle are more sensitive to T-2 toxin than to DON. Ruminal bacteria such as *Butyrivibrio fibrisolvens* and *Selenomodos ruminatium* can partially degrade this mycotoxin, mostly to HT-2 toxin and T-2-triol, by using this T-2 toxin as an energy and carbon source (Dohnal et al., 2008). Degradation of T-2 toxin was also reported in the intestine, where deepoxy-HT-2 was detected (Cavret and Lecoeur, 2006). The T-2 toxin has been reported to cause various diseases in dairy cows, such as gastroenteritis, abomasal and ruminal ulcers, gastrointestinal lesions, and intestinal bleeding (Whitlow and Hagler, 2005). Consumption of feed contaminated with T-2 toxin resulted in irritation in the upper respiratory tract of cows (Fink-Gremmels, 2008), immunosupression via decreasing the cellular immune response (0.3 mg/kg/day during 56 days or 0.5mg/kg/day during 28 days; Mann et al., 1984) and necrosis in lymphoid tissues in calves (0.6 mg/kg/day during 43 days; Buening et al., 1982).

Ferreras et al. (2013) studied both acute and chronic mycotoxicosis resulting from natural intermittent T-2 exposure within a flock of 1000 sheep. The acute phase of exposure was when 190 animals died from ulcerative abomasitis and ruminitis, lymphocyte decline within lymphoid organs, necrosis in the exocrine pancreas, myocarditis, and skin and brain oedema. The remaining animals, while still alive, showed numerous effects such as reduced water consumption, anorexia, ruminal atony, soft stools, and apathy. During the chronic phase of exposure the animals showed loss of weight and impaired reproductive efficiency. Compared to the acute phase only a few animals died, in this case the deaths were associated with gastrointestinal inflammation, myocardial fibrosis, necrotic and suppurative lesions in the oral cavity and opportunistic infections (such as mastitis or parasitic pneumonia).

6.3 *Fumonisins*

FUM are characterized by poor bioavailability; in particular, the mycotoxin is rapidly distributed to the liver and kidneys before being excreted with faeces. FUM possess half-lives of only a few hours, and consequently, their rapid elimination occurs via the systemic circulation. FB_1 excretion in cattle occurs primarily through faeces, which also contain the hydrolyzed forms, and secondarily through urine. FB_1, FB_2, and FB_3 in cattle cause bile duct hyperplasia, hepatocellular lesions, and liver toxicity, as evidenced by increased serum enzymes and bilirubin. Because of their similar structure, FUM may also inhibit ceramide synthases (CerS), which are key enzymes involved in the biosynthesis of ceramide and complex sphingolipids. This causes a major impairment of cellular signalling mechanisms (Knutsen et al., 2018). FUM causes cytotoxic, hepatotoxic, and nephrotoxic effects. To date, there is limited literature regarding the actual effect of mycotoxins on digestive, physiological, and pathophysiological parameters in dairy cows (Fink-Gremmels, 2008; Gallo et al., 2015).

Calves fed FUM-contaminated diets for 31 days refused the feed and presented impaired liver and immune functions when exposed to a mixture of 105 mg/kg FB_1, 32 mg/kg FB_2, and 11 mg/kg FB_3, but neutrophil toxicity was already observed after exposure to 26 mg/kg FB_1 and 5 mg/kg FB_2 (Osweiler et al., 1993). Absorption of FUM by ruminants is low (Kemboi et al., 2020), and these animals are therefore considered less sensitive than other animal species (Voss et al., 2007; Mostrom and Jacobsen, 2011). Adverse effects in ruminants are observed only after exposure to a high FUM dietary intake. When dairy cows fed a FUM-contaminated diet (100 mg/kg) from 7 days before calving to 70 days after calving (Diaz and Smith, 2005), milk production decreased, and calves fed 1 mg/kg/BW of FUM showed nephrotoxic effects (Mathur, 2001). Lambs intraruminally exposed to FUM (11.1 mg/kg BW daily during 4 days) presented decreased FI as well as liver and kidney damage, and when this dose was increased to 45.5 mg/kg BW, the animals died 3 to 7 days after exposure started (Edrington et al., 1995). These levels of contamination tested in cows and lambs, however, rarely occur in the field, and FUM will probably have a harmful effect when combined with other mycotoxins.

6.4 *Zearalenone*

After ingestion, ZEN is rapidly metabolised. Microbial populations colonizing the rumen degrade ZEN mainly into α-ZEL, while β-ZEL is produced subsequent to hepatic biotransformation. As mentioned in the previous chapters, α-ZEL possesses a 60-fold higher potency than ZEN, whereas β-ZEL possesses a potency of 0.2 compared to ZEN. *In vivo* experiments report that β-ZEL is the metabolite produced the most in cattle. A trial showed that after 28 weeks consuming a ZEN-contaminated diet, the ratios of ZEN, α-ZEL, and β-ZEL in the bile of the dairy cows were 12-20%, 6-13% and 68-76%, respectively (Dänicke et al., 2014). Likewise, the excreta of goats administered ZEN intravenously showed a ZEN : α-ZEL : β-ZEL ratio of 1.6 : 1.0 : 3.6, respectively (Dong et al., 2010). This evidence suggests that these ruminants are less sensitive to ZEN. However, this percentage distribution is affected by the influence of both ruminal pH and FI (Seeling et al., 2005). Importantly, sheep resemble pigs in the ZEN biotransformation, where ZEN is mostly converted into α-ZEL regardless of rumen or liver function (Zinedine et al., 2007).

ZEN is characterized by oestrogenic activity, which it exerts by binding to oestrogen receptors and inducing an oestrogen-like response that interferes with the reproductive function of animals (Weaver et al., 1986a). The ZEN causes oestrous cycle disorders, pathological changes in the reproductive system in both sexes, decreased fertility, altered and impaired spermatogenesis, and decreases neonatal survival. Generally, female individuals are more susceptible at this mycotoxin (EFSA, 2011). Symptoms of hyperoestrogenism include enlargement of the vulva, uterus and mammary glands; it causes ovarian cysts, vulvovaginitis, atrophy of the testes and ovaries, altered oocyte maturation and may results in spontaneous abortion (CAST, 2003; Minervini and Dell'Aquila, 2008). Oxidative stress is a consequence of the action of ZEN on the reduction of cell proliferation and modulation of the immune response (Frisvad and Samson, 2004; Marin et al., 2013). The oral administration of 25 mg/kg ZEN to lactating cows had no visible effect on animals other than a reduction in the size of corpora lutea and a 25% reduction in the conception rate of animals fed ZEN (Weaver et al., 1986b). It was previously reported that herds of low fertility cows had higher levels of ZEN metabolites in their blood (Knutsen et al., 2017).

7 Main *Aspergillus* and *Penicillium* mycotoxins affecting ruminants

The two predominant toxigenic field *Aspergillus* species in crops and silages are *A. flavus* and *A. parasiticus*. Silages are often contaminated by *A. fumigatus* related toxins (Storm et al., 2008; Frisvad et al., 2009; Alonso et al., 2015). Among 226 potentially bioactive secondary metabolites produced by this mycotoxigenic fungus, gliotoxin is the most toxic present in maize and other silages (Pereyra et al., 2008b; Richard et al., 2009). Additional mycotoxins produced by *A. flavus* and other *Aspergillus* species are kojic, cyclopiazonic and β-nitropropionic acids (Chang et al., 2009), but are only sporadically present in silages (Storm et al., 2014). Boudra and Morgavi (2005) reported that gliotoxin, helvolic acid and verruculogen are stable during forage storage phases, whereas fumagillin was unstable under ensiling conditions. Agroclavine and festuclavine are other mycotoxins potentially produced by *A. fumigatus* (Frisvad et al., 2009). Lastly, maltoryzine can be produced by *A. clava-*

tus. However, data on its occurrence in forages are still poorly available (Gallo et al., 2015).

7.1 *Aflatoxins*

A portion of ingested AFB_1 can be transformed by ruminal microbes, resulting in the formation of aflatoxicol (AFL) (Upadhaya et al., 2010). A significant amount of AFs can also be sequestered by rumen fluid constituents, such as chlorophyllin, bacteria and yeast cell walls (Gallo and Masoero, 2009; Spanghero et al., 2023). The transfer of AFB_1 to milk as AFM_1, which poses a potential hazard for human health, is the primary concern associated with AFs exposure in ruminants. Hence, the regulatory limit of AFB_1 in dairy cattle feed in Europe is 5 μg/kg, and in the feed of beef cattle, sheep, and goats the regulatory limit of AFB_1 is 50 μg/kg (see details in Chapter 17). Exposure to significantly higher levels of AFB_1 in the diet are necessary to observe clinical symptoms. Applebaum et al. (1982) conducted a 7-day trial involving exposure to 13 mg/kg/day of AFB_1 through the ruminal orifice. The treated animals had lower milk production and at the same time the metabolite of AFB_1 in milk, i.e. AFM_1, was detectable in the milk with values ranging from 1.05 to 10.58 ng/l. In order to evaluate chronic exposure over a long period of time, Helferich et al. (1986) conducted a 155-day trial revealing that the administration of AFB_1 (600 μg/kg) resulted in a decrease in feed efficiency and in the rates of BWG in steers. In 60-day-old lambs, on the other hand, a halving of BWG was reported following the administration of AFB_1 at much lower level (100 μg/kg), but without the occurrence of increased mortality (Wang et al., 2019).

7.2 *Ochratoxin A*

Silages are generally low in OTA, relative to other major mycotoxins produced by *Fusarium* strains, such as DON, ZEN, FUM, and T-2 toxin (Kosicki et al., 2016). The fungi that produce this toxin cannot tolerate high concentrations of acetic acid and carbon dioxide that characterize silage during the conservation phase (Frisvad and Samson, 2004; Gallo et al., 2015). Ruminants are relative tolerant to these toxins because OTA is extensively and rapidly degraded in the rumen to a less toxic product, ochratoxin-α (OTα) (Fink-Gremmels, 2008). This activity is, however, only present in those animals that have a developed rumen. Calves with limited rumen functionality are more exposed to the negative effects of ingestion of OTA than calves that already have a developed rumen. Bioavailability of dietary OTA may be increased with high-grain diets, probably because they predispose to reduced ruminal pH (pH 5.5 to 5.8) (Höhler et al., 1999). To reduce the detoxifying capacity of the rumen, an extremely high and unrealistic intake of this toxin (13 mg/kg BW) is required, which could potentially reduce milk production and transfer the toxin into milk (Ribelin et al., 1978). These authors reported that goats were more sensitive to OTA than cows, where consumption of 3 mg/kg BW OTA resulted in diarrhoea, dehydration, and death.

7.3 *Mycophenolic acid and roquefortine C*

Species belonging to *Penicillium* section *Roquefortorum,* such as *P. roqueforti* and *P. paneum* (Nielsen et al., 2006), are considered some of the most prevalent post-harvest fungi found in silages (Nout et al., 1993; Driehuis, 2013; Gallo et al., 2016a,b). Different critical factors, such as unfavourable weather or storage conditions, could promote fungal growth and mycotoxin production (Boudra and Morgavi, 2005). A list of main mycotoxins produced by *Penicillium* strains was reported by Nielsen et al. (2006). These mycotoxins are certainly the most researched and detected in forages such as silage. Mycophenolic acid (MPA) and roquefortines are considered the most studied *Penicillium*-derived compounds in ensiled products.

MPA, produced by *P. roqueforti* and *Byssochlamys nivae* (Storm et al., 2008), was detected in maize silage with variable incidences, i.e. about 40% (Mansfield et al., 2008), 30% (Schneweis et al., 2000), 10% (Gallo et al., 2016a) or lower than 3% (Storm et al., 2010, 2014). Furthermore, MPA concentrations higher than 20 mg/kg were reported mainly in altered silage sample (Auerbach et al., 1998; Schneweis et al., 2000; Medina et al., 2014; Gallo et al., 2015).

Among the roquefortines produced by different strains of *P.* section *Roquefortorum*, the most studied was roquefortine C and this toxin was detected in more than 40% of sampled silages (Auerbach et al., 1998; O'Brien et al., 2006; Mansfield et al., 2008). Other *Penicillium*-derived exometabolites have been detected in silages, such as andrastin A, citreoisocoumarin, agroclavine, festuclavine and the *P. paneum* biomarker marcfortine A (O'Brien et al., 2006, Storm et al., 2014). Specially, O'Brien et al., (2006) reported that 2 of 5 collected maize silages were contaminated by agroclavine and festuclavine produced by *Penicillium* strains, with concentrations ranging from 100 to 1000 µg/kg. *Penicillium crutosum* produces not only roquefortine C, but also the tremorgenic mycotoxin Penitrem A. Although this mycotoxin is observed in pets with concerning results (see Chapter 8), neurotoxic effects were reported in grazing ruminant in New Zealand (DiMenna et al., 1976). When 5.5 mg/kg penitrem A was intravenously administered to sheep, only mild tremours were observed in the first 20 min, which then disappeared 2 hours later, whereas a strong inhibition of the reticulum and rumen skeletal muscle was observed (McLeay et al., 1999).

The influence of mycotoxins on the parameters of ruminal fermentation was examined *in vitro* by Gallo et al. (2014), who studied the influence of MPA and roquefortine C at the level of microbial populations, suggesting that these two mycotoxins interfere with digestive processes and remained stable even after 48 hours of rumen incubation.

7.4 *Ergot alkaloids*

EAs are found in the sclerotia of the fungus, which actually present a cemented compact mass of hyphae, very similar to a large and dark wheat grain (Stoev, 2023). The exposure to fungal alkaloids can unfortunately not be reduced by the activity of the rumen microflora. In this sense, it is understandable what was found by Schumann et al. (2009) who observed that the administration of EAs affects the levels of isovalerate, propionate and ammoniacal nitrogen thus indicating a change within the microbial population. These authors reported that the rumen absorbs or transforms about 30% of the ingested EAs. However, the absorption and metabolism of EAs in cattle remain unclear. Ergotism refers to the disease conditions associated with the ingestion of toxic levels of EAs. The leading chronic symptoms of the disease are the gangrene of peripheral parts of the extremities and subsequent loss of the same peripheral parts of the limbs, as well as some gastrointestinal ailments. The mycotoxins involved in the ergotism (ergometrine, ergotamine, and ergocristine) also provoke increased contractility of the uterus and many subsequent spontaneous abortions or prolapse of the uterus (Schrenk et al., 2024). As reported by these authors, the experimental studies failed to show any effect in cattle and future studies need to elucidate mode of action of the EAs induced effects, as related to a highly variable individual animal response to the exposure to these toxins. This is due mainly to the complex plant–fungus–animal–microbe–environmental interactions that result in changing alkaloid concentrations, availability, and distribution of various isomeric forms throughout the animal. The primary modes of action of the EAs, whether of tall fescue endophytic or ergot origin, involve vasoconstriction or hyperprolactinemia due to reduced secretion of prolactin by the anterior pituitary (Schrenk et al., 2024). In ruminants, the vasoconstrictive effects of EAs, whether of endophytic or ergot origin, cause both the so-called gangrenous and hyperthermic forms of EAs intoxication. The gangrenous syndrome generally occurs at low environmental temperatures and is characterized by diminished blood flow to the extremities, ultimately leading to dry gangrene. Hyperthermic syndrome is usually observed under heat-stress conditions and develops due to an animal's impaired ability to thermoregulate. Ruminal administration of ergotamine (105 µg/kg BW) in goats during the last third of gestation resulted in hyperthermia, feed refusal of the goats, and abortion or weak offspring in 50% of the pregnancies (Engeland et al., 1998).

8 Other mycotoxins

8.1 *Enniatins and Beauvericin*

Due to their diverse toxic activities, ENNs and BEA have been attracting increasing attention. Their occurrence has been reported in cereals such as wheat, barley, maize, and oats. A high prevalence of ENNs and BEA was found in a survey conducted on Norwegian grains (Uhlig et al., 2006), where 100% and 32% of the 228 grain samples (oats, barley and wheat) were contaminated by ENNs and BEA, respectively. In Denmark, 100% and 98% of the maize harvested in 2006 was contaminated by ENNs and BEA, respectively (Sørensen et al., 2008). In both reports,

ENNs levels above 1 mg/kg were found, and BEA levels above 100 μg/kg were detected. In later studies, even higher contamination levels of these mycotoxins have been reported. Although no maximum levels of ENNs and BEA have been set until now, their toxic effects in combination with high contamination levels could pose possible hazards to health. For more details, see Chapter 17.

9 Other fungal toxins

In *Monascus ruber* infected silage, CIT has been detected (Schneweis et al., 2001; Garon et al., 2006; Richard et al., 2009). Among other exometabolites produced by *M. ruber*, Schneweis et al. (2001) reported that monacolin (statin family, cholesterol lowering) was detected in about 20% of collected silages. Some zygomycetes can, via endophytic bacteria, produce secondary metabolites and toxic rhizonins (Jennessen et al., 2005). Jensen et al. (1994) reported these fungi could cause zygomycosis in immunosuppressed animals.

Lolitrem B is a tremorgenic mycotoxin produced by *Acremonium lolii*, and it is responsible for ryegrass staggers in livestock (Gallagher et al., 1984). It was shown before that sheep grazing on endophyte-infected pasture may be exposure to this mycotoxin, and severe disruption of digestion may take place 12 hours after exposure (McLeay et al., 1999).

Phomopsin A is highly toxic for sheep, with an LD_{50} of 40 μg/kg, which is over 100 times more sensitive than rats. The clinical symptoms of contamination are anorexia, jaundice, and liver dysfunction characterized by fat accumulation (Jago et al., 1982). This micotoxicosis is denominated lupinosis because it occurs when animals consume lupins contaminated with *Phomopsis leptostromiformis* (now denominated is *Diaporthe toxica*). Although cattle are less susceptible to lupinosis, researchers have reported cases of fatty liver syndrome in cows during pregnancy or after calving, resulting in clinical symptoms such as inappetence and ketosis (Allen, 1981). In Australia, outbreaks were reported when 9–12-month-old heifers consumed contaminated lupins during 3 weeks. After presenting the symptoms of staggering, jaundice, and bleeding from body orifices, the heifers were prevented from consuming the contaminated lupins, but 44% of the heifers died up to 2 weeks later, and only 8% did not present symptoms (Makie et al., 1992). These authors, however, did not exclude the possibility of co-contamination with phomopsin and other hepatotoxins.

Stenocarpella maydis, often found in maize ears, produces neurotoxins such as diplonine and chaetoglobosin K. They damage myelin in newborn animals by stopping actin polymerization, resulting in nervous disorders affecting grazing cattle, goats, and sheep in South Africa and South America (Kellerman et al., 1985, 1991; Odriozola et al., 2005; Wicklow et al., 2011). A high mortality rate was reported in lambs of ewes exposed to *S. maydis* during the second (66% mortality) or third trimester (87% mortality) of gestation, where 44% of the dams presented some symptoms of intoxication, i.e., ataxia, arched back, stiff-hindlegged gait with short steps, hopping with the hindlegs together like a rabbit, recumbency, and animals can lay laterally or sternally with the legs abnormally disposed (Kellerman et al., 1985; 1991). Goats exhibit similar symptoms. Cattle sometimes keep their heads low, arched back, and tails tucked in; when induced to move, they usually walk tottering crab-wise, and may develop a involuntary high-step gait of the hindlimbs (Kellerman et al., 1985). Morbidity and mortality rates in cattle (heifers) may reach 5.4 and 37%, respectively (Odriozola et al., 2005).

10 Unspecified fungi toxins

Illicolins are fungal metabolites produced by different strains of *Fusarium* spp. and other pathogenic fungi of cereal crops, such as *Ascochyta* and *Cylindrocladium* spp. Illicolin A and B are not often detected in maize silages. These compounds were detected in maize silages as secondary metabolites of different fungal strains, but their effects on animal health are still unknown.

11 Multi-mycotoxin exposure

Figure 6.1 aims to summarize and classify within macro-categories the results obtained in 31 trials, published in specialist journals, conducted to evaluate the effect of mycotoxins produced by *Fusarium* spp. (i.e. DON, ZEN and FUM) (Gallo et al., 2022). Some of the trials focused on a single mycotoxin, while others were aimed at assessing the effect of several mycotoxins taken at the same time. In the latter case, i.e. when mycotoxins with different effects were studied in the same trial, each effect found was attributed only to the mycotoxin responsible for that effect (in some cases only to the mycotoxins responsible) and not to the entire set of mycotoxins examined in the trial for data processing.

Numerous trials have been conducted to identify the effects of co-exposure to mycotoxins, most of them considering the *Fusarium* toxins DON and ZEN, which may be often present simultaneously in ruminant diets (Gallo et la., 2022). In Table 6.1 the mycotoxins effects are classified and reported as effects on milk yield and quality, nutrient digestibility, as well as immune and metabolic status, and reproduction.

Catellani et al. (2023) carried out a trial to assess the effects following a long period of exposure (54 days) to *Fusarium* mycotoxins. The results showed that DM intake remained constant in control and treated animals but the treated group had lower milk production (37.6 kg/day) than the control group (38.2 kg/day). Other differences were found in milk quality as the treated group had a higher milk fat content (3.7%) than the control group (3.5%).

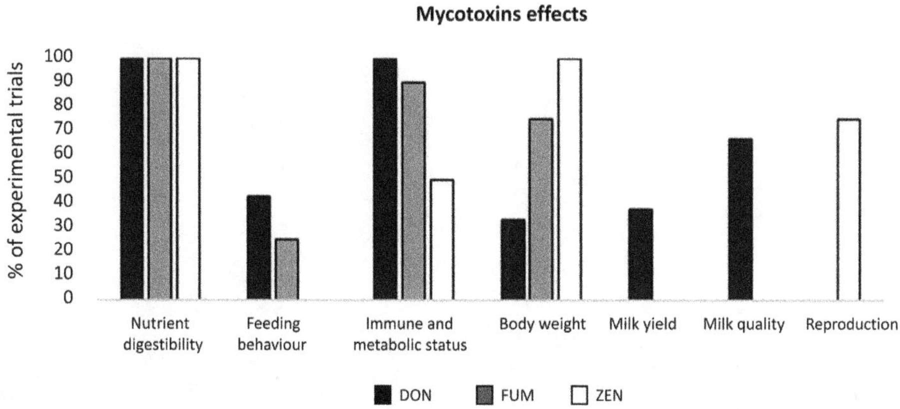

FIGURE 6.1 Effects of main *Fusarium* mycotoxins on nutrient digestibility, feeding behaviour, immune state, body weight, milk yield and quality, and reproduction function in dairy cows.

In addition to the known effects of ZEN on the reproductive function, there are also other mycotoxins that affect animal reproduction by first impairing reproductive cell function and then embryo development, such as DON, DOM-1, FUM, T-2/HT-2, and BEA (Albonico et al., 2016; Pizzo et al., 2016). The administration of diets co-contaminated by ZEN (1.5 mg/kg) and DON (1 mg/kg) causes an increase in reproductive tract infections, total reproductive failure, diarrhoea and a decrease in milk production (Coppock et al., 1990).

Gallo et al. (2021b) studied how the potential of fermentation fluid could vary in animals that ingested a diet contaminated by DON and FUM. It was observed that gas production kinetics were adversely affected by the presence of these secondary metabolites. Although mycotoxins possess a deleterious action at the ruminal level, it is indisputable that the rumen possesses a detoxifying action against these secondary metabolites. This detoxifying action occurs, however, in organisms that already have a well-developed rumen and, above all, an adequate microbial colonization. Individuals that are still young and do not have a fully developed rumen suffer in a similar way to monogastrics from the deleterious effect of mycotoxins. This conclusion is supported by a trial conducted on ewes that were administered 50 µg AFB_1 and 100 µg OTA/kg DM during the peripartum period. It was observed that the survival rate of the lambs went from 100% to 50% (Giorni et al., 2019) due to the fact that the rumen of these animals is not yet fully developed. Oxidative indices and antioxidant enzyme activity in this trial behaved similarly to what was also observed in another trial that reported the negative effect of AFB_1 and OTA on the oxidative balance of small ruminants (Peng et al., 2018).

TABLE 6.1 Survey on the effects of *Fusarium* toxin ingestion in ruminants based on experimental, field, and *in vitro* trials

Animal category	Mycotoxins	Milk	Nutrient digestibility	Immune and metabolic status	Reproduction
Experimental trials					
Lactating and non-lactating dairy cows[a]	DON: 4.4 mg/kg diet	–	↓ NDF digestibility ↓ in microbial CP	–	–
Lactating dairy cows[b]	DON: 4.4 or 5.3 mg/kg DM	↓ Milk Fat	↑ DMI	–	–
Lactating dairy cows[c]	DON: 4.4 or 5.3 mg/kg DM	–	↓ pH ↑ valerate ↓acetate ↓isobutyrate	–	–
Lactating dairy cows[d]	DON: 42, and 104 mg/kg BW	–	–	–	–
Lactating dairy cows[e]	DON: 8 mg/kg diet	–	–	–	–
Non-lactating cows[f]	DON: 8 or 35 mg/kg BW/day	–	↓ dietary consumption	–	–
Non-lactating cows[g]	DON: 4.0 or 3.6 mg/kg diet ZEN: 0.13 or 0.05 mg/kg diet	–	↓ microbial protein and ↑rumen NH_3-N	–	–
Lactating dairy cows[h,i]	DON: 3.5 mg/kg diet ZEN: 0.24 mg/kg diet	= milk production	= DMI	immunosuppression	–
Lactating dairy cows[j]	DON: 2.6 mg/kg DM ZEN: 0.3 mg/kg DM or DON: 5.2 mg/kg DM ZEN: 0.7 mg/kg DM	–	–	–	–
Lactating dairy cows[k]	DON: 14.3 mg/kg diet ZEN: 0.68 mg/kg diet	–	–	↑ AST and LDH ↑ VFAs	–
Lactating dairy cows[l]	FUM: 75 mg/kg BW FB_1: 3 mg/kg BW	–	–	–	–
Steers[m]	FB_1: 94 mg/kg diet	–	–	↑ AST, GGT, liver injury, and biliary hyperplasia	–
Calves[n]	FUM: 148 mg/kg diet	–	= Feed ingestion	↑AST, GGT, LDH, bilirubin,	–

Milk-fed calves[o]	FB_1: 1mg/kg BW intravenous	–	–	Liver and kidney lesions ↑ AST, ALP, GGT, and SDH	–
Lambs[p]	FUM: 45.5 mg/kg BW	–	–	↑ALP, GGT, AST, cholesterol, triglyceride, urea nitrogen and creatinine, death	–
Heifers[q]	ZEN: 250 mg/heifer	–	–	–	↓ Conception rate
Dairy cow[r]	ZEN: 500 mg/cow	–	–	–	No effect
Ewes[s]	ZEN: 3 mg/ewe/day (10 days)	–	–	–	↓ ovulation ↓lambing infertility
Beef cattle[t]	DON: 1.7 mg/kg TMR FUM: 3.5 mg/kg TMR	–	↓↓ pH Rumen	inflammation	
Beef cattle[u]	DON: 1.7 mg/kg TMR FUM: 3.5 mg/kg TMR	–	↓↓ CP	immunotoxicity	–
Dairy cows[v]	DON: 733 or 897 μg/kg DM FUM: 994 or 1247 μg/kg DM	↓↓↓ production and curd firmness	↓ DM and NDF digestibility	↑ ALT and AST immunosuppression	–
Dry cow[w]	FUM: 22 mg/kg TMR	–	Altered rumen microflora ↑ fibre breakdown	Hepatotoxicity ↑ liver enzymes	–
Red dairy cows[x]	DON: 455 μg/kg DM ZEN: 550 μg/kg DM AFB_1: 10.0 μg/kg DM	↓↓Fat	–	immunosuppression	–
Holstein cows[y]	DON: 4.61 mg/kg DM ZEN: 0.58 mg/kg DM	–	–	↑ AST ↑ bilirubin	–
Holstein Friesian cows[z]	DON: 2 mg/kg TMR DM ZEN: 366μg/kg TMR DM	↓ milk yield, milk fat and protein kg	↓DM		–

TABLE 6.1 Survey on the effects of *Fusarium* toxin ingestion in ruminants based on experimental, field, and *in vitro* trials (*cont.*)

Animal category	Mycotoxins	Milk	Nutrient digestibility	Immune and metabolic status	Reproduction
Dry cow[w]	ZEN: 5.9 mg/kg TMR	–	Altered rumen microflora ↓ rumen pH ↑ fibre breakdown	–	–
Cow[aa]	ZEN: 1 mg/kg diet	–	Altered protein metabolism		↑ ovarian antral follicle population ↑ synthesis of AMH by granulosa cells No effect on fertility
Heifer[bb]	ZEN: 0.3 mg/kg diet	–	–	–	↓ oocyte quality No effect on oestradiol
Heifer[cc]		ZEN: 10 mg/kg diet	–	–	–
Bulls[dd]	ZEN: 0.32 mg/kg diet	–	–	–	↓ testicular weight (tendency)
Cows[ee]	DOM-1: 100 ng/ml (intra ovarian injection)	–	–	–	↓ follicular size
	DOM-1: 6 mg/kg diet	–	–	–	No effect on the number of viable embryos Serum level of DON: 1.37 ng/ml Serum level of DOM1: 20.4 ng/ml
Bulls[ff]	T2-toxin: 0.6 mg/kg diet	–	–	–	↓ sperm quality ↓ motility poor morphology

Heifers[gg]	DON: 0.5 mg /kg diet ZEN: 0.75 mg/kg diet	–	–	–	unsynchronized ovarian cycles, vaginitis and early development of mammary gland in the prepubertal heifers
Steers[hh]	FUM: 15, 30, 60, 90 mg/kg	–	–	↑ SA:SO ratio	–
In vitro					
Rumen donor animal[ii]	DON: 0.3, 0.7, 0.9 mg/kg DM FUM: 0.1, 1.0, 1.3 mg/kg DM	–	↓↓ gas production ↓ acetic acid	–	–
Ruminal liquid fermentation[jj]	12 mg/kg DON, 60 mg/kg NIV, 1 mg/kg ENNB, 6 mg/kg MPA, 2 mg/kg ROQ-C, 3 mg/kg ZEN	–	No effect on VFAs	–	–

ALP: alkaline phosphatase; AMH: Anti-Müllerian hormone; AST: aspartate aminotransferase; CP: crude protein; DM: dry matter; DMI: dry matter intake; GGT: gamma-glutamyl transferase; LDH: lactate dehydrogenase; NDF: neutral detergent fibre; NH_3-N: ammonia nitrogen; SA:SO: sphingnanine:sphingosine ratio; SDH: sorbitol dehydrogenase; TMR: total mixed ration; VFAs: volatile fatty acids. In general, no major effects on growth performance.
a Hildebrand et al., 2012; b Keese et al., 2008a; c Keese et al., 2008b; d Charmley et al., 1993; e Ingalls, 1996; f Trenholm et al., 1985; g Dänicke et al., 2005;
h Korosteleva et al., 2007; i Korosteleva et al., 2009; j Winkler et al., 2014; k Hochsteiner et al., 2000; l Richard et al., 1996; m Baker and Rottinghaus, 1999;
n Osweiler et al., 1993; o Mathur, 2001; p Edrington et al., 1995; q Weaver et al., 1986a; r Weaver et al., 1986b; s Smith et al., 1990; t Duringer et al., 2020;
u Roberts et al., 2021; v Gallo et al., 2020; w Hartinger et al., 2022; x Jovaišienė et al., 2016; y Kinoshita et al., 2015; z McKay et al., 2019; aa Fushimi et al., 2015;
bb Silva et al., 2021; cc Kallela and Ettala, 1984; dd Dänicke et al., 2002; ee Guerrero-Netro et al., 2020; ff Alm et al., 2002; gg Coppock et al., 1990;
hh Jennings et al., 2020; ii Gallo et al., 2021b; jj Debevere et al., 2020. Source: Adapted by Gallo et al., 2015, 2022.

12 Carry-over of mycotoxins into animal-derived food products

Numerous mycotoxins and their metabolites may be transferred into animal tissues and fluids, including edible animal products like milk. The carry-over of DON and its metabolites is negligible (Keese et al., 2008c). More specifically, DOM-1 was found in milk instead of DON because, at ruminal level, DON is rapidly biotransformed into DOM-1, which is the less toxic form (Winkler et al., 2014). The administration of a single dose of 1.7 mg/kg BW of DON to lactating cows resulted in a carry-over in milk of 1-3 ng/ml at 8 h of free DON and <1-2 ng/ml at 20 h of conjugated DON (Prelusky et al., 1984).

The carry-over of FUM within milk is negligible. One experiment identified that FB_1 is able to cross the mammary barrier, but no further evidence of the presence of this molecule in this organ is currently known (Spotti et al., 2001).

The carry-over of ZEN in cattle occurs in the milk and liver (Mirocha et al., 1981; Kleinova et al., 2002); the latter in particular can be explained by the enterohepatic recirculation of ZEN (Dänicke and Winkler, 2015). ZEN has been detected in both milk as such and milk powder (Huang et al., 2014). Despite the presence of ZEN and its metabolites in milk, the consumption of these products does not pose a risk to human health, unlike cereals which leave individuals exposed to increased levels of these substances (EFSA, 2011; Dänicke and Winkler, 2015).

With regard to OTA, this mycotoxin was detected in milk in quantities between 0.07 and 0.11 μg/l and in 20% of the hand-made cheeses analysed, which contained OTA at concentrations ranging from 1 to 262 μg/kg in the rind and 18 to 146 μg/kg in the interior (Pattono et al., 2011; 2013). The assessed hand-made cheese also contained PAT, where 3% of the samples had this mycotoxin in the inner part of the cheese (26.6 μg/kg), and 25% of the samples contained 15.4 to 460.8 μg/kg PAT (Pattono et al., 2013).

AFs are of great concern for human health because they can be transformed by lactating animals into the hydroxylated derivatives AFM_1 and AFM_2, which can be found in milk products (Masoero et al., 2007). This ability to transfer AFB_1 from contaminated feed to milk represents a significant safety risk for milk and milk products, because AFM_1 is a group 1 human carcinogen (IARC, 2002). For this reason, AFB_1 is the only mycotoxin for which there is government legislation concerning its maximum permissible levels in feedstuffs and, consequently, in milk products (see Chapter 17). The carry-over of AFB_1 varies according to the feeding regimen, health status of the animals, biotransformation capacity, lactation period, and yield (EFSA, 2004b). It was reported that highly productive cows (~ 40 kg milk per day) in early to mid-lactation will have a higher level of AFM_1 in milk (up to 6.8%) than those with a lower milk yield (~ 20-25 kg milk per day) and in late lactation (about 2%), and this is not related to the milk volume but is probably associated with liver metabolism (Veldman et al., 1992). These authors suggested that, although the maximum level of AFB_1 in dairy cattle feed is set at 5 μg/kg, the maximum dietary level of AFB_1 for highly yielding cows should be kept below 3.4 μg/kg. Estimates indicated a 2% transfer in animals with average performance and a 6% transfer in animals with high performance. This pattern of AFM_1 transfer to milk is not unique to cows, but applies to all ruminants in general based on their milk yield, i.e., sheep (2 vs. 4 kg milk per day), goats (3 vs. 6 kg milk per day), camels (10 vs. 20 kg milk per day), and buffalos (5 vs. 10 kg milk per day) (EFSA, 2004b).

When analysing the carry-over of mycotoxins within edible tissues (e.g. in meat), it would appear that there is no carry-over, except for AFs, for which the carry-over is low due to their partial degradation in the rumen and their rapid metabolization in the liver. No carry-over of EAs in milk is reported in dairy cows (Schumann et al., 2009). Lolitrem B can be detected in the fat tissue of animals consuming the contaminated pasture, but only residues in muscle, liver, kidney, or brain have been detected (Miyazaki et al., 2007).

13 Conclusions

In general, ruminants are considered less sensitive to mycotoxins than monogastrics, due to the presence of the ruminal microbiota that can successfully degrade and/or inactivate several mycotoxins. However, the detoxifying capacity of the ruminal microflora is limited and varies with changes in the dietary composition or as a consequence of metabolic diseases. Besides this, mycotoxins cause different effects depending on various factors, such as the diet, the physiological state of the animals or the environmental conditions in which they live. The transition period is an extremely difficult time for animals that are in a negative energy balance and therefore show increased sensitivity to feed contaminated by fungi or mycotoxins. Often matrices are multi-contaminated by several mycotoxins that can be produced by the same fungus or even by different fungi. The simultaneous presence of different mycotoxins is very important as these substances can interact with each other causing additive, antagonistic or synergistic effects (see chapter 3). In order to provide more information on the risk of mycotoxin contamination in silage and cattle diets and to verify the effects of mycotoxin ingestion in animals, included some emerging mycotoxins, an international network should be created to monitor mycotoxin occurrence in silages and other forage crops worldwide, thus enabling the verification of exposure levels to mycotoxin ingestion of ruminants and the real effect of mycotoxin ingestion in livestock.

References

Albonico, M., Schütz, L.F., Caloni, F., Cortinovis, C. and Spicer, L.J., 2016. Toxicological effects of fumonisin B1 alone and in combination with other fusariotoxins on bovine granulosa cells. Toxicon 118: 47–53. https://doi.org/10.1016/j.toxicon.2016.04.041.

Allen, J.G., 1981. An evaluation of lupinosis in cattle in Western Australia. Australian Veterinary Journal 57: 212-215. https://doi.org/10.1111/j.1751-0813.1981.tb02660.x.

Alm, K., Dahlbom, M., Säynäjärvi, M., Andersson, M.A., Salkinoja-Salonen, M.S. and Andersson, M.C., 2002. Impaired semen quality of AI bulls fed with moldy hay: a case report. Theriogenology 58: 1497–1502. https://doi.org/10.1016/S0093-691X(02)01079-8.

Alonso, V., Díaz Vergara, L., Aminahuel, C., Pereyra, C., Pena, G., Torres, A., Dalcero, A. and Cavaglieri, L., 2015. Physiological behaviour of gliotoxigenic *Aspergillus fumigatus sensu stricto* isolated from maize silage under simulated environmental conditions. Food Additives and Contaminants Part A 32: 236–244. https://doi.org/10.1080/19440049.2014.996256.

Alonso, V.A., Pereyra, C.M., Keller, L.A.M., Dalcero, A.M., Rosa, C.A.R., Chiacchiera, S.M. and Cavaglieri, L.R., 2013. Fungi and mycotoxins in silage: an overview. Journal of Applied Microbiology. 115: 637–643. https://doi.org/10.1111/jam.12178.

Antonissen, G., Martel, A., Pasmans, F., Ducatelle, R., Verbrugghe, E., Vandenbroucke, V., Li, S., Haesebrouck, F., Van Immerseel, F. and Croubels, S., 2014. The impact of *Fusarium* mycotoxins on human and animal host susceptibility to infectious diseases. Toxins 6: 430-452. https://doi.org/10.3390/toxins6020430.

Applebaum, R.S., Brackett, R.E., Wiseman, D.W. and Marth, E.H., 1982. Responses of dairy cows to dietary aflatoxin: feed intake and yield, toxin content, and quality of milk of cows treated with pure and impure aflatoxin. Journal of Dairy Science 65: 1503–1508. https://doi.org/10.3168/jds.S0022-0302(82)82374-6.

Auerbach, H., Oldenburg, E. and Weissbach, F., 1998. Incidence of *Penicillium roqueforti* and roquefortine C in silages. Journal of the Science of Food and Agriculture 76: 565–572. https://doi.org/10.1002/(SICI)1097-0010(199804)76:4<565::AID-JSFA990>3.0.CO;2-6.

Baker, D.C. and Rottinghaus, G.E. 1999. Chronic experimental fumonisin intoxication of calves. Journal of Veterinary Diagnostic Investigation 11: 289–292. https://doi.org/10.1177/104063879901100315.

Boudra, H., and Morgavi, D.P., 2005. Mycotoxin risk evaluation in feeds contaminated by *Aspergillus fumigatus*. Animal Feed Science and Technology 120: 113–123. https://doi.org/10.1016/j.anifeedsci.2005.01.006.

Buening, G.M., Mann, D.D., Hook, B. and Osweiler, G.D., 1982. The effect of T-2 toxin on the bovine immune system: Cellular factors. Veterinary Immunology and Immunopathology 3: 411–417. https://doi.org/10.1016/0165-2427(82)90023-X.

CAST (Council for Agricultural Science and Technology), 2003. Mycotoxins: risks in plant, animal, and human systems. Task Force Report 139, Ames, Iowa, USA.

Catellani, A., Ghilardelli, F., Trevisi, E., Cecchinato, A., Bisutti, V., Fumagalli, F., Swamy, H.V.L.N., Han, Y., van Kuijk, S. and Gallo, A., 2023. Effects of supplementation of a mycotoxin mitigation feed additive in lactating dairy cows fed fusarium mycotoxin-contaminated diet for an extended period. Toxins 15: 546. https://doi.org/10.3390/toxins15090546.

Cavret, S. and Lecoeur, S., 2006. Fusariotoxin transfer in animal. Food Chemistry and Toxicology 44: 444-453. https://doi.org/10.1016/j.fct.2005.08.021.

Chang, P.-K., Ehrlich, K.C. and Fujii, I., 2009. Cyclopiazonic acid biosynthesis of *Aspergillus flavus* and *Aspergillus oryzae*. Toxins 1: 74–99. https://doi.org/10.3390/toxins1020074.

Charmley, E., Trenholm, H.L., Thompson, B.K., Vudathala, D., Nicholson, J.W., Prelusky, D.B. and Charmley, L.L., 1993. Influence of level of deoxynivalenol in the diet of dairy cows on feed intake, milk production, and its composition. Journal of Dairy Science 76: 3580–3587. https://doi.org/10.3168/jds.S0022-0302(93)77697-3.

Colombatto, D., Morgavi, D.P., Furtado, A.F. and Beauchemin, K.A., 2003. Screening of exogenous enzymes for ruminant diets: Relationship between biochemical characteristics and in vitro ruminai degradation. Journal of Animal Science 81: 2628–2638. https://doi.org/2003.81102628x.

Cook, W.O., Richard, J.L., Osweiler, G.D. and Trampel, D.W., 1986. Clinical and pathologic changes in acute bovine aflatoxicosis: rumen motility and tissue and fluid concentrations of aflatoxins B1 and M1. American Journal of Veterinary Research 47: 1817–1825.

Coppock, R.W., Mostrom, M.S., Sparling, C.G., Jacobsen, B. and Ross, S.C., 1990. Apparent zearalenone intoxication in a dairy herd from feeding spoiled acid-treated corn. Veterinary and Human Toxicology. 32: 246–248.

Dänicke, S. and Winkler, J., 2015. Invited review: diagnosis of zearalenone (ZEN) exposure of farm animals and transfer of its residues into edible tissues (carry over). Food and Chemical Toxicology 84: 225-249. https://doi.org/10.1016/j.fct.2015.08.009.

Dänicke, S., Gädeken, D., Ueberschär, K.H., Meyer, U. and Scholz, H., 2002. Effects of *Fusarium* toxin contaminated wheat and of a detoxifying agent on performance of growing bulls, on nutrient digestibility in wethers and on the carry over of zearalenone. Archives in Animal Nutrition 56: 245–261. https://doi.org/10.1080/00039420214347.

Dänicke, S., Keese, C., Meyer, U., Starke, A., Kinoshita, A., & Rehage, J. (2014). Zearalenone (ZEN) metabolism and residue concentrations in physiological specimens of dairy cows exposed long-term to ZEN-contaminated diets differing in concentrate feed proportions. Archives of Animal Nutrition, 68: 492-506.. https://doi.org/10.1080/1745039X.2014.973236.

Dänicke, S., Valenta, H., Gareis, M., Lucht, H.W. and Von Reichenbach, H., 2005. On the effects of a hydrothermal treatment of deoxynivalenol (DON)-contaminated wheat in the presence of sodium metabisulphite (Na 2S 2O 5) on DON reduction and on piglet performance. Animal Feed Science and Technology 118: 93-108. https://doi.org/10.1016/j.anifeedsci.2004.09.011.

Dänicke, S., Winkler, J., Meyer, U., Kersten, S., Wernike, K., Beer, M. and Frahm, J., 2018. Antibody response of growing German Holstein bulls to a vaccination against bovine viral diarrhea virus (BVDV) is influenced by Fusarium toxin exposure in a non-linear fashion. Mycotoxin Research 34: 123-139. https://doi.org/10.1007/s12550-018-0307-4.

Devebere, S., Cools, A., De Baere, S., Haesaert, G., Rychlik, m., Croubels, S. And Fievez, V., 2020. *In vitro* rumen simulations show a reduced disappearance of deoxynivalenol, nivalenol and enniatin B at conditions of rumen acidosis and lower microbial activity. Toxins 12: 101. https://doi.org/10.3390/toxins12020101.

Diaz, D.E. and Smith, T.K., 2005. Mycotoxin sequestering agents: practical tools for the neutralisation of mycotoxins. The Mycotoxin Blue Book Nottingham University Press 323-340.

DiMenna, M.E., Mantle, P.G. and Mortimer, P.H., 1976. The role of penicillia in rye-grass staggers. Research in Veterinary Science 24: 347.

Dohnal, V., Jezkova, A., Jun, D. And Kuca, K., 2008. Metabolic pathways of T-2 toxin. Current Drug Metabolism 9: 77-82. https://doi.org/10.2174/138920008783331176.

Dong, M., He, X.J., Tulayakul, P., Li, J.Y., Dong, K.S., Manabe, N., Nakayama, H. and Kumagai, S., 2010. The toxic effects and fate of intravenously administered zearalenone in goats. Toxicon 55: 523-530. https://doi.org/10.1016/j.toxicon.2009.10.004.

Driehuis, F. 2013. Silage and the safety and quality of dairy foods: a review. Agricultural and Food Science 22: 16–34. https://doi.org/10.23986/afsci.6699.

Duringer, J.M., Roberts, H.L., Doupovec, B., Faas, J., Estill, C.T., Jiang, D. and Schatzmayr, D., 2020. Effects of deoxynivalenol and fumonisins fed in combination on beef cattle: health and performance indices. World Mycotoxin Journal 13: 533–543. https://doi.org/10.3920/WMJ2020.2567.

Edrington, T.S., Harvey, R.B., Kubena, L.F., Elissalde, M.H. and Rottinghaus, G.E., 1995. Acute hepatic and renal toxicity in lambs dosed with fumonisin-containing culture material. Journal of Animal Science 73: 508–515. https://doi.org/102527/1995.732508x.

EFSA Panel on Contaminants in the Food Chain (CONTAM), 2011. Scientific Opinion on the risks for public health related to the presence of zearalenone in food. EFSA Journal 2011 9: 2197. https://doi.org/10.2903/j.efsa.2011.2197.

EFSA Panel on Contaminants in the Food Chain, 2004a. Opinion of the Scientific Panel on contaminants in the food chain [CONTAM] related to Deoxynivalenol (DON) as undesirable substance in animal feed. EFSA Journal 2: 73. https://doi.org/102903/j.efsa.2004.73.

EFSA Panel on Contaminants in the Food Chain, 2004b. Opinion of the Scientific Panel on contaminants in the food chain [CONTAM] related to Aflatoxin B1 as undesirable substance in animal feed. EFSA Journal 2: 39. https://doi.org/102903/j.efsa.2004.39.

Elgioushy, M.M., Elgaml, S.A., El-Adl, M.M., Hegazy, A.M. and Hashish, E.A., 2020. Aflatoxicosis in cattle: clinical findings and biochemical alterations. Environmental Science and Pollution Research 27: 35526-35534. https://doi.org/10.1007/s11356-020-09489-3.

Engeland, I.F., Andresen, O., Ropstad, E., Kindahl, H., Waldeland, H., Daskin, A. and Eik, L.O., 1998. Effect of fungal alkaloids on the deelopment of pregnancy and endocrine foetal-placental function in the goat. Animal Reproduction Science 52: 289-302. https://doi.org/10.1016/S0378-4320(98)00107-9.

Fernandez, A., Hernandez, M., Sanz, M.C., Verde, M.T. and Ramos, J.J., 1997. Serological serum protein fraction and responses to *Brucella melitensis* in lambs fed aflatoxins. Veterinary and Human Toxicology 39: 137-40.

Ferreras, M.C., Benavides, J., García-Pariente, C., Delgado, L., Fuertes, M., Muñoz, M., García-

Marín, J.F. and Pérez, V., 2013. Acute and chronic disease associated with naturally occurring T-2 mycotoxicosis in sheep. Journal of Comparative Pathology. 148: 236-242. https://doi.org/10.1016/J.JCPA.2012.05.016.

Fink-Gremmels, J. 2008. The role of mycotoxins in the health and performance of dairy cows. Veterinary Journal 176: 84–92. https://doi.org.10.1016/j.tvjl.2007.12.034.

Frisvad, J.C., and Samson, R.A., 2004. Polyphasic taxonomy of *Penicillium subgenus Penicillium:* a guide to identification of food and air-bone teverticilliate penicillia and their mycotoxins. Studies in Mycology 49: 1-52.

Frisvad, J.C., C. Rank, K.F. Nielsen, and T.O. Larsen. 2009. Metabolomics of *Aspergillus fumigatus*. Medical Mycology 47: S53–S71. https://doi.org/10.1080/13693780802307720.

Fu, Y.R., Jin, Y.C., Zhao, Y., Shan, A.S., Fang, H.T., Shen, J.L., Zhou, C.H., Yu, H., Zhou, Y.F., Wang, X., Wang, J.M., Li, R.H. Wang, R. and Zhang, J., 2019. Zearalenone induces apoptosis in bovine mammary epithelial cells by activating endoplasmic reticulum stress. Journal of Dairy Science. 102: 10543-10553. https://doi.org/10.3168/jds.2018.16216.

Fushimi, Y., Takagi, M., Monniaux, D., Uno, S., Kokushi, E., Shinya, U., Kawashima, C., Otoi, T., Deguchi, E. and Fink-Gremmels, J., 2015. Effects of dietary contamination by zearalenone and its metabolites on serum Anti-Müllerian hormone: impact on the reproductive performance of breeding cows. Reproduction in Domestic Animals 50: 834-839. https://doi.org/10.1111/rda.12599.

Gallagher, R.T., Hawkes, A.D., Steyn, P.S. and Vleggaar, R., 1984. Tremorgenic neurotoxins from perennial ryegrass causing ryegrass staggers disorder of livestock: structure elucidation of lolitrem B. Journal of the Chemical Society, Chemical Communications. 9: 614–616. https://doi.org/10.1039/C39840000614.

Gallo, A. and Masoero, F., 2009. *In vitro* models to evaluate the capacity of different sequestering agents to adsorb aflatoxins. Italian Journal of Animal Science 9: 109–116. https://doi.org/10.4081/ijas.2010.e21.

Gallo, A., Bertuzzi, T., Giuberti, G., Moschini, M., Bruschi, S., Cerioli, C. and Masoero, F., 2016a. New assessment based on the use of principal factor analysis to investigate corn silage quality from nutritional traits, fermentation end products and mycotoxins. Journal of Science Food and Agriculture 96: 437–448. https://doi.org/10.1002/jsfa.7109.

Gallo, A., Ghilardelli, F., Atzori, A.S., Zara, S., Novak, B., Faas, J. and Fancello, F., 2021a. Co-occurrence of regulated and emerging mycotoxins in corn silage: relationships with fermentation quality and bacterial communities. Toxins 13: 232. https://doi.org/10.3390/toxins13030232.

Gallo, A., Ghilardelli, F., Doupovec, B., Faas, J., Schatzmayr, D. and Masoero, F., 2021b. Kinetics of gas production in the presence of *Fusarium* mycotoxins in rumen fluid of lactating dairy cows. Journal of Dairy Science Communications 2: 243–247. https://doi.org/10.3168/jdsc.2021-0100.

Gallo, A., Giuberti, G., Bertuzzi, T., Moschini, M. and Masoero, F., 2014. Study of the effects of PR toxin, mycophenolic acid and roquefortine C on in vitro gas production parameters and their stability in the rumen environment. The Journal of Agricultural Science 153: 163-176. https://doi.org/10.1017/S0021859614000343.

Gallo, A., Giuberti, G., Bruschi, S., Fortunati, P. and Masoero, F. 2016b. Use of principal factor analysis to generate a corn silage fermentative quality index to rank well- or poorly preserved forages. Journal of Science Food and Agriculture 96: 1686–1696. https://doi.org/10.1002/jsfa.7272.

Gallo, A., Giuberti, G., Frisvad, J., Bertuzzi, T. and Nielsen, K., 2015. Review on mycotoxin issues in ruminants: occurrence in forages, effects of mycotoxin ingestion on health status and animal performance and practical strategies to counteract their negative effects. Toxins 7: 3057–3111. https://doi.org/10.3390/toxins7083057.

Gallo, A., Minuti, A., Bani, P., Bertuzzi, T., Cappelli, F.P., Doupovec, B., Faas, J., Schatzmayr, D. and Trevisi, E., 2020. A mycotoxin-deactivating feed additive counteracts the adverse effects of regular levels of *Fusarium* mycotoxins in dairy cows. Journal of Dairy Science. 103: 11314-11331. https://doi.org/10.3168/jds.2020-18197.

Gallo, A., Mosconi, M., Trevisi, E. and Santos, R.R., 2022. Adverse effects of *Fusarium* toxins in ruminants:a review of *in vivo* and *in vitro* studies. Dairy 3: 474–499. https://doi.org/10.3390/dairy3030035.

Garon, D., Richard, E., Sage, L., Bouchart, V., Pottier, D. and Lebailly, P., 2006. Mycoflora and multimycotoxin detection in corn silage: experimental study. Journal of Agriculture and Food Chemistry 54: 3479–3484. https://doi.org/10.1021/jf060179i.

Giorni, P, Bertuzzi, T. and Battilani, P., 2019. Impact of fungi co-occurrence on mycotoxin contamination in maize during the growing season. 10: 2019. http://doi.org/10.3389/fmicb.2019.01265.

Grenier, B., and Applegate, T.J., 2013. Modulation of intestinal functions following mycotoxin ingestion: Meta-analysis of published experiments in animals. Toxins 5: 396–430. https://doi.org/10.3390/toxins5020396.

Guerrero-Netro H.M., Barreta, M.H., Costa, E., Goetten, A., Dupras, R., Mills, L., Koch, J., Portela, V.M., Price, C.A. and Chorfi, Y., 2020. Effects of the mycotoxin metabolite de-epoxy-deoxynivalenol (DOM-1) on embryo development and sperm motility in cattle. Journal of Applied Toxicology 41: 1180–1187. https://doi.org/10.1002/jat.4102.

Guo, L., Yao, D., Li, D., Lin, Y, Bureenok, S., Ni, K. and Yang, F., 2020. Effects of lactic acid bacteria isolated from rumen fluid and feces of dairy cows on fermentation quality, microbial community, and *in vitro* digestibility of alfalfa silage. Frontiers in Microbiology 10: 2998. https://doi.org/10.3389/fmicb.2019.02998.

Hartinger, T., Grabher, L., Pacífico, C., Angelmayr, B., Faas, J. and Zebeli, Q., 2022. Short-term exposure to the mycotoxins zearalenone or fumonisins affects rumen fermentation and microbiota, and health variables in cattle. Journal of Food Chemistry and Toxicology 162: 112900. https://doi.org/10.1016/j.fct.2022.112900.

Helferich, W.G., Garrett, W.N., Hsieh, D.P.H. and Baldwin, R.L., 1986. Feedlot performance and tissue residues of cattle consuming diets containing aflatoxins. Journal of Animal Science 62: 691-696. https://doi.org/10.2527/jas1986.623691x.

Hildebrand, B., Boguhn, J., Dänicke, S. and Rodehutscord, M., 2012. Effect of Fusarium toxin-contaminated triticale and forage-to-concentrate ratio on fermentation and microbial protein synthesis in the rumen. Journal of Animal Physiology and Animal Nutrition 96: 307-318. https://doi.org/10.1111/j.1439-0396.2011.01143.x.

Hochsteiner, W., Schuh, M., Luger, K. and Baumgartner, W., 2000. Effect of mycotoxin contaminated feed on production parameters of dairy cows. Berliner und Münchener Tierärztliche Wochenschrift 113: 14-21.

Höhler, D., Südekum, K.H., Wolffram, S., Frohlich, A., Marquardt, R.R. and Ho, D., 1999. Metabolism and excretion of ochratoxin A fed to sheep. Journal of Animal Science 77: 1217-1223. https://doi.org/10.2527/1999.7751217x.

Hou, Y.J., Zhao, Y.Y., Xiong, B., Cui, X.S., Kim, N.H., Xu, Y.X. and Sun, S.C., 2013. Mycotoxin-containing diet causes oxidative stress in the mouse. PLoS One 8: e60374. https://doi.org/10.1371/JOURNAL.PONE.0060374.

Huang, L.C., Zheng, N., Zheng, B.Q., Wen, F., Cheng, J.B., Han, R.W., Xu, X.M., Li, S.L. and Wang, J.Q., 2014. Simultaneous determination of aflatoxin M-1, ochratoxin A, zearalenone and alpha-zearalenol in milk by UHPLC-MS/MS. Food Chemistry 146: 242–249. https://doi.org/10.1016/j.foodchem.2013.09.047.

Ingalls, J.R. 1996. Influence of deoxynivalenol on feed consumption by dairy cows. Animal Feed Science and Technology 60: 297–300. https://doi.org/10.1016/0377-8401(96)00984-4.

International Agency for Research on Cancer (IARC). 2002. Aflatoxins. IARC Summary and Evaluation 82: 1–171.

Jago, M.V., Peterson, J.E., Payne, A.L., Campbell, D.G., 1982. Lupinosis: response of sheep to different doses of phomopsin. Australian Journal of Experimental Biology and Medical Sciences 60: 239-251. https://doi.org/10.1038/icd.1982.29.

Jennessen, J., Nielsen, K.F., Houbraken, J., Lyhne, E.K., Schnorer, J., Frisvad, J.C. and Samson, R.A., 2005. Secondary metabolite and mycotoxin production by the *Rhizopus microsporus* group. Journal of Agriculture and Food Chemistry 53: 1833–1840. https://doi.org/10.1021/jf048147n.

Jennings, J.S., Ensley, S.M., Smith, W.N., Husz, T.C. and Lawrence, T.E., 2020. Impact of increasing levels of fumonisin on performance, liver toxicity, and tissue histopathology of finishing beef steers. Journal of Animal Science 98: 1–8. https://doi.org/10.1093/JAS/SKAA390.

Jensen, H.E., Olsen, S.N. and Aalbaek, B., 1994. Gastrointestinal aspergillosis and zygomycosis of cattle. Veterinary Pathology 31: 28–36. https://doi.org/10.1177/030098589403100104.

Jovaišienė, J., Bakutis, B., Baliukonienė, V. and Gerulis, G., 2016. *Fusarium* and *Aspergillus* mycotoxins effects on dairy cow health, performance and the efficacy of anti-mycotoxin additive. Polish Journal of Veterinary Science 19: 79–87. https://doi.org/10.1515/pjvs-2016-0011.

Kallela, K. and Ettala, E., 1984. The oestrogenic *Fusarium* toxin (zearalenone) in hay as a cause of early abortions in the cow. Nordisk Veterinaermedicin 36: 305–309.

Keese, C., Meyer, U., Rehage, J., Spilke, J., Boguhn, J., Breves, G. and Dänicke, S., 2008a. On the effects of the concentrate proportion of dairy cow rations in the presence and absence of a *Fusarium* toxin-contaminated triticale on cow performance. Archives in Animal Nutrition 62: 241–262. https://doi.org/10.1080/17450390802066435.

Keese, C., Meyer, U., Rehage, J., Spilke, J., Boguhn, J., Breves, G. and Dänicke, S., 2008b. Ruminal fermentation patterns and parameters of the acid base metabolism in the urine as influenced by the proportion of concentrate in the ration of dairy cows with and without *Fusarium* toxin-contaminated triticale. Archives in Animal Nutrition 62: 287–302. https://doi.org/10.1080/17450390802066443.

Kellerman, T.S., Prozesky, L., Schultz, R.A., Rabie, C.J., Van Ark, H., Maartens, B.P. and Lubben, A., 1991. Perinatal mortality in lambs of ewes exposed to cultures of Diplodiamaydis (*Stenocarpella maydis*) during gestation. Onderstepoort Journal of Veterinary Research 58: 297-308.

Kellerman, T.S., Rabie, C.J., van der Westhuizen, G.C., Kriek, N.P. and Prozesky, L., 1985. Induction of diplodiosis, a neuromycotoxicosis, in domestic ruminants with cultures of indigenous and exotic isolates of *Diplodia maydis*. Onderstepoort Journal of Veterinary Research 52: 35e42.

Kemboi, D.C., Antonissen, G., Ochieng, P.E., Croubels, S., Okoth, S., Kangethe, E.K., Faas, J., Lindahl, J.F. and Gathumbi, J.K., 2020. A review of the impact of mycotoxins on dairy cattle health: challenges for food safety and dairy production in sub-Saharan Africa. Toxins 12: 222. https://doi.org/103390/toxins12040222.

Kiessling, K.H., Pettersson, H., Sandholm, K. and Olsen, M., 1984. Metabolism of aflatoxin, ochratoxin, zearalenone, and three trichothecenes by intact rumen fluid, rumen protozoa, and rumen bacteria. Applied Environmental Microbiology 47: 1070–1073. https://doi.org/10.1128/aem.47.5.1070-1073.1984.

Kinoshita, A., Keese, C., Beineke, A., Meyer, U., Starke, A., Sauerwein, H., Dänicke, S. and Rehage, J., 2015. Effects of *Fusarium* mycotoxins in rations with different concentrate proportions on serum haptoglobin and hepatocellular integrity in lactating dairy cows. Journal of Animal Physiology and Animal Nutrition 99: 887-892. https://doi.org/10.1111/jpn.12293.

Kleinova, M., Zollner, P., Kahlbacher, H., Hochsteiner, W. and Lindner, W., 2002. Metabolic profiles of the mycotoxin zearalenone and of the growth promoter zeranol in urine, liver, and muscle of heifers. Journal of Agricultural and Food Chemistry 50: 4769–4776. https://doi.org/10.1021/jf020160p.

Knutsen, H.K., Alexander, J., Barregård, L., Bignami, M., Brüschweiler, B., Ceccatelli, S., Cottrill, B., Dinovi, M., Edler, L., Grasl-Kraupp, B., Hogstrand, C., Hoogenboom, L.R., Nebbia, C.S., Petersen, A., Rose, M., Roudot, C., Schwerdtle, T., Vleminckx, C., Vollmer, G., Wallace, H., Dall'Asta, C., Eriksen, G.S., Taranu, I., Altieri, A., Roldán-Torres, R., and Oswald, I.P., 2018b. Risks for animal health related to the presence of fumonisins, their modified forms and hidden forms in feed. EFSA Journal. 16: 5242. https://doi.org/10.2903/j.efsa.2018.5242.

Knutsen, H.K., Alexander, J., Barregård, L., Bignami, M., Brüschweiler, B., Ceccatelli, S., Cottrill, B., Dinovi, M., Edler, L., Grasl-Kraupp, B., Hogstrand, C., Hoogenboom, L.R., Nebbia, C.S., Petersen, A., Rose, M., Roudot, C., Schwerdtle, T., Vleminckx, C., Vollmer, G., Wallace, H., Dall'Asta, C., Dänicke, S., Eriksen, G.S., Altieri, A., Roldán-Torres, R. and Oswald. I.P., 2017. Risks for animal health related to the presence of zearalenone and its modified forms in feed. EFSA Journal 15: 4851. https://doi.org/10.2903/j.efsa.2017.4851.

Korosteleva, S.N., Smith, T.K. and Boermans, H.J., 2007. Effects of feedborne *Fusarium* mycotoxins on the performance, metabolism, and immunity of dairy cows. Journal of Dairy Science 90: 3867-3873. https://doi.org/10.3168/jds.2007-0162.

Korosteleva, S.N., Smith, T.K. and Boermans, H.J., 2009. Effects of feed naturally contaminated with *Fusarium* mycotoxins on metabolism and immunity of dairy cows. Journal of Dairy Science 92: 1585–1593. https://doi.org/10.3168/jds.2008-1267.

Kosicki, R., Błajet-Kosicka, A., Grajewski, J. and Twaruzek, M., 2016. Multiannual mycotoxin survey in feed materials and feedingstuffs. Animal Feed Science Technology 215: 165–180. https://doi.org/10.1016/j.anifeedsci.2016.03.012.

Makie, J.T., Rahaley, R.S. and Bennett, R., 1992. Lupinosis in yearling cattle. Australian Veterinary Journal 69: 172-173. https://doi.org/10.1111/j.1751-0813.tb07507.x.

Makowska, K., Obremski, K. and Gonkowski, S., 2018. The impact of T-2 toxin on vasoactive intestinal polypeptide-like immunoreactive (VIP-LI) nerve structures in the wall of the porcine stomach and duodenum. Toxins 10: 138. https://doi.org/10.3390/toxins10040138.

Mann, D.D., Buening, G.M., Osweiler, G.D. and Hook, B.S., 1984. Effect of subclinical levels of T-2 toxin on the bovine cellular immune system. Canadian Journal of Comparative Medicine 48: 308–312.

Mansfield, M.A., Jones, A.D. and Kuldau, G.A., 2008. Contamination of fresh and ensiled maize by multiple *Penicillium* mycotoxins. Phytopathology 98: 330–336. https://doi.org/10.1094/PHYTO-98-3-0330.

Marczuk, J., Obremski, K., Lutnicki, K., Gajecka, M. and Gajecki, M., 2012. Zearalenone and deoxynivalenol mycotoxicosis in dairy cattle herds. Polish Journal of Veterinary Sciences 15: 365-372. https://doi.org/10.2478/v10181-012-0055-x.

Marin, S., Ramos, A.J., Cano-Sancho, G. and Sanchis, V., 2013. Mycotoxins: occurrence, toxicology, and exposure assessment. Food and Chemistry Toxicology 60: 218–237. https://doi.org/10.1016/j.fct.2013.07.047.

Masoero, F., Gallo, A., Moschini, M., Piva, G. and Diaz, D., 2007. Carryover of aflatoxin from feed to milk in dairy cows with low or high somatic cell counts. Animal 1: 1344–1350. https://doi.org/10.1017/S1751731107000663.

Mathur, S. 2001. Fumonisin B1 is hepatotoxic and nephrotoxic in milk-fed calves. Toxicological Sciences 60: 385–396. https://doi.org/10.1093/toxsci/60.2.385.

May, H.D., Wu, Q. and Blake, C.K., 2000. Effects of the *Fusarium* spp. mycotoxins fusaric acid and deoxynivalenol on the growth of *Ruminococcus albus* and *Methanobrevibacter ruminantium*. Canadian Journal of Microbiology 46: 692–699. https://doi.org/10.1139/w00-045.

McKay, Z.C.C., Averkieva, O., Rajauria, G. and Pierce, K.M.M., 2019. The effect of feedborne *Fusarium* mycotoxins on dry matter intake, milk production and blood metabolites of early lactation dairy cows. Animal Feed Science and Technology 253: 39–44. https://doi.org/10.1016/j.anifeedsci.2019.05.004.

McLeay, L.M., Smith, B.L. and Munday-Finch, S.C., 1999. Tremorgenic mycotoxins paxilline, penitrem and lolitrem B, the non-tremorgenic 31-epilolitrem B and electromyographic activity of the reticulum and rumen of sheep. Research in Veterinary Science 66: 119-127. https://doi.org/10.1053/rvsc.1998.0253.

Medina, A., Rodriguez, A. and Magan, N., 2014. Effect of climate change on *Aspergillus flavus* and aflatoxin B1 production. Frontiers in Microbiology 5: 348–366. https://doi.org/10.3389/fmicb.2014.00348.

Minervini, F. and Dell'Aquila, M.E., 2008. Zearalenone and reproductive function in farm animals. International Journal of Molecular Science 9: 2570–2584. https://doi.org/10.3390/ijms9122570.

Mirocha, C.J., Pathre, S.V. and Robinson, T.S., 1981. Comparative metabolism of zearalenone and transmission into bovine milk. Food and Cosmetics Toxicology 19: 25-30. https://doi.org/10.1016/0015-6264(81)90299-6.

Miyazaki, S., Ikeda, T., Hanazumi, M., Fukumoto, Y., Yamata, T., Mikami, O., Yamanaka, N., Murata, H., Shimada, N. and Ishikuro, E., 2007. Toxicological evaluation of endophyte-infected perennial ryegrass straw to Japanese Black steers. Journal of New Zealand Grasslands Research and Practice Series 13: 415-418. https://doi.org/10.33584/rps.13.2006.3113.

Morgavi, D.P., Martin, C. and Boudra, H., 2013. Fungal secondary metabolites from Monascus spp. reduce rumen methane production *in vitro* and *in vivo*. Journal of Animal Science 91: 848–860. https://doi.org/10.2527/jas.2012-5665.

Mostrom, M.S. and Jacobsen, B.J., 2011. Ruminant mycotoxicosis. Veterinary Clinics of North America: Food Animal Practice 27: 315–344. https://doi.org/10.1016/j.cvfa.2011.02.007.

Nayakwadi, S., Ramu, R., Kumar Sharma, A., Kumar Gupta, V., Rajukumar, K., Kumar, V., Shirahatti, P.S., Kanthesh, R.L. and Basalingappa, M., 2020. Toxicopathological studies on the effects of T-2 mycotoxin and their interaction in juvenile goats. PLOS ONE 15: e0229463. https://doi.org/10.1371/journal.pone.0229463.

Nielsen, K.F., Sumarah, M.W., Frisvad, J.C. and Miller, J.D., 2006. Production of metabolites from the *Penicillium roqueforti* complex. Journal of Agriculture and Food Chemistry 54: 3756–3763. https://doi.org/10.1021/jf060114f.

Nout, M.J.R., Bouwmeester, H.M., Haaksma, J. and van Dijk, H., 1993. Fungal growth in silages of sugarbeet press pulp and maize. Journal of Agriculture Science 121: 323–326. https://doi.org/10.1017/S0021859600085506.

O'Brien, M., Nielsen, K.F., O'Kiely, P., Forristal, P.D., Fuller, H.T. and Frisvad, J.C., 2006. Mycotoxins and other secondary metabolites produced in vitro by *Penicillium paneum* and *Penicillium roqueforti* isolated from baled grass silage in Ireland. Journal of Agriculture and Food Chemistry 54: 9268–9276. https://doi.org/10.1021/jf0621018.

Odriozola, E., Odeon, A., Canton, G., Clemente, G. and Escande, A., 2005. *Diplodia maydis*: a cause of death of cattle in Argentina. New Zealand Veterinary Journal 53: 160-161. https://doi.org/10.1080/00480169.2005.36496.

Osweiler, G.D., Kehrli, M.E., Stabel, J.R., Thurston, J.R., Ross, P.F. and Wilson, T.M., 1993. Effects of fumonisin-contaminated corn screenings on growth and health of feeder calves. Journal of Animal Science 71: 459–466. https://doi.org/10.2527/1993.712459x.

Overton, T.R., McArt, J.A.A. and Nydam, D.V., 2017. A 100-year review: metabolic health indicators and management of dairy cattle. Journal of Dairy Science 100: 10398–10417. https://doi.org/10.3168/jds.2017-13054.

Pattono, D., Gallo, P.F. and Civera, T., 2011. Detection and quantification of Ochratoxin A in milk produced in organic farms. Food Chemistry 127: 374–377. https://doi.org/10.1016/j.foodchem.2010.12.051.

Pattono, D., Grosso, A., Stocco, P.P., Pazzi, M. And Zeppa, G., 2013. Survey of the presence of patulin and ochratoxin A in traditional semi-hard cheeses. Food Control 33: 54-57. https://doi.org/10.1016/j.foodcont.2013.02.019.

Peng, K., Jin, L., Niu, Y.D., Huang, Q., McAllister, T.A., Yang, H.E., Denise, H., Xu, Z., Acharya, S., Wang, S. and Wang, Y., 2018. Condensed tannins affect bacterial and fungal microbiomes and mycotoxin production during ensiling and upon aerobic exposure. Applied Environmental Microbiology 84: 1–20. https://doi.org/10.1128/AEM.02274-17.

Pereyra, C.M., Alonso, V.A., Rosa, C.A.R., Chiacchiera, S.M., Dalcero, A.M. and Cavaglieri, L.R., 2008b. Gliotoxin natural incidence and toxigenicity of *Aspergillus fumigatus* isolated from corn silage and ready dairy cattle feed. World Mycotoxin Journal 1: 457-462. https://doi.org/10.3920/WMJ2007.1012.

Pereyra, M.L.G., Alonso, V.A., Sager, R., Morlaco, M.B., Magnoli, C.E., Astoreca, A.L., Rosa, C.A.R., Chiacchiera, S.M., Dalcero, A.M. and Cavaglieri, L.R., 2008a. Fungi and selected mycotoxins from pre- and postfermented corn silage. Journal of Applied Microbiology 104: 1034–1041. https://doi.org/10.1111/j.1365-2672.2007.03634.x.

Pestka, J.J., 2007. Deoxynivalenol: Toxicity, mechanisms and animal health risks. Animal Feed Science and Technology 137: 283–298. https://doi.org/10.1016/j.anifeedsci.2007.06.006.

Pestka, J.J., Zhou, H.-R., Moon, Y. and Chung, Y.J., 2004. Cellular and molecular mechanisms for immune modulation by deoxynivalenol and other trichothecenes: unraveling a paradox. Toxicology Letters 153: 61–73. https://10.1016/j.toxlet.2004.04.023.

Phillips, T.D., Clement, B.A., Kubena, L.F. and Harvey, R.B., 1990. Detection and detoxification of aflatoxins: prevention of aflatoxicosis and aflatoxin residues with hydrated sodium calcium aluminosilicate. Veterinary and Human Toxicology 32 Suppl: 15–19.

Pirestani, A. and Toghyani, M., 2010. The effect of aflatoxin levels on milk production, reproduction and lameness in high production Holstein cows. African Journal of Biotechnology 9: 7905–7908. https://doi.org/10.5897/AJB10.250.

Pizzo, F., Caloni, F., Schreiber, N.B., Cortinovis, C. and Spicer, L.J., 2016. *In vitro* effects of deoxynivalenol and zearalenone major metabolites alone and combined, on cell proliferation, steroid production and gene expression in bovine small-follicle granulosa cells. Toxicon 109: 70–83. https://doi.org/10.1016/j.toxicon.2015.11.018.

Prelusky, D.B., Trenholm, H.L., Lawrence, G.A. and Scott, P.M., 1984. Nontransmission of deoxynivalenol (vomitoxin) to milk following oral administration to dairy cows. Journal of Environmental Science and Health Part B 19: 593–609. https://doi.org/10.1080/03601238409372453.

Prelusky, D.B., Vieira, D.M., Trenholm, H.L. and Hartin, K.E., 1986. Excretion profiles of the mycotoxin deoxynivalenol, following oral and intravenous administration to sheep. Fundamental and Applied Toxicology 6: 356-363. https://doi.org/10.1016/0272-0590(86)90251-4.

Przybylska-Gornowicz, B., Lewczuk, B., Prusik, M., Hanuszewska, M., Petrusewicz-Kosińska,

M., Gajęcka, M., Zielonka, Ł. and Gajęcki, M., 2018. The effects of deoxynivalenol and zearalenone on the pig large intestine. A light and electron microscopy study. Toxins 10: 1–18. https://doi.org/10.3390/toxins10040148.

Reisinger, N., Schürer-Waldheim, S., Mayer, E., Debevere, S., Antonissen, G., Sulyok, M. and Nagl. V., 2019. Mycotoxin occurrence in maize silage-a neglected risk for bovine gut health? Toxins 11: 577. https://doi.org/10.3390/toxins11100577.

Ribelin, W.E., Fukushima, K. and Still P.E., 1978. The toxicity of ochratoxin A to ruminants. Canadian Journal of Comparative Medicine 42: 172–176.

Richard, E., Heutte, N., Bouchart, V. and Garon, D., 2009. Evaluation of fungal contamination and mycotoxin production in maize silage. Animal Feed Science and Technology 148: 309–320. https://doi.org/10.1016/j.anifeedsci.2008.02.004.

Richard, J.L., Meerdink, G., Maragos, C.M., Tumbleson, M., Bordson, G., Rice, L.G. and Ross, P.F., 1996. Absence of detectable fumonisins in the milk of cows fed *Fusarium proliferatum* (Matsushima) Nirenberg culture material. Mycopathologia 133: 123–126. https://doi.org/10.1007/BF00439124.

Roberts, H.L., Bionaz, M., Jiang, D., Doupovec, B., Faas, J., Estill, C.T., Schatzmayr, D. and Duringer, J.M., 2021. Effects of deoxynivalenol and fumonisins fed in combination to beef cattle: Immunotoxicity and gene expression. Toxins 13: 714. https://doi.org/10.3390/toxins13100714.

Rodrigues, I. 2014. A review on the effects of mycotoxins in dairy ruminants. Animal Production Science 54: 1155–1165. https://doi.org/10.1071/AN13492.

Rodrigues, I., and Naehrer, K., 2012. A three-year survey on the worldwide occurrence of mycotoxins in feedstuffs and feed. Toxins 4: 663–675. https://doi.org/10.3390/toxins4090663.

Rodrigues, M.A.M., Cone, J.W., Ferreira, L.M.M., Blok, M.C. and Guedes, C.V.M., 2009. Relationship between in situ degradation kinetics and *in vitro* gas production fermentation using different mathematical models. Animal Feed Science and Technology 151: 86–96. https://doi.org/10.1016/j.anifeedsci.2008.12.004.

Schneweis, I., K. Meyer, S. Hörmansdorfer, and J. Bauer. 2001. Metabolites of *Monascus ruber* in silages. Journal of Animal Physiology and Animal Nutrition 85: 38–44. https://doi.org/10.1046/j.1439-0396.2001.00300.x.

Schneweis, I., Meyer, K., Hormansdorfer, S. and Bauer, J., 2000. Mycophenolic acid in silage. Applied Environmental Microbiology 66: 3639–3641. https://doi.org/10.1128/AEM.66.8.3639-3641.2000.

Schrenk, D., Bignami, M., Bodin, L., Chipman, J.K., del Mazo, J., Grasl-Kraupp, B., Hogstrand, C., Leblanc, J.C., Nebbia, C.S., Nielsen, E., Ntzani, E., Petersen, A., Sand, S., Schwerdtle, T., Vleminckx, C., Wallace, H., Gropp, J., Mulder, P., Oswald, I.P., Woutersen, R., Gómez Ruiz, J.Á., Rovesti, E. and Hoogenboom, L., 2024. Risks for animal health related to the presence of ergot alkaloids in feed. EFSA Journal 22: 8496. https://doi.org/10.2903/j.efsa.2024.8496.

Schumann, B., Lebzien, P., Ueberschär, K.H. and Dänicke, S., 2009. Effects of the level of feed intake and ergot contaminated concentrate on ergot alkaloid metabolism and carry over into milk. Molecular Nutrition and Food Research. 53: 931-938. https://doi.org/10.1002/MNFR.200800319.

Seeling, K., Dänicke, S., Ueberschär, H., Lebzien, P. and Flachowsky, G., 2005. On the effects of *Fusarium* toxin-contaminated wheat and the feed intake level on the metabolism and carry over of zearalenone in dairy cows. Food Additives and Contaminants: Part A 22: 847–855. https://doi.org/10.1080/02652030500163708.

Seeling, K., Dänicke, S., Valenta, H., van Egmond, H.P., Schothorst, R.C., Jekel, A.A., Lebzien, P., Schollenberger, M., Razzazi-Fazeli, E. and Flachowsky, G., 2006. Effects of *Fusarium* toxin-contaminated wheat and feed intake level on the biotransformation and carry-over of deoxynivalenol in dairy cows. Food Additives and Contaminants 23: 1008-1020. https://doi.org/10.1080/02652030600723245.

Silva, L.A., Mello, M.R.B., Piao, D.O., Silenciato, L.N., Quadros, T.C.O., Souza, A.H. and Barbero, R.P., 2021. Effects of experimental exposure to zearalenone on reproductive system morphometry, plasma oestrogen levels, and oocyte quality of beef heifer. Reproduction in Domestic Animals 56: 775-782. https://doi.org/10.1111/rda.13917.

Smith, J.F., DiMenna, M.E. and McGowan, L.T., 1990. Reproductive performance of Coopworth ewes following oral doses of zearalenone before and after mating. Journal of Reproduction and Fertility 89: 99–106. https://doi.org/10.1530/jrf.0.0890099.

Sørensen, J.L., Nielsen, K.F., Rasmussen, P.H., and Thrane, U., 2008. Development of a LC-MS/MS method for analysis of enniatins and beauvericin in whole fresh and ensiled maize. Journal of Agricultural and Food Chemistry 56: 10439-10443. https://doi.org/10.1021/jf802038b.

Spanghero, M., Braidot, M., Sarnataro, C., Fabro, C., Piani, B. and Gallo, A., 2023. *In vitro* aflatoxins recovery after changing buffer or protozoa concentrations in the rumen fermentation fluid. Journal of Animal Physiology and Animal Nutrition 107: 1311–1319. https://doi.org/10.1111/JPN.13818.

Speijers, G.J.A. and Speijers, M.H.M., 2004. Combined toxic effects of mycotoxins. Toxicology Letters 153: 91–98. https://doi.org/10.1016/j.toxlet.2004.04.046.

Spotti, M., Caloni, F., Fracchiolla, L., Pompa, G., Vigo, D. and Maffeo, G., 2001. Fumonisin B1 carry-over into milk in the isolated perfused bovine udder. Veterinary and Human Toxicology 43: 109–111.

Stoev, S.D., 2023. Foodborne diseases due to underestimated hazard of joint mycotoxin exposure at low levels and possible risk assessment. Toxins 15: 464. https://doi.org/10.3390/toxins15070464.

Storm, I.M.L.D., Kristensen, N.B., Raun, B.M.L., Smedsgaard, J. and Thrane, U., 2010. Dynamics in the microbiology of maize silage during whole-season storage. Journal of Applied Microbiology 109: 1017–1026. https://doi.org/10.1111/j.1365-2672.2010.04729.x.

Storm, I.M.L.D., Rasmussen, R.R. and Rasmussen, P.H., 2014. Occurrence of pre- and post-harvest mycotoxins and other secondary metabolites in Danish maize silage. Toxins 6: 2256–2269. https://doi.org/10.3390/toxins6082256.

Storm, I.M.L.D., Sørensen, J.L., Rasmussen, R.R., Nielsen, K.F. and Thrane, U., 2008. Mycotoxins in silage. Stewart Postharvest Reviews. 4: 1-12. https://doi.org/10.2212/spr.2008.6.4.

Sulzberger, S.A., Kalebich, C.C., Melnichenko, S. and Cardoso, F.C., 2016. Effects of clay after a grain challenge on milk composition and on ruminal, blood, and fecal pH in Holstein cows. Journal of Dairy Science 99: 8028–8040. https://doi.org/10.3168/jds.2016-11030.

Trenholm, H.L., Thompson, B.K., Martin, K.E., Greenhalgh, R. and McAllister, A.J., 1985. Ingestion of vomitoxin (Deoxynivalenol)-contaminated wheat by nonlactating dairy cows. Journal of Dairy Science 68: 1000–1005. https://doi.org/10.3168/jds.S0022-0302(85)80921-8.

Uhlig, S., Torp, M. and Heier, B.T., 2006. Beauvericin and enniatins A, A1, B and B1 in Norwegian grain: a survey. Food Chemistry 94: 193–201. https://doi.org/10.1016/j.foodchem.2004.11.004.

Upadhaya, S.D., Park, M.A. and Ha, J.K., 2010. Mycotoxins and their biotransformation in the rumen: a review. Asian-Australian Journal of Animal Science 23: 1250–1260. https://doi.org/10.5713/ajas.2010.r.06.

Veldman, A., Meijs, J.A.C., Borggreve, G.J. and Heeres-van der Tol, J.J., 1992. Carry-over of aflatoxin from cow's food to milk. Animal Science 55: 163. https://doi.org/10.1017/s0003356100037417.

Voss, K.A., Smith, G.W. and Haschek, W.M., 2007. Fumonisins: toxicokinetics, mechanism of action and toxicity. Animal Feed Science and Technology 137: 299–325. https://doi.org/10.1016/j.anifeedsci.2007.06.007.

Wang, J., Lin, L., Jiang, Q. and Huang, W., 2019. Effect of supplemental lactic acid bacteria on growth performance, glutathione turnover and aflatoxin B1 removal in lambs. Czech Journal in Animal Science 64: 272–278. https://doi.org/10.17221/5/2019-CJAS.

Weaver, G.A., Kurtz, H.J., Behrens, J.C., Robison, T.S., Seguin, B.E., Bates, F.Y. and Mirocha, C.J., 1986a. Effect of zearalenone on the fertility of virgin dairy heifers. American Journal of Veterinary Research 47: 1395–1397.

Weaver, G.A., Kurtz, H.J., Behrens, J.C., Robison, T.S., Seguin, B.E., Bates, F.Y. and Mirocha, C.J., 1986b. Effect of zearalenone on dairy cows. American Journal of Veterinary Research 47: 1826–1828.

Whitlow, L. and Hagler, W.M., 2005. Mycotoxins in dairy cattle: Occurrence, toxicity, prevention and treatment. Proceedings of the Southwest Nutrition Conference 124–138.

Wicklow, D.T., Rogers, K.D., Dowd, P.F. and Gloer, J.B., 2011. Bioactive metabolites from *Stenocarpella maydis*, a stalk and ear rot pathogen of maize. Fungal Biology 115: 133-142. https://doi.org/10.1016/j.funbio.2010.11.003.

Winkler, J., Kersten, S., Meyer, U., Engelhardt, U. and Dänicke, S., 2014. Residues of zearalenone (ZEN), deoxynivalenol (DON) and their metabolites in plasma of dairy cows fed *Fusarium* contaminated maize and their relationships to performance parameters. Food Chemistry and Toxicology 65: 196-204. https://doi.org/10.1016/j.fct.2013.12.020.

Yang, Y., Ferreira, G., Teets, C.L., Corl, B.A., Thomason, W.E., Griffey, C.A., Candy, L., Bauman, D.E., Sohn, M. and Hicks, K., 2017. Effects of feeding hull-less barley on production performance, milk fatty acid composition, and nutrient digestibility of lactating dairy cows. Journal of Dairy Science 83: 554-568 https://doi.org/10.3168/jds.2016-12241.

Zain, M.E., 2011. Impact of mycotoxins on humans and animals. Journal of Saudi Chemical Society 15: 129–144. https://doi.org/10.1016/j.jscs.2010.06.006.

Zinedine, A., Soriano, J.M., Moltó, J.C. and Mañes, J., 2007. Review on the toxicity, occurrence, metabolism, detoxification, regulations and intake of zearalenone: an oestrogenic mycotoxin. Food Chemistry and Toxicology. 45: 1-18. https://doi.org/10.1016/j.fct.2006.07.030.

CHAPTER 7

Effects of mycotoxins in fish

C. Pietsch

ProFishCare GmbH, Switzerland
pietsch@profishcare.com

1 Introduction

As shown in the previous chapters, the contamination of feed ingredients is unavoidable. The contamination of fish feeds has not been investigated in great detail, but evidence is increasing that a number of mycotoxins are frequently occurring in commercial fish feeds (Pietsch et al., 2013; Marijani et al., 2019; Mwihia et al., 2020; Koletsi et al., 2021). Similar to the differences in the global distribution of mycotoxins in feed ingredients, finished fish feeds also show differences in mycotoxin levels (Marijani et al., 2019; Koletsi et al., 2021). In addition, storage conditions can be unsuitable for fish feeds and typical storage-related mycotoxins such as AFs and ochratoxins can occur.

In summary, the mycoflora on crops, feed ingredients and finished fish feeds determines the set of mycotoxins and the levels that are transferred to farmed fish (Marijani et al., 2017; Pietsch, 2020). Accordingly, it has been observed that especially DON is frequently found in fish feeds in Europe (Koletsi et al., 2021). Most commonly, fish feed contains a mixture of different mycotoxins at the same time. Consequently, the high prevalence of mycotoxin contamination leads to frequent exposure of farmed fish to a variety of mycotoxins often with detrimental effects on the animals.

Aquaculture is increasing continuously and more dry feeds are needed every year, while at the same time fishmeal and fish oil are increasingly limited resources for feed production. Consequently, more plant-based ingredients are used for feed production, which increases the probability of mycotoxin transfer into the farmed fish. But also non-plant-based ingredients, such as insect larvae meals, have been proposed as promising alternatives for fishmeal. Accordingly, meal worms feeding on ZEN-contaminated materials did not show accumulation of the parent compound and indicated metabolisation of the mycotoxins (Niermans et al., 2019). For further details see also Chapter 10.

 | DOI: 10.1163/9789004712720_8

2 Occurrence of mycotoxins in fish diets

A high number of different mycotoxins are known so far, and our increasing knowledge about their metabolisation and degradation adds further to the number of different chemical substances that have to be taken into account if we would like to estimate the number of relevant mycotoxins fish are exposed to. Initial investigations on the mycotoxin exposure for fish (and also other vertebrates) concentrated on the most toxic ones as well as the ones that were more easily detectable. Further development of the extractions and detection methods allowed more detailed surveys to be conducted, which confirmed that fish are commonly exposed to a broad combination of mycotoxins (Koletsi et al., 2021). We consider the effects mycotoxins have on the animals to be of paramount importance.

Generally, exposure to mycotoxins may occur via two different routes: water and feed. However, water contamination does not usually account for high levels of exposure in natural water bodies, since, for example, in agricultural regions, DON and ZEN levels did not exceed the toxic levels for fish (Schenzel et al., 2010; Wettstein and Bucheli, 2010). In contrast, feed-borne exposure contributes a lot to the exposure of farmed fish to toxic mycotoxins and consequently only feed-related contaminations have been used for the calculations within this article. The information on mycotoxin occurrence in commercial fish feeds is increasing. For the feed commodities it is known that there are regional differences in the occurrence of mycotoxins which, for example, are shown by the typical contamination of cereals with DON in Europe and Northern America. However, the global trade in feed commodities leads to less pronounced regional mycotoxin occurrence patterns in commercial fish feeds.

Figure 7.1 gives an overview of the mycotoxin contamination of commercial fish feeds. It shows the potential mycotoxin contamination levels if each of the mycotoxins is present. However, the prevalence of each mycotoxin varies considerably, with DON often being present in 50 to 80 % of the feed samples. The frequency of occurrence for ZEN is more variable with detectable levels in commercial feed between 2.4 to 100 % (Pietsch et al., 2013; personal data of the author). Moreover, the local differences in the occurrence of AFB_1 in commercial fish feeds ranges from approximately 5 % in Europe to 55 % in a study from Brazil (personal data of the author; Nogueira et al., 2020). Consequently, the data used for Figure 7.1 would be more conclusive if the prevalence of each mycotoxin in fish feeds were known more precisely. For this, more detailed studies on the occurrence of mycotoxins in commercial fish feeds would be needed in the future.

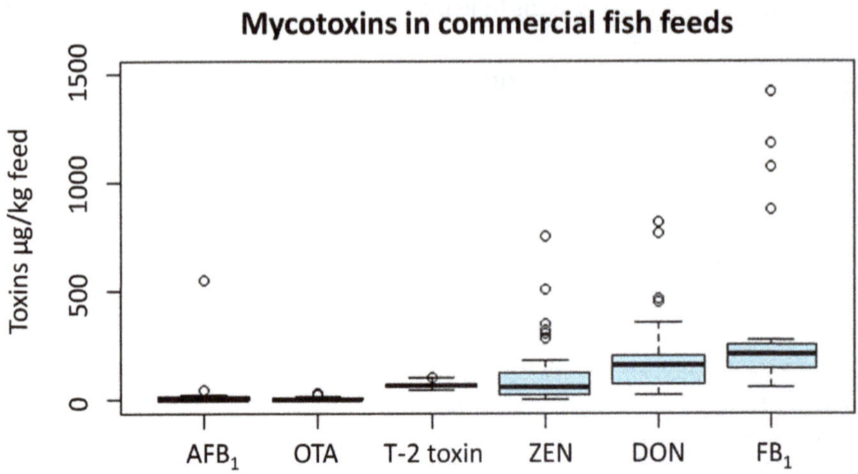

FIGURE 7.1 The contamination of commercial fish feeds with the most commonly investigated mycotoxins: AFB_1 with data for 25 feed samples, OTA with data for 25 feed samples, T-2 toxin and ZEN with data for 53 feed samples, DON with data for 50 feed samples, and FB_1 with data for 27 feed samples (outliers are indicated by the circles with one outlier for FB_1 not included).

3 Main mycotoxins affecting fish

In order to learn more about the consequences for fish, the first step would be to estimate the uptake and retention of mycotoxins in different fish species and in different parts of the fish. This requires conducting a number of controlled feeding studies. Such feeding trials have been performed for a limited number of mycotoxins. The following sections will review the current state of knowledge about these experiments and investigations. To increase the calculations that can be performed with the data from these studies, the models and predictions were based on data sets that have been derived by using the adjusted bootstrap percentile (BCa) method. On the one hand, the models aimed at gaining insight into the definition of mycotoxin concentrations which may pose a risk to the fish. These concentrations have been derived from a comparison of the lowest feed contamination levels that showed an effect on the fish (LOAEL) and the concentration below that, which showed no more detectable effects on the animals (NOAEL). Where possible, the prediction of hazardous concentrations for 5 % of a fish population (CC5) was also used, according to the methods published previously (Pietsch, 2020). On the other hand, the calculations also take into account that acute toxicity may be relevant if the fish are exposed to high mycotoxin concentrations. More relevant may be the chronic exposure to mycotoxins; fish in aquaculture are commonly farmed for several months and may be exposed to mycotoxins for quite a long time, but experimental studies rarely take this into account (Woźny et al., 2019, 2020).

3.1 *Deoxynivalenol*

Due to its common occurrence in feed ingredients and feeds at considerable concentrations, fish frequently have to deal also with DON. Hence, DON has been investigated in fish feeding studies more often compared with most of the other mycotoxins. In total 65 data points from the different feeding studies have been compiled and used to estimate the lowest DON level in feed (LOAEL) that would potentially lead to a negative effect on the fish. The mean LOAEL for DON is 2,845 µg/kg feed (with 95 % confidence intervals of 2,311 and 3,532 µg/kg feed and a standard error of 308 derived from the bootstrap method). But this value includes a number of different biological endpoints in different fish species. Consequently, it makes sense to have a look at the endpoints individually.

For effects on fish behaviour and body malformation only one study was found for each of these endpoints (Šišperová et al., 2015; Huang et al., 2018), hence these endpoints cannot be reviewed in detail here. In contrast, most feeding studies with DON have documented FI and growth performance and a mean LOAEL for effects on fish growth. A level of 3,617 µg/kg feed has been observed, which is a lower value than that currently recommended value by the EU (Commission Recommendation 576/2006/EC; European Commission, 2006) as indicated by the orange line in Figure 7.2.

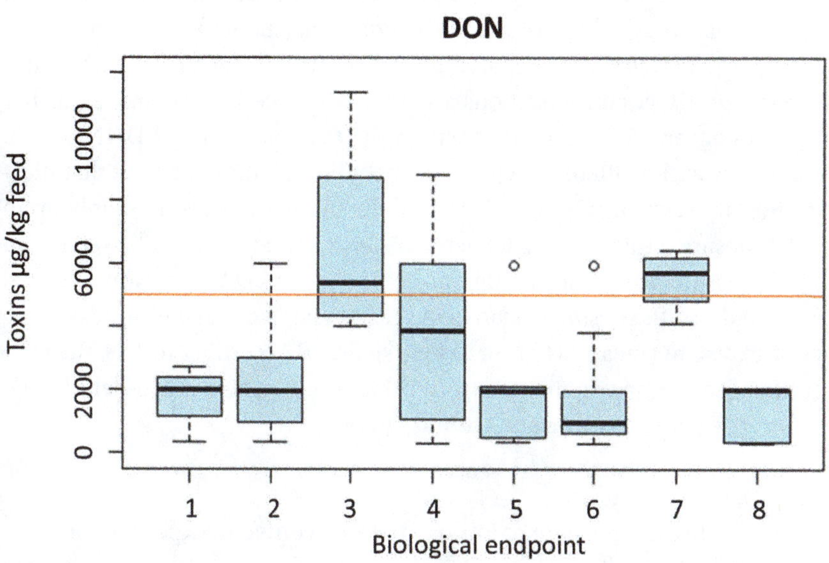

FIGURE 7.2 This graph represents the lowest DON concentrations (LOAEL) in fish that show negative effects on different endpoints, with 3 data points showing effects on *1: biotransformation*, 7 data points collected for the endpoint *2: blood*, 4 data points collected for the endpoint *3: body composition*, 20 data points showing effects on *4: growth* performance, 10 data points showing changes in *5: histology*, 9 data points from studies investigating *6: immune system*, 3 data points showing increased *7: mortality*, and 6 data points for the endpoint *8: oxidative stress*. The orange line indicates the recommendation level for feedstuffs set by the European Commission (2006). Outliers are indicated by the circles.

As can be seen from Figure 7.2, DON is also one of the rare mycotoxins for which a relationship between the FI and reduction in growth has been shown (Gonçalves et al., 2018a,b, 2019). The difference between naturally occurring or artificially added DON was further investigated by Koletsi et al. (2021), indicating that purified DON is tendentially less toxic than naturally contaminated fish feed. This is thought to be due to the mixture of mycotoxins that occur in naturally contaminated feeds which contribute to the resulting toxicity in the fish. Most importantly, as indicated by the orange line in Figure 7.2, a number of endpoints are sensitive to DON even at lower levels than recommended as the maximum DON level in feedstuffs (Commission Recommendation 576/2006/EC; European Commission, 2006).

A number of studies on DON have tried to reveal the mechanisms that lead to decreased growth performance in fish and it has been observed that DON also affects FI and appetite (Koletsi et al., 2021). The reason for this might be related to DON effects on the brain serotonin pathways as has been observed in other vertebrates (Shen et al., 2021). In contrast to growth performance, oxidative stress and immunity appear to be sensitive endpoints for DON exposure with LOAELs of 1,218 and 1,684 µg/kg feed, which is in closer proximity to the potential levels that can be expected in commercial fish feeds as shown in Figure 7.1. A recent study also shows that effects on isolated carp leukocytes involve oxidative stress and subsequent cell death (Ding et al., 2021), which supports the effects on the leukocyte function of DON-treated carp *in vivo* (Pietsch et al., 2014a). The study of Szabó et al. (2021) also showed that high DON concentrations are able to induce DNA damage, further adding to our concerns over underestimating the toxic potential of DON in fish. However, looking at the different DON feeding studies it is quite obvious that most of the studies investigating DON toxicity in different fish species used only up to 60 days of exposure. Studies using longer exposures are rare. As can be seen from Figure 7.3 there is no pronounced difference in the mean LOAEL between studies using up to 60 days of exposure (mean LOAEL 2,749 µg/kg feed) and studies using >60 days of exposure (mean LOAEL 1,835 µg/kg feed). According to this, the 67% difference between these two means is lower than the expected one order of magnitude difference between acute and chronic studies.

3.2 *T-2 toxin*

For effects of T-2 toxin some data points have been recorded in experimental feeding studies and these are displayed in Figure 7.4. The effects on histology appear to contain quite some uncertainty since only two studies could be obtained for this. In contrast, the remaining effects occur at levels of less than 5 mg/kg. What is raising questions about defining safe toxin levels in the future is the observation that a concentration of 4.11 mg/kg T-2 toxin already induced DNA damage in carp. In addition, especially the endpoint blood appeared to be sensitive even to comparable low T-2 levels (Szabó et al., 2021). Taking blood samples in future T-2 feeding studies might therefore be a promising approach to detecting effects on the fish, and the analysis of blood parameters is also less invasive than organ dissection.

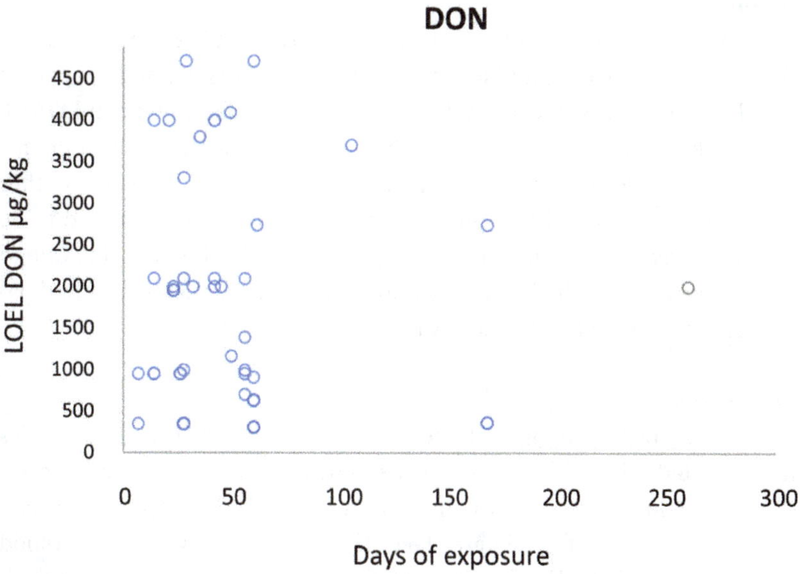

FIGURE 7.3 The knowledge gap between duration of exposure and the lowest DON concentrations (LOAEL) that show negative effects in the fish.

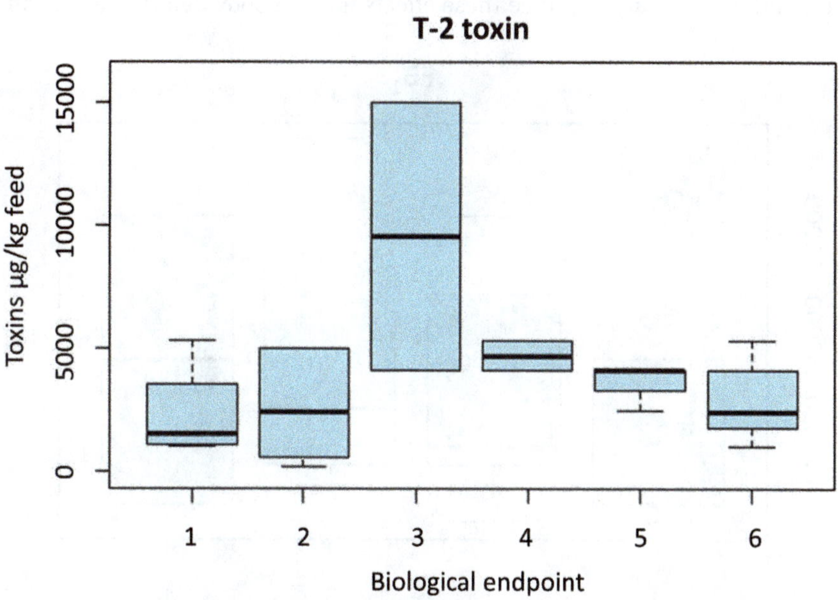

FIGURE 7.4 The lowest sublethal T-2 toxin concentrations (LOAEL) that show negative effects on different endpoints in the fish with 4 data points collected for the endpoint *1: blood*, 5 data points collected for the endpoint *2: growth*, 2 data points showing changes in *3: histology*, 2 data points from studies investigating *4: immune system*, 3 data points showing increased *5: mortality*, and 5 references for the endpoint oxidative stress *6: oxidative stress*.

3.3 *Fumonisins*

FUM are a group of mycotoxins with FB_1 being the most prominent member. The effects of FB_1 have been investigated less frequently and in less detail. FB_1 appears to be toxic only at higher toxin levels in feeds. In line with that, the mean LOAEL for all 42,482 mg/kg feed was determined by bootstrapping (with 95% confidence values of 26,550 to 74,489 mg/kg and a standard error of 10,706). Similar to other vertebrates (Chen et al., 2021), FUM and especially FB_1 have been reported to be immunotoxic and organo-toxic in fish (Figure 7.5). Especially, the mortality due to FB_1 exposure occurs at rather high concentrations. In contrast, effects on blood parameters and growth occur at considerably lower FB_1 concentrations.

3.4 *Zearalenone*

Although ZEN often occurs along with DON, considerably fewer feeding studies have been conducted with ZEN. Thus, fewer biological endpoints have been investigated, with the immune system being the most sensitive endpoint for ZEN effects with a mean LOAEL of 587.8 µg/kg feed. This is followed by effects on blood parameters with a mean LOAEL of 768.7 µg/kg feed (Figure 7.6). Effects on growth and the endocrine system appear at higher ZEN concentrations. However, effects that have been less frequently investigated raise even more concerns about ZEN contamination in aquafeeds, since these effects include body deformations, dam-

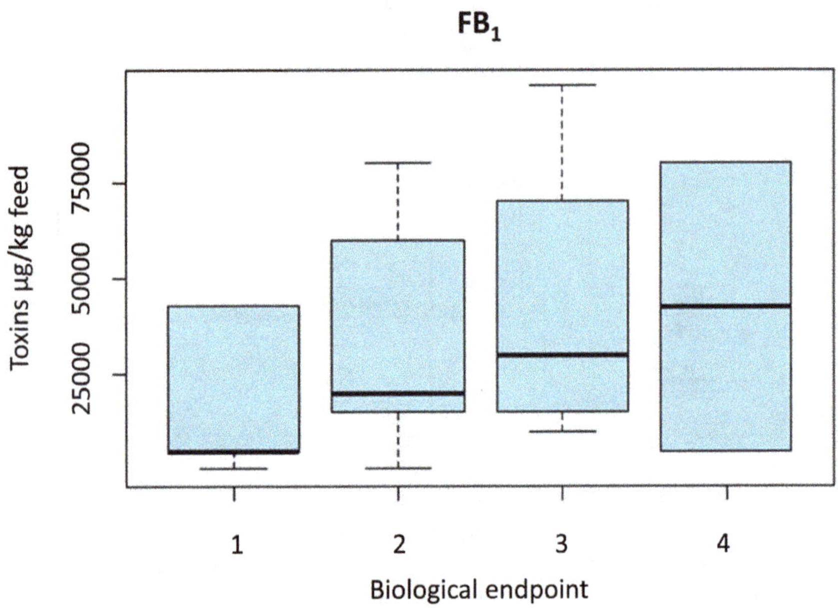

FIGURE 7.5 The lowest sublethal FB_1 concentrations (LOAEL) that show negative effects on different endpoints in the fish with 5 data points collected for the endpoint *1: blood*, 12 data points collected for the endpoint *2: growth*, 4 data points showing effects on *3: histology performance*, and 2 data points showing changes in *4: immune system*; outliers are indicated by circles.

ages in the intestine related to increased oxidative stress and damage to the DNA (Wang et al., 2019). The latter is especially concerning, since the mycotoxin ZEN is listed as a pro-carcinogenic substance at this time (Ouanes et al., 2003), but may also have other serious consequences in farm animals. The majority of these effects occurred at concentrations lower than the level still currently recommended by the European Union (Commission Recommendation 576/2006/EC; European Commission, 2006) as indicated by the orange line in Figure 7.6.

Exact estimations of chronic effects of dietary ZEN exposure cannot be derived from the currently available data. The studies that showed effects of ZEN at concentrations up to 1,041 µg/kg feed for 28 to 70 days included effects on survival, growth and immune parameters (Pietsch et al., 2015a,b; Pietsch and Junge, 2016; Abdel-Tawwab, et al., 2020; Ghafarifarsani et al., 2021a,b; Wang et al., 2019; Wu et al., 2021), whereas the study of Woźny et al. (2019) exposed fish for 72 weeks to more than 2 mg/kg ZEN in the feeds (Figure 7.7). If we assume that exposure to higher (but not extreme) ZEN concentrations realistically occurs for 1 -2 months until a bad feed batch is used up and replaced by new, potentially less contaminated feeds, the ZEN concentrations that may be problematic for the fish are > 439 µg/kg (with 95% confidence values of 359 and 544 µg/kg and a standard error of 46.5 derived

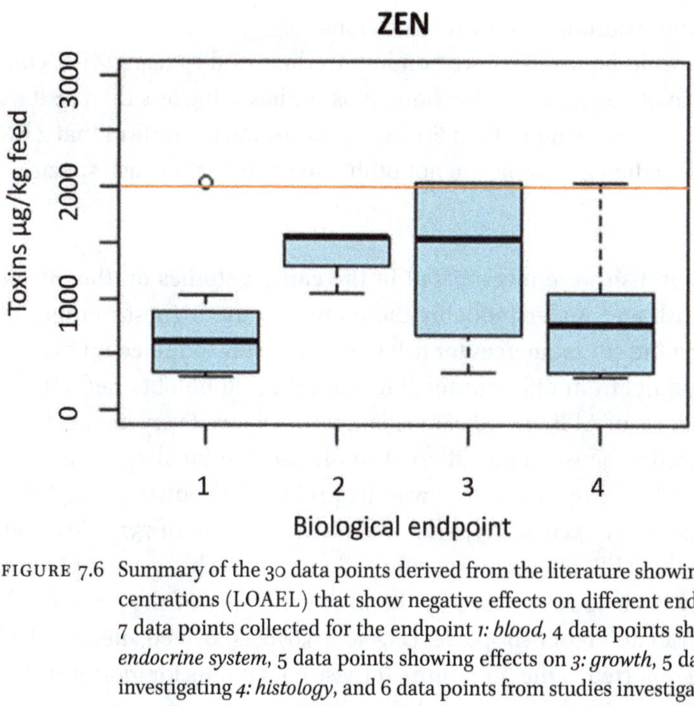

FIGURE 7.6 Summary of the 30 data points derived from the literature showing the lowest ZEN concentrations (LOAEL) that show negative effects on different endpoints in the fish with 7 data points collected for the endpoint *1: blood*, 4 data points showing effects on the *2: endocrine system*, 5 data points showing effects on *3: growth*, 5 data points from studies investigating *4: histology*, and 6 data points from studies investigating the *5: immune system*. The orange line indicates the recommended guidance level for finished feedstuff (EU, 2006); outliers are indicated by the circles.

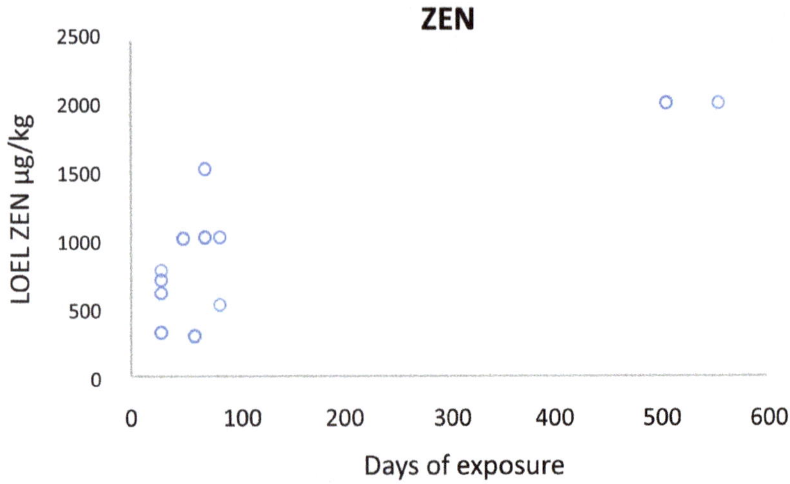

FIGURE 7.7 Each circle represents a study reporting the lowest ZEN concentrations (LOAEL) that show negative effects on the fish. This graph shows the knowledge gap between duration of exposure and effects in different fish species.

from the bootstrap method). In this respect, the observed effect of ZEN on the oxygen demand of common carp treated with 621 µg/kg ZEN for 4 weeks (Pietsch and Junge, 2016) may be a problem if fish were exposed even longer than that and had to deal with ZEN under chronic exposure conditions.

In the future it would be necessary to conduct feeding studies using ZEN concentrations of less than 500 µg/kg feed in both, plus studies using less than 28 days of exposure and studies using more than 60 days of exposure, to confirm that ZEN concentrations of less than 454 µg/kg are not problematic for farmed fish species.

3.5 *Aflatoxins*

The effects of AFs on fish were investigated in the earliest studies on the effects of mycotoxins on fish and understandably these are also the highest number of studies investigating the consequences for fish that have been compiled up to now. In total, 208 data points from different feeding studies could be obtained for the evaluation of the effects of AFB_1 on fish. Some endpoints showed very high LOAEL values and high variation between the different studies, so that for all the effects together a mean LOAEL of 2,392 µg/kg feed was determined by bootstrapping (with 95% confidence values of 1,374 to 4,833 µg/kg and a standard error of 757). However, if we take a look at the AFB_1 effects on specific endpoints this changes the picture. The high variability is obviously caused by the high variability of the AFB_1 levels that cause mortality in different fish species (14,550 ± 9,676 µg/kg feed, mean ± SEM from 14 data points reported in the literature). If these data points for mortality due to AFB_1 exposure are omitted from the calculations, a number of more sensitive endpoints can be identified (Figure 7.8).

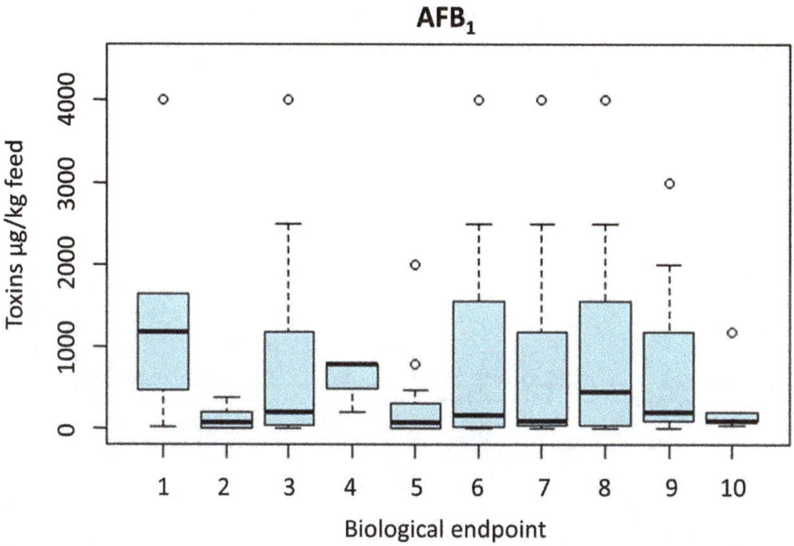

FIGURE 7.8 The lowest sublethal AFB_1 concentrations (LOAEL) that show negative effects on different endpoints in the fish can be seen in 5 studies for the endpoint: *1: behaviour*, 10 data points showing effects on *2: biotransformation*, 24 data points collected for the endpoint *3: blood*, 3 data points showing alteration of *4: body colouration*, 15 data points collected for the endpoint *5: body composition*, 19 data points reported for *6: cancer* development, 61 data points showing effects on *7: growth* performance, (and one outlier removed from the graph) and 27 data points showing changes in the *8: histology*, 22 data points from studies investigating *9: immune system* and 5 references for the endpoint *10: oxidative stress*; outliers are indicated by the circles.

For the endpoints shown in Figure 7.8, a mean LOAEL of 2,137 µg/kg feed was determined by bootstrapping (with 95 % confidence values of 979 to 4,980 µg/kg and a standard error of 821, which still indicates some variation in the data). The differentiation of distinct endpoints in the fish is thus very useful. Figure 7.8 reveals that oxidative stress measured in the intestine or in the liver belongs to the most sensitive markers of AFB_1 exposure in fish, together with the measurement of detoxifying enzymes such as CYP in liver tissue. AFB_1 is known to cause cancer in vertebrates (Li and Liu, 2019), which is also why most of the early investigations of AFB_1 in fish addressed carcinogenic effects. Taking all studies on carcinogenic effects in fish together produces a mean LOAEL of 1,448 µg/kg feed, but the studies still indicate varying outcomes. One potential reason for variation in the data sets might be the presence of species-dependent differences in the reaction to AFB_1 intoxication. This assumption is supported by Figure 7.9. Although the majority of the studies on AFB_1 concern tilapia species, Figure 7.9 still indicates that other species may be more suitable for toxicity studies. For example, rainbow trout shows a high sensitivity to AFB_1 in a number of experimental feeding studies.

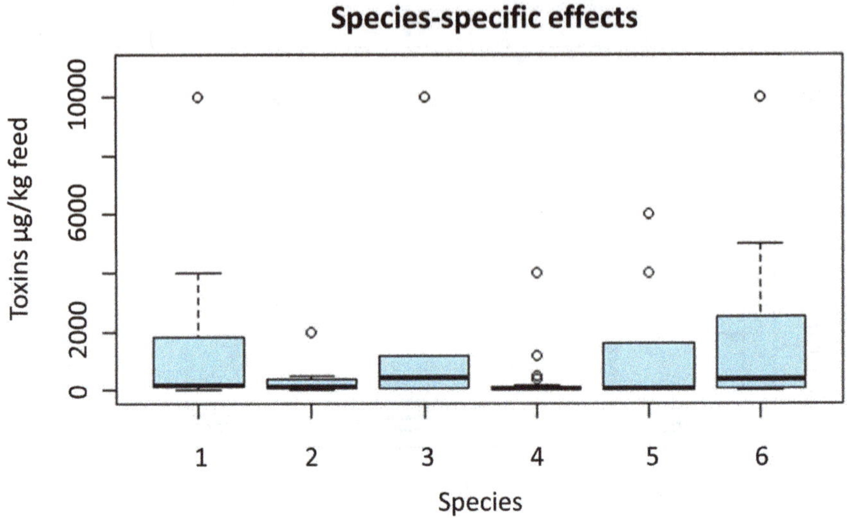

FIGURE 7.9 In total, 166 data points for the lowest sublethal AFB_1 concentrations (LOAEL) with negative impact on different fish species were retrieved from the literature, all showing negative effects on *1: tilapia species* with 71 data points, 7 data points collected for the endpoint *2: seabass and red drum* (*Sciaenops ocellatus*), 27 data points showing effects on *3: catfish species*, 24 data points showing changes in *4: rainbow trout*, 28 data points from studies investigating effects on *5: carp species*, including *Cyprinus*, *Catla* and *Labeo* species, and 9 data points showing effects on *6: different fish species*, such as beluga, medaka, guppy, matrinxã fish, walleye and pearlspot; outliers are indicated by the circles.

With respect to the individual endpoint that can be affected by AFB_1 exposure, changes in the body composition appear to commonly occur after feeding AFB_1. However, these markers are invasive markers that require killing fish before analysis. In comparison blood sampling is less invasive, but this endpoint is unfortunately also less sensitive in AFB_1 feeding studies. What makes AFB_1 very dangerous is not only its carcinogenic potential, but also the ability to cross the blood-brain barrier and thereby impair fish behaviour (Baldissera et al., 2018). Although behaviour may be a parameter that can be assessed often without disturbing the fish, the literature up to now does not provide much information on the behavioural effects of AFB_1. Similarly, a suitable non-invasive marker could also be the assessment of skin colouration. But changes in body colouration have been investigated in tilapia species in only two different studies (Deng et al., 2010; Hegazi et al., 2013). In addition, only one study reported increased levels of the stress hormone in the blood circulation after ABF_1 exposure. Hence, it would be advantageous for a better endpoint sensitivity estimation if more feeding studies were to be performed in the future, investigating a broader variety of different biological endpoints. The relationship between negative effects on the fish and the duration of the experiments is shown in Figure 7.10, revealing no clear correlation between the exposure time to AFB_1 and the concentrations that are expected to result in negative effects in the fish. Consequently, even small amounts of AFB_1 in fish feeds should be avoided since safe AFB_1 levels cannot precisely be predicted.

This is also supported by the estimation of CC5 as shown in Figure 7.11. The median CC5 for AFB_1 intoxication was calculated to be 1.48 mg/kg with 95 % confidence intervals of 0.60 and 3.06 mg/kg. These values are located in close proximity to the limit of 2 mg/kg currently set for human food consumption by the European Union (Van Egmont & Jonker, 2004), so consequently a risk for at least 5 % of a fish population can be assumed under realistic conditions.

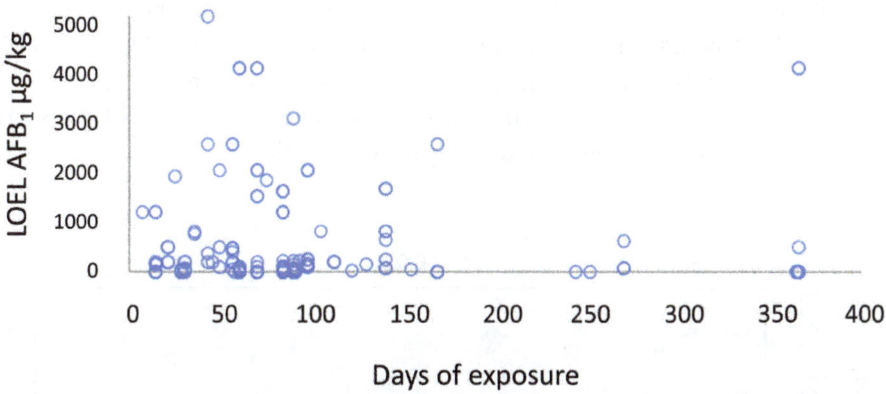

FIGURE 7.10 The relationship between duration of exposure and the lowest AFB_1 concentrations (LOAEL) that show negative effects in fish.

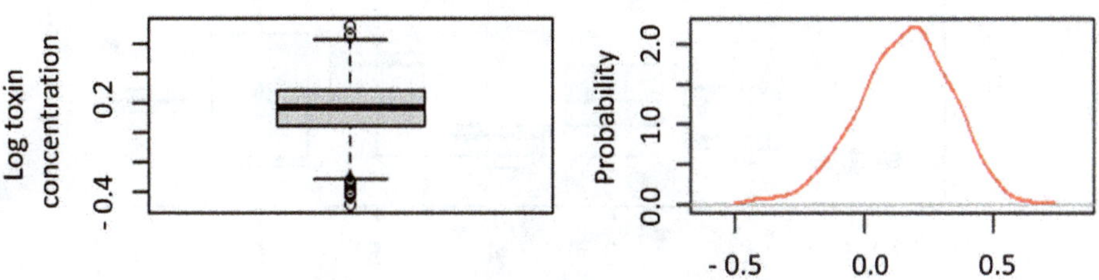

FIGURE 7.11 The predicted hazard concentrations for 5 % of a fish population (CC5) on the left side, and on the right side the density plot of the predicted CC5 concentrations for AFB_1 intoxication in fish.

3.6 *Ochratoxin A*

From the group of ochratoxins, only OTA toxicity has been investigated in fish in some detail. In total 42 data points could be derived from feeding studies using OTA contaminated fish feeds. For all endpoints, a mean LOAEL of 1,395 µg/kg feed was determined by bootstrapping (with 95% confidence values of 988 to 2,072 µg/kg and a standard error of 262). Only two of these studies investigated behaviour and enzyme activities in muscle of fish after OTA exposure (Shalaby, 2004; El-Sayed et al., 2009) and have therefore not been used in the calculations for Figure 7.12.

Hepatotoxicity of OTA is thought to compromise the detoxification capacity and antioxidant pathways, which further affects energy metabolism (Baldissera et al., 2020). Similarly, OTA exposure was reported to increase ROS production and cellular damages in the intestine of grass carp (Liu et al., 2021). The disruption of the integrity of the intestinal structure was also assumed to lead to the known negative effects of OTA on growth performance (Liu et al., 2021). The mean LOAEL for effects on growth was 2,191 ± 720 µg/kg (mean ± SEM). A promising result is the low mean LOAEL for changes that can be observed in blood (mean LOAEL 471 ± 104 µg/kg) since blood sampling is a minimally invasive method.

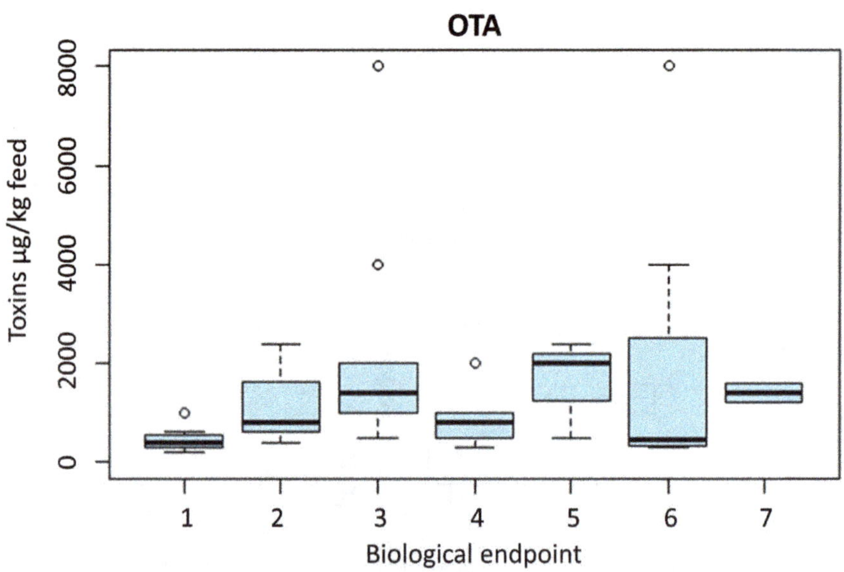

FIGURE 7.12 In total 42 data points for the lowest sublethal OTA concentrations (LOAEL) with negative impact on fish were retrieved from the literature, showing negative effects on different endpoints in the fish with 7 data points collected for the endpoint *1: blood*, 5 data points collected for the endpoint *2: body composition*, 10 data points showing effects on *3: growth* performance, 5 data points showing changes in *4: histology*, 3 data points from studies investigating *5: immune system*, 8 data points showing increased *6: mortality*, and 2 references for the endpoint oxidative stress *7: oxidative stress*; outliers are indicated by circles.

3.7 *Other mycotoxins*

So-called emerging mycotoxins commonly occur in feed ingredients and fish feeds, but only recently has research focused on their importance (Tolosa et al., 2014). One of the mycotoxins that has not received much attention is moniliformin (MON) which is a mycotoxin that is thought to mainly exert its effects via disruption of the conversion of pyruvate to acetyl-CoA in cells with profound consequences for the body's metabolism. With respect to its toxicity in fish, not many feeding studies have been performed within the last 20 years. It follows that the estimation of its toxicity probably contains high uncertainties. Accordingly, the endpoint blood shows a mean LOAEL of 46,667 µg/kg, but only three data points have been compiled for this endpoint. Similarly, for the endpoint growth a mean LOAEL of 71,429 µg/kg was calculated from 5 data points. Last, but not least, for the endpoint histology only the study of Yildrim et al. (2000) showed hepatocellular changes after the fish had been fed 60 mg/kg MON for 10 weeks.

Another group of mycotoxins that are currently getting more attention are the ENNs, which show a number of effects in farm animals including cytotoxicity (Křížová et al., 2021). For some time, it was assumed that the high levels of ENNs in raw feed materials are degraded sufficiently during the feed processing. In addition, the ENNs concentrations are rather low in fish feeds produced by extrusion. Nevertheless, ENNs and especially ENNA, ENNA1, ENNB, and ENNB1 are frequently found at low concentrations in fish feeds, which raises the question of their toxicity. According to Nácher-Mestre et al. (2020), ENNB in fish feed at average concentrations of 19.9 µg/kg is comparable to the ENNB level found in a commercial trout feed (up to 18.5 µg/kg) in the study of Pietsch (2020). Other studies confirmed that ENNB, ENNB1, and ENNA1 can also be present in fish samples (Tolosa et al., 2017). However, feeding trials with relevant ENNs concentrations are lacking and should be conducted in the future.

Instead, water-borne exposure with zebrafish embryos has been performed by Pietsch (with up to now unpublished results) according to the methods described in the study of Tschirren et al. (2018). These experiments show that high ENNA concentrations can rapidly impair the normal development of the fish embryos. Accordingly, a concentration that impaired 50 % of the development (EC_{50}) of the embryos after 24 h of exposure was found at 983 ng/ml ENNA (Figure 7.13). The EC_{50} for the occurrence of detrimental effects on the embryos after 48 h of exposure to ENNA was found to be higher with 1,496 ng/ml and especially embryos exposed to 2,000 ng/ml did not show any further development compared with the status observed at 24 h of exposure (Figure 7.14). In addition, the developmental impairment of the zebrafish embryos was also obvious after 72 h of exposure to 1,000 ng/ml, showing underdevelopment of organs and curvature of the spine (Figure 7.14). In contrast, exposure to 1,000 ng/ml ENNB did not significantly impair the normal development of the zebrafish, for example after 72 h of exposure (Figure 7.14F).

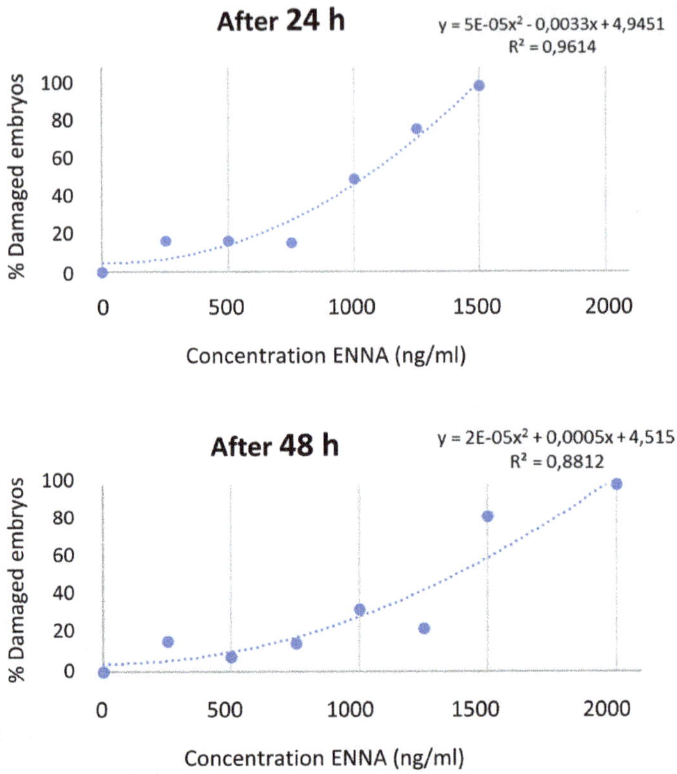

FIGURE 7.13 The concentration-dependent effects of water-borne exposure of zebrafish embryos to ENNA for 24 h and 48 h, n=12 for each toxin concentration. Damaged embryos either showed retarded development, oedema or were dead.

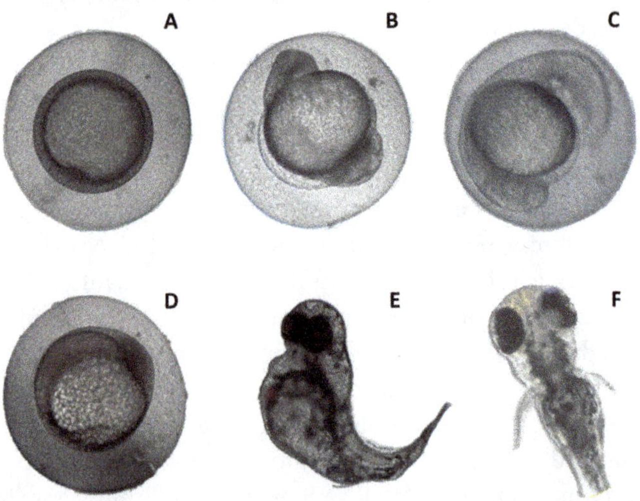

FIGURE 7.14 The effects on zebrafish embryos of water-borne exposure to A: 2,000 ng/ml ENNA for 24h, B: 1,000 ng/ml ENNA for 24 h, C: the solvent ethanol at a concentration of 0.01% for 24 h, D: 2,000 ng/ml ENNA for 48 h, E: 1,000 ng/ml ENNA for 72 h, F: 1,000 ng/ml ENNB for 72 h.

Exact staging of the development of each embryo revealed that already after 48 h of exposure, the ENNA treated embryos were lagging behind with respect to their development, especially at concentrations of 1,000 ng/ml or higher (Figure 7.15).

BEA is also an emerging mycotoxin that is mainly formed by *Fusarium* species and less frequently by *Beauveria* or *Isaria* species (Hamill, 1969, Luangsaard et al., 2009). Intestinal effects in non-fish models have been reviewed by Bertero et al. (2020) and cytotoxic effects have frequently been observed. According to García-Herranz et al. (2019) the cellular targets and the endocrine regulating potential of BEA appear to be a bit different for two fish cell lines derived respectively from a cold-water species and a warm-water species. This confirms that the effects of BEA and also other mycotoxins should in the best case be investigated in a number of fish species for comparison. So far only Nácher-Mestre et al. (2020) have reported average BEA levels of 30 µg/kg in fish feed, which shows that our knowledge on the prevalence and importance of this mycotoxin is limited.

3.8 *Multi-mycotoxin exposure*

The effects of natural and artificial contamination of fish are different (Koletsi et al., 2021) probably due to the fact that natural contamination of feed ingredients rarely includes only one single mycotoxin. The set of mycotoxins that may occur in fish feeds is determined by the fungi present in the feeds and by environmental conditions such as temperature and humidity during feed storage (Marijani et al., 2019; Pietsch et al., 2020). DON frequently co-occurs with its acetylated forms as well as with ZEN (Koletsi et al., 2021), which is also supported by our previous study (Pietsch et al., 2013). But the co-occurrence of mycotoxins also strongly depends on the geographical region. This determines the mycoflora on crops and the environmental parameters for mycotoxin formation. Consequently, the potential mixtures of mycotoxins are diverse, but scientific studies so far have often concentrated on

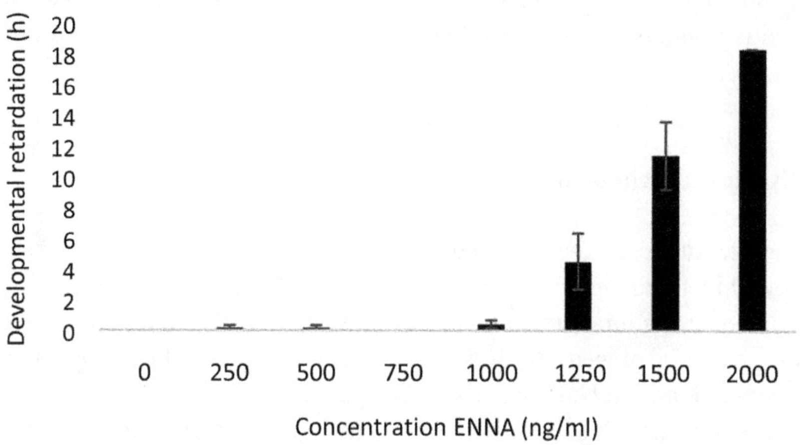

FIGURE 7.15 The concentration-dependent effects of water-borne ENNA exposure on the developmental stage of zebrafish embryos after 48h, n=12 for each toxin concentration; exact staging of the embryos has been performed according to Braunbeck and Lammer (2006).

the combination of only two mycotoxins. For example, the effects of AFB_1 and FB_1 on mosquitofish, *Gambusia affinis,* have been investigated by McKean et al. (2006) and strong additive modes of action have been observed similar to those in rats. Similarly, FB_1 added to AFB_1 also increased the detrimental effects in Nile tilapia (Pumipuntu et al., 2013). In some studies, AFB_1 has also been combined with other mycotoxins. For example, mixtures of AFB_1 and ZEN were fed to trout by Ghafari-farsani et al. (2021a,b) resulting in synergistic effects on the fish. The effects of AFB_1 and DON in combination have been investigated *in vitro* using carp hepatocytes (He et al., 2010). The results of this study showed that additive effects have to be assumed for the combination of AFB_1 and DON.

Most studies on other farm animals investigating the combined effects of FB_1 and MON assumed an interactive toxicity of these two mycotoxins, although the exact type of interaction still needs to be elucidated (Fremy et al., 2019). In addition, the combined adverse effects exerted by FB_1 and MON have not so far been proven to be synergistic. Similarly, an interaction of FB_1 and MON in channel catfish was also not indicated by Yildrim et al. (2000). Also turkeys give no clear evidence of synergistic mechanisms of action of these two mycotoxins (Fremy et al., 2019). In contrast, chickens appear to show additive effects of MON and DON. Although DON is one of the most frequently occurring mycotoxins, effects on fish of DON in combination with other mycotoxins still need to be investigated. Interestingly, a study on the effects of DON and ZEN together in humans indicates that early exposure is characterised by synergistic effects whereas later also antagonistic interactions can be observed (Thapa et al., 2021). This shows how much more still needs to be investigated in different fish species until we can understand the modes of action of each mycotoxin and their potential interactions with each other.

In summary, the effects from mycotoxin mixtures frequently occur and need to be considered. Often synergistic or additive effects of mycotoxins can be assumed. But since commercial fish feeds commonly contain two or more mycotoxins, the effects resulting from mycotoxin combinations still need more attention in future studies.

3.9 Mycotoxin decontamination

Labour-intensive aflatoxin decontamination procedures for feed ingredients have been reviewed by Nazhand et al. (2020). However, feed additives are more often used to counteract the impairment by mycotoxins. Most studies investigating the ameliorative potential of feed additives on the impairment of fish health by mycotoxin contamination used Nile tilapia and AFB_1-contaminated diets. In addition, the most commonly used feed additive is clay particles. One example for this is the study of Abdelaziz et al. (2010) showing that clay prevented the effects on growth performance of Nile tilapia caused by a diet containing AFB_1 and OTA. But adding adsorbing compounds to feeds does not reduce the toxic potential of DON and ZEN. Hence, it is useful to differentiate mycotoxin decontaminants from different modes of action in order to prevent the negative effects of mycotoxins with different chemical structures and modes of action in the fish. See also chapter 16.

Several decontaminants were, for example, investigated by Diab et al. (2018) and are included in Figures 7.16 to 7.18, summarising the ameliorative potential of different feed additives. Figure 7.16 shows the potential improvement with respect to the feed conversion of fish, calculated as the FCR, i.e., the amount of feed that is needed to let the fish grow by a certain amount. If less feed is used to achieve comparable growth of the fish, this is advantageous for the farmers.

The studies used for calculating the numbers for Figure 7.16 lasted between 28 and 120 days, with a mean duration of the trials of 77 days. Most of the studies that have been compiled used AFB_1 (57.3% of the data points) as the mycotoxin and tilapia as the test subject (77% of the data points). In addition to clay and bentonite, nanoparticles of hydrated sodium aluminium silicates and the potential of copper oxide to ameliorate the effects of the mycotoxins have also been investigated (Fadl et al., 2020). Copper oxide and other chemical compounds have been included as chemicals in Figure 7.16 if these have been applied to the mycotoxin contaminated feeds individually and have been used as separate treatments in the feeding trials. From Figure 7.16 it can also be seen that the clay particles, chemicals and com-

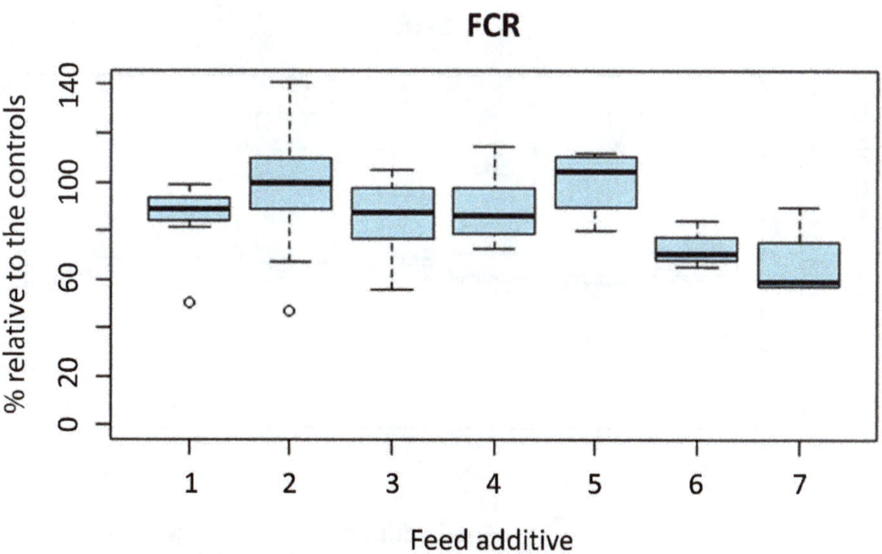

FIGURE 7.16 Feed conversion ratios (FCR) calculated from feeding studies using mycotoxin contaminated feeds and the same feeds with the addition of feed additives, with 10 data points collected for the endpoint *1: chemicals*, 20 data points collected for *2: clay* as a feed additive, 30 data points for the performance of *3: commercial mixtures*, 9 data points from studies investigating *4: herbs, including herbal extracts*, 4 data points showing the advantageous effects of *5: experimental mixtures*, 3 data points showing increased *6: prebiotics*, and 3 references for *7: probiotics*; outliers are indicated by circles.

mercial mixtures did not influence the FCR of the fish profoundly. Commercial mycotoxin effect-reducing mixtures often contain a combination of clay particles, probiotics and herbal extracts. However, probiotics and prebiotics alone showed the best result on the FCR values in the different mycotoxin feeding studies.

Consequently, it is not surprising that the use of microorganisms has attracted some attention. Accordingly, Bata-Vidács et al. (2020) investigated the ability of different bacterial strains to detoxify AFB_1 and sterigmatocystin with promising results for the use of *Pediococcus* strains. Another study using *Bacillus* ssp. as a detoxifying component against aflatoxicosis showed improvement in the growth performance of Yellow river carp (*Cyprinus carpio haematopterus* Temminck et Schlegel; Fan et al., 2018). But also microorganisms derived from the rainbow trout intestine may show AFB_1-binding capacity (Martinez et al., 2018). In the same studies that had been used for the FCR calculations, the specific growth rates (SGR) of the fish (expressed as percentage growth per day) have been reported in most cases, and the effect of the mycotoxin decontaminants on the SGR are shown in Figure 7.17.

An increased SGR in a feeding trial would be preferred; however, this could not be achieved by using clay particles as mycotoxin binders. While clay particles are used to avoid uptake of mycotoxins in the fish intestine, herbal extracts should reduce the negative effects of the toxic substances after the uptake into the body. But also the effects of the herbs are different. For example, basil was the best detoxify-

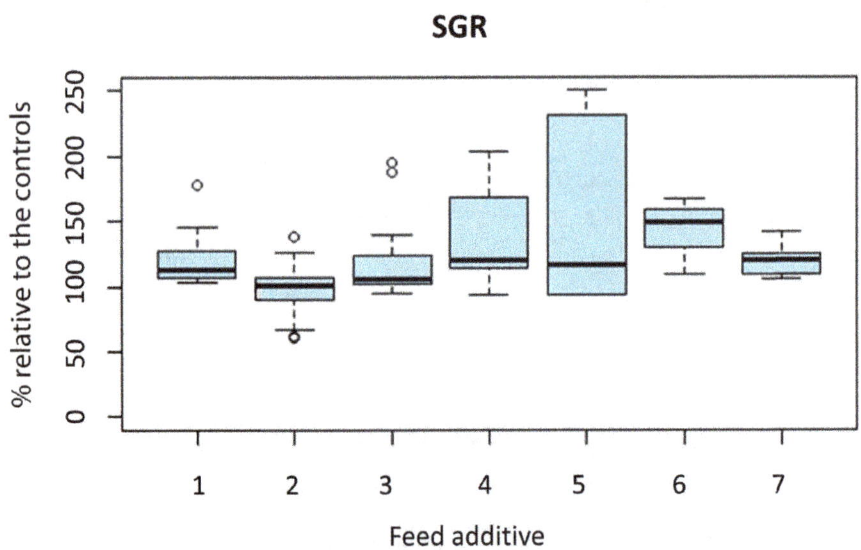

FIGURE 7.17 Specific growth rates (SGR) calculated from feeding studies using mycotoxin contaminated feeds and the same feeds with the addition of feed additives, with 12 data points collected for the endpoint *1: chemicals*, 27 data points collected for *2: clay* as a feed additive, 33 data points for the performance of *3: commercial mixtures*, 10 data points from studies investigating *4: herbs, including herbal extracts*, 5 data points showing advantageous effects of *5: experimental mixtures*, 3 data points showing increased *6: prebiotics*, and 5 references for *7: probiotics*; outliers are indicated by the circles.

ing agent of OTA, followed by black seed and curcuma. And finally, dried ginger had the lowest positive effect in Nile tilapia (El-Marakby et al., 2018). Consequently, and in contrast to the effects observed when clay particles were used, herbal extracts, prebiotics and in some cases also some experimental mixtures of mycotoxin decontaminants showed positive effects on the growth rates of the fish exposed to mycotoxins.

Finally, the survival rates in a number of mycotoxin feeding studies have been compared and summarised in Figure 7.18. Within this graph, the 100% point means that the mortality observed in the mycotoxin-treated group was not changed by the addition of feed additives. From this chart, it can be seen that almost all feed additives improved the survival of the fish in the different experimental studies, which is a promising result. The exception is again the clay particles that did not counteract the mycotoxin-caused mortality in most clay particle-using studies. Nevertheless and as mentioned earlier, most of the studies on mycotoxin binders have been conducted with tilapia species and AFB_1 as the mycotoxin. Hence, broader surveys on mycotoxin alleviation by mycotoxin decontaminants should be conducted in the future to yield more realistic information on the potential advantage of feed additives used against a variety of mycotoxins.

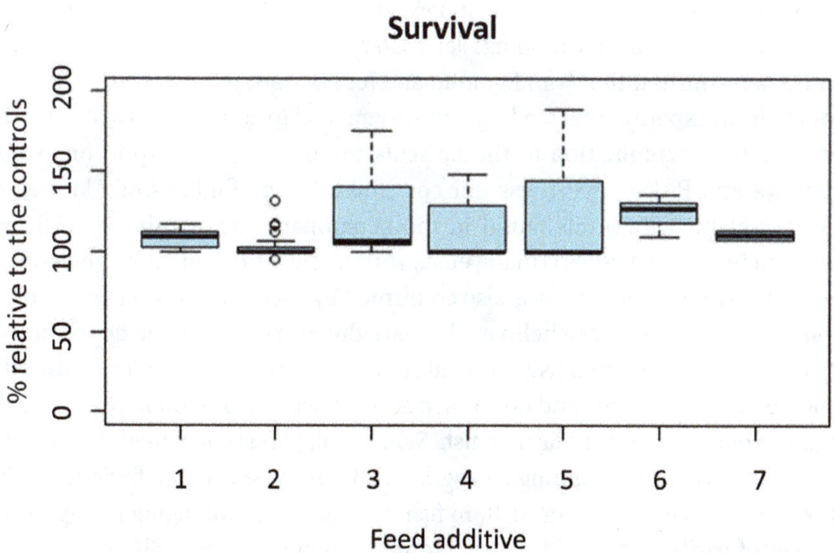

FIGURE 7.18 Survival calculated from feeding studies using mycotoxin contaminated feeds and the same feeds with the addition of feed additives, with 11 data points collected for the endpoint *1: chemicals*, 21 data points collected for *2: clay* as a feed additive, 23 data points for the performance of *3: commercial mixtures*, 6 data points from studies investigating *4: herbs, including herbal extracts*, 4 data points showing advantageous effects of *5: experimental mixtures*, 3 data points showing increased *6: prebiotics*, and 2 references for *7: probiotics*.

4 Mycotoxins carry-over in fish food products

The bioaccessibility of the mycotoxins determines their potential effects in the fish (Nogueira et al., 2020). But the chemical characteristics of each mycotoxin determine its potential for accumulation in fish parts that are destined to be used as a food source. Although mycotoxins occur in feeds frequently and are transferred to the fish, in most cases they are not reaching significant levels in edible parts (Rokvic et al., 2020). Thus, and probably due to rapid metabolisation of a number of mycotoxins, low retention levels of the mycotoxins do not reflect their toxic potential for fish. Nevertheless, the retention of mycotoxins in edible parts of fish has consequences for human nutrition and health, which is why exposure assessments have been conducted and daily intake levels estimated by assuming an adult human BW of 60 kg. In addition, tolerable weekly intake levels for humans have been derived for a number of mycotoxins (JECFA, 2001; European Committee Scientific Committee on Food, 2002) with the exception of AFs, which are listed as human carcinogens and should be avoided completely.

In an optimal case, the actual contamination levels found in fish products should not exceed the tolerable intake levels. It was assumed that generally less than 10% of meat and fish samples are contaminated with mycotoxins, but Carballo et al. (2018) reported DON contamination in 17% of the samples, with 29 fish investigated. For AFB_1 the European Union has set a safety level of 2 μg/kg for food designated for human consumption (van Egmond and Jonker, 2004).

Results from experimental feeding studies can lead to contradictory results. The absence of AFs accumulation in the musculature of common carp in one study (Svobodova and Piskac, 1980) was not confirmed by the findings of Akter et al. (2010). Typically, AFB_1 levels found in the hepatopancreas in fish fed AFB_1 are higher than in muscle tissues (Huang et al., 2011). A clear time and dose-dependent accumulation of AFB_1 in liver was also confirmed by a feeding study using Pacu, *Piaractus mesopotamincus* (Michelin et al., 2021). But also extrahepatic deposition of AFB_1 has also been observed (Ngethe et al., 1992, Larsson et al., 1992). This raises the question of how much time and how much AFB_1 is needed to result in relevant contamination of edible parts of farmed fish. Selim et al. (2014) confirmed that feeding Nile tilapia with a diet containing 200 μg/kg AFB_1 for 2 weeks was sufficient to raise the toxin residues to >20 μg/kg AFB_1 in fish musculature. Prolonging the exposure for 8 weeks further increased this level to more than 90 μg/kg AFB_1. However, in the study of Mehrim et al. (2016) AFB_1 in muscle of tilapia only increased to values close to 5.5 mg/kg after feeding the fish a diet containing 150 mg/kg feed.

But AFs accumulation obviously also depends on the fish species and the study design, since dietary exposure of European seabass (*Dicentrarchus labrax*) to 18 μg/kg BW of AFB_1 for 28 days resulted in 2.5 μg/kg AFB_1 in the musculature and even higher levels (4.25 μg/kg AFB_1) after 42 days of exposure (El-Sayed and Kahlil, 2009). Compared to this, feeding lambari fish (*Astyanax altiparanae*) with AFB_1 resulted in detectable body residues after 90 days of feeding. Feeding them with 50 µg/kg feed for 120 days caused muscle and liver tissues to show AFB_1 concentrations as high as in the feed (Michelin et al., 2017). In contrast, the AFB_1 content in Nile tilapia reached only 2.24-2.99 µg/kg in the body carcass after feeding them AFB_1-contaminated feeds (3.49 to 4.38 mg/kg feed) for 12 weeks (Adeniji et al., 2014). Only traces of AFB_1 were detected in muscle of matrinxã fish (*Brycon cephalus*) after feeding them diets contaminated with 50 mg/kg AFB_1 feed, whereas liver tissue showed slightly higher values (Bedoya-Serna et al., 2018).

But mycotoxin contamination of fish products can also occur when mycotoxin-producing fungi are present during food processing. Accordingly, a study from Kenya and Zambia showed AFs in dried fish but not in fresh fish (Orony et al., 2016; Singh and Nsokolo, 2020). A study from Egypt showed predominantly xerophilic moulds such as *Aspergillus* species (58.2 %) and also *Penicillium* species (32.7 %) in salted products and smoke-cured fish (Ahmed et al., 2005). Similarly, several studies from Nigeria showed mycotoxin-producing fungi and AFs in smoke-dried fish (Adebayo-Tayo et al., 2008; Akinyemi et al., 2012; Olajuyigbe et al., 2014; Job et al. 2016).

Feeding other fish species with experimental diets, resulted in detectable AFB_1 residues in the liver but not in fish musculature (Deng et al., 2010; Huang et al., 2011). Taken together, AFs contamination can be a threat to human health if the fish have been exposed to AFB_1-contaminated diets for a certain duration. A promising result was described by Hussain et al. (1993) who showed that walleye (*Sander vitreus*) exposed to considerable AFB_1 levels are able to reverse the accumulation of AFB_1 in the musculature when fed mycotoxin-free diets for 2 weeks (Huang et al., 2011). However, in food samples AFs can be surprisingly stable, as has been shown in muscle of crucian carp (*Carassius carassius*); AFB_2 (1.2 µg/kg) was still detectable even after storage for 3 months at room temperature (Sun et al., 2015). Results from an Egyptian study revealed that for smoked fish the moisture and salt amounts during food processing influence the OTA and AFB_1 levels in the products, which can consequently exceed the permissible limits for both mycotoxins (Farag et al., 2011).

For mycotoxins other than AFB_1 the information on potential accumulation in fish is rather scarce. However, the International Agency for Research on Cancer (IARC, 1993) categorised OTA as potentially carcinogenic to humans under Group 2B. Nevertheless, according to a Polish food study, fish samples did not contain OTA (Wiśniewska-Dmytrow et al., 2003). In another study, musculature of seabass and seabream (*Sparus aurata*) contained only low OTA levels (Guardone et al., 2018). The tissue distribution and elimination of OTA in dietary-exposed Atlantic salmon (*Salmo salar*) has been investigated by Bernhoft et al. (2017). In that study, OTA residues were detected only in the liver and kidney tissues of the fish (at levels of 3.0 µg/kg in liver tissue and 4.6 µg/kg in kidney tissue) after 6 weeks of feeding and after two further weeks of OTA withdrawal (with levels of 1.7 and 1.9 µg/kg in liver and kidney tissue, respectively). Not surprisingly the detectable amounts of OTA in the intestinal samples of grass carp increased continuously as the OTA level in the diet increased (Liu et al., 2021), but dietary OTA levels of 406 μg/kg resulted in no detectable mycotoxin levels in the intestine. From the studies of Hagelberg et al. (1989) and Fuchs and Hult (1992) it was concluded that in fish OTA is mainly excreted via the liver and bile rather than via the urine, which obviously prevents OTA accumulation in fish to a certain extent. The daily human intake via direct consumption of cereal-based products accounts for up to 22.2 ng OTA/kg BW (Huong et al., 2016). Compared to this the contribution from fish products to the human toxicological burden appears to be rather low. Nevertheless, the potential risk of OTA intoxication should also not be underestimated since OTA was assumed to be even more stable in the environment than AFs (Moss, 2002; Duarte et al., 2010).

Two mycotoxins that are found to be rapidly metabolised in fish and therefore not accumulating in edible fish parts, are DON and ZEN. Consequently, only a low risk to humans can be assumed from the traces of DON or ZEN that could be observed in rainbow trout and common carp fillets (Pietsch et al., 2014b; Pietsch et al., 2015a; Woźny et al., 2015). Similarly, Nácher-Mestre et al. (2015) reported no detectable mycotoxin levels in sea bream or Atlantic salmon after feeding of fish for 8 months with DON concentrations of up to 79.2 μg/kg and FUM levels of up to 754 μg/kg. The low risk for DON or ZEN intoxication of humans via consumption of fish parts is feasible since there is a considerably higher risk of exceeding the toxicological reference values by consuming cereal products directly (Caldas and Jardim, 2012; Raad et al., 2014). But this does not mean that accumulation of DON and ZEN is completely absent in fish. Woźny et al. (2013) showed that ZEN exposure resulted in retention in the ovaries of farmed trout and also in food samples small amounts may be observed (e.g., 1.19 μg/kg DON in the study of Carballo et al., 2018). Similarly, the ZEN residues in seabass after feeding them ZEN-contaminated diets for 4 weeks reached levels of 1.53 μg/kg dry weight of muscle (Abdel-Tawwab et al., 2020).

Mycotoxins have also been reported for sun-dried fish products that are typically used in tropical and subtropical regions in which high humidity and high temperatures can lead to considerable fungal growth and toxin formation. Accordingly, samples of dried seafood contained high levels of ZEN and OTA (317.3 and 1.9 μg/kg, respectively) in the study of Sun et al. (2015). Similarly, ZEN concentrations of up to nearly 54 μg/kg have been detected in sun-dried kapenta fish (Gonkowski et al., 2018). Moreover, small amounts of mycotoxins such as AFB_1, T2 toxin, OTA and DON in dried fish products have been observed in a survey in China (Deng et al., 2021).

Rarely, other mycotoxins are also investigated in food samples: for example, 24 % of fish samples in the study by Carballo et al. (2018), with mean ENNA concentrations of 0.89 μg/kg, mean concentrations of the trichothecene mycotoxins neosolaniol and DAS of 0.86 μg/kg and 0.55 μg/kg, respectively. Furthermore, ENNs were also detectable in 20 % of the salmon samples and 10% of the rainbow trout samples in the study of Tolosa et al. (2017), but further processing such as cooking or smoking appeared to mitigate the mycotoxin content (Tolosa et al., 2019).

Compared to this, FUM levels in fish appear not to be problematic for human consumption since they rarely occur in edible parts of farmed fish (e.g., a survey in Switzerland reported for only one fillet sample a concentration of <0.06 μg/kg FB_1+FB_2, (personal communication C. Pietsch). Taken together, it is not surprising that, for example, mycotoxins are detectable in human breast milk, which has already been related to the consumption of fish, besides cereals, coffee, dried fruits, nuts, and spices (Dinleyici et al., 2018; Mehta et al., 2021).

5 Conclusions

We are becoming more and more aware that mycotoxins are a real threat to the health of farmed fish. However, to have a better understanding of their toxic potential more research is needed, including feeding studies with single mycotoxins and with combinations. This would improve the research that is needed to work on strategies to prevent detrimental effects of mycotoxins in fish. In addition, the present book chapter indicates that some biological endpoints are more useful in some studies than others. Besides the already mentioned biomarkers, the development of new methods to detect so far unidentified effects in fish is also highly recommended.

References

Abdelaziz, M., Answer, W. and Abdelrazek, A.H., 2010. Field study on the mycotoxin binding effects of clay in *Oreochromis niloticus* feeds and their impacts on the performance as well as the health status throughout the culture season. Interdisciplinary Bio Central 2: 1 – 6. https://doi.org/10.4051/ibc.2010.2.2.0010.

Abdel-Tawwab, M., Khalifa, E., Diab, A.M., Khallaf, M.A., Abdel-Razek, N. and Khalil, R.H., 2020. Dietary garlic and chitosan alleviated zearalenone toxic effects on performance, immunity, and challenge of European sea bass, *Dicentrarchus labrax*, to *Vibrio alginolyticus* infection. Aquaculture International 28: 493-510. https://doi.org/10.1007/s10499-019-00477-0.

Adebayo-Tayo, B.C., Onilude, A.A. and Patrick, U.G., 2008. Mycofloral of smoke-dried fishes sold in Uyo, Eastern Nigeria. World Journal of Agricultural Sciences 4: 346-350.

Adeniji, C.A., Okiki, P.A., Rasheed, A.M. and Bolaji, R., 2014. Mouldy groundnut cake and hydrated sodium calcium aluminosilicate in practical diet for African catfish *Clarias gariepinus* (Burchell, 1822). Fisheries and Aquaculture Journal 4: 1. https://doi.org/10.4172/2150-3508.1000083.

Ahmed, A.M., Ismail, S.A. and Abd-El-Rahman, H.A.E.L., 2005. Quantitative, qualitative and toxigenic evaluations of xerophilic mold in traditional Egyptian salted fish, Molouha. Journal of Food Safety 25: 9-18. https://doi.org/10.1111/j.0149-6085.2005.25546.x.

Akinyemi, A.A., Adejola, A.Q., Obasa, S.O. and Ezeri, G.N.O., 2012. Aflatoxins in smoked-dried fish sold in Abeokuta, Ogun State, South-west Nigeria. COLERM Proceedings 2: 478-487.

Akter, A., Rahman, M. and Hasan, M., 2010. Effects of aflatoxin B_1 on growth and bioaccumulation in common carp fingerling in Bangladesh. Asia-Pacific Journal of Rural Development 20: 1-14. https://doi.org/10.1177/1018529120100201.

Baldissera, M.D., Souza, C.F., Zeppenfeld, C.C., Descovi, S.N., Moreira, K.L.S., da Rocha, M.I.U.M., da Veiga, M.L., da Silva, A.S. and Baldisserotto, B., 2018. Aflatoxin B_1-contaminated diet disrupts the blood–brain barrier and affects fish behavior: Involvement of neurotransmitters in brain synaptosomes. Environmental Toxicology and Pharmacology 60: 45-51. https://doi.org/10.1016/j.etap.2018.04.003.

Baldissera, M.D., Souza, C.F., da Silva, J.A., Barroso, D.C., Glória, E.M., Mesadri, J., Wagner, R., Baldisserotto, B. and Val, A.L., 2020. Dietary ochratoxin A (OTA) decreases growth performance and impairs muscle antioxidant system and meat fatty acid profiles in juvenile tambaqui (*Colossoma macropomum*), Comparative Biochemistry and Physiology Part C: Toxicology and Pharmacology 236: 108803. https://doi.org/10.1016/j.cbpc.2020.108803.

Bata-Vidács, I., Kosztik, J., Mörtl, M., Székács, A. and Kukolya, J., 2020. Aflatoxin B_1 and sterigmatocystin binding potential of non-lactobacillus LAB strains. Toxins 12: 799. https://doi.org/10.3390/toxins12120799.

Bedoya-Serna, C.M., Michelin, E.C., Massocco, M.M., Carrion, L.C.S., Godoy, S.H.S., Lima, C.G., Ceccarelli, P.S., Yasui, G.S., Rottinghaus, G.E., Sousa, R.L.M. and Fernandes, A.M., 2018. Effects of dietary aflatoxin B1 on accumulation and performance in matrinxã fish (*Brycon cephalus*). PLoS ONE 13: e0201812. https://doi.org/10.1371/journal.pone.0201812.

Bernhoft, A., Høgåsen, H.R., Rosenlund, G., Ivanova, L., Bernfssen, M.H.G., Alexander, J., Eriksen G.S. and Fæste, C.K., 2017. Tissue distribution and elimination of deoxynivalenol and ochratoxin A in dietary-exposed Atlantic salmon (*Salmo salar*). Food Additives and Contaminants: Part A 3: 1211-1224. https://doi.org/10.1080/19440049.2017.1321149.

Bertero, A., Fossati, P., Tedesco, D.E.A. and Caloni, F., 2020. Beauvericin and enniatins: *In vitro* intestinal effects. Toxins 12: 686. https://doi.org/10.3390/toxins12110686.

Braunbeck, T. and Lammer, E., 2006. Background paper on fish embryo toxicity assays. German Federal Environment Agency, UBA Contract Number 203 85 422, Heidelberg, Germany, 40.

Caldas, E.D. and Jardim, A.N.O., 2012. Exposure to toxic chemicals in the diet: Is the Brazilian population at risk. Journal of Exposure Science and Environmental Epidemiology 22: 1-15. https://doi.org/10.1038/jes.2011.35.

Carballo, D., Moltó, J.C., Berrada, H. and Ferrer, E., 2018. Presence of mycotoxins in ready-to-eat food and subsequent risk assessment. Food and Chemical Toxicology 121: 558-565. https://doi.org/10.1016/j.fct.2018.09.054.

Chen, J., Wen, J., Tang, Y., Shi, J., Mu, G., Yan, R., Cai, J. and Long, M., 2021. Research progress on fumonisin B_1 contamination and toxicity: A review. Molecules 26: 5238. https://doi.org/10.3390/molecules26175238.

Deng, S.-X., Tian, L.-X., Liu, F.-J., Jin, S.-J., Liang, G.-Y., Yang, H.-J., Du, Z.-Y. and Liu, Y.-J., 2010. Toxic effects and residue of aflatoxin B1 in tilapia (*Oreochromis niloticus* × *O. aureus*) during long-term dietary exposure. Aquaculture 307: 233-240. https://doi.org/10.1016/j.aquaculture.2010.07.029.

Deng, Y., Wang, Y., Deng, Q., Sun, L., Wang, R., Ye, L., Tao, S., Liao, J. and Gooneratne, R., 2021. Fungal diversity and mycotoxin contamination in dried fish products in Zhanjiang market, China. Food Control 121: 107614. https://doi.org/10.1016/j.foodcont.2020.107614.

Diab, A.M., Salem, RM., Abeer, E.M.S., Ali, G.I.E. and El-Habashi, N., 2018. Experimental ochratoxicosis A in Nile tilapia and its amelioration by some feed additives. International Journal of Veterinary Science and Medicine 6: 149-158. https://doi.org/10.1016/j.ijvsm.2018.09.004.

Ding, C., Shi, X., Guan, Y. and Li, X., 2021. Deoxynivalenol induces carp neutrophil apoptosis and necroptosis via CYP450s/ROS/PI3K/AKT pathway. Aquaculture 545: 737182. https://doi.org/10.1016/j.aquaculture.2021.737182.

Dinleyici, M., Aydemir, O., Yildirim, G.K., Kaya, T.B. and Carman, K.B., 2018. Human mature milk zearalenone and deoxynivalenol levels in Turkey. Neuroendocrinology Letters 39: 325-330.

Duarte, S.C., Pena, A. and Lino, C.M., 2010. A review on ochratoxin A occurrence and effects of processing of cereal and cereal derived food products. Food Microbiology 27: 187-198. https://doi.org/10.1016/j.fm.2009.11.016.

El-Marakby, H.I., Helal, G. and Fawzy, R.M., 2018. Hazard effect of ochratoxin on tilapia fish and ways for reduction. Research Journal of Pharmaceutical, Biological and Chemical Sciences 9: 1713.

El-Sayed, Y.S. and Khalil, R.H., 2009. Toxicity, biochemical effects and residue of aflatoxin B_1 in marine water-reared sea bass (*Dicentrarchus labrax* L.). Food and Chemical Toxicology 47: 1606-1609. https://doi.org/10.1016/j.fct.2009.04.008.

El-Sayed, Y.S., Khalil, R.H. and Saad, T.T., 2009. Acute toxicity of ochratoxin-A in marine water-reared sea bass (*Dicentrarchus labrax* L.). Chemosphere 75: 878-882. https://doi.org/10.1016/j.chemospehre.2009.01.049.

European Commission. 2006. Commission recommendation 2006/576/EC of 17 August 2006 on the presence of deoxynivalenol, zearalenone, ochratoxin A, T-2 and HT-2 and fumonisins in products intended for animal feeding. Off J L 229: 7-9.

European Committee Scientific Committee on Food. 2002. Opinion on *Fusarium* toxins. Part 6: Group evaluation of T-2 toxin, HT-2 toxin, nivalenol and deoxynivalenol. European Commission 2002 1-12.

Fadl, S.E., El-Shenawy, A.M., Gad, D.M., El Daysty, E.M., El-Sheshtawy, H.S. and Abdo, W.S., 2020. Trial for reduction of ochratoxin A residues in fish feed by using nano particles of hydrated sodium aluminum silicates (NPsHSCAS) and copper oxide. Toxicon 184: 1-9. https://doi.org/10.1016/j.toxicon.2020.05.014.

Fan, Y., Liu, L., Zhao, L., Wang, X., Wang, D., Huang, C., Zhang J., Ji, C. and Ma, Q., 2018. Influence of *Bacillus subtilis* ANSB060 on growth, digestive enzyme and aflatoxin residue in Yellow River carp fed diets contaminated with aflatoxin B_1. Food and Chemical Toxicology 113: 108-114. https://doi.org/10.1016/j.fct.2018.01.033.

Farag, H.E.M., El-Taiby, A.A. and Hassan, H.M. 2011. Assessment of ochratoxin A and aflatoxin B_1 levels in the smoked fish with special reference to the moisture and sodium chloride content. Research Journal of Microbiology 6: 813-825. https://doi.org/10.3923/jm.2011.813.825.

Fremy, J.-M., Alassane-Kpembi, I., Oswald, I.P., Cottrill, B. and van Egmond, H.P., 2019. A review on combined effects of moniliformin and co-occurring *Fusarium* toxins in farm animals. World Mycotoxin Journal 12: 281-291. https://doi.org/10.3920/WMJ2018.2405.

Fuchs, R., Hult, K. 1992. Ochratoxin A in blood and its pharmacokinetic properties. Food and Chemical Toxicology 30: 201-204.

García-Herranz, V., Valdehita,, A., Navas, J.M. and Fernández-Cruz, M.L., 2019. Cytotoxicity against fish and mammalian cell lines and endocrine activity of the mycotoxins beauvericin, deoxynivalenol and ochratoxin-A. Food and Chemical Toxicology 127: 288-297. https://doi.org/10.1016/j.fct.2019.01.036.

Ghafarifarsani, H., Imani, A., Niewold, T.A., Pietsch-Schmied and C. Kourosh, S., 2021. Histopathological effects of aflatoxin B_1 and zearalenone on liver tissue of rainbow trout (*Oncorhynchus mykiss*). Journal of Fisheries 74: 97-106. https://doi.org/10.22059/jfisheries.2021.317750.1224.

Ghafarifarsani, H., Niewold, T.A., Pietsch-Schmied, C. Moghanlou, K.S. and Imani, A., 2021. Synergistic toxicity of dietary aflatoxin B_1 (AFB_1) and zearalenone (ZEN) in rainbow trout (*Oncorhynchus mykiss*) is attenuated by anabolic effects. Aquaculture 541: 736793. https://doi.org/10.1016/j.aquaculture.2021.736793.

Gonçalves, R.A., Navarro-Guillén, C., Gilannejad, N., Dias, J., Schatzmayr, D., Bichl, G., Czabany, T., Moyano, F.J., Rema, P., Yúfera, M., Mackenzie, S. and Martínez-Rodríguez, G., 2018a. Impact of deoxynivalenol on rainbow trout: Growth performance, digestibility, key gene expression regulation and metabolism. Aquaculture 490: 362-372. https://doi.org/10.1016/j.aquaculture.2018.03.001.

Gonçalves, R.A., Engrola, S., Aragão, C., Mackenzie, S., Bichl, G., Czabany, T. and Schatzmayr, D., 2018b. Fate of [^{3}H]-deoxynivalenol in rainbow trout (*Oncorhynchus mykiss*) juveniles: Tissue distribution and excretion. Journal of Aquaculture Research and Development 9: 7. https://doi.org/10.4172/2175.9546.1000540.

Gonçalves, R.A., Menanteau-Ledouble, S., Schöller, M., Eder, A., Schmidt-Posthaus, H., Mackenzie, S. and El-Matbouli, M., 2019. Effects of deoxynivalenol exposure time and contamination levels on rainbow trout. Journal of the World Aquaculture Society 50: 137-154. https://doi.org/10.1111/jwas.12542.

Gonkowski, S., Obremski, K., Makowska, K., Rytel, L. and Mwaanga, E.S., 2018. Levels of zearalenone and its metabolites in sun-dried kapenta fish and water of Lake Kariba in Zambia — A preliminary study. Science of the Total Environment 637–638: 1046-1050. https://doi.org/10.1016/j.scitotenv.2018.05.091.

Guardone, L., Tinacci, L., Luci, G., Meucci, V., Intorre, L. and Armani, A., 2018. Determination of ochratoxin A in European seabass and gilthead seabream in Italy. In: Poster by the European College of Veterinary Public Health. https://doi.org/10.13140/RG.2.2.31611.77604.

Hagelberg, S.K., Hult, K. and Fuchs, R., 1989. Toxicokinetics of ochratoxin A in several species and its Plasma-binding properties. Journal of Applied Toxicology 9: 91-96.

Hamill, R.L., Higgens, C.E., Boaz, H.E. and Gorman, M., 1969. The structure op beauvericin, a new depsipeptide antibiotic toxic to artemia salina. Tetrahedron Letters 10: 4255-4258.

He, C.-H., Fan, Y.-H., Wang, Y., Huang, C.-Y., Wang, X.-C. and Zhang, H.-B., 2010. The individual and combined effects of deoxynivalenol and aflatoxin B_1 on primary hepatocytes of *Cyprinus carpio*. International Journal of Molecular Sciences 2010: 11. https://doi.org/10.3390/ijms11103760.

Hegazi, S.M., El-Sabagh, M.R., El-Keeidy, A. and Zein El-Dein, A.I., 2013. Aflatoxin in feed and its effect on fish health. Kafrelsheikh Veterinary Medicine Journal 11: 317-329.

Huang, Y., Han, D., Zhu, X., Yang, Y., Jin, J., Chen, Y. and Xie, S., 2011. Response and recovery of gibel carp from subchronic oral administration of aflatoxin B_1. Aquaculture 319: 89-97. https://doi.org/10.1016/j.aquaculture.2011.06.024.

Huang, C., Wu, P., Jiang, W.D., Liu, Y., Zeng, Y.Y., Jiang, J., Kuang, S.Y., Tang, L., Zhang, Y.A., Zhou, X.Q. and Feng, L., 2018. Deoxynivalenol decreased the growth performance and impaired intestinal physical barrier in juvenile grass carp (*Ctenopharyngodon idella*). Fish and Shellfish Immunology 80: 376-391.

Huong, B.T.M., Tuyen, L.D., Tuan, D.H., Brimer, L. and Dalsgaard, A., 2016. Dietary exposure to aflatoxin B1, ochratoxin A and fumonisins of adults in Lao Cai province, Vietnam: A total dietary study approach. Food and Chemical Toxicology. 98: 127-133. https://doi.org/10.1016/j.fct.2016.10.012.

Hussain, M., Gabal, M.A., Wilson, T., Summerfelt, R.C. 1993. Effect of aflatoxin-contaminated feed on morbidity and residues in walleye fish. Veterinary and Human Toxicology 35: 396-398.

IARC Monographs on the Evaluation of Carcinogenic Risks to Humans, Volume 56: Some naturally occurring substances: Some food items and constituents, heterocyclic aromatic amines and mycotoxins, IARC, Lyon, France 1993.

JECFA, 2001. Joint FAO/WHO Expert Committee on Food Additives. Safety evaluation of certain mycotoxins in food. WHO Food Additives Ser. 47.

Job, M., Agina, S. and Dapiya, H., 2015. Occurrence of aflatoxigenic fungi in smoke-dried fish sold in Jos Metropolis. British Microbiology Research Journal 11: 1-7. http://sciencedomain.org/review-history/11795.

Koletsi, P., Schrama, J.W., Graat, E.A.M., Wiegertjes, G.F., Lyons, P. and Pietsch, C., 2021. Occurrence of mycotoxins in feedstuffs and fish feeds in Europe and the potential effects of deoxynivalenol DON on important farmed fish species. Toxins 13: 403. https://doi.org/10.3390/toxins13060403.

Křížová, L., Dadáková, K., Dvořáčková, M. and Kašparovský, T., 2021. Feedborne mycotoxins beauvericin and enniatins and livestock animals. Toxins 13: 32. https://doi.org/10.3390/toxins13010032.

Larsson, P., Ngethe, S., Ingebrigtsen, K. and Tjälve, H., 1992. Extrahepatic disposition of ^{3}H-aflatoxin B_1 in the rainbow trout (*Oncorhynchus mykiss*). Pharmacology and Toxicology 71: 262-271.

Li, J. and Liu, M., 2019. The carcinogenicity of aflatoxin B_1. In: Aflatoxin B_1 occurrence, detection and toxicological effects (Long, X.-D., ed.). InTech Open https://doi.org/10.5772/intechopen.88353.

Liu, X., Wu, P., Jiang, W.-D., Liu, Y., Jiang, J., Kuang, S.-Y., Tang, L., Zhou, X.-Q. and Feng, L., 2021. Effects of dietary ochratoxin A on growth performance and intestinal apical junctional complex of juvenile grass carp (*Ctenopharyngodon idella*). Toxins 13: 11. https://doi.org/10.3390/toxins13010011.

Luangsaard, J.J., Berkaew, P., Ridkaew, R., Hywel-Jones, N.L. and Isaka, M., 2009. A beauvericin hot spot in the genus *Isaria*. Mycology Research 113: 1389-1395. https://doi.org/10.1016/j.mycres.2009.08.017.

Marijani, E., Wainaina, J.M., Charo-Karisa, H., Nzayisenga, L., Munguti, J., Gnonlonfin, G.J.B., Kigadye, E. and Okoth, S., 2017. Mycoflora and mycotoxins in finished fish feed and feed ingredients from smallholder farms in East Africa. The Egyptian Journal of Aquatic Research 43: 169-176. https://doi.org/10.1016/j.ejar.2017.07.001.

Marijani, E., Kigadye, E., Okoth, S., 2019. Occurrence of fungi and mycotoxins in fish feeds and their impact on fish health. International Journal of Microbiology Article ID 6743065 https://doi.org/10.1155/2019/6743065.

Martinez, M.P., Gonzalez Pereyra, M.L., Fernandez Juri, M.G., Poloni, V. and Cavaglieri, L., 2018. Probiotic characteristics and aflatoxin B_1 binding ability of *Debaryomyces hansenii* and *Kazaschtania exigua* from rainbow trout environment. Aquaculture Research 49: 1588-1597.

McKean, C., Tang, L., Tang, M., Billam, M., Wang, Z., Theodorakis, C.W., Kendall, R.J. and Wang, J.-S., 2006. Comparative acute and combinative toxicity of aflatoxin B_1 and fumonisin B_1 in animals and human cells. Food and Chemical Toxicology 44: 868-876. https://doi.org/10.1016/j.fct.2005.11.011.

Mehrim, A.I., Refaey, M.M. and Elmeleigy, K.M., 2016. Glutathione-Enhancer™ against foodborne aflatoxicosis of *Oreochromis niloticus* (Linnaeus, 1758). Journal of Fisheries and Aquatic Science 11: 131-146. https://doi.org/10.3923/jfas.2016.131.146.

Mehta, R.V., Wenndt, A.J., Girard, A.W., Taneja, S., Ranjan, S., Ramakrishnan, U., Martorell, R., Ryan, P.B., Rangiah, K. and Young, M.F. 2021., Risk of dietary and breastmilk exposure to mycotoxins among lactating women and infants 2–4 months in northern India. Maternal and Child Nutrition 17: e13100. https://doi.org/10.1111/mcn.13100.

Michelin, E.C., Massocco, M.M., Godoy, S.H.S., Baldin, J.C., Yasui, G.S., Lima, C.G., Rottinghaus, G.E., Sousa, R.LM. and Fernandes, A.M., 2017. Carryover of aflatoxins from feed to Lambari fish (*Astyanax altiparanae*) tissues. Food Additives and Contaminants: Part A: Chemistry, Analysis, Control, Exposure and Risk Assessment. 34: 265-272. https://doi.org/10.1080/19440049.2016.1266097.

Michelin, E.C., Bedoya-Serna, C.M., Carrion, L.C.S., Godoy, S.H.S., Baldin, J.C., Lima, C.G., Yasui, G.S., Rottinghaus, G.E., Sousa, R.L.M. and Fernandes, A.M., 2021. Long-term exposure of Pacu (*Piaractus mesopotamincus*) fish to dietary aflatoxin B1: residues in tissues and performance. World Mycotoxin Journal 14: 411-419. https://doi.org/10.3920/WMJ2020.2659.

Moss, M.O., 2002. Risk assessment for aflatoxins in foodstuffs. International Biodeterioration and Biodegradation. 50: 137-142. https://doi.org/10.1016/S0964-8305(02)00078-1.

Mwihia, E.W., Lyche, J.L., Mbuthia, P.G., Ivanova, L., Uhlig, S., Gathumbi, J.K., Maina, J.G., Eshitera, E. E. and Eriksen, G.S., 2020. Co-occurrence and levels of mycotoxins in fish feeds in Kenya. Toxins 12: 627. https://doi.org/10.3390/toxins12100627.

Nácher-Mestre, J., Serrano, R., Beltrán, E., Pérez-Sánchez, J., Silva, J., Karalazos, V., Hernández, F. and Berntssen, M.H.G., 2015. Occurrence and potential transfer of mycotoxins in gilthead sea bream and Atlantic salmon by use of novel alternative feed ingredients. Chemosphere 128: 314-320. https://doi.org/10.1016/j.chemosphere.2015.02.021.

Nácher-Mestre, J., Beltrán, E., Strachan, F., Dick, J.R., Pérez-Sánchez, J., Berntssen, M.H.G. and Tocher, D.R., 2020. No transfer of the non-regulated mycotoxins, beauvericin and enniatins, from feeds to farmed fish reared on plant-based diets. Food Chemistry 323: 126773. https://doi.org/10.1016/j.foodchem.2020.126773.

Nazhand, A., Durazzo, A., Lucarini, M., Souto, E.B. and Santini, A. 2020. Characteristics, occurrence, detection and detoxification of aflatoxins in foods and feeds. Foods 9: 644. https://doi.org/10.3390/foods9050644.

Ngethe, S., Horsberg, T.E. and Ingebrigtsen, K., 1992. The disposition of ^{3}H-aflatoxin B1 in the rainbow trout (*Oncorhynchus mykiss*) after oral and intravenous administration. Aquaculture 108: 323-332.

Niermans, K., Woyzichovski, J., Kröncke, N., Benning, R. and Maul, R., 2019. Feeding study for the mycotoxin zearalenone in yellow mealworm (*Tenebrio molitor*) larvae—investigation of biological impact and metabolic conversion. Mycotoxin Research 35: 231-242. https://doi.org/10.1007/s12550-019-00346-y.

Nogueira, W.V., Kerstner de Oliveira, F., Marimón Sibaja, K.V., de Oliveira Garcia, S., Kupski, L., de Souza, M.M., Tesserband, M.B. and Garda-Buffon, J., 2020. Occurrence and bioacessibility of mycotoxins in fish feed. Food Additives and Contaminants: Part B 13: 244-251. https://doi.org/10.1080/19393210.2020.1766577.

Olajuyigbe, O.O., Akande, G.R., Ezekiel, C.N. and Ezekiel, M.O. 2014. Aflatoxigenic moulds and aflatoxin contamination of retailed fishery products in Lagos markets. Mycotoxicology Society of Nigeria Mycotoxicology 1: 57-63.

Orony, D.N.A., Lalah, J.O. and Jondiko, I.O., 2016. Determination of carcinogenic polycyclic aromatic hydrocarbons (PAHs), aflatoxins, and nitrosamines in processed fish from the Winam Gulf area of Kenya and estimated potential exposure in human. Polycyclic Aromatic Compounds 36: 295-317. https://doi.org/10.1080/10406638.2014.980433.

Ouanes, Z., Abid, S., Ayed, I., Anane, R., Mobio, T., Creppy, E.E. and Bacha, H., 2003. Induction of micronuclei by Zearalenone in Vero monkey kidney cells and in bone marrow cells of mice: protective effect of Vitamin E. Mutation Research 538: 63-70. https://doi.org/10.1016/s1383-5718(03)00093-7.

Pietsch, C., Kersten, S., Burkhardt-Holm, P., Valenta, H. and Dänicke, S., 2013. Occurrence of deoxynivalenol and zearalenone in commercial fish feed: An initial study. Toxins 5: 184-192. https://doi.org/10.3390/toxins5010184.

Pietsch, C., Michel, C., Kersten, S., Valenta, H., Dänicke, S., Schulz, C., Kloas, W. and Burkhardt-Holm, P., 2014a. *In vivo* effects of deoxynivalenol (DON) on innate immune responses of carp (*Cyprinus carpio* L.). Food and Chemical Toxicology 68: 44-52. https://doi.org/10.1016/j.fct.2014.03.012.

Pietsch, C., Schulz, C., Rovira, P., Kloas, W. and Burkhardt-Holm, P. 2014b. Organ damage and hepatic lipid accumulation in carp (*Cyprinus carpio* L.) after feed-borne exposure to the mycotoxin, deoxynivalenol (DON). Toxins 6: 756-778. https://doi.org/10.3390/toxins6020756.

Pietsch, C., Kersten, S., Valenta, H., Dänicke, S., Schulz, C., Burkhardt-Holm, P. and Junge, R., 2015a. Effects of dietary exposure to zearalenone (ZEN) on carp (*Cyprinus carpio* L.). Toxins 7: 3465-3480. https://doi.org/10.3390/toxins7093465.

Pietsch, C., Junge, R., Burkhardt-Holm, P. 2015b. Immunomodulation by zearalenone (ZEN) in carp (*Cyprinus carpio* L.). BioMed Research International ID 420702 https://doi.org/10.1155/2015/420702.

Pietsch, C. and Junge, R., 2016. Physiological responses of carp (*Cyprinus carpio* L.) to dietary exposure to zearalenone (ZEN). Comparative Biochemistry and Physiology Part C 188: 52-59. https://doi.org/10.1016/j.cbpc.2016.06.004.

Pietsch, C., 2020. Risk assessment for mycotoxin contamination in fish feeds in Europe. Mycotoxin Research 36: 41-62. https://doi.org/10.1007/s12550-019-00368-6.

Pietsch, C., Müller, G., Mourabit, S., Carnal, S. and Bandara, K., 2020. Occurrence of fungi and fungal toxins in fish feed during storage. Toxins 12: 171. https://doi.org/10.3390/toxins12030171.

Pumipuntu, N., Tengjaroenkul, B., Pimpukdee, K. 2013. Effects of diet containing aflatoxin B_1 and fumonisin B_1 on hematology, immune response and histopathology in nile tilapia. KKU Veterinary Journal 23: 201-217. http://vmj.kku.ac.th.

Raad, F., Nasreddine, L., Hilan, C., Bartosik, M. and Parent-Massin, D., 2014. Dietary exposure to aflatoxins, ochratoxin A and deoxynivalenol from a total diet study in an adult urban Lebanese population. Food and Chemical Toxicology 73: 35-43. https://doi.org/10.1016/j.fct.2014.07.034.

Rokvić, N., Aksentijević, K., Kureljušić, J., Vasiljević, M., Todorović, N., Zdravković, N. and Stojanac, N., 2020. Occurrence and transfer of mycotoxins from ingredients to fish feed and fish meat of common carp (*Cyprinus carpio*) in Serbia. World Mycotoxin Journal 13: 545-552. https://doi.org/10.3920/WMJ2020.2580.

Schenzel, J., Schwarzenbach, R.P. and Buchel, T.D., 2010. Multi-residue screening method to quantify mycotoxins in aqueous environmental samples. Journal of Agricultural Food Chemistry 58: 11207-11217. https://doi.org/10.1021/jf102737.

Selim, K.M., El-Hofy, H. and Khalil, R.H., 2014. The efficacy of three mycotoxin adsorbents to alleviate aflatoxin B_1-induced toxicity in *Oreochromis niloticus*. Aquaculture International. 22: 523-540. https://doi.org/10.1007/s10499-013-9661-6.

Shalaby, A.M.E. 2004. The opposing effect of ascorbic acid (vitamin C) on ochratoxin toxicity in Nile tilapia (*Oreochromis niloticus*). In: Proceedings of the 6th International Symposium on Tilapia in Aquaculture. 209-221.

Shen, Y.-B., Weaver, A.C. and Kim, S.W., 2021. Physiological effects of deoxynivalenol from naturally contaminated corn on cerebral tryptophan metabolism, behavioral response, gastrointestinal immune status and health in pigs following a pair-Feeding model. Toxins 13: 393. https://doi.org/10.3390/toxins13060393.

Singh, I.S. and Nsokolo, E., 2020. Prevalence of aflatoxins in smoked-dried and fresh fish in Zambia. Journal of Environmental Protection 11: 13-21. https://doi.org/10.4236/jep.2020.111002.

Šišperová, E., Modrá, H., Ziková, A., Kloas, W., Blahová, J., Matejová, I., Živná, D. and Svobodová, Z., 2015. The effect of mycotoxin deoxynivalenol (DON) on the oxidative stress markers in rainbow trout (*Oncorhynchus mykiss*, Walbaum 1792). Journal of Applied Ichthyology. 31: 855-861. https://doi.org/10.1111/jai.12809.

Sun, W., Han, Z., Aerts, J., Nie, D., Jin, M., Shi, W., Zhao, Z., De Saeger, S., Zhao, Y. and Wu, A., 2015. A reliable liquid chromatography–tandem mass spectrometry method for simultaneous determination of multiple mycotoxins in fresh fish and dried seafoods. Journal of Chromatography A 1387: 42-48. https://doi.org/10.1016/j.chroma.2015.01.071.

Svobodova, Z. and Piskac, A., 1980. Effect of feeds with a low content of aflatoxin B_1 on the health of carp *Cyprinus carpio*. Zivocisna Vyroba–UVTIZ 25: 809-814.

Szabó, R.T., Kovács-Weber, M., Balogh, K.M., Mézes, M. and Kovács, B., 2021. Changes of DNA damage effect of T-2 or deoxynivalenol toxins during three weeks exposure in common carp (*Cyprinus carpio* L.) revealed by LORD-Q PCR. Toxins 13: 576. https://doi.org/10.3390/toxins13080576.

Thapa, A., Horgan, K.A., White, B. and Walls, D., 2021. Deoxynivalenol and zearalenone—synergistic or antagonistic agri-food chain co-contaminants? Toxins 13: 561. https://doi.org/10.3390/toxins13080561.

Tolosa, J., Font, G., Mañes, J. and Ferrer, E., 2014. Natural occurrence of emerging *Fusarium* mycotoxins in feed and fish from aquaculture. Journal of Agricultural and Food Chemistry 62: 12462-12470. https://doi.org/10.1021/jf5036838.

Tolosa, J., Font, G., Mañes, J. and Ferrer, E., 2017. Mitigation of enniatins in edible fish tissues by thermal processes and identification of degradation products. Food and Chemical Toxicology 101: 67-74. https://doi.org/10.1016/j.fct.2016.12.039.

Tolosa, J., Barba, F.J., Font, G. and Ferrer, E., 2019. Mycotoxin incidence in some fish products: QuEChERS methodology and liquid chromatography linear ion trap tandem mass spectrometry approach. Molecules 24: 527. https://doi.org/10.3390/molecules24030527.

Tschirren, L., Siebenmann, S. and Pietsch, C., 2018. Toxicity of ochratoxin to early life stages of zebrafish (*Danio rerio*). Toxins 10: 264. https://doi.org/10.3390/toxins10070264.

van Egmond, H. and Jonker, M., 2004. Worldwide Regulations for Mycotoxins in Food and Feed in 2003. 81. Rome, Italy: FAO Food and Nutrition Paper.

Wang, Y.-L., Zhou, X.-Q., Jiang, W.-D., Wu, P., Liu, Y., Jiang, J., Wang, S.-W., Kuang, S.-Y., Tang, L. and Feng, L. 2019. Effects of dietary zearalenone on oxidative stress, cell apoptosis, and tight junction in the intestine of juvenile grass carp (*Ctenopharyngodon idella*). Toxins 11: 333. https://doi.org/10.3390/toxins11060333.

Wettstein, F. and Bucheli, T., 2010. Poor elimination rates in waste-water treatment plants lead to continuous emission of deoxynivalenol into the aquatic environment. Water Research 44: 4137-4142. https://doi.org/10.1016/j.waters.2010.05.038.

Wiśniewska-Dmytrow, H., Zmudzki, J., Burek, O. and Pietruszka, K., 2013. Official control of ochratoxin A in food of animal origin in Poland between 2003 and 2012. Bulletin of the Veterinary Institute in Pulawy 57: 519-523. https://doi.org/10.2478/bvip-2013-0090.

Woźny, M., Obremski, K., Jakimiuk, E., Gusiatin, M. and Brzuzan, P., 2013. Zearalenone contamination in rainbow trout farms in North-Eastern Poland. Aquaculture 416-417: 209-211. https://doi.org/10.1016/j.aquaculture.2013.09.030.

Woźny, M., Dobosz, S., Obremski, K., Hliwa, P., Gomułka, P., Łakomiak, A., Różyński, R., Zalewski, T. and Brzuzana, P. 2015. Feed-borne exposure to zearalenone leads to advanced ovarian development and limited histopathological changes in the liver of premarket size rainbow trout. Aquaculture 448: 71-81. https://doi.org/10.1016/j.aquaculture.2015.05.032.

Woźny, M., Obremski, K., Hliwa, P., Gomułka, P., Różyński, R., Wojtacha, P., Florczyk M., Segner, H. and Brzuzan, P. 2019. Feed contamination with zearalenone promotes growth but affects the immune system of rainbow trout. Fish and Shellfish Immunology 84: 680-694. https://doi.org/10.1016/j.aquaculture.2020.735522.

Woźny, M., Dobosz, S., Hliwa, P., Gomułka, P., Królc, J., Obremski, K., Blahova, J., Svobodova, Z., Michalik, O., Ocalewicz, K. and Brzuzan, P. 2020. Feed-borne exposure to zearalenone impairs reproduction of rainbow trout. Aquaculture 528: 735522. https://doi.org/10.1016/j.aquaculture.2020.735522.

Wu, J., Du, H., Wu, J., Luo, J., Fu, P., Qiao, X. and Wie, Q. 2021. Effects of dietary zearalenone on the serum biochemistry, hepatic and intestinal histology, and intestinal microbiota of juvenile Dabry's sturgeon (*Acipenser dabryanus*). Journal of Applied Ichthyology 37: 932-941. https://doi.org/10.1111/jai.14270.

Yildirim, M., Manning, B.B., Lovell, R.T., Grizzle, J.M. and Rottinghaus, G.E., 2000. Toxicity of moniliformin and fumonisin B_1 fed singly and in combination in diets for young channel catfish *Ictalurus punctatus*. Journal of the World Aquaculture Society 31: 599-608. https://doi.org/10.1111/j.1749-7345.2000.tb00909.x.

CHAPTER 8

Effects of mycotoxins in pets

J.H. Tegzes

College of Veterinary Medicine, Western University of Health Sciences, Pomona, USA
jtegzes@westernu.edu

1 Introduction

Pets are increasingly at risk from exposure to mycotoxins through dietary sources (Singh and Chuturgoon, 2017). While the risks of exposure and development of disease are, perhaps, better known in livestock and production animals, research continues to reveal the risks of exposure to pets such as dogs and cats through clinical signs and disease. This chapter will describe the reported incidence of these clinical signs and disease and the effects of mycotoxins in dogs and cats.

Mycotoxin exposure in pets can be through their regular diet and through the consumption of treats (Kępinska-Pacelik and Biel, 2021a; Macías-Montes et al., 2020). Mycotoxin exposure risk is higher through the consumption of grains and starches, although there is still some risk associated with the consumption of meats, particularly organ meats (Alaboudi et al., 2022; Tegzes et al., 2019). In a study examining the differences between wet and dry commercial dog feeds containing grains (starches) and those without, mycotoxin contamination was most common and showed the highest degree of contamination in dry, kibbled types of dog feed containing grains (starches) (Tegzes et al., 2019). In the study, dry, extruded dog feeds contained measurable contamination with DON, FB_1, FB_2, and ZEN, while freshly prepared wet-type feeds did not contain any. However, others have found that organ meats from commercial poultry operations can also contain measurable levels of mycotoxins. This is important since organ meats are sometimes used in the formulations of commercial pet feeds. Alaboudi et al. (2022) found multiple mycotoxins in organ meats from poultry, with all samples analysed containing AFB_1, OTA, ZEN, FB_1, and DON. Others have also noted that plant-based products are commonly contaminated with mycotoxins. Products of animal origin, such as milk and meat, may also be contaminated (Kępinska-Pacelik and Biel, 2021b; Mohammedi-Ameur et al., 2020; Muaz et al., 2021; Nazhand et al., 2020).

In 2022 it was projected that the worldwide market value of extruded, dry, kibbled feed for dogs and cats would reach over $70 billion (Corsato Alvarenga et al., 2021), and it is expected to reach more than $100 billion by 2028 (Research and Markets, 2024). It is important to note that ingredients such as starches and veg-

 | DOI: 10.1163/9789004712720_9

etable proteins promote binding, homogenisation, and structure in the dough of extruded dog and cat feeds. Of the starches used, ground maize or maize flour are the most common (Corsato Alvarenga et al., 2021). This helps to explain the susceptibility of extruded kibbled types of feed to contamination with mycotoxins. Fungal growth and mycotoxin production are related to methods of handling and storage of products post-harvest, and after manufacture (Bryden, 2012; Singh and Chuturgoon, 2017; Tulpule, 1981). It is estimated that, globally, most pet dogs and cats are fed dry, extruded, kibbled types of feeds, and that approximately 80% of pet diets contain grains and starches. The proportion of starches and grains contained can be as high as between 30% and 60% of the dry matter ingredients (Corsato Alvarenga and Aldrich, 2020).

2 Occurrence of mycotoxin contamination in pet diets

Throughout the years there have been reports in the literature about the relative prevalence of mycotoxin contamination in dog and cat feeds. It is uncertain whether incidence is actually increasing, or is further revealed by increased investigations. From 1980 to 2017, mycotoxins were the most common notification to the Rapid Alert System for Food and Feed (RASFF), accounting for 21.3% of reports (Pigłowski, 2019). The largest number of these notifications related to AFB_1 (Kępinska-Pacelik and Biel, 2021a; Pigłowski, 2019).

A report in 2017 noted that the protein content of many dog feeds is largely sourced from vegetables and minimally from meat sources. This could help to explain why mycotoxins are more studied and reported in recent years (Singh and Chuturgoon, 2017).

In South Africa, a survey of commercial pet feeds readily available to consumers through supermarkets, as well as premium pet feeds available through specialty stores and veterinary clinics was performed. The survey reported that all pet feeds sampled contained fungi that produce mycotoxins (i.e., *Aspergillus flavus, A. fumigatus,* and *A. parasiticus*) as well as the mycotoxins themselves, most frequently AFs and FUM. It is worthwhile noting that the price of the feed was not a factor in differentiating the presence of mycotoxins, and that most of these feeds were produced and imported from North America and Europe (Singh et al., 2017; Singh and Chuturgoon, 2017). This is in contrast to a study in the USA that found diets containing higher quality grains did not contain mycotoxins while feeds that contained lower quality starches and grains did (Tegzes et al., 2019).

Figure 8.1 summarizes some surveys performed of dog and cat feeds. Dry dog feeds were analysed for mycotoxins in Austria applying two different methods: a) enzyme linked immunosorbent assay (ELISA) and b) high performance liquid chromatography (HPLC). For more details about these methods of analysis, see Chapter 12. Of the 76 dog feed samples analysed with ELISA test kits, 97% contained very high concentrations of DON (more than 250 µg/kg); 47% contained ZEN (up to 298 µg/kg); 42% contained FUM (up to 568 µg/kg); and 5% contained OTA (up to 4.7 µg/kg) (Böhm et al., 2010). Some of these feed contaminated samples were submitted to confirmatory analysis with HPLC, and 83% contained DON (up to 1,390 µg/kg); 68% contained ZEN (up to 549 µg/kg); and 67% contained FUM (up to 119 µg/kg). It was remarkable that up to 47% of the samples were co-contaminated with two or more mycotoxins. A study conducted in Spain was based on LC-MS/MS. The study sampled the best-selling dog and cat feed brands and found mycotoxins in all samples analysed, often with multiple mycotoxins contaminating the same feed sample. Of note, AFB_1, DON, FB_1, FB_2, and HT-2 were detected in 100% of the 34 grocery dog feed brands sampled, while AFB_1, DON, ZEN, FB_1, FB_2, HT-2 and T-2 were detected in 100% of the 26 premium dog feed brands. Of the 62 cat feed brands sampled, 100% contained measurable AFB_1, HT-2, DON, FB_1, and FB_2. Although there was a low risk of acute toxicity, all dog and cat feeds were co-contaminated with at least six different mycotoxins, and 30% of the feed samples had nine different mycotoxins. These results indicate not only chronic exposure to mycotoxins, but also chronic, simultaneous exposure to multiple mycotoxins. The researchers raised the possibility of hidden mycotoxins that might not have been detected in the samples. They concluded that further research on the effects of chronic exposure to multiple mycotoxins is needed (Macías-Montes et al., 2020).

An investigation into the occurrence of mycotoxins in dry feeds intended for cats in Italy found that a relatively high number of samples exceeded the European regulatory limits established for AFB_1, ZEN, FB_1, FB_2, T-2, and OTA. Two samples exceeded the guidance values for FB_1 and FB_2 (5,000 µg/kg), containing both 5,190 µg/kg FB_1 and 8,800 µg/kg FB_2 (Grandi et al., 2019). A similar study in South Africa analysing both grocery store brands and premium brands of dry cat feeds found that all of the 12 brands sampled contained both fungi producing mycotoxins and mycotoxins. All of the samples submitted to HPLC analysis tested positive for AFB_1, ZEN, FB_1, FB_2, and OTA. The levels of AFB_1 in the premium brands were much higher (125 µg/kg) than in the grocery brands (41.57 µg/kg), whereas the levels of FB_1 + FB_2 were higher in the grocery brands (321µg/kg) than in the premium brands (15.43 µg/kg) (Singh et al., 2017).

Dry dog and cat feed samples from 16 producers in Poland were analysed. A total of 38 samples were obtained comprising 12 feeds for cats and 26 for dogs. These samples were purchased from pet shops and veterinary clinics. The samples were further classified as those containing maize; those containing wheat; those containing both maize and wheat; and those without any grains. The results of the analyses revealed that the dog feed samples containing maize, and maize and wheat, had higher concentrations of all mycotoxins when compared to cat feeds containing the same grains. In contrast, when analysing both dog and cat feeds that were labelled as grain-free, the cat feeds tended to contain higher concentrations of ZEN, DON,

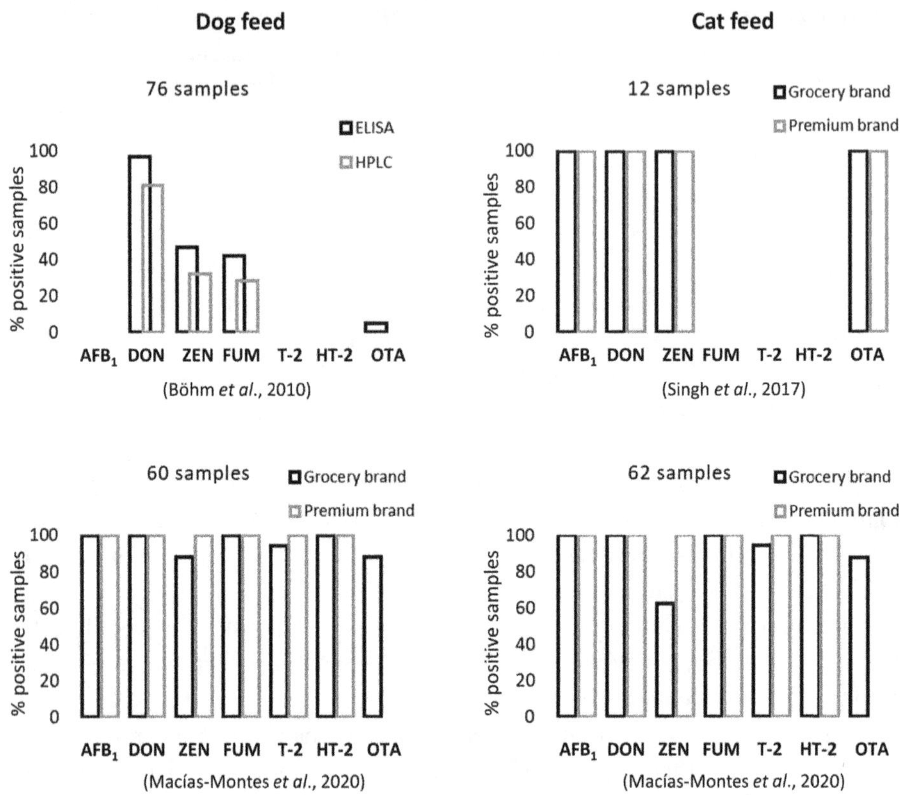

FIGURE 8.1 Summary of surveys performed in dog and cat feed.

and NIV (Witaszak et al., 2020). In grain-free diets, cereals are replaced by potatoes and beet pulp, which are also sources of mycotoxin contamination. The presence of these *Fusarium* mycotoxins has previously been detected in both sugar beet pulp and potato (Boudra et al., 2015; Qi et al., 2024).

Others have noted the importance of masked and hidden mycotoxins in pet feeds. As described in chapter 2, masked mycotoxins are undetectable by conventional analytical methods because their structures are modified within the plants that they contaminate (Freire et al., 2018), and these modified mycotoxins (metabolites of fungi) are undetected when testing for the parent mycotoxin. These modified forms can be produced or generated as part of an infected plant's defence mechanism. They may also be produced during feed processing (Kępinska-Pacelik and Biel, 2021b). Because of their hidden state, they may be underestimated as part of a risk assessment performed on pet feeds (Kovač et al., 2018). To assess the total risk to pets, both modified and hidden mycotoxins should be considered as potential contaminants (Kępinska-Pacelik and Biel, 2021b).

Because such a high number of pets are fed feeds that are susceptible to the growth of fungi which produce mycotoxins, there is ongoing risk of both acute and chronic exposures. It is understandable that many studies have revealed mycotoxin levels that often exceed the tolerable levels for dogs. The soil fungi that create mycotoxins are ubiquitous in the soils where agricultural crops of maize, nuts and grains are grown (Leung et al., 2006). Because multiple mycotoxins may be present within the same feed ingredient, multiple mycotoxin exposures are probably occurring at the same time, making it quite difficult to evaluate affected animals. Some researchers have referenced a multi-etiological syndrome after exposure to pet feeds contaminated with multiple mycotoxins (Arnot et al., 2012; Mulunda et al., 2013; Singh and Chuturgoon, 2017).

3 Main mycotoxins affecting pets

While there are many reports of mycotoxin contamination in pet feeds, there are far fewer describing disease in dogs and cats attributable to mycotoxin exposure. This does not imply that mycotoxin exposure is unimportant in dogs and cats. Rather, it is more likely that mycotoxin disease is not readily recognized by treating veterinarians and may not be actively uncovered in the diagnostic process. While acute, high dose exposures to mycotoxins can result in significant morbidity and mortality, chronic low-dose exposures can lead to a multitude of chronic conditions that are not easily connected to mycotoxin exposures such as liver and kidney fibrosis, infections resulting from immunosuppression, and cancer. These findings suggest that mycotoxin contamination in pet feed poses a serious health threat to pets (Boermans and Leung, 2007). Table 1 summarizes the clinical findings associated with exposure to mycotoxins.

3.1 *Deoxynivalenol and Diacetoxyscirpenol*

Common mechanisms of the trichothecenes include inhibition of protein and DNA synthesis, interference with normal cellular metabolic activities leading to cell death. Rapidly dividing cells of the GIT and the immune system are most vulnerable to the toxic effects of the trichothecenes (Boermans and Leung, 2007; Rotter et al., 1996; Smith, 1992). Several studies in dogs have demonstrated the effects of trichothecenes on FI and vomiting in dogs and cats. In a two-week feeding trial in dogs and cats it was estimated that 4.5 mg of DON per kg feed produced significant decreases in FI. Vomiting was observed with exposures of 8 mg of DON per kg feed. Cats were slightly more resistant to these effects than dogs, but were still affected (Hughes et al., 1999). Interestingly, dogs in the study were able to preferentially select an uncontaminated diet when given the opportunity. Another study investigating similar effects found that 3 mg of DON per kg feed caused weight loss and reduced feed intake in dogs (Leung et al., 2007).

TABLE 8.1 Clinical findings associated with mycotoxin exposure in pets

Myco-toxins	Species	Exposure doses (mg/kg BW)		Clinical signs	Clinical pathology (acute and chronic)	Post-mortem find
		Acute	Chronic			
DON	Dogs Cats			Decreased feed intake, vomiting, weight loss, bloody diarrhoea, immunosuppression		
FUM	Dogs Cats			Insidious liver failure, anorexia, vomiting, lethargy, oedema, icterus, and enlarged liver. Later signs include coagulopathy and hepatic encephalopathy. Increased susceptibility to infections with poor response to treatment. Heart failure with fluid overload and pulmonary edema. Possibly carcinogenic	Increased serum ALT, AST, ALP; coagulopathies	
ZEN	Dogs	0.2	0.025-0.05 (50 days)	Pyometra and changes in the reproductive tracts of sexually intact dogs. GIT inflammation, changes in immune function with inflammatory reactions, evidence of damage to kidneys and liver insufficiency		
AFs	Dogs	>0.5-1.0	0.05-0.3 (6-8 weeks)	Vomiting, anorexia, haematemesis, haematochezia, melena, lethargy, icterus, enlarged livers, disseminated intravascular coagulation, internal haemorrhage, and death in 2-3 weeks. Chronic exposures lead to carcinogenesis, hepatic encephalopathy evidenced by depression, stupor, seizures, and coma	Increased serum ALT, AST, ALP; coagulopathies	Hemorrhage, hepatomegaly, oe ma and congestio GI lumen, splenor aly, dark red kidne with subcapsular depressions and multifocal hemorrhages
OTA	Dogs Cats	7.8	0.2 (2 weeks)	Anorexia, weight loss, vomiting, tenesmus, bloody diarrhoea, fever, polydipsia, polyuria, dehydration, and carcinogenesis. Acute kidney injury progressing to chronic renal failure		Degeneration of re proximal tubules interstitial fibrosis
Tremor-genic	Dogs			Hyperexcitability, panting, tachycardia, generalized muscle fasciculations, hyperextension of limbs, blepharospasms		

ALP: alkaline phosphatase; ALT: alanine aminotransferase; AST: aspartate aminotransferase; GI: gastrointestinal; GIT: gastrointestinal tract.

The clinical effects typically seen in dogs and cats exposed to trichothecene mycotoxins include feed refusal, decreased FI, vomiting, weight loss, bloody diarrhoea, and immunosuppression (Boermans and Leung, 2007). Where pets present with similar signs but are recalcitrant to traditional therapies for these conditions, ongoing trichothecene exposure should be suspected. Suspect dog and cat feeds can be submitted for mycotoxin analyses. If a trichothecene or other mycotoxin is revealed, elimination of the suspect feed is indicated. Therapy, following cessation of exposure, can include the use of appetite stimulants, antiemetics, fluid and electrolyte replacement in cases of dehydration and traditional treatments for any infectious diseases when they are present. In cases of anaemia or severe immunosuppression, the use of blood products may be indicated until the effects of the mycotoxins on the bone marrow have subsided – the use of immunoglobulins through plasma products may be helpful.

Experimentally, an intravenous dose of DAS (0.5 mg/kg BW) damaged the haematopoietic cells in the bone marrow of dogs (Coppock et al., 1989).

3.2 *Fumonisins*

While many studies have demonstrated the contamination of pet feeds with FUM, most of the toxicology investigations reported are on animal species other than pets. This is important, as the clinical effects of FUM vary between animal species (Boermans and Leung, 2007). FB_1 was found in higher concentrations where dog and cat feeds contained both maize and wheat, although in some studies these remained within the acceptable levels recommended by regulatory bodies (Witaszak et al., 2019, 2020). However, it is often concluded that pets' health may be affected by these low exposure levels, especially when considering that ingestion of FUM by pets across their lifespan could ultimately impact the development of chronic disease (Witaszak et al., 2020). Most animal studies have concluded that the toxic effects of FUM are cumulative (Kępinska-Pacelik and Biel, 2021b).

The structure of FB_1 is similar to that of the sphingolipids, which are important components of the plasma membrane of cells and are regulators of cell-to-cell interactions and cell-to-cell recognition (Hannun and Obeid, 2018; Witaszak et al., 2020). FB_1 can compete with the sphingolipids sphinganine and sphingosine, altering the integrity of the plasma membrane and interrupting cellular processes. As FB_1 exposure increases, leakage of transaminases from hepatocytes can be detected (Chen et al., 2021; Witaszak et al., 2020). In addition to impacting the integrity of the plasma membrane, FB_1 causes oxidative stress, immunotoxicity and organ toxicity in various species of animals including the liver, kidneys, heart, lungs, brain, reproductive tract, and GIT (Witaszak et al., 2020).

Considering these exposure risks and the mechanisms of toxicity, clinical effects in exposed dogs and cats could include a variety of clinical presentations. In all likelihood, progressive deterioration in organ function might be expected, with insidious liver failure possible as the presenting clinical effect. As with the AFs, evidence of liver insult might first present with elevations in the serum concentrations of the transaminases ALT, AST, and ALP. Evidence of liver insult can present clinically with signs of anorexia, vomiting, lethargy, oedema, icterus, and enlarged liver on palpation. These signs can progress to include coagulopathy and hepatic encephalopathy. The effects on immune functioning may first present as increased susceptibilities to infectious organisms, or poor response to treatment for infections. Effects on the cardiovascular and pulmonary systems might include the development of heart failure, with accumulations in fluid load on the vascular system and development of pulmonary oedema. While this is more likely to occur in pigs, it is possible to see these effects in other species including pets. Considering the renal system, evidence of renal tumours might be an indication of FUM exposure in pets (Chen et al., 2021).

FUM are classified as group 2B (possibly carcinogenic) by the International Agency for Research on Cancer (IARC). While the carcinogenic mechanism of FUM is not fully understood, they are known to affect sphingolipid metabolism, leading to intracellular accumulation of sphingoid bases. Changes in sphinganine and sphingosine bases disturb the cell cycle, mainly DNA replication and cell proliferation (Frehse et al., 2015; Riley et al., 2001). A study associating the development of mammary tumours in dogs exposed to mycotoxins found an association between the development of cancer and exposure to AFs. However, it is worthwhile to note that FUM were also contaminating these diets and the association between carcinogenesis and FUM needs to be studied further to evaluate the risks (Frehse et al., 2015).

As with the treatment of pets exposed to AFs, the clinical care plan for dogs and cats exposed to FUM is focused primarily on eliminating sources of exposure and supporting the animals based on the presentation of clinical signs.

3.3 *Zearalenone*

In intact female dogs, ZEN can induce oviduct and uterine hyperplasia and oedema. In intact male dogs it results in reduced spermatogenesis (Boermans and Leung, 2007; Mahato et al., 2021; Witaszak et al., 2020).

Surveys of commercial dog and cat feeds have found many samples contaminated with ZEN (Leung et al., 2006; Singh et al., 2017; Singh and Chuturgoon, 2017; Witaszak et al., 2020). This high incidence highlights the need for periodic monitoring of pet feeds and for examination of pets for potential clinical signs. Pathological changes can be seen within the reproductive tracts of dogs after 7 days of exposure to 200 μg/kg BW daily or after 50 to 100 days of exposure to 25 to 50 μg/kg BW (Böhm et al., 2010; Gajęcka et al., 2007).

The clinical effects of ZEN are poorly understood in dogs and cats. For many years the oestrogenic effects were well known and characterized across animal species. However, because most pet dogs and cats have had their reproductive organs and tissues removed, the oestrogenic effects were not considered significant. However, there are other mechanisms associated with both acute and chronic exposures. Oxidative stress occurs in the intestines and this can cause structural changes to the intestinal villi. It has also been shown to cause genotoxicity and oxidative injuries to DNA. It can also lead to DNA fragmentation, cell cycle arrest, micronuclei formation and chromosomal aberrations. In other tissues it exerts nephrotoxic and hepatotoxic effects and can lead to the development of hepatocarcinoma (Abid-Essefi et al., 2003; Cheng et al., 2020; Fleck et al., 2012; Grenier and Applegate, 2013; Mahato et al., 2021). Immune response effects have also been described, showing that ZEN has a negative impact on white blood cells, B-cells, immunoglobulins, T-cells, and cytokines that are produced in response to infection (Mahato et al., 2021; Marin et al., 1996). A 50-day dietary exposure to ZEN at 25 and 50 μg/kg BW per day resulted in a reduction in serum antibodies and white blood cell counts in dogs (Boermans and Leung, 2007; Gajęcka et al., 2007, 2015).

For sexually intact dogs, diagnosis and treatment for ZEN toxicosis includes examination of suspect feeds for the presence and concentration of ZEN and elimination of any sources of exposure. Cessation of exposure may be enough for signs referrable to the reproductive tract to subside. However, surgical sterilization may be indicated in some cases. Antibiotics and surgical sterilization may be indicated in bitches with evidence of both pyometra and changes in the reproductive tract.

For dogs and cats with evidence of alterations in immune functioning, oxidative and inflammatory reactions, or evidence of organ damage such as kidney or liver insufficiency, treatments would be aimed at addressing the clinical signs in individual animals consistent with general standards of care.

3.4 *Aflatoxins*

While only a small number of mycotoxin exposure outbreaks in dogs and cats have been published, outbreaks relating to AFs have been the most reported following exposure to commercial dog feed, with maize being the usual source in these cases. After ingestion, AFs are absorbed in the duodenum. Upon absorption they bind to plasma albumin and are transported to various tissues, especially the liver and the kidneys. AFs undergo biotransformation to highly electrophilic forms. These can bind to DNA forming adducts, to RNA and to protein enzymes resulting in damage to liver cells (Boermans and Leung, 2007; Cullen and Newberne, 1994; Martínez-Martínez et al., 2021). The primary clinical effects of aflatoxicosis are related to liver damage. In acute cases of AFs exposure, dogs ingesting > 0.5-1 mg/kg typically die within days, showing enlarged livers, disseminated intravascular coagulation and internal haemorrhage. Exposures of 0.5 to 1 mg of AFs per kg feed cause a subacute syndrome characterized by vomiting, anorexia, haematemesis, haematochezia, melena, lethargy, icterus, disseminated intravascular haematemesis, and death within 2 to 3 weeks of exposure. Chronic AFs exposure to doses ranging from 0.05 to 0.3 mg of AFs per kg of pet feed can produce similar clinical effects over exposure lengths of 6 to 8 weeks. AFs are also carcinogenic. They bind to DNA causing genotoxicity and mutation in cells. These chronic changes occur at doses much lower than the doses associated with acute illness (Boermans and Leung, 2007; Bohn and Razzai-Fazeli, 2005; Martínez-Martínez et al., 2021). Clinical signs associated with central nervous system disturbances are the result of hepatic encephalopathy, which can be evidenced by depression, stupor, seizures, and coma. Because of the clinical signs associated with acute and chronic changes in liver function and the development of certain cancers over time, the actual numbers of dogs affected by AFs is thought to be far more than the numbers reflected in published case reports associated with acute poisoning cases.

A study investigating the impact of AFs on the development of mammary tumours in intact female dogs demonstrated an association with an increased chance of developing mammary tumours in dogs exposed chronically to AFB_1, AFG_1, and AFG_2 (Frehse et al., 2015).

Laboratory findings observed with AFs exposure in dogs include serum elevations in liver transaminases, including ALT, AST, and ALP. These changes in serum/plasma concentrations of liver enzymes are biomarkers for damage to the liver induced by the epoxides generated by AFs biotransformation reactions. Blood tests of coagulation parameters can also demonstrate a poor ability of the blood to clot as evidence of liver damage induced by AFs (Martínez-Martínez et al., 2021). Post-mortem findings in dogs can include gross lesions demonstrating haemorrhage, hepatomegaly, oedema and congestion of the gastrointestinal lumen, splenomegaly, and dark red colouration with subcapsular depressions in the kidneys with multifocal haemorrhages (Martínez-Martínez et al., 2021).

The function of the GST enzymes is an important detoxification mechanism after AFs exposure. These enzymes bind the reactive epoxides generated after AFs are biotransformed in the liver. These enzymes allow the reactive epoxides to bind glutathione, which is subsequently eliminated through bile or urine. Dogs have reduced GST activity, which is the reason why they are especially susceptible to the damaging effects of AFs after ingestion (Bruchim et al., 2012; Martínez-Martínez et al., 2021; Rawal et al., 2010). Administration of N-acetylcysteine has been successful in dogs because it provides L-cysteine for the intracellular production of glutathione, which interacts with GST in the detoxification process of AFs. In addition, N-acetylcysteine acts as a free radical scavenger and has anti-inflammatory effects after administration in dogs (Martínez-Martínez et al., 2021; Viviano and VanderWielen, 2013).

Treatment of dogs suffering from aflatoxicosis is aimed at correcting coagulation abnormalities, and haemodynamic stabilization. Whole blood products as well as plasma products can support anaemic dogs while replacing depleted clotting factors and replacing platelets. Vitamin K1 can also help to support the production of activated clotting factors. Fluid therapy with intravascular crystalloid fluids can help to maintain haemodynamic stability. Gastrointestinal effects can be treated with the administration of gastric protectants such as sucralfate and famotidine, and antiemetics such as maropitant and ondansetron (Martínez-Martínez et al., 2021).

While there is evidence in the literature of exposure to AFs in cats through commercial cat feeds, there is a scarcity of reports on the acute and chronic effects in cats. Since cats are obligate carnivores, they may lack the biotransformation capacity to metabolise substances that are mainly of plant origin. Because the detrimental effects of AFs are mainly due to production of reactive metabolites in the liver, cats may not biotransform as efficiently as in dogs. More research examining the acute and chronic effects in cats is needed. In particular, the risk that chronic exposure to mycotoxins in cats with already poor health or concurrent disease might impact their susceptibility to toxic injuries (Singh et al., 2017).

3.5 *Ochratoxin A*

Where ochratoxins are present, initial clinical signs observed in all species include anorexia, polydipsia, polyuria, and dehydration and are associated with renal damage. After ingestion, ochratoxins are absorbed and enter the systemic circulation, bind tightly to serum proteins and distribute to and accumulate within the kidneys. There they are disruptors of protein synthesis and interfere with the function of the proximal tubular cells. Ultimately, there is degeneration of the proximal tubules with interstitial fibrosis. There is also some evidence of carcinogenesis, in that they have been noted to induce renal tumours in some animal models. However the mechanism for this remains controversial (Boermans and Leung, 2007; Faucet et al., 2004; Krogh, 1992; Mally et al., 2004; Meucci et al., 2017).

Dogs are highly susceptible to the toxic effects of OTA. A daily dose of 0.2 mg/kg BW for two weeks, or a single dose of 7.8 mg/kg BW were fatal in experiments with beagle dogs (Boermans and Leung, 2007; Szczech et al., 1973). Others have established the chronic toxic dose in dogs to be 0.2 mg/kg feed (Razzazi et al., 2001; Singh and Chuturgoon, 2017).

While both chronic renal disease and the occurrence of OTA in pet diet are relatively common, some have speculated about the relationship between chronic exposures to ochratoxins and the development of renal failure. One study attempted to correlate plasma OTA concentrations in dogs with chronic renal failure. The study compared plasma OTA concentrations between dogs with known renal disease with those without any clinical signs or laboratory findings associated with renal disease. The study found a statistically significant difference between the concentrations of OTA in the plasma of healthy dogs versus those with evidence of chronic kidney disease. The results suggest that OTA may have a role in the onset of kidney disease in dogs, although the concentration of OTA found in the plasma of dogs in renal failure did not correlate with the degree of failure (Meucci et al., 2017).

Because many dog and cat feeds have been found to contain ochratoxins, and by comparison the prevalence of renal disease is lower, it is difficult to make an association between exposure to OTA and the development of disease in dogs and cats (Macías-Montes et al., 2020; Razzazi et al., 2001; Singh et al., 2017; Singh and Chuturgoon, 2017). In addition to the effects on the kidneys, OTA exposure leads to oxidative damage to lipids, proteins and DNA through the over-production of ROS, and the reduction of antioxidant defence mechanisms of cells by reducing glutathione availability. The extent of these mechanisms on animal health is not clearly understood. Exposure to OTA has been associated with neurotoxic and teratogenic effects and immune impairment (Kępinska-Pacelik and Biel, 2021b; Stoev et al., 2022).

Clinical signs of toxic exposures to OTA in dogs include anorexia, weight loss, vomiting, tenesmus, bloody diarrhoea, fever, and dehydration (Boermans and Leung, 2007). Dogs and cats presenting evidence of ochratoxicosis, such as renal disease not attributable to other causes, and evidence of oxidative injuries, with or without immunosuppression, can benefit from the elimination of potential sources of exposure from the diet. Further treatments should be directed toward standards of care for other causes of renal disease and can include the use of protein-modified diets, fluid replacement, phosphate binders, appetite stimulants, antiemetics, and frequent monitoring by the treating veterinarian.

3.6 *Tremorgenic Mycotoxins*

Penitrem A and roquefortine C are tremorgenic mycotoxins produced by *Penicillium* fungi. Dogs can be exposed to these mycotoxins through the ingestion of spoiled food wastes or fallen nuts and fruits. It can also be found in the grass *Poa huecu* (Scuteri et al., 1992). In certain regions, exposures in dogs have been associated with over-wintered fallen walnuts with the dog only needing to hold the shell-encased walnut within its mouth for a short period of time. One clinical case report indicated that a dog was exposed through a mouldy cream cheese wrapper and macaroni and cheese that the dog ate out of the garbage (Young et al., 2003). Once the mycotoxin is ingested, it is absorbed into the systemic circulation where it passes the blood-brain barrier into the central nervous system. While not completely understood, absorption is believed to affect GABAergic neurotransmission. It may also inhibit glycine transmission in inhibitory neurons (Bradford et al., 1990). In some cells it also interferes with potassium channels and causes cell death (Moldes-Anaya et al., 2009, 2011; Uhlig et al., 2020). Ultimately, there is a loss of coordination of neural mechanisms that control muscle action and brain stem reflexes.

Clinical signs in exposed dogs include hyperexcitability, panting, tachycardia, and generalized muscle fasciculations. Hyperextension of the limbs and blepharospasms can also be observed. Treatment is aimed at controlling the muscle tremours, which, in case reports, were resolved after administration of phenobarbital. Interestingly, the administration of a benzodiazepine alone did not resolve the muscle tremours but was later useful when administered in combination with phenobarbital (Young et la., 2003). Clinical signs generally resolve within 24 to 48 hours following appropriate treatments. In severe, untreated cases, muscle fasciculations may progress to seizures.

3.7 Conclusions

Prevention of exposure is key to controlling mycotoxin disease. It is important that mouldy food wastes be disposed in areas where pets cannot gain access. Dogs and cats are susceptible to mycotoxin diseases when exposed to dangerous mycotoxins through pet feeds. For all mycotoxins, limiting further exposure is paramount in recovery from mycotoxin ill-effects and disease. Treatment is aimed primarily at controlling the clinical signs observed in individual animals, providing care accordingly, and supporting their recovery. Most mycotoxin exposures are associated with the feeding of dry, extruded kibbled types of pet feed. Switching the pet from a dry, extruded kibble diet and replacing it with wet-type feeds would be a significant way to minimize, or even eliminate further mycotoxin exposure. A further consideration would be to focus the diet on freshly prepared feeds containing only high quality grains which are less susceptible to mould growth and that do not contain any added starches. Diets such as these have been found to be at lowest risk of mycotoxin accumulation (Tegzes et al., 2019).

References

Abid-Essefi, S., Baudrimont, I., Hassen, W., Ouanes, Z., Mobio, T.A., Anane, R., Creppy, E.E. and Bacha, H., 2003. DNA fragmentation, apoptosis and cell cycle arrest induced by zearalenone in cultured DOK, Vero and Caco-2 cells: prevention by Vitamin E. Toxicology 192: 237–248. https://doi.org/10.1016/s0300-483x(03)00329-9.

Alaboudi, A.R., Osaili, T.M. and Otoum, G., 2022. Quantification of mycotoxin residues in domestic and imported chicken muscle, liver and kidney in Jordan. Food Control 132: 108511. https://doi.org/10.1016/j.foodcont.2021.108511.

Arnot, L.F., Duncan, N.M., Coetzer, H. and Botha, C.J., 2012. An outbreak of canine aflatoxicosis in Gauteng Province, South Africa. Journal of the South African Veterinary Association 83: 2. https://doi.org/10.4102/jsava.v83i1.2.

Boermans, H.J., Leung and M.C.K., 2007. Mycotoxins and the pet food industry: toxicological evidence and risk assessment. International Journal of Food Microbiology 119: 95–102. https://doi.org/10.1016/j.ijfoodmicro.2007.07.063.

Böhm, J., Koinig, L., Razzazi-Fazeli, E., Blajet-Kosicka, A., Twaruzek, M., Grajewski, J. and Lang, C., 2010. Survey and risk assessment of the mycotoxins deoxynivalenol, zearalenone, fumonisins, ochratoxin A, and aflatoxins in commercial dry dog food. Mycotoxin Research 26: 147–153. https://doi.org/10.1007/s12550-010-0049-4.

Böhn, J. and Razzai-Fazeli, E., 2005. Effects of mycotoxins on domestic pet species, in: Diaz, D. (Ed.), The Mycotoxin Blue Book. Nottingham University Press, Nottingham, pp. 77–92.

Boudra, H., Rouillé, B., Lyan, B. and Morgavi, D., 2015. Presence of mycotoxins in sugar beet pulp silage collected in France. Animal Feed Science and Technology 205: 131-135. https://hal.science/hal-01901484/document.

Bradford, H.F., Norris, P.J. and Smith, C.C., 1990. Changes in transmitter release patterns in vitro induced by tremorgenic mycotoxins. Journal of Environmental Pathology, Toxicology and Oncology 10: 17–30.

Bruchim, Y., Segev, G., Sela, U., Bdolah-Abram, T., Salomon, A. and Aroch, I., 2012. Accidental fatal aflatoxicosis due to contaminated commercial diet in 50 dogs. Research in Veterinary Science 93: 279–287. https://doi.org/10.1016/j.rvsc.2011.07.024.

Bryden, W.L., 2012. Mycotoxin contamination of the feed supply chain: Implications for animal productivity and feed security. Animal Feed Science and Technology 173: 134–158. https://doi.org/10.1016/j.anifeedsci.2011.12.014.

Chen, J., Wen, J., Tang, Y., Shi, J., Mu, G., Yan, R., Cai, J. and Long, M., 2021. Research Progress on Fumonisin B1 Contamination and Toxicity: A Review. Molecules 26: 5238. https://doi.org/10.3390/molecules26175238.

Cheng, Q., Jiang, S., Huang, L., Wang, Y., Yang, W., Yang, Z. and Ge, J., 2020. Effects of zearalenone-induced oxidative stress and Keap1-Nrf2 signaling pathway-related gene expression in the ileum and mesenteric lymph nodes of post-weaning gilts. Toxicology 429: 152337. https://doi.org/10.1016/j.tox.2019.152337.

Coppock, R.W., Hoffmann, W.E., Gelberg, H.B., Bass, D. and Buck, W.B., 1989. Hematologic changes induced by intravenous administration of diacetoxyscirpenol in pigs, dogs, and calves. American journal of Veterinary Research 50: 411–415.

Corsato Alvarenga, I. and Aldrich, C.G., 2020. Starch characterization of commercial extruded dry pet foods. Translational Animal Science 4: 1017-1022. https://doi.org/10.1093/tas/txaa018.

Corsato Alvarenga, I., Dainton, A.N. and Aldrich, C.G., 2021. A review: nutrition and process attributes of corn in pet foods. Critical Reviews in Food Science and Nutrition 62: 8567-8576. https://doi.org/10.1080/10408398.2021.1931020.

Cullen, J. and Newberne, P., 1994. Acute hepatotoxicity of aflatoxins, in: Eaton, D., Groopman, J. (Eds.), The Toxicology of Aflatoxins: Human Health, Veterinary and Agricultural Significance. Academic Press, San Diego, CA, pp. 3–26.

Faucet, V., Pfohl-Leszkowicz, A., Dai, J., Castegnaro, M. and Manderville, R.A., 2004. Evidence for covalent DNA adduction by ochratoxin A following chronic exposure to rat and subacute exposure to pig. Chemical Research in Toxicology 17: 1289–1296. https://doi.org/10.1021/tx049877s.

Fleck, S.C., Hildebrand, A.A., Müller, E., Pfeiffer, E. and Metzler, M., 2012. Genotoxicity and inactivation of catechol metabolites of the mycotoxin zearalenone. Mycotoxin Research 28: 267–723. https://doi.org/10.1007/s12550-012-0143-x.

Frehse, M.S., Martins, M.I.M., Ono, E.Y.S., Bracarense, A.P.F.R.L., Bissoqui, L.Y., Teixeira, E.M.K., Santos, N.J.R. and Freire, R.L., 2015. Aflatoxins ingestion and canine mammary tumors: There is an association? Food and Chemical Toxicology 84: 74–78. https://doi.org/10.1016/j.fct.2015.08.004.

Freire, L., Guerreiro, T.M., Caramês, E.T.S., Lopes, L.S., Orlando, E.A., Pereira, G.E., Lima Pallone, J.A., Catharino, R.R. and Sant'Ana, A.S., 2018. Influence of maturation stages in different varieties of wine grapes (*Vitis vinifera*) on the production of Ochratoxin A and its modified forms by *Aspergillus carbonarius* and *Aspergillus niger*. Journal of Agricultural and Food Chemistry 66: 8824–8831. https://doi.org/10.1021/acs.jafc.8b02251.

Gajęcka, M., Janowski, T., Jakimiuk, E., Polak, M., Podhalicz-Dziegielewska, M., Rotkiewicz, T., Otrocka-Domagała, I., Obremski, K., Zielonka, L. and Gajęcki, M., 2007. Histopathological and immunohistochemical examinations, and changes in proliferation activity of the uterus in bitches following zearalenone micotoxicosis. Polish Journal of Veterinary Sciences 10: 143–151.

Gajęcka, M., Zielonka, Ł. and Gajęcki, M., 2015. The effect of low monotonic doses of zearalenone on selected reproductive tissues in pre-pubertal female dogs – A review. Molecules 20: 20669–20687. https://doi.org/10.3390/molecules201119726.

Grandi, M., Vecchiato, C.G., Biagi, G., Zironi, E., Tondo, M.T., Pagliuca, G., Palmonari, A., Pinna, C., Zaghini, G. and Gazzotti, T., 2019. Occurrence of mycotoxins in extruded commercial cat food. American Chemical Society Omega 4: 14004–14012. https://doi.org/10.1021/acsomega.9b01702.

Grenier, B. and Applegate, T.J., 2013. Modulation of intestinal functions following mycotoxin ingestion: meta-analysis of published experiments in animals. Toxins 5: 396–430. https://doi.org/10.3390/toxins5020396.

Hannun, Y.A. and Obeid, L.M., 2018. Sphingolipids and their metabolism in physiology and disease. Nature Reviews Molecular Cell Biology 19: 175–191. https://doi.org/10.1038/nrm.2017.107.

Hughes, D.M., Gahl, M.J., Graham, C.H. and Grieb, S.L., 1999. Overt signs of toxicity to dogs and cats of dietary deoxynivalenol. Journal of Animal Science 77: 693–700. https://doi.org/10.2527/1999.773693x.

Kępińska-Pacelik, J. and Biel, W., 2021a. Microbiological hazards in dry dog chews and feeds. Animals 11: 631. https://doi.org/10.3390/ani11030631.

Kępińska-Pacelik, J. and Biel, W., 2021b. Alimentary risk of mycotoxins for humans and animals. Toxins 13: 822. https://doi.org/10.3390/toxins13110822.

Kovač, M., Šubarić, D., Bulaić, M., Kovač, T. and Šarkanj, B., 2018. Yesterday masked, today modified; what do mycotoxins bring next? Archives of Industrial and Toxicology 69: 196–214. https://doi.org/10.2478/aiht-2018-69-3108.

Krogh, P., 1992. Role of ochratoxin in disease causation. Food and Chemical Toxicology 30: 213–224. https://doi.org/10.1016/0278-6915(92)90036-k.

Leung, M.C.K., Díaz-Llano, G. and Smith, T.K., 2006. Mycotoxins in pet food: a review on worldwide prevalence and preventative strategies. Journal of Agricultural and Food Chemistry 54: 9623–9635. https://doi.org/10.1021/jf062363+.

Leung, M.C.K., Smith, T.K., Karrow, N.A. and Boermans, H.J., 2007. Effects of foodborne *Fusarium* mycotoxins with and without a polymeric glucomannan mycotoxin adsorbent on food intake and nutrient digestibility, body weight, and physical and clinicopathologic variables of mature dogs. American Journal of Veterinary Research 68: 1122–1129. https://doi.org/10.2460/ajvr.68.10.1122.

Macías-Montes, A., Rial-Berriel, C., Acosta-Dacal, A., Henríquez-Hernández, L.A., Almeida-González, M., Rodríguez-Hernández, Á., Zumbado, M., Boada, L.D., Zaccaroni, A., Luzardo, O.P., 2020b. Risk assessment of the exposure to mycotoxins in dogs and cats through the consumption of commercial dry food. The Science of the Total Environment 708: 134592. https://doi.org/10.1016/j.scitotenv.2019.134592.

Mahato, D.K., Devi, S., Pandhi, S., Sharma, B., Maurya, K.K., Mishra, S., Dhawan, K., Selvakumar, R., Kamle, M., Mishra, A.K. and Kumar, P., 2021. Occurrence, impact on agriculture, human health, and management strategies of zearalenone in food and feed: A review. Toxins 13: 92. https://doi.org/10.3390/toxins13020092.

Mally, A., Zepnik, H., Wanek, P., Eder, E., Dingley, K., Ihmels, H., Völkel, W. and Dekant, W., 2004. Ochratoxin A: lack of formation of covalent DNA adducts. Chemical Research in Toxicology 17: 234–242. https://doi.org/10.1021/tx034188m.

Marin, M.L., Murtha, J., Dong, W. and Pestka, J.J., 1996. Effects of mycotoxins on cytokine production and proliferation in EL-4 thymoma cells. Journal of Toxicology and Environmental Health 48: 379–396. https://doi.org/10.1080/009841096161267.

Martínez-Martínez, L., Valdivia-Flores, A.G., Guerrero-Barrera, A.L., Quezada-Tristán, T., Rangel-Muñoz, E.J. and Ortiz-Martínez, R., 2021. Toxic effect of aflatoxins in dogs fed contaminated commercial dry feed: A review. Toxins 13: 65. https://doi.org/10.3390/toxins13010065.

Meucci, V., Luci, G., Vanni, M., Guidi, G., Perondi, F. and Intorre, L., 2017. Serum levels of ochratoxin A in dogs with chronic kidney disease (CKD): a retrospective study. The Journal of Veterinary Medical Science 79: 440–447. https://doi.org/10.1292/jvms.16-0226.

Mohammedi-Ameur, S., Dahmane, M., Brera, C., Kardjadj, M. and Ben-Mahdi, M.H., 2020. Occurrence and seasonal variation of aflatoxin M1 in raw cow milk collected from different regions of Algeria. Veterinary World 13: 433–439. https://doi.org/10.14202/vetworld.2020.433-439.

Moldes-Anaya, A., Wilkins, A.L., Rundberget, T. and Faeste, C.K., 2009. *In vitro* and *in vivo* hepatic metabolism of the fungal neurotoxin penitrem A. Drug and Chemical Toxicology 32: 26–37. https://doi.org/10.1080/01480540802416232.

Moldes-Anaya, A.S., Fonnum, F., Eriksen, G.S., Rundberget, T., Walaas, S.I. and Wigestrand, M.B., 2011. In vitro neuropharmacological evaluation of penitrem-induced tremorgenic syndromes: importance of the GABAergic system. Neurochemistry International 59: 1074–1081. https://doi.org/10.1016/j.neuint.2011.08.014.

Muaz, K., Riaz, M., Oliveira, C.A.F. de, Akhtar, S., Ali, S.W., Nadeem, H., Park, S. and Balasubramanian, B., 2021. Aflatoxin M_1 in milk and dairy products: global occurrence and potential decontamination strategies. Toxin Reviews 41: 2022. https://doi.org/10.1080/15569543.2021.1873387.

Mulunda, M., Ndou, R. v, Dzoma, B., Nyirenda, M. and Bakunzi, F., 2013. Canine aflatoxicosis outbreak in South Africa (2011): a possible multi-mycotoxins aetiology. Journal of the South African Veterinary Association 84: E1-5. https://doi.org/10.4102/jsava.v84i1.133.

Nazhand, A., Durazzo, A., Lucarini, M., Souto, E.B. and Santini, A., 2020. Characteristics, occurrence, detection and detoxification of aflatoxins in foods and feeds. Foods 9: 644. https://doi.org/10.3390/foods9050644.

Pigłowski, M., 2019. Pathogenic and non-pathogenic microorganisms in the rapid alert system for food and feed. International Journal of Environmental Research and Public Health 16: 477. https://doi.org/10.3390/ijerph16030477.

Qi, T., He, F., Zhang, X., Wang, J., Zhang, Z., Jiang, H., Zhao, B., Du, C., Che, Y., Feng, X., Wang, Y. and Li F., 2024. Genome-wide identification and expression profiling of potato (*Solanum tuberosum* L.) universal stress proteins reveals essential roles in mechanical damage and deoxynivalenol stress. International Journal in Molecular Science. 25: 1341. https://doi.org/10.3390/ijms25021341.

Rawal, S., Kim, J.E. and Coulombe, R., 2010. Aflatoxin B1 in poultry: toxicology, metabolism and prevention. Research in Veterinary Science 89: 325-331. https://doi.org/10.1016/j.rvsc.2010.04.011.

Razzazi, E., Böhm, J., Grajewski, J., Szczepaniak, K., Kübber-Heiss, A.J. and Iben, C.H., 2001. Residues of ochratoxin A in pet foods, canine and feline kidneys. Journal of Animal Physiology and Animal Nutrition 85: 212–216. https://doi.org/10.1046/j.1439-0396.2001.00331.x.

Research and Markets, 2024. Global pet food extrusion market size and growth. Pet Food Extrusion Global Market Report 2024. https://www.researchandmarkets.com/reports/5895366/pet-food-extrusion-global-market-report.

Riley, R.T., Enongene, E., Voss, K.A., Norred, W.P., Meredith, F.I., Sharma, R.P., Spitsbergen, J., Williams, D.E., Carlson, D.B. and Merrill, A.H., 2001. Sphingolipid perturbations as mechanisms for fumonisin carcinogenesis. Environmental Health Perspectives 109: 301–308. https://doi.org/10.1289/ehp.01109s2301.

Rotter, B.A., Prelusky, D.B. and Pestka, J.J., 1996. Toxicology of deoxynivalenol (vomitoxin). Journal of Toxicology and Environmental Health 48: 1–34. https://doi.org/10.1080/009841096161447.

Scuteri, M., Sala de Miguel, M.A., Blanco Viera, J. and Planes de Banchero, E., 1992. Tremorgenic mycotoxins produced by strains of Penicillium spp. isolated from toxic *Poa huecu parodi*. Mycopathologia 120: 177–182. https://doi.org/10.1007/BF00436396.

Singh, S.D., Baijnath, S. and Chuturgoon, A.A., 2017. A comparison of mycotoxin contamination of premium and grocery brands of pelleted cat food in South Africa. Journal of the South African Veterinary Association 88: e1-e4.

Singh, S.D. and Chuturgoon, A.A., 2017. A comparative analysis of mycotoxin contamination of supermarket and premium brand pelleted dog food in Durban, South Africa. Journal of the South African Veterinary Association 88: a1488. https://doi.org/10.4102/jsava.v88i0.1488.

Smith, T.K., 1992. Recent advances in the understanding of *Fusarium* trichothecene mycotoxicoses. Journal of Animal Science 70: 3989–3993. https://doi.org/10.2527/1992.70123989x.

Stoev, S.D., Paskalev, M., MacDonald, S. and Mantle, P.G., 2002. Experimental one year ochratoxin A toxicosis in pigs. Experimental and Toxicologic Pathology 53: 481–487. https://doi.org/10.1078/0940-2993-00213.

Szczech, G.M., Carlton, W.W. and Tuite, J., 1973. Ochratoxicosis in Beagle dogs. I. Clinical and clinicopathological features. Veterinary Pathology 10: 135–154. https://doi.org/10.1177/030098587301000207.

Tegzes, J.H., Oakley, B.B. and Brennan, G., 2019. Comparison of mycotoxin concentrations in grain versus grain-free dry and wet commercial dog foods. Toxicology Communications 3: 61–66. https://doi.org/10.1080/24734306.2019.1648636.

Tulpule, P.G., 1981. Aflatoxins-experimental studies. Journal of Cancer Research and Clinical Oncology 99: 137–142. https://doi.org/10.1007/BF00412449.

Uhlig, S., Ivanova, L., Voorspoels, P. and Fæste, C.K., 2020. *In vitro* toxicokinetics and phase I biotransformation of the mycotoxin Penitrem A in dogs. Toxins 12: 293. https://doi.org/10.3390/toxins12050293.

Viviano, K.R. and VanderWielen, B., 2013. Effect of N-acetylcysteine supplementation on intracellular glutathione, urine isoprostanes, clinical score, and survival in hospitalized ill dogs. Journal of Veterinary Internal Medicine 27: 250–258. https://doi.org/10.1111/jvim.12048.

Witaszak, N., Stępień, Ł., Bocianowski, J. and Waśkiewicz, A., 2019. *Fusarium* species and mycotoxins contaminating veterinary diets for dogs and cats. Microorganisms 7: 26. https://doi.org/10.3390/microorganisms7010026.

Witaszak, N., Waśkiewicz, A., Bocianowski, J. and Stępień, Ł., 2020. Contamination of pet food with mycobiota and *Fusarium* mycotoxins-focus on dogs and cats. Toxins 12: 130. https://doi.org/10.3390/toxins12020130

Young, K.L., Villar, D., Carson, T.L., Ierman, P.M., Moore, R.A. and Bottoff, M.R., 2003. Tremorgenic mycotoxin intoxication with penitrem A and roquefortine in two dogs. Journal of the American Veterinary Medical Association 222: 52–53. https://doi.org/10.2460/javma.2003.222.52.

CHAPTER 9

Effects of mycotoxins in horses

J. Fink-Gremmels and J.C. Vendrig**

Department of Population Health Sciences, Institute for Risk Assessment Sciences (IRAS), Faculty of Veterinary Medicine, Utrecht University, Utrecht, the Netherlands
**correspondence: J.Fink@uu.nl; J.C.Vendrig@uu.nl*

1 Introduction

Mycotoxins are considered the most frequently occurring undesirable contaminants of animal food and feed materials as indicated by global surveys. These overviews included many feed materials that are also included in diets for horses such as cereal grains, maize and roughages. However, in contrast to the many investigations that established a relationship between the dietary intake and the adverse health effects in pigs and poultry, fish and ruminants, and even pets, systematic investigations in horses are scarce. The current knowledge about the adverse effects of individual mycotoxins in horses is largely based on case reports, which often lack a precise estimate of the ingested dose (Miyazaki et al., 2001).

Horses (and other domestic solipeds like donkeys and mules) are strict herbivores, and the typical diet of horses consists of grass, mixed pastures, hay and straw. These roughages are enriched with cereal grains, maize and other protein-rich components to balance energy needs, palatability, and digestive functions of (sport-) horses. Mycotoxin contamination of feed materials may comprise toxins formed already during plant growth at the preharvest stage. Examples include the large group of *Fusarium* toxins, in line with the global distribution of the genus *Fusarium* in the soil. *Fusarium* toxins are structurally highly diverse and include the FUM, the mycoestrogen ZEN, as well as the large group of trichothecenes, including more than 160 different toxins with DON, T-2 toxin, NIV and fusaric acid as the most prominent examples and the ENNs and BEA as emerging mycotoxins. Straw, which is used as feed and bedding in horse accommodation may also contain significant amounts of *Fusarium* toxins, and particularly DON in straw can reach high concentrations and contribute significantly to animal exposure. A special group, the macrocyclic trichothecenes, including *Stachybotrus chartarum* toxins, is frequently found in straw and hay and has been related to the first major fatal cases of mycotoxicosis in horses in the 1930s. In contrast to these detrimental effects of epiphytic fungi, which are often also plant pathogens, common pasture grasses may also benefit from the mutualistic symbiosis with endophytes, in

 | DOI: 10.1163/9789004712720_10

particular the *Epichloë* species, conveying via their secondary metabolites an increased resilience to insects, helminths and abiotic stressors such as drought and heat stress. These *Epichloë* species, however, also produce secondary metabolites such as ergovaline (*Neotyphodium coenophialum*) and lolitrems (*N. lolii*) which are toxic for horses and other herbivorous species. Moreover, following harvest and storage of grass (and grains) as hay or haylage and straw, storage moulds such as *Penicillium, Aspergillus* and *Alternaria* species can contribute to the multi-mycotoxin exposure pattern of animal diets. Of special relevance in equine medicine is the exposure to the myriad of spores produced *by Aspergillus fumigatus* which are a major causative agent of equine asthma (heaves) and related airway inflammation and obstruction.

The aim of this chapter is to present a brief overview of the sensitivity of horses to various mycotoxins. Special emphasis is given to toxins that are less often described in other animal species but could be measured in typical diets of horses. This overview also aims to show the gaps in the current knowledge as well as the uncertainties about the reliable description of health-based guidance levels for horses and the impact of mycotoxins in the aetiology of multifactorial diseases.

2 Fungal species and toxins in roughages

2.1 Fusarium *species and their toxins*

Fusarium spp. invade numerous plants, often acting as phytopathogens, responsible for significant economic losses in the production of cereal and other crops. Fusaria in cereal grains have been historically related to fatal outbreaks of human intoxications such as alimentary toxic aleukia in Russia resembling all clinical signs of a T-2 toxin intoxication. The first fusariosis in horses was described as Stachybotryosis and associated with ingestion of satratoxin, a macrocyclic trichothecene. Signs of intoxication include irritation of the skin and finally a severe neurological (and often fatal) syndrome. *Fusarium* toxins have been studied particularly in maize (corn), resulting for example in the description of ZEN as the maize (*Zea mays* L.) metabolite and the identification of FB_1 and ELEM (Equine leukoencephalomalacia), one of the most prominent forms of fusariosis in horses as described below.

Fusarium species infest pasture grasses as well, and hay and haylage have regularly been found to be contaminated with high amounts of DON as well as ZEN and FUM, albeit at rather low concentrations. Of increasing concern is the contamination of grass and cereals (monocotyledons) with so-called emerging mycotoxins, such as the group of ENNs and BEA, but also emodin and aurofusarin. Advancement of analytical techniques allows the simultaneous detection of a wide array (almost 200) of fungal and plant metabolites simultaneously. The assessment of the clinical relevance of these complex exposure patterns remains an unmet challenge, particularly for horses. Even for prominent mycotoxins there are limited data available with regard to horses, whilst numerous investigations have been conducted in, for example, pigs and poultry.

2.1.1 Trichothecenes: deoxynivalenol and T-2 toxin as examples

Major representatives of the structurally related group of epoxy-trichothecenes, from which adverse effects have been documented in horses, are DON and T-2 toxin.

2.1.1.1 *Deoxynivalenol*

DON is a trichothecene that interacts with the dopaminergic system of the central nervous system, which can induce nausea, feed refusal and vomiting in pigs and was therefore also described as vomitoxin in the clinical literature. The effect of DON in horses has been studied as well, but with controversial outcomes. For example, Johnson et al. (1997) showed no adverse effects in horses after exposure to 36-44 mg DON/kg diet. However, when horses were exposed to feed that was naturally contaminated with several *Fusarium* mycotoxins (DON, fusaric acid, ZEN, and 15-ADON) adverse effects were observed at lower concentrations, including loss of appetite and weight loss (Raymond et al., 2003, 2005). Such a decrease ultimately resulted in an impairment of condition and reduced performance.

The current concerns regarding the adverse effects of DON in other animal species, focus on the intestinal tract, as the first site of contact. In pigs and poultry (and rodents), an impairment of the intestinal barrier leads to a cascade of adverse effects such as intestinal inflammation, liver toxicity, alterations in the intestinal microbiota and subsequently poor feed utilization and weight gain, and even neurological disturbances (Akbari et al., 2014; Knutsen et al., 2017b). No comparable detailed investigations have been conducted in horses. However, it should be noted that in an investigation comparing blood concentrations of DON in colic horses (n=102) to a control group (n=55), almost all positive samples were found in the colic horses, while there was no significant difference regarding the height of the blood concentrations of DON between the two groups (Dänicke et al., 2021). Another retrospective study identified mycotoxin contamination (i.e. AFs, ochratoxins and/or *Fusarium* toxins) as a risk factor for equine colic (Gomaa et al., 2022).

DON has also been recognised as a modulator of various immune functions in monogastric animal species, such as pigs and poultry. Subsequently, in an experiment in horses, lymphocyte subsets ($CD4^{+}$, $CD8^{+}$, $CD2^{+}$, $CD21^{+}$, $MHCII^{+}$) and the lymphocyte proliferation rate after stimulation with different mitogens was tested. As no significant differences were observed, the authors concluded that a daily DON intake up to 9.5 mg/100 kg BW appears to have no major impact on the measured immune response in horses (Khol-Parisini et al., 2012). This would indicate that horses have a rather high tolerance for the systemic effects of DON, which might be associated with limited oral bioavailability, as DON is known to be a substrate for ABC-efflux transporters (Videmann et al., 2007) which have a higher capacity in herbivorous species compared to omnivores.

In the most recent EFSA evaluation (Schrenk et al., 2023a) an adverse effect level of 5.6 mg DON/kg feed for horses was established based on studies showing reduced FI, with a reference point for adverse animal health effects of 3.5 mg/kg feed (see Chapter 17). The overall conclusion was that despite some knowledge gaps, the risk for adverse health effects from feed containing DON in horses is low when compared to other monogastric animal species such as pigs and poultry. The current EU recommendation suggests that the concentration of DON in feedstuff for horses should not exceed 0.9 mg/kg feed (see Chapter 17).

2.1.1.2 *T-2 toxin*

T-2 toxin and HT-2 toxin belong to the group A of epoxy-trichothecenes (epoxy-sesquiterpenoids). These closely related toxins, which are often reported together or even as a sum of T-2/HT-2, are produced by various *Fusarium* species including *F. langsethiae* as well as *F. poae* and *F. sporotrichioides,* which invade various feed ingredients such as oats, in which often the highest toxin concentration can be measured, as well as barley (and bakery products thereof), and maize, rice, beans, and several herbs and spices which are sometimes added to equine diets to improve palatability and digestive functions. Comprehensive reviews on the occurrence and analysis of these toxins are presented in the EFSA Opinion on T-2/HT-2 in food and feed (Arcella et al., 2017; Knutsen et al., 2017c).

T-2 toxin is readily absorbed after oral intake and inhalation. Exposure of horses to T-2 toxin may cause oral lesions, as also observed in other animal species. Skin lesions around the mouth were also the only clinical signs observed in a field trial in which mares were exposed to 7 mg purified T-2 toxin for up to 40 days (Juhász et al., 1997). The rate of absorption through the intact skin has not been studied in horses but has been demonstrated in experiments with rodent species and rabbits. After absorption, T-2 and HT-2 toxin are rapidly metabolised (Wu et al., 2020) and plasma levels can be measured only for a short period (around 20 minutes) after exposure, which often makes it impossible to establish a correlation between toxin exposure and clinical signs in daily equine practice. In addition to its direct cytotoxicity (*in vitro* cytotoxicity is observed already at a concentration of 1 ng/ml toxin in the cell culture medium), and dermatotoxicity with necrotic lesions, T-2 toxin is a prominent haematotoxin affecting circulating blood cells as well as inducing apoptosis in their progenitor cells in the bone marrow and lymphoid tissues. Clinically this is described as pancytopenia, which is particularly prominent in cats after exposure to T-2 toxin in the feed even at low concentrations ranging between 0.06 to 0.10 mg/kg BW (Lutsky and Mor, 1981). Pancytopenia was also described as a main clinical finding in a case study in which horses accidentally consumed a high concentration of T-2 toxin with the feed (240 mg/kg in a grain/maize concentrate, a concentration that would equate to an exposure of approximately 4.3 mg/kg BW).

In this accident 12/30 horses died within 4 weeks after exposure (Gabal et al., 1986). As T-2 and HT-2 toxin inhibit protein and DNA synthesis in all actively dividing cells in bone marrow, lymph nodes, spleen, and thymus, the resulting leukopenia also makes them potent immunotoxins supressing cellular and humoral immune responses. Due to these general and species-independent mechanisms, the immune system is therefore considered as the main target of T-2/HT-2 toxin in all animal species. However, specific (experimental) data for horses are lacking and a NOAEL cannot currently be defined.

2.1.2 Fumonisin and equine leukoencephalomalacia

FUM were described for the first time in 1998 due to their uncommon structure and poor solubility in several common organic solvents used for the extraction of other mycotoxins in feed materials. FB_1, the main toxin of *Fusarium verticilloides* (previously denoted *F. moniliforme*) has been implicated in the aetiology of a neurological condition in horses described as ELEM or mouldy maize poisoning. FB_1 has a sphingolipid-like structure, explaining one of its main mechanisms of action, the inhibition of the enzymes sphingosine-sphinganine-transferases and ceramide synthases. Subsequently, the concentrations of free sphinganine and sphingosine increase (Sa/So ratio used as biomarker of effect), resulting in cellular oxidative damage, while at the same time the synthesis of complex sphingolipids decreases, impairing cell metabolism and regeneration. Despite these common mechanisms of action, the clinical signs differ significantly between animal species. In horses, early signs of ELEM are paralysis of the tongue, forelimb incoordination and hind limb ataxia, which may occur hours or days before the more severe clinical signs. Other characteristic signs of ELEM are incoordination, apparent blindness, circling, paresis, ataxia, head pressing, depression and hyperexcitation. Mortality due to ELEM is high and death can occur within hours or days after onset of the clinical signs, or even without obvious clinical signs (Caloni and Cortinovis, 2010, 2011). Horses seem to be among the most sensitive animal species, and clinical signs are observed following exposure to >10 mg/kg feed for horses, compared to for example 160 mg/kg feed in pigs in which the clinical symptoms are different and cardiovascular symptoms followed by severe pulmonary oedema prevail.

Several outbreaks of severe and fatal FUM intoxications have been reported in the literature. One of the first was a major outbreak of ELEM in the fall of 1989 involving 18 of 66 purebred Arabian horses at a breeding/training stable in Arizona (USA). Of the 18 horses affected, 14 cases were fatal. These horses, as well as 48 unaffected horses, had been fed a diet containing a substantial amount of white maize screenings. Gross pathological findings included liquefactive necrosis in parts of the cerebral white matter and haemorrhagic foci of various sizes in the brain stem. Histopathological findings included rarefied white matter with pyknotic nuclei and eosinophilic cytoplasm. Concentrations of FB_1 ranged from 37 to 122 mg/kg. FB_2 was also detected (Wilson et al., 1990). While the clinical as well as the histopathological findings clearly suggested FB_1 as the major cause of disease, the study did not allow for the determination of a threshold level for FB_1. A major obstacle in the interpretation of the many case reports describing the clinical signs and even fatal outcome in horses following ingestion of FUM-contaminated feed materials is the lack of a precise description of the amount of toxin ingested. This was also the main conclusion in the EFSA report, following a comprehensive review of all published data (Knutsen et al., 2018). Moreover, in many reports only single components of the diet (often only concentrates) were tested and the analytical methods applied ranged from screening methods to quantitative analyses with advanced methods. Finally, in many cases only the concentration of FB_1 is given, while it cannot be excluded that the other (modified) forms of FUM such as FB_2, FB_3 and FB_4 contribute to the observed clinical signs. Despite the obvious uncertainties due to incomplete data, for a conclusive dose-response relationship regarding the dietary intake of FUM and the risk of developing the clinical signs of ELEM a concentration of 1 mg/kg feed was proposed as a reference point for adverse animal health effects. This guidance level refers to the complete diet, which was estimated to consist of 50% roughages and 50% concentrates. The estimates are very conservative and might result in undesirable (or subclinical) effects such as an increase in serum concentrations of liver enzymes, observed in horses, even at lower doses.

2.1.3 Zearalenone

ZEN is a phenolic resorcylic acid lactone mycotoxin produced by various *Fusarium* species, including *F. graminearum,* and is found as a contaminant of maize and maize products around the world (see also Chapter 2). In feed materials for horses the ZEN metabolites α-ZEL and β-ZEL, α-ZAL and β-ZAL, ZAN and plant-derived conjugates such as glucose and sulphate conjugates can be measured and should be included in an exposure assessment (for review see Knutsen et al., 2017a).

Although the chemical conformation of ZEN and its modified forms is non-steroidal, all non-conjugated forms interact with oestrogen receptors. Initial *in vitro* studies with equine hepatocytes demonstrated a rapid transformation of ZEN into α-ZEL, which was immediately glucuronidated (Malekinejad, 2013). In contrast, in *in vivo* studies measuring the levels of ZEN and its metabolites in plasma, urine and faeces of horses fed with naturally *Fusarium* toxin-contaminated oats (calculated exposure rate approx. 0.003 mg ZEN/kg BW), suggest that β-ZEL might be the main metabolite in horse, as in blood samples taken after 10 days of exposure the ration of α-ZEL/β-ZEL was 1:20 (Songsermsakul et al., 2013). Again, a high rate of glucuronidation was observed, and in urine samples ZEN as well as α-ZEL were mainly detected. Given that in the same study a prolonged absorption was also seen with a T_{max} of 8-12 h after the beginning of the administration, the ultimate formation of the very potent α-ZEL might be more dependent on the pattern of ZEN and its modified form and on the dose ingested than previously considered. This hypothesis is supported by the results of an *in vivo* screening of blood samples collected from patients sent to a horse clinic for orthopaedic reasons (Schumann et al., 2016). Out of a total of 49 blood samples, 24 were positive for ZEN with a median concentration of 0.11 ng/ml (0.02–0.38 ng/ml), 3 samples were positive for α-ZEL with a median of 0.19 ng/ml (0.14–0.28 ng/ml), and 37 samples were positive for β-ZEL with a median of 1.24 ng/ml (0.28–7.65 ng/ml). While in most cases β-ZEL was clearly the dominant metabolite, the presence of ZEN and α-ZEL may refer to differences between ZEN exposure originating from pasture grass versus short-term exposure to ZEN contaminating maize products.

The well-established interaction of ZEN with oestrogen receptors resulted in various *in vitro* and *in vivo* studies aiming to predict potential adverse effects on fertility. For example, *in vitro* experiments with equine ovarian granulosa cells confirmed the oestrogenic potency and cellular disturbances induced by ZEN and its metabolites (Minervini et al., 2006). In a similar approach, ZEN and its hydroxy-metabolites were shown to affect equine sperm chromatin structure (Minervini et al., 2010).

The main concern regarding dietary exposure to ZEN is its potential adverse effect on female reproduction. The clinical studies conducted in mares, however, often gave inconsistent or equivocal results due to a small number of animals and a limited observation period. For example, in an on-farm study, 3 groups of 7 mares were fed 0, 2, and 8 mg/day crystalline (>95% pure) ZEN for the duration of three oestrous cycles, followed by artificial insemination (Vance et al., 2019). No significant differences regarding reproductive hormone levels or pregnancy rate were observed, although minor changes were observed in the length of the oestrus cycle. Moreover, the effect of ZEN was studied in 6 cycling trotter mares in the summer period. After an entire oestrous cycle (cycle 1), mares were given 7 mg purified ZEN *per os* daily (dissolved in alcohol), beginning on day 10 of cycle 2. Toxin exposure, calculated to be comparable to natural exposure rates, was continued until the subsequent ovulation. No adverse effects of ZEN on reproduction were reported (Juhász et al., 2001).

In a feeding trial, mares were exposed to two different mixtures of ZEN and DON (1.5 or 3 mg/day ZEN, and 18 or 36 mg/day DON, respectively) for the duration of two oestrous cycles in a cross-over design (Aurich et al., 2006). No clear adverse effects on hormone levels and the oestrous cycle were recorded, but an increased incidence of haemorrhagic corpora lutea and follicular haematomas was observed.

The main difference between these experiments is the nature of ZEN, with pure toxin in the first study and natural contaminated feed material in the older studies. ZEN is occurring in plants in different forms including plant-derived modified forms, with higher oestrogenic activity as mentioned above (Knutsen et al., 2017a). This could explain the apparent differences in the outcome of clinical trials, as natural contaminated feed materials contain the sum of ZEN and its modified forms and therefore induce a stronger biological effect.

In consideration of the lack of sufficient data to establish a conclusive dose-response curve, no reference point for adverse animal health effect was set for horses (solipeds) in Europe by the EFSA and national organisations. This implies that the general recommendation for feed materials, i.e. 2 mg/kg for cereals and cereal products except for maize by-products, and 3 mg/kg for maize by-products, also applies to feeds prepared for horses.

2.2 *Endophytes in cool season grasses: lolitrem and ergovaline intoxications*

Endophytes occur invisibly in grass and in the seeds of grasses and small grains, and are sometimes made visible by the discolouration of seeds. As grass, hay and straw are among the major dietary components for horses, clinical intoxication by mycotoxins produced by endophytes has been described worldwide.

Common mutualists living in cool-season grass are *Epichloë* (formerly denoted *Neotyphodium*) species, a large genus of endophytes that often reside invisibly in plant tissue. The relationship between endophyte and grass is a true symbiosis in which the plant provides the fungus with water and nutrients, while the fungus produces a variety of secondary metabolites that prevent insect damage, infections with phytopathogenic microorganisms and helminths, thereby protecting the host plant against multiple stressors (Bastias et al., 2017). Early surveys of the grasses in North and South America indicated *Epichloë* infectivity in 60-100% of all fescue grasses. An increase in the prevalence of endophytes was observed in agronomic grasses compared to the wild type flora, which might also explain the apparent increase in endophyte positive roughage samples over the last decades. A review (Krauss et al., 2020) revisited the current prevalence of *Epichloë* infections in Europe, mainly found in *Lolium perenne*. Again, an increase in prevalence was noted, which might be attributable to the abundant use of endophyte-positive grass seeds. In Europe, the Netherlands had been identified earlier as a hot-spot for *Epichloë*-positive *Lolium perenne* grass on the dunes and sandy grounds.

Endophytes produce a variety of secondary metabolites, of which the alkaloids (ergovaline) and a variety of indole terpenoids, particularly Lolitrem B, are recognised as a health risk for horses (Philippe, 2016; Porras-Alfaro and Bayman, 2011; Schardl, 2001).

2.2.1 Fescue toxicosis and ergovaline

Tall Fescue (*Festuca arundinacea*) is a perennial pasture grass that is often infected with endophytes (*Epichloë festucae*, previously described as *Neotyphodium coenophialum*). Endophyte-infected tall fescue can cause equine fescue toxicosis, which is mostly found in the USA, due to the widespread occurrence of tall fescue in natural grassland. The effect of tall fescue on horses differs from symptoms observed in other animal species such as cattle. For example, unlike cattle, the body temperature of horses does not increase after exposure to the endophyte toxins. Also, the well-known 'Fescue Foot', a gangrenous condition of the claws, has not been reported in a similar way in horses. The most prominent signs of equine fescue toxicosis are observed in pregnant mares and include prolonged gestation, abortion, weakness, dysmaturity and mortality, thickened or retained placentas and attenuated lactation or agalactia (Brendemuehl et al., 1994; Blodgett, 2001; Cross et al., 1995; Putnam et al., 1991).

Most of these symptoms can be attributed to a stimulation of dopamine-2 receptors. Due to the negative effects on the reproductive system, removal of pregnant mares from the endophyte-infected pasture 60 to 90 days (before day 300 of gestation) prior to the expected foaling date has been recommended in order to prevent these symptoms and to prevent adverse effects on foals, which are often weak and oversized but underdeveloped (Richards et al., 2021).

2.2.2 Lolitrem and ryegrass staggers

Lolitrem B is one of the most prominent mycotoxins of *Epichloë festucae* var. *lolii* (formerly described as *Neotyphodium lolii*) infecting *Lolium perenne* (perennial ryegrass) (for details see Chapter 2). Horses are exposed to lolitrem on the pastures under condition of drought (and deep grazing) or via consumption of hay, produced by late-harvested grass, including flowering grass plants. In equine practice the intoxication is known as ryegrass staggers. Lolitrem has been found to bind to calcium-activated K^+ channels (BK channels), thereby inhibiting K^+ currents exiting the cells (Imlach et al., 2011). Subsequent clinical symptoms include tremours, muscle fasciculation, ataxia, truncal swaying, and tetanic convulsions, which increase with stress or exercise (Blythe et al., 2007; Fink-Gremmels, 2005; Johnstone et al., 2012; Tor-Agbidye et al., 2001). These severe clinical signs are not correlated with typical (pathological) lesions in the central nervous system. Following ingestion of contaminated feed materials, lolitrem is easily absorbed from the GIT and distributed into fatty tissues, including the central nervous system (Johnstone et al., 2012). As it apparently also accumulated in fatty tissues for a certain period, the measurable plasma concentrations did not correlate with the clinical symptoms (Johnstone et al., 2012). There is no known effective treatment against lolitrem B intoxications in horses, but the clinical signs of a lolitrem intoxication are completely reversible and once removed from the contaminated feeds, animals generally recover within five days (Blythe et al., 2007). In an attempt to provide guidance to horse owners and equine practitioners and based on multiple clinical cases, levels below 0.8 mg/kg were considered safe for adult horses, levels between 0.8 and 1.2 mg/kg hay indicated concentrations at which already mild symptoms may occur, whereas lolitrem B concentrations from 1.8–2.0 mg/kg must be considered as severely toxic (Craig et al., 2014; Fink-Gremmels, 2005). Due to a lack of more complete data, no statutory limits are set for lolitrem in the animal feed regulations.

2.3 Slaframine *and* swainsonine

Slafractonia leguminicola (formally denoted *Rhizoctonia leguminicola*) is a fungal species found in red clover and other clover species (described as black patch disease of legumes). These fungal diseases of clover and the associated toxins were also attributed to the species *Botrytis fabae Sardiña* (Kagan, 2016). The two most common toxins found on infected clover leaves are the indolizidine alkaloids, slaframine (l-acetoxy-8-aminooctohydroindolizine) and swainsonine (1,2,8-trihydroxyoctahydroindolizine). In horses, slaframine causes excessive salivation, which gave the toxic syndrome the popular name Slobbers syndrome. Excessive salivation can be the only clinical sign found (Wijnberg et al., 2009), but longer exposure to slaframine will result in anorexia, bloating, diarrhoea, urination and lacrimation, and even death (Hibbard et al., 1995). Slaframine intoxications can be treated with atropine, an acetylcholine muscarinic receptor antagonist.

A related indole derivative in locoweeds (*Astragalus* sp., *Oxytropis* sp.) is swainsonine (named after the plant *Swainsona canescensas,* in which it was first described). Swainsonine causes different clinical signs including ataxia, staggering gait, depression, muscular incoordination, appetite suppression and severe loss of condition (Croom et al., 1995). The toxin inhibits Golgi α-mannosidase II, which is an enzyme that is involved in N-linked glycosylation, ultimately resulting in neuronal swelling and vacuolation due to accumulations of mannose in the lysosomes (van den Elsen et al., 2001). Young animals have a higher risk of irreversible damage than adults. Intoxications in horses (referred to as locoism) are regularly described following the ingestion of spotted locoweed (*Astragalus lentiginosus*), common in North America. An investigation in China describes an increasing incidence of Swainsonine producing strains of *Alternaria* (sect. *Undifilum*), considered to form a commensal association with the two related genera of *Astragalus* and *Oxytropis.* Subsequently, an increasing number of intoxications have been described in cattle and horses (Guo et al., 2022).

2.4 *Storage fungi in hay:* Aspergillus fumigatus *and gliotoxin*

The most common spoilage moulds in hay are Aspergilli, and particularly *Aspergillus fumigatus.* This fungal species that cleaves fibres produces very high numbers of conidia (spores), which form a substantial fraction of the dust in horse stables. Inhalation of such spores is associated with equine asthma, one of the most common inflammatory respiratory diseases in horses (Laan et al., 2005, 2006). In more complicated cases, for instance in cases with concomitant viral or bacterial infections, inhalation of fungal spores might result in colonization of the fungus in the bronchoalveolar space, causing a chronic fungal pulmonary infection. Invasive pulmonary aspergillosis was confirmed in a study in horses, as well as the internal production of several fungal metabolites including gliotoxin, which was found both in lung tissue samples and bronchoalveolar lavage fluid (Dobiás et al., 2023).

2.5 *Storage fungi in haylage and grass silage*

Major fungal species found in silage are Penicillia (*Penicillium roqueforti* group), *Monascus* species (*Monascus purpureus*), as well as Aspergilli, particularly *A. fumigatus,* and *Alternaria* species (for more details see Chapter 2).

Next to fungal species, silage may also contain undesirable yeasts of which *Byssochlamys* species are known as mycotoxin producers (PAT, MPA), while *Trichoderma* spp., *Geotrichum* spp. and *Paecilomyces* spp. seem not to produce toxic metabolites in silage. As silage may also contain mycotoxins from the pre-harvest stage, it is the complex mixture of mycotoxins that needs to be considered when potential adverse effects on animal health and performance are assessed (van Pamel et al., 2011). In a comprehensive study, Andersen et al., 2020) described the fungal and chemical diversity of fungal species and their toxins in wrapped haylages and conventional hay. The results show that *Sordaria fimicola* was common in both hay and haylage, while *Penicillium* spp. was prevalent in haylage and *Aspergillus* spp. in hay. In haylage the *Penicillium* metabolites andrastins, fumigaclavines, isofumigaclavines, marcfortines, MPA, PR toxins, and roquefortines were detected in hot spots, whereas roquefortines and fumigaclavines were detected also in haylage samples with no visible mould growth. In hay, *Aspergillus* spp. were the most prominent fungal species. The authors pointed to the lack of data on the clinical effects of these toxins and the fact that multi-toxin exposure of horses via common diets remains a matter of concern.

2.6 *Patulin in silage and fruits*

PAT is produced by *Penicillium, Aspergillus,* and *Byssochlamys* species occurring in silages, but also in fruits (apple and apple products), vegetables (tomato), and grains. PAT exerts a large diversity of toxic effects in rodents and due to these findings a provisional maximum tolerable daily intake (PMTDI) of 0.4 µg/kg BW/day for fruits and fruit products has been set for humans within the EU. These regulations were set in consideration of experimental data in rodents showing that patulin may exert nausea, vomiting and other gastrointestinal symptoms such as gastric ulcers, intestinal haemorrhages and lesions, and act as a disrupter of the intestinal barrier, followed by hepato- and renal toxicity (Vidal et al., 2019). As mechanisms of action the binding to thiol-containing macromolecules, including GSH and the induction of oxidative stress, are discussed (for review, see Wei et al. (2020)). Of note is also the strong antimicrobial effect of PAT (Malekijenad and Fink-Gremmels, 2020; Wei et al., 2020), but its intrinsic toxicity prevented the clinical use of PAT as an antibiotic. While studies in horse are lacking, the typical anatomical features of the equine stomach and the basic molecular mechanism of PAT toxicity makes it likely that PAT can induce gastric ulcers in horses, as well as colic and diarrhoea due to the broad antibacterial activity of PAT, which is regularly found in equine diets.

3 Mycotoxins in concentrates used in the diet of horses

3.1 *Aflatoxins*

AFs produced by *Aspergillus flavus* or *A. parasiticus* are found in maize, soybeans, and diverse oilseeds and nuts. The production of AFs, which can start at the pre-harvest stage, is highest at temperatures between 25 to 30 degrees Celsius and high humidity. The toxicity of AFs and particularly AFB_1 has been described in numerous articles (Caloni and Cortinovis, 2011; Malekinejad and Fink-Gremmels, 2020; Osweiler, 2001) and is not discussed here in detail. AFs are typical hepatotoxins, inducing bile duct proliferation and centrilobular necrosis resulting in an impairment of liver function (including the synthesis of clotting factors with subsequent haemorrhaging as observed in the intestines of exposed horses), reduced FI and weight loss. Other symptoms of aflatoxin intoxication are depression, fever, tremour, ataxia and cough. A total dietary concentration of 500–1000 μg/kg has been shown to induce significant clinical findings and acute and chronic liver damage, depending on the duration of exposure (for review, see Caloni and Cortinovis, 2011).

3.2 *Ochratoxins and citrinin*

In contrast to AFs, which affect the liver, the target organs of ochratoxins are the kidneys. OTA, the most prominent representative of the structurally related ochratoxins, is produced by various Aspergilli (*Aspergillus ochraceus, A. carbonarius*) and *Penicillium* species. OTA intoxications are characterized by proximal tubule damage and renal karyomegaly. Initially clinical symptoms are mild, including an increase in water uptake and excretion. Mortality is usually low. Like AFs, OTA is known for its immunosuppressive effects, which may reduce the health status of the horse. CIT, produced by *Penicillium* and *Monascus* species and regularly co-occurring with ochratoxins, is also a renal toxin causing tubular lesions. The effects of OTA have been widely studied in pigs and poultry, but data for horses are lacking. Considering the typical accumulation of OTA in renal tissues, it cannot be excluded that particularly in older horses OTA presents a risk for the development of renal insufficiency (Schrenk et al., 2023b).

4 Conclusions

Knowledge about the potential adverse effects of mycotoxins in horses is scarce and often confined to clinical reports and case descriptions which do not allow for the definition of a reliable no-effect level or a reliable dose-response estimate. The sensitivity of horses in comparison to other animal species varies. For example, both a higher sensitivity (FUM, lolitrems) as well as an apparent lower sensitivity (DON) are observed. Clinical experiments conducted with horses are limited and focus mainly on reproduction. Global surveys on the rate of contamination of feed materials regularly indicate that under current agricultural practice, fungal invasion and mycotoxin formation in crops cannot be entirely prevented and are very likely to increase in the face of global climate change. Surveys of equine feed materials also indicate that not only concentrates but also roughages such as grass, hay and haylage are often invaded by various fungal species, resulting in the involuntary and uncontrolled exposure of horse to multiple mycotoxins. Hence, in an overall assessment of possible harmful effects of mycotoxins, different exposure scenarios and the co-exposure to different toxins remain major factors of uncertainty. However, advanced analytical techniques requiring only a droplet of blood or a small urine sample will hopefully soon fill various knowledge gaps between exposure scenarios and adverse outcomes or disease conditions, such as colic or liver disease.

As yet, clinical studies in horses have focused on organ-specific damage and typical clinical signs of intoxication. However, as in other animal species, such as pigs, poultry and ruminants, the major concern is the subclinical impairment of immune competence and resilience to (infectious) disease as well as the impact of mycotoxin exposure on metabolic processes, reproduction, and performance of (sports)horses. The increasing prevalence of mycotoxins in feed materials has resulted in professional risk management and risk mitigation strategies in (commercial) farm animal species and these are often routinely used to avoid production losses. Awareness among horse owners and trainers is still much lower, but the increasing risk of exposure to mycotoxins deserves the attention of horse owners, trainers and veterinarians, as the adverse effects of mycotoxins have become an important differential diagnosis in equine practice.

References

Akbari, P., Braber, S., Gremmels, H., Koelink, P.J., Verheijden, K.A.T., Garssen, J. and Fink-Gremmels, J., 2014. Deoxynivalenol: a trigger for intestinal integrity breakdown. FASEB journal: official publication of the Federation of American Societies for Experimental Biology 28: 2414-2429. https://doi.org/10.1096/fj.13-238717.

Andersen, B., Phippen, C., Frisvad, J.C., Emery, S. and Eustace, R.A., 2020. Fungal and chemical diversity in hay and wrapped haylage for equine feed. Mycotoxin Research 36: 159-172. https://doi.org/10.1007/s12550-019-00377-5.

Arcella, D., Gergelova, P., Innocenti, M.L. and Steinkellner, H., 2017. Human and animal dietary exposure to T-2 and HT-2 toxin. EFSA journal 15: e04972. https://doi.org/10.2903/j.efsa.2017.4972.

Aurich, J.E., Hoppen, H., Trampler, R., Zentek, J., Boehm, J., Razzazi-Fazeli, E. and Aurich, C., 2006. Effects of mycotoxins on reproductive function in mares. Animal Reproduction Science 2006: 238-241. https://doi.org/10.1016/anireprosci.2006.03.037.

Bastias, D.A., Martínez-Ghersa, M.A., Ballaré, C.L. and Gundel, P.E., 2017. Epichloë Fungal Endophytes and Plant Defenses: Not Just Alkaloids. Trends in Plant Science 22: 939-948. https://doi.org/10.1016/j.tplants.2017.08.005.

Blodgett, D.J., 2001. Fescue toxicosis. The Veterinary Clinics of North America. Equine Practice 17: 567-577. https://doi.org/10.1016/s0749-0739(17)30052-4.

Blythe, L., Estill, C., Males, J. and Craig, A.M., 2007. Determination of the toxic threshold of lolitrem B in cattle eating endophyte-infected perennial ryegrass. In: Popay A.J. and Thom, E.R., eds, New Zealand Grassland Association: Endophyte Symposium. Christchurch, New Zealand: 399-402.

Brendemuehl, J.P., Boosinger, T.R., Pugh, D.G. and Shelby, R.A., 1994. Influence of endophyte-infected tall fescue on cyclicity, pregnancy rate and early embryonic loss in the mare. Theriogenology 42: 489-500. https://doi.org/10.1016/0093-691x(94)90687-e.

Caloni, F. and Cortinovis, C., 2010. Effects of fusariotoxins in the equine species. Veterinary Journal 186: 157-161. https://doi.org/10.1016/j.tvjl.2009.09.020.

Caloni, F. and Cortinovis, C., 2011. Toxicological effects of aflatoxins in horses. Veterinary Journal 188: 270-273. https://doi.org/10.1016/j.tvjl.2010.06.002.

Craig, A.M., Blythe, L.L. and Duringer, J.M., 2014. The role of the Oregon State University Endophyte Service Laboratory in diagnosing clinical cases of endophyte toxicoses. Journal of Agricultural and Food Chemistry 62: 7376-7381. https://doi.org/10.1021/jf5027229.

Croom, W.J., Hagler, W.M., Froetschel, M.A. and Johnson, A.D., 1995. The involvement of slaframine and swainsonine in slobbers syndrome: a review. Journal of Animal Science 73: 1499-1508. https://doi.org/10.2527/1995.7351499x.

Cross, D.L., Redmond, L.M. and Strickland, J.R., 1995. Equine fescue toxicosis: signs and solutions. Journal of Animal Science 73: 899-908. https://doi.org/10.2527/1995.733899x.

Dänicke, S., Saltzmann, J., Liermann, W., Glatter, M., Hüther, L., Kersten, S., Zeyner, A., Feige, K. and Warnken, T., 2021. Evaluation of Inner Exposure of Horses to Zearalenone (ZEN), Deoxynivalenol (DON) and Their Metabolites in Relation to Colic and Health-Related Clinical-Chemical Traits. Toxins 13: 588. https://doi.org/10.3390/toxins13080588.

Dobiás, R., Jahn, P., Tóthová, K., Dobesová, O., Visnovská, D., Patil, R., Škríba, A., Jaworská, P., Škoric, M., Podojil, L., Kantorová, M., Mrázek, J., Krejcí, E., Stevens, D.A. and Havlícek, V., 2023. Diagnosis of Aspergillosis in Horses. Journal of Fungi 9: 161. https://doi.org/10.3390/jof9020161.

Fink-Gremmels, J., 2005. Mycotoxins in forages. In: DIAZ, D.E. ed, The Mycotoxin Blue Book. Nottingham, UK: Nottingham University Press, 249-268.

Gabal, M.A., Awad, Y.L., Morcos, M.B., Barakat, A.M. and Malik, G., 1986. Fusariotoxicoses of farm animals and mycotoxic leucoencephalomalacia of the equine associated with the finding of trichothecenes in feedstuffs. Veterinary and Human Toxicology 28: 207-212.

Gomaa, N., Elemiri, M., Hegazy, Y., Zeineldin, M., Nassif, M., Alacal-Canto, Y., Barbarosa-Pliego, A., Rivas-Caceres, R.R. and Abdelmegeid, M., 2022. Clinical and pathological examination of mycotoxicosis as an associated risk factor for colic in equine. Microbial pathogenesis 163: 105377. https://doi.org/10.1016/j.micpath.2021.105377.

Guo, C., Zhang, L., Zhao, Q., Beckmann, M., Phillips, H., Meng, H., Mo, C., Mur, L.A.J. and He, W., 2022. Host-species variation and environment influence endophyte symbiosis and mycotoxin levels in chinese oxytropis species. Toxins 14: 181. https://doi.org/10.3390/toxins14030181.

Hibbard, B., Peters, J.P., Chester, S.T., Robinson, J.A., Kotarski, S.F., Croom, W.J. and Hagler, W.M., 1995. The effect of slaframine on salivary output and subacute and acute acidosis in growing beef steers. Journal of Animal Science 73: 516-525. https://doi.org/10.2527/1995.732516x.

Imlach, W.L., Finch, S.C., Zhang, Y., Dunlop, J. and Dalziel, J.E., 2011. Mechanism of action of lolitrem B, a fungal endophyte derived toxin that inhibits BK large conductance Ca^{2+}-activated K^+ channels. Toxicon 57: 686-694. https://doi.org/10.1016/j.toxicon.2011.01.013.

Johnson, P.J., Casteel, S.W. and Messer, N.T., 1997. Effect of feeding deoxynivalenol (vomitoxin)-contaminated barley to horses. Journal of Veterinary Diagnostic Investigation 9: 219-221. https://doi.org/10.1177/104063879700900224.

Johnstone, L.K., Mayhew, I.G. and Fletcher, L.R., 2012. Clinical expression of lolitrem B (perennial ryegrass) intoxication in horses. Equine Veterinary Journal 44: 304-309. https://doi.org/10.1111/j.2042-3306.2011.00439.x.

Juhász, J., Nagy, P., Huszenicza, G., Szigeti, G., Reiczigel, J. and Kulcsár, M., 1997. Long term exposure to T-2 *Fusarium* mycotoxin fails to alter luteal function, follicular activity and embryo recovery in mares. Equine Veterinary Journal 25: 17-21. https://doi.org/10.1111/j.2042-3306.1997.tb05093.x.

Juhász, J., Nagy, P., Kulcsár, M., Szigeti, G., Reiczigel, J. and Huszenicza, G., 2001. Effect of low-dose zearalenone exposure on luteal function, follicular activity and uterine oedema in cycling mares. Acta Veterinaria Hungarica 49: 211-222. https://doi.org/10.1556/004.49.2001.2.11.

Kagan, I.A., 2016. Blackpatch of clover, cause of slobbers syndrome: a review of the disease and the pathogen, Rhizoctonia leguminicola. Frontiers in Veterinary Science 3: 3. https://doi.org/10.3389/fvets.2016.00003.

Khol-Parisini, A., Hellweg, P., Razzazi-Fazeli, E., Saalmüller, A., Strasser, A., Tichy, A. and Zentek, J., 2012. Highly deoxynivalenol contaminated oats and immune function in horses. Archives of Animal Nutrition 66: 149-161. https://doi.org/10.1080/1745039x.2012.672220.

Knutsen, H., Alexander, J., Barregård, L., Bignami, M., Brüschweiler, B., Ceccatelli, S., Cottrill, B., Dinovi, M., Edler, L., Grasl-Kraupp, B., Hogstrand, C., Hoogenboom, L.R., Nebbia, C.S., Petersen, A., Rose, M., Roudot, A., Schwerdtle, T., Vleminckx, C., Vollmer, G., Wallace, H., Dall'asta, C., Eriksen, G., Taranu, I., Altieri, A., Roldán-Torres, R. and Oswald, I.P., 2018. Risks for animal health related to the presence of fumonisins, their modified forms and hidden forms in feed. EFSA journal 16: e05242. https://doi.org/10.2903/j.efsa.2018.5242.

Knutsen, H., Alexander, J., Barregård, L., Bignami, M., Brüschweiler, B., Ceccatelli, S., Cottrill, B., Dinovi, M., Edler, L., Grasl-Kraupp, B., Hogstrand, C., Hoogenboom, L.R., Nebbia, C.S., Petersen, A., Rose, M., Roudot, A., Schwerdtle, T., Vleminckx, C., Vollmer, G., Wallace, H., Dall'asta, C., Dänicke, S., Eriksen, G., Altieri, A., Roldán-Torres, R. and Oswald, I.P., 2017a.

Risks for animal health related to the presence of zearalenone and its modified forms in feed. EFSA journal 15: e04851. https://doi.org/10.2903/j.efsa.2017.4851.

Knutsen, H., Alexander, J., Barregård, L., Bignami, M., Brüschweiler, B., Ceccatelli, S., Cottrill, B., Dinovi, M., Grasl-Kraupp, B., Hogstrand, C., Hoogenboom, L.R., Nebbia, C.S., Oswald, I.P., Petersen, A., Rose, M., Roudot, A., Schwerdtle, T., Vleminckx, C., Vollmer, G., Wallace, H., de Saeger, S., Eriksen, G.S., Farmer, P., Fremy, J., Gong, Y.Y., Meyer, K., Naegeli, H., Parent-Massin, D., Rietjens, I., van Egmond, H., Altieri, A., Eskola, M., Gergelova, P., Ramos Bordajandi, L., Benkova, B., Dörr, B., Gkrillas, A., Gustavsson, N., van Manen, M. and Edler, L., 2017b. Risks to human and animal health related to the presence of deoxynivalenol and its acetylated and modified forms in food and feed. EFSA journal 15: e04718. https://doi.org/10.2903/j.efsa.2017.4718.

Knutsen, H., Barregård, L., Bignami, M., Brüschweiler, B., Ceccatelli, S., Cottrill, B., Dinovi, M., Edler, L., Grasl-Kraupp, B., Hogstrand, C., Hoogenboom, L.R., Nebbia, C.S., Oswald, I.P., Petersen, A., Rose, M., Roudot, A., Schwerdtle, T., Vleminckx, C., Vollmer, G., Wallace, Dall'Asta, C., Gutleb, A., Metzler, M., Oswald, I., Parent -Massin, Binaglia, M., Steinkellner, H. and Alexander, J., 2017c. Appropriateness to set a group health based guidance value for T2 and HT2 toxin and its modified forms. EFSA journal. 15: e04655. https://doi.org/10.2903/j.efsa.2017.4655.

Krauss, J., Vikuk, V., Young, C.A., Krischke, M., Mueller, M.J. and Baerenfaller, K., 2020. *Epichloë* endophyte infection rates and alkaloid content in commercially available grass seed mixtures in Europe. Microorganisms 8: 498. https://doi.org/10.3390/microorganisms8040498.

Laan, T.T.J.M., Bull, S., Pirie, R.S. and Fink-Gremmels, J., 2005. Evaluation of cytokine production by equine alveolar macrophages exposed to lipopolysaccharide, *Aspergillus fumigatus*, and a suspension of hay dust. American Journal of Veterinary Research 66: 1584-1589. https://doi.org/10.2460/ajvr.2005.66.1584.

Laan, T.T.J.M., Bull, S., van Nieuwstadt, R.A. and Fink-Gremmels, J., 2006. The effect of aerosolized and intravenously administered clenbuterol and aerosolized fluticasone propionate on horses challenged with *Aspergillus fumigatus* antigen. Veterinary Research Communications 30: 623-635. https://doi.org/10.1007/s11259-006-3346-9.

Lutsky, I. and Mor, N., 1981. Experimental alimentary toxic aleukia in cats. Laboratory Animal Science 31: 43-47.

Malekinejad, H. and Fink-Gremmels J., 2020. Mycotoxicoses in veterinary medicine: Aspergillosis and penicilliosis. Veterinary Research Forum 11: 97-103. https://doi.org/10.30466/vfrf.2020.112820.2686.

Malekinejad, H., 2013. Zearalenone is converted to a potent oestrogenic metabolite by the equine hepatic subcellular fractions and hepatocytes. Bulgarian Journal of Veterinary Medicine 16: 260-270.

Minervini, F., Giannoccaro, A., Fornelli, F., Dell'Aquila, M.E., Minoia, P. and Visconti, A., 2006. Influence of mycotoxin zearalenone and its derivatives (alpha and beta zearalenol) on apoptosis and proliferation of cultured granulosa cells from equine ovaries. Reproductive biology and endocrinology 4: 62. https://doi.org/10.1186/1477-7827-4-62.

Minervini, F., Lacalandra, G.M., Filannino, A., Nicassio, M., Visconti, A. and Dell'Aquila, M.E., 2010. Effects of in vitro exposure to natural levels of zearalenone and its derivatives on chromatin structure stability in equine spermatozoa. Theriogenology 73: 392-403. https://doi.org/10.1016/j.theriogenology.2009.09.023.

Miyazaki, S., Fukumura, M., Yoshioka, M. and Yamanaka, N., 2001. Detection of endophyte toxins in the imported perennial ryegrass straw. The Journal of Veterinary Medical Science 63: 1013-1015. https://doi.org/10.1292/jvms.63.1013.

Osweiler, G.D., 2001. Mycotoxins. The Veterinary Clinics of North America. Equine Practice, 17: 547-566, viii.

Philippe, G., 2016. Lolitrem B and indole diterpene alkaloids produced by endophytic fungi of the genus Epichloë and their toxic effects in livestock. Toxins 8: 47. https://doi.org/10.3390/toxins8020047.

Porras-Alfaro, A. and Bayman, P., 2011. Hidden fungi, emergent properties: endophytes and microbiomes. Annual Review of Phytopathology 49: 291-315. https://doi.org/10.1146/annurev-phyto-080508-081831.

Putnam, M.R., Bransby, D.I., Schumacher, J., Boosinger, T.R., Bush, L., Shelby, R.A., Vaughan, J.T., Ball, D. and Brendemuehl, J.P., 1991. Effects of the fungal endophyte *Acremonium coenophialum in* fescue on pregnant mares and foal viability. American Journal of Veterinary Research 52: 2071-2074.

Raymond, S.L., Smith, T.K. and Swamy, H.V.L.N., 2003. Effects of feeding a blend of grains naturally contaminated with *Fusarium* mycotoxins on feed intake, serum chemistry, and hematology of horses, and the efficacy of a polymeric glucomannan mycotoxin adsorbent. Journal of Animal Science 81: 2123-2130. https://doi.org/10.2527/2003.8192123x.

Raymond, S.L., Smith, T.K. and Swamy, H.V.L.N., 2005. Effects of feeding a blend of grains naturally contaminated with *Fusarium* mycotoxins on feed intake, metabolism, and indices of athletic performance of exercised horses. Journal of Animal Science 83: 1267-1273. https://doi.org/10.2527/2005.8361267x.

Richards, N., Nielsen, B.D. and Finno, C.J., 2021. Nutritional and non-nutritional aspects of forage. The Veterinary clinics of North America. Equine practice 37: 43-61. https://doi.org/10.1016/j.cveq.2020.12.002.

Schardl, C.L., 2001. *Epichloë festucae* and related mutualistic symbionts of grasses. Fungal Genetics and Biology 33: 69-82. https://doi.org/10.1006/fgbi.2001.1275.

Schrenk, D., Bignami, M., Bodin, L., Chipman, J.K., Del Mazo, J., Grasl-Kraupp, B., Hogstrand, C., Hoogenboom, L.R., Leblanc, J., Nielsen, E., Ntzani, E., Sand, S., Schwerdtle, T., Vleminckx, C., Wallace, H., Gropp, J., Antonissen, G., Rychen, G., Gómez Ruiz, J.Á, Innocenti, M.L., Rovesti, E. and Petersen, A., 2023a. Risks for animal health related to the presence of ochratoxin A (OTA) in feed. EFSA journal 21: e08375. https://doi.org/10.2903/j.efsa.2023.8375.

Schrenk, D., Bignami, M., Bodin, L., Del Mazo, J., Grasl-Kraupp, B., Hogstrand, C., Leblanc, J., Nielsen, E., Ntzani, E., Petersen, A., Sand, S., Schwerdtle, T., Vleminckx, C., Wallace, H., Dänicke, S., Nebbia, C.S., Oswald, I.P., Rovesti, E., Steinkellner, H. and Hoogenboom, L.R.,

2023b. Assessment of information as regards the toxicity of deoxynivalenol for horses and poultry. EFSA journal. 21: e07806. https://doi.org/10.2903/j.efsa.2023.7806.

Schumann, B., Winkler, J., Mickenautsch, N., Warnken, T. and Dänicke, S., 2016. Effects of deoxynivalenol (DON), zearalenone (ZEN), and related metabolites on equine peripheral blood mononuclear cells (PBMC) in vitro and background occurrence of these toxins in horses. Mycotoxin Research 32: 153-161. https://doi.org/10.1007/s12550-016-0250-1.

Songsermsakul, P., Böhm, J., Aurich, C., Zentek, J. and Razzazi-Fazeli, E., 2013. The levels of zearalenone and its metabolites in plasma, urine and faeces of horses fed with naturally, *Fusarium* toxin-contaminated oats. Journal of Animal Physiology and Animal Nutrition 97: 155-161. https://doi.org/10.1111/j.1439-0396.2011.01253.x.

Tor-Agbidye, J., Blythe, L.L. and Craig, A.M., 2001. Correlation of endophyte toxins (ergovaline and lolitrem B) with clinical disease: fescue foot and perennial ryegrass staggers. Veterinary and Human Toxicology 43: 140-146.

van den Elsen, J.M., Kuntz, D.A. and Rose, D.R., 2001. Structure of golgi alpha-mannosidase II: a target for inhibition of growth and metastasis of cancer cells. The EMBO journal 20: 3008-3017. https://doi.org/10.1093/emboj/20.3008.

van Pamel, E., Verbeken, A., Vlaemynck, G., de Boever, J. and Daeseleire, E., 2011. Ultra-high-performance liquid chromatographic-tandem mass spectrometric multimycotoxin method for quantitating 26 mycotoxins in maize silage. Journal of Agricultural and Food Chemistry 59: 9747-9755. https://doi.org/10.1021/jf202614h.

Vance, C.K., King, E.H., Bowers, S.D., Ryan, P.L., Walters, K. and Shappell, N.W., 2019. Reproductive performance of mares fed dietary zearalenone. Frontiers in veterinary science 6: 423. https://doi.org/10.3389/fvets.2019.00423.

Vidal, A., Ouhibi, S., Ghali, R., Hedhili, A., de Saeger, S. and de Boevre, M., 2019. The mycotoxin patulin: An updated short review on occurrence, toxicity and analytical challenges. Food and Chemical Toxicology 129: 249-256. https://doi.org/10.1016/j.fct.2019.04.048.

Videmann, B., Tep, J., Cavret, S. and Lecoeur, S., 2007. Epithelial transport of deoxynivalenol: involvement of human P-glycoprotein (ABCB1) and multidrug resistance-associated protein 2 (ABCC2). Food and Chemical Toxicology 45: 1938-1947. https://doi.org/10.1016/j.fct.2007.04.011.

Wei, C., Yu, L., Qiao, N., Zhao, J., Zhang, H., Zhai, Q., Tian, F. and Chen, W., 2020. Progress in the distribution, toxicity, control, and detoxification of patulin: A review. Toxicon 184: 83-93. https://doi.org/10.1016/j.toxicon.2020.05.006.

Wijnberg, I.D., van der Ven, P.J. and Gehrmann, J.F., 2009. Outbreak of salivary syndrome on several horse farms in the Netherlands. The Veterinary Record 164: 595-596. https://doi.org/10.1136/vr.164.19.595.

Wilson, T.M., Ross, P.F., Rice, I.G., Osweiler, G.D., Nelson, H.A., Owens, D.L., Plattner, R.D., Reggiardo, C., Noon, T.H. and Pickrell, J.W., 1990. Fumonisin B1 levels associated with an epizootic of equine leukoencephalomalacia. Journal of Veterinary Diagnostic Investigation 2: 213-216. https://doi.org/10.1177/104063879000200311.

Wu, Q., Qin, Z., Kuca, K., You, L., Zhao, Y., Liu, A., Musilek, K., Chrienova, Z., Nepovimova, E., Oleksak, P., Wu, W. and Wang, X., 2020. An update on T-2 toxin and its modified forms: metabolism, immunotoxicity mechanism, and human exposure assessment. Archives of Toxicology 94: 3645-3669. https://doi.org/10.1007/s00204-020-02899-9.

CHAPTER 10

Mycotoxins in insects

M. Mézes

Hungarian University of Agriculture and Life Science, Institute of Physiology and Nutrition, Department of Feed Safety, H-2103 Gödöllő, Páter Károly u. 1., Hungary
mezes.miklos@uni-mate.hu

1 Introduction

Edible insects or processed insect proteins are important as a novel food (Commission Regulation EU No. 2015/2283), and edible insects have the potential to become a major global future food due to their high nutrient content as well as economic and environmental benefits (Lange and Nakamura, 2021). In addition, insect larvae are an important alternative protein source in animal nutrition (DiGiacomo and Leury, 2019).

Insect larvae and processed insect proteins were first allowed in aquaculture (Commission Regulation EU No. 2017/893) and, more recently, poultry and pigs (Commission Regulation EU No. 2021/1372). However, insects or insect proteins may contain food or feed safety hazards, for instance, undesirable contaminants such as heavy metals, herbicides, and pesticide residues, antinutritive factors, pathogens, and mycotoxins (van Huis et al., 2013; EFSA, 2015; van der Fels-Klerx, 2018; Mézes, 2018) because insects can grow on a different substrate, which could be naturally contaminated by undesirable substances, for instance, mycotoxins. Although mycotoxin occurrence in food and feed is regulated in the EU, some rearing substrates, for instance, vegetable waste on small-scale farms, are not covered by these regulations. However, the accumulation of mycotoxins is low in most edible insect species, possibly due to the metabolisation of mycotoxins within the insect body (Niermans et al., 2021).

2 Mould infection of insects

Pathogenic moulds, such as *Aspergillus, Penicillium, Mucor, Rhizopus, Alternaria,* and yeast, such as *Candida,* may infect insects via the feed materials. These moulds on the surface and gut of insects can be directly contagious to animals and humans due to their potentially toxic or allergenic effects (Banjo et al., 2005; Suh et al., 2008). More than 1000 mould species are known to infect insects (Vega et al.,

 | DOI: 10.1163/9789004712720_11

TABLE 10.1 Edible insects as animal feeds

Insect species	Rearing larvae period[a]	Production scale	Target animal feed
Black soldier fly (*Hermetia illucens*)	3-5 weeks	Industrial	fish, poultry, pig, pets
Yellow mealworm (*Tenebrio molitor*)	5-7 weeks	Industrial	fish, poultry, pig, pets
Lesser mealworm (*Alphitobius diaperinus*)	6-8 weeks	Industrial	fish, poultry, pig, pets
Housefly (*Musca domestica*)	4-6 weeks	Small	poultry, pets

a Rearing period depends on temperature and developmental stage at harvesting

2012), but only a few have toxic secondary metabolic products, such as mycotoxins. Furthermore, moulds have different optimal temperatures and humidity necessary for mycotoxin production. Still, in general, high humidity and high temperature increase the risk of mould growth, particularly in the case of insect meal, which is highly susceptible to moisture absorption (Amadi et al., 2005). For instance, a high *Aspergillus* mould infection was found in grasshopper (*Imbrasia belina*) in Botswana due to tropical environmental conditions (Mpuchane et al. 1996). The pathogenesis of mould infection in insects follows a series of events comprising conidial adhesion to the insect host cuticle, followed by penetration of the cuticle via fungal enzymes, and after released the mycotoxins from the moulds to haemocoel of insects (Khachatorians, 1996). The exact mechanism of mycotoxin toxicosis in insects or insect larvae is not fully elucidated, but it has been proposed that mycotoxins suppress the immune system (Vilcinskas, 1999).

Heat treatment is beneficial against toxicogenic mould infection but not against mycotoxins because of their high heat stability (Bullerman and Bianchini, 2007). There are some heat-resistant mould species (Tournas, 1994), but those have low importance regarding the mycotoxin contamination of insect larvae.

3 Mycotoxins in insects

Mycotoxins in insects may originate from those produced by pathogenic fungi, such as *Aspergillus* spp., *Penicillium* spp., and *Fusarium* spp. in the substrates and from mycotoxin production in the gut of insects (Schabel, 2010). Mycotoxins can also be produced in insect hosts as the effect of entomophagous moulds proposed for use as alternative insecticides. However, entomophagous moulds are insignificant in edible insects, and their mycotoxins as contaminants through the feed have little chance of contaminating insects as feed or food materials (Qiongbo et al., 2016). Mycotoxins of other toxinogenic moulds can enter the interior of the insect body through the digestive tract with contaminated feedstuffs (Dowd, 1992). However, the concentrations of mycotoxins in the larvae are generally much lower than in the feedstuffs as was found in the study by Guo et al. (2014) with mealworm (*Tenebrio molitor*) larvae reared on wheat kernels infected with BEA, ENNA, ENNA1, ENNB, and ENNB1. The same result was found by Charlton et al. (2015), who analysed the presence of mycotoxins in the housefly (*Musca domestica*) larvae. Although BEA (6.9 μg/kg), ENNA (12.5 μg/kg), and ENNA1 (7.3 μg/kg) were detected in the samples, their levels were not considered to pose a risk for animal health.

There is relatively little information on mycotoxin contamination and its effect on insects. Therefore, an *in vitro* study was performed with 65 mycotoxins using the Sf-9 cell line from fall armyworm (*Spodoptera frugiperda*) pupal ovarian tissue (Boguś et al., 2021). The results showed that all tested mycotoxins significantly reduced the *in vitro* proliferation and displayed disintegrating cells and varying degrees of cell fragmentation compared to the untreated control cells. The most potent effect on Sf-9 cells (low proliferation rate, 13-30%, compared to control) was the effect of 19 mycotoxins. The following are important from feed or food safety aspects: AFB_1, AFB_2, 3-ADON, T-2 tetraol, FB_2, OTA, AOH, AME, BEA.

In contrast, the weakest effect (reduction of cell proliferation by 9-30% compared to control) was detected in the case of 13 mycotoxins, such as AFL, AFM_1, CIT, HT-2-toxin, roquefortine C, T-2 triol, ZEN, and α-ZEL. However, these *in vitro* results cannot be directly extrapolated to *in vivo* conditions or to other insect species because there are species differences, mainly due to the different metabolic activity of insect larvae which influenced the response to the ingested mycotoxins, for instance, through xenobiotic transformation (Schrögel and Wätjen, 2019). This was proven by the results of Dowd (1990), who found that DON and T-2 toxin changes the activity of some midgut detoxifying enzymes in the larvae of corn earworm (*Helicoverpa zea*) and fall armyworm. O-demethylation was induced by 1.6 times in corn earworm and 6.1 times in fall armyworm after 48 h rearing on 250 mg/kg DON contaminated substrate. GST activity and glutathione conjugation was induced in corn earworm by exposure to DON at 25 or 250 mg/kg 25 mg/kg T-2 toxin. Still, it was inhibited by 10–20% in fall armyworm larvae. High tolerance to AFB_1 in some insect larvae was proposed by efficient conversion of AFB_1 to less toxic AFL, AFB_2, or AFM_1, but not to epoxidation products, such as AFBO. This hypothesis was proven by an *in vitro* study (Lee et al., 2000). De Zutter et al. (2016) described a novel mechanism of detoxification; it was they who detected the phase II metabolite DON-3G for the first time in pea aphids. This defence mechanism against the cytotoxic effect of mycotoxins was previously only known in plants (Tucker et al., 2019).

The insecticidal activity of mycotoxins *in vivo* was investigated with many insect species. For instance, AFB_1, rubratoxin B, PAT, and DAS were found to be toxic in common fruit fly (*Drosophila melanogaster*) (Reiss, 1975) and BEA in blowfly (*Calliphora erythrocephala*) (Wang and Xu, 2012). In another study, a range of 22 mycotoxins was found to strongly influence the growth and mortality of corn earworm and fall armyworm, e.g. AFB_1, DON, DAS, T-2 toxin, ZEN, OTA, sterigmatocystin,

CIT, CPA, penicillic acid, MON, and penitrem A (Dowd, 1992). However, it is important to mention that these tests were performed with extremely high mycotoxins levels needed to observe an insecticidal activity (above 10 mg/kg and sometimes 100 mg/kg). The effects of several mycotoxins on edible insect larvae are summarised in Table 2.

Although mycotoxins are usually not harmful to insect larvae, they can be metabolised to higher or less toxic metabolites in the insect larvae, which may affect farm animals' production traits and health, as summarised in Table 3 and described below.

TABLE 10.2 Effect of mycotoxins on edible insect larvae

Insect species	Mycotoxins	Dose (µg/kg substrate)	Effects	
Yellow mealworm (*Tenebrio molitor*)	DON	2000-12,000[a]	ADWG	Ø
			Survival (%)	>96
	DON	4900- 25,000[b]	Larval weight	↓
			Locomotion	↓
	T-2/HT-2 toxin	100-250[c]	Survival (%)	↓
	ZEN	500-2000[d]	Larval development	Ø
			Larval weight	↑
	AFB_1	1070[e]	Weight gain	↓
Lesser mealworm (*Alphitobius diaperinus*)	DON	5000-125,000[f]	Larval development	Ø
	ZEN	500-12500[f]	Larval development	Ø
	AFB_1	20-500[f]	Larval development	Ø
	OTA	100-2500[f]	Larval development	Ø
Black soldier fly (*Hermetia illucens*)	DON	5000-125,000[f]	Larval development	Ø
	ZEN	500-12,500[f]	Larval development	Ø
	AFB_1	10 – 500[g]	Larval weight	Ø
			Survival (%)	Ø
	AFB_1	20-500[g]	Larval development	Ø
	OTA	100-2500[g]	Larval development	Ø
	AFB_1/AFB_2/AFG_2/DON/ZEN/OTA	4600/88/17/46/260/860[h]	Larval weight	Ø
			Survival (%)	Ø
			Feed intake	Ø

ADWG = Average daily weight gain; Ø = no effect; ↑ = increased; ↓ = decreased. a Ochoa-Sanabria et al. (2019); b Janković-Tomanić et al. (2019); c Piacenza et al. (2021); d Niermans et al. (2019); e Gützkow et al. (2021); f Camenzuli et al. (2018); g Bosch et al. (2017); h Purschke et al. (2017).

TABLE 10.3 Metabolism of mycotoxins in several insect larvae

Insect	Mycotoxins	Metabolites	Effect of metabolites on farm animals
Yellow mealworm (*Tenebrio molitor*)	DON	3-ADON, NIV	3-ADON causes the same intestinal damage as DON[1] NIV is haematotoxic in long-term exposure[a]
	ZEN	α- and β-ZEL	60/0.2 times more oestrogenic than ZEN[b]
	AFB_1	AFM_1	AFM_1 is 7× less toxic than AFB_1[c]
Lesser mealworm (*Alphitobius diapernus*)	DON	Unknown	–
	ZEN	α- and β-ZEL	60/0.2 times more oestrogenic than ZEN[b]
	AFB_1	AFL, AFM_1	AFL is 18 × less hepatotoxicity than AFB1[d] AFM_1 is 7× less toxic than AFB_1[c]
Black soldier fly (*Hermetia illucens*)	ZEN	α- and β-ZEL	60/0.2 times more oestrogenic than ZEN[b]
	AFB_1	AFL	18 × less hepatotoxicity than AFB_1[d]
Corn earworm (*Helicoverpa zea*)	AFB1	AFP1	16× less genotoxic than AFB_1[e]

a Alizadeh et al. (2016)
b Eriksen and Pettersson (2004)
c Li et al. (2018)
d Karabulut et al. (2014)
e Buchi et al. (1973). Source: Modified from Evans and Shao (2022).

3.1 *Deoxynivalenol*

Camenzuli et al. (2018) investigated the accumulation of DON in larvae of both black soldier flies and lesser mealworms. They found that DON (5, 50, or 125 mg/kg substrate) did not accumulate in the larvae and was present only at a minimal concentration or below the LOD (10 µg/kg). Purschke et al. (2017) also investigated the accumulation of DON (4600 ± 1400 µg/kg substrate) in black soldier fly larvae, and this mycotoxin was at concentrations lower than the LOD (10 µg/kg) in the larvae after 24 hours of fasting before harvesting. DON accumulation and excretion was investigated in yellow mealworm larvae. The larvae were reared on experimentally contaminated wheat flour with DON (8000 µg/kg), but no detectable levels of the parent compound (DON and two derivatives (DON-3G and 15-ADON) analysed) were measured in the larvae after 14 days (LOD 10 µg/kg). The excreta contained up to 41% of the DON content of the feed, which suggests that the mealworm larvae may have metabolised the remaining amount to unknown metabolites (van Broekhoven et al., 2017). Corn earworm larvae and fall armyworm larvae were sensitive to DON only at a high concentration of 25 mg/kg, as suggested by the reduced growth (Dowd, 1990). Another study proved that yellow mealworm larvae could utilise

DON-contaminated wheat as a nutrient source when fed wheat contaminated with 2,000, 10,000, or 12,000 μg/kg. The DON content of the larvae in different groups was the same (121.8±19.3 to 136.4±40.5 μg/kg), which means that less than 1.2% of the ingested DON was accumulated in the larvae when they consumed DON-contaminated wheat up to 12,000 μg/kg. Excretion of DON ranged from 131.0 to 742.1 μg/kg in the experimental groups (Ochoa-Sanabria, 2019).

3.2 *T-2 toxin*

There is experimental evidence about the different sensitivity of insects to T-2 toxin toxicity (Dowd, 1992). German cockroach (*Blattella germanica*) and honeybee (*Apis mellifera*) are resistant to T-2 toxin, while the larvae of domestic silk moth (*Bombyx mori*), corn earworm (*Helicoverpa zea*), and fall armyworm showed sensitivity. The flour beetle (*Tribolium confusum*) was fed wheat flour contaminated with an extremely high level (100 mg/kg) of T-2 toxin. It increased the fecundity of the flour beetle, but the larval hatch was reduced by 10–12% compared to the control. Progeny of beetles fed T-2 toxin-contaminated wheat flour showed high egg production. However, the high dose of T-2 toxin resulted in a slow development of the larvae and moderate mortality (Wright et al., 1976). In a study by van Broekhoven et al. (2017) with lesser mealworm larvae, the T-2 toxin accumulation was 45 μg/kg when larvae were fed with a substrate of 500 μg/kg T-2 toxin content. Still, no significant detrimental effects on larval growth and survival were observed. However, the T-2 concentration of the larvae decreased quickly below the LOD (10 μg/kg) when they fasted or consumed control feed 24 hours before harvesting. The cause of this rapid decrease is possibly due to the excretion through the frass. In another study by Piacenza et al. (2021), T-2 and HT-2 toxins at the dose of 100 and 250 μg/kg substrate decreased the survival rate of yellow mealworm larvae. T-2 and HT-2 toxins were not detected in the surviving larvae, but a high level of T-2 toxin was found in the dead larvae from the high dose group (44.2 μg/kg). The cause of this difference would be the metabolisation of trichothecenes to unknown metabolites or conjugates in the living larvae up to harvesting.

3.3 *Fumonisins*

Physiological effects such as growth performance and respiratory metabolism of yellow mealworm larvae were assessed by Abado-Becognee et al. (1998) after exposure to a high dose of FB_1 (450 mg/kg feed). Reduced growth was observed only after consuming FB_1 for several weeks; there was a reduced rate of CO_2 production, reduced feed consumption, and reduced protein metabolism, whereas the mortality was not affected. According to the results of this study, approximately 40% of the ingested FB_1 was excreted with the frass. Exposure of lesser mealworm larvae with lower FB_1 and FB_2 concentrations (FB_1: 727±6 μg/kg; FB_2: 294±5 μg/kg) only FB_1 (127± 6 μg/kg) but not FB_2 was detected in the larvae at the end of the feeding trial (LOD 25 μg/kg) (Leni et al., 2019).

3.4 *Zearalenone*

ZEN-contaminated wheat flour (100 mg/kg) as substrate did not significantly affect the flour beetle. The beetles showed a prolonged reproductive life, as was proven by the increased egg production with age. The fecundity and fertility of progeny were the same as those of the parents (Wright et al., 1976). In the study of Niermans et al. (2019) with yellow mealworm larvae, no changes in the mortality were observed, and the growth rate increased by 37% and 62% in the low (600 μg/kg) and high (900 μg/kg) ZEN contaminated wheat flour fed groups as compared to the control in a long-term (8 weeks) feeding trial.

ZEN did not accumulate in the larvae of black soldier fly or lesser mealworm larvae at the dose range of 0.5-12.5 mg/kg substrate, and it was recovered at minimal concentrations or below the LOD (2 μg/kg). However, the concentration of the analysed metabolites showed that the parent compound (ZEN) was metabolised in the larvae, to a large extent, into two metabolites, α-and β-ZEL in black soldier fly larvae, and mainly to α-ZEL in lesser mealworm larvae (Camenzuli et al., 2018). Purschke et al. (2017) investigated the accumulation of ZEN (860±260 μg/kg substrate) in black soldier fly larvae, and no ZEN was detected. The same effect, no accumulation but nearly 100% excretion, was found using 173±4 μg/kg natural contamination of the substrate in black soldier fly larvae (Leni et al., 2019). A study with lesser mealworms showed that the level of 500 μg/kg ZEN in the substrate did not cause detrimental effects on larval growth and development and did not increase the mortality in the feeding trial (van Broekhoven et al., 2017). The ZEN accumulation was 42 μg/kg at harvesting, but ZEN concentration decreased quickly below the LOD (2 μg/kg) when the larvae fasted or consumed control feed 24 hours before harvesting. The cause of this rapid decrease is possibly due to the excretion through the frass. Approximately 30% of the total ZEN is metabolised to reductive metabolites, α- and β-ZEL. It is important to consider that the toxicity of α-ZEL is much higher than ZEN, and this risk should be considered when these larvae enter the animal feed chain. Camenzuli et al. (2018) observed the same metabolism of ZEN in black soldier fly larvae. Phase II ZEN metabolites, such as ZEN-14S and ZEL-sulphate, were also detected in the larval faeces but not in the larvae, which means that these metabolites were rapidly excreted.

3.5 *Aflatoxins*

The toxic effects of AFB_1 in insect species have been widely studied. A recent literature review showed that insects of the order Coleoptera (e.g. yellow mealworm) have lower susceptibility to AFB_1 than Lepidoptera (e.g. butterfly) and Diptera (e.g. black soldier fly, housefly) (Niermans et al., 2021). For instance, it was found that German cockroach and honeybee are relatively resistant to AFB_1 (Dowd, 1992), possibly due to the effective xenobiotic transformation based on CYP enzymes (Guodong, 2010). Larvae of black soldier fly (*Hermetia illucens*) and yellow mealworm also showed that AFB_1 up to 0.5 mg/kg dry feed, which is 25 times higher than the maximum limits (0.02 mg/kg) set by the European Commission for complete feed (Commission Recommendation 576/2006/EC), did not affect the survival rate and weight gains in both species (Bosch et al., 2017). However, larvae of lesser mealworm showed decreased weight gain by 15% in the presence of a higher dose of AFB_1 (10.7 mg/kg) in feed. Other insect species, such as common fruit fly, domestic silk moth, corn earworm, and fall armyworm, are more sensitive to AFB_1 (Dowd, 1992). A high mortality rate was detected when aflatoxin-contaminated (10 µg/kg) substrate was used on tobacco budworm larvae (*Heliothis virescens*). Still, the sensitivity of larvae to AFs decreases with age (Gudauskas, 1967). Aflatoxin contamination of insects or insect meals would be a food and feed safety problem. Edible grasshopper was checked, and the level of total AFs varied from 0 to 50 µg/kg product (Mpuchane et al., 1996).

Larvae of black soldier fly and yellow mealworm were grown on poultry feed experimentally contaminated with AFB_1 (25 µg/kg). It was found that black soldier fly and yellow mealworm larvae did not accumulate AFB_1, and the larvae had a high tolerance to this mycotoxin (Bosch et al. 2017). Other studies (Purschke et al. 2017; Camenzuli et al. 2018) also revealed that black soldier fly and lesser mealworm (*Alphitobius diaperinus*) larvae did not accumulate AFs (AFB_1, AFB_2 and AFG_2). Yellow mealworm larvae fed with a substrate contaminated 10.7 mg/kg aflatoxin for four weeks accumulate less than 0.5% of AFB_1. In addition, a novel, possibly monohydroxylated AFB_1 metabolite was found in the larvae after harvesting, which showed no toxicity in the genotoxicity assay (Gützkow et al., 2021).

3.6 *Ochratoxin A*

The accumulation of OTA was studied in black soldier fly and lesser mealworm larvae. The substrate was experimentally contaminated with 0.1, 1.0, or 2.5 mg/kg OTA. It was found that OTA did not accumulate in the larvae over the LOD concentration (20 µg/kg), even at the highest exposure level (Camenzuli et al., 2018). Purschke et al. (2017) also studied the accumulation of OTA (260 µg/kg substrate) in black soldier fly larvae, and no mycotoxin was detected in the larvae over the LOD value (20 µg/kg).

3.7 *Other mycotoxins*

No research is available regarding the effects of other mycotoxins on edible insects. However, studies using the globally found *Galleria mellonella* (the Greater Wax moth) as a model to understand the interaction between larvae and fungal pathogens, could be used as a starting point. Exposure of *G. mellonella* larvae to 100 µg/g ENNB did not affect the survival or growth of the larvae, but 50 µg/g DAS negatively affected the larvae (Mulè et al., 1992). After oral administration, a significant decrease in *G. mellonella* larvae biomass was observed after two days' exposure to 2.5 mg/g TeA and after six days' exposure to 0.25 mg/g TeA, whereas no effect was observed at a concentration of 0.025 mg/g (Salimova et al., 2021). Interestingly, when these authors offered a substrate with 25 mg/g TeA, feed refusal was observed, which may explain a mortality that did not exceed 20%. A 48-hour exposure of *G. mellonella* to a clinical isolate of *A. fumigatus* with the ability to produce high levels of gliotoxin (~ 2.5 µg/g) resulted in a high mortality rate probably because of the combined effect of the fungal infection and the immunosuppressive effect of this mycotoxin. However, no information is available when this larvae is exposed to lower gliotoxin levels or tissue accumulation (Reeves et al., 2004). Fungi of the genus *Metarhizium* are usually applied as biological control against insects. However, inoculation of *G. mellonella* larva with fungi of this genus resulted in the synthesis of EAs and accumulation in the insect (Leadmon et al., 2020). Furthermore, the presence of EAs contributes to an increase in the virulence of *A. fumigatus* (Jones and Panaccione, 2023).

4 Conclusion

According to the available literature, mycotoxins adversely affected the survival rate in insects only at extremely high concentrations. However, no accumulation of mycotoxins was observed with different edible insect species, even up to a mycotoxin concentration in the rearing substrate of several times the maximum European limits and guidance values. It was also demonstrated that the gut content significantly impacts the insect body burden of mycotoxins; therefore, a fasting period of at least 24 h before or after harvesting to empty their gut would be suitable in industrial rearing procedures, even for those fed with a mycotoxin-contaminated substrate. However, the larvae or the residues could not recover a substantial portion of the mycotoxin ingested. There are some species-specific metabolic processes in insect larvae that may detoxify mycotoxins. However, there is a lack of studies with different mycotoxins, and further *in vitro* and *in vivo* studies are required to investigate the potential metabolism pathways in insect larvae to gain more insight into them. In conclusion, edible insect larvae contain no more hazards than other animal products, and insects can be considered safe when properly managed. However, it should be noted that farmed insects may be contaminated with other toxic materials, e.g. heavy metals, herbicides, pesticides, etc., and accumulate them to a much greater extent than mycotoxins.

References

Abado-Becognee, K., Fleurat-Lessard, F., Creppy, E.E. and Melcion, D., 1998. Effects of fumonisin B1 on growth and metabolism of larvae of the yellow mealworm, *Tenebrio molitor*. Entomologia Experimentalis et Applicata 86: 135–143. https://doi.org/10.1046/j.1570-7458.1998.00274.x.

Alizadeh, A., Braber, S., Akbari, P., Kraneveld, A., Garssen, J. and Fink-Gremmels, J., 2016. Deoxynivalenol and its modified forms: are there major differences? Toxins 8: 334. https://doi.org/10.3390/toxins8110334.

Amadi, E.N., Ogbalu, O.K, Barimalaa, I.S. and Pius, M., 2005. Microbiology and nutritional composition of an edible insect (*Bunaea alcinoe* Stoll) of the Niger Delta. Journal of Food Safety 25: 193–197. https://doi.org/10.1111/j.1745-4565.2005.00577.x.

Banjo, A.D., Lawal, O.A. and Adeduji, O.O., 2005. Bacteria and fungi isolated from housefly (*Musca domestica* L.) larvae. African Journal of Biotechnology 4: 780–784.

Boguś, M.I., Wrońska, A.K., Kaczmarek, A. and Boguś-Sobocińska, M., 2021. *In vitro* screening of 65 mycotoxins for insecticidal potential. PLoS ONE 16: e0248772. https://doi.org/10.1371/journal.pone.0248772.

Bosch, G., van der Fels-Klerx. H.J., de Rijk, T.C. and Oonincx, D.G.A.B., 2017. Aflatoxin B1 tolerance and accumulation in black soldier fly larvae (*Hermetia illucens*) and yellow mealworms (*Tenebrio molitor*). Toxins 9: 185. https://doi.org/10.3390/toxins9060185.

Buchi, G., Spitzner, D., Paglialunga, S. and Wogan, G.N., 1973. Synthesis and toxicity evaluation of afatoxin P1. Life Sciences 13: 1143–1149. https://doi.org/10.1016/0024-3205(73)90381-0.

Bullerman, L.B. and Bianchini, A., 2007. Stability of mycotoxins during food processing. International Journal of Food Microbiology 119: 140–146. https://doi.org/10.1016/j.ijfoodmicro.2007.07.035.

Camenzuli, L., van Dam, R., De Rijk, T. C., Andriessen, R., van Schelt, J. and van der Fels-Klerx, H. J., 2018. Tolerance and excretion of mycotoxins aflatoxin B1, zearalenone, deoxynivalenol and ochratoxin A by *Alphitobius diaperinus* and *Hermetia illucens* from a contaminated substrate. Toxins 10: 91. https://doi.org/10.3390/toxins10020091.

Charlton, A. J., Dickinson, M., Wakefield, M. E., Fitches, E., Kenis, M., Han, R. Zhu, F., Kone, N., Grant, M., Devic, E., Bruggeman, G., Prior, R. and Smith, R., 2015. Exploring the chemical safety of fly larvae as a source of protein for animal feed. Journal of Insects for Food and Feed 1: 7–16. https://doi.org/10.3920/JIFF2014.0020.

Commission Regulation EU No. 2021/1372 of 18 August 2021 on the amending Annex IV to Regulation (EC) No. 999/2001 of the European Parliament and of the Council as regards the prohibition to feed non-ruminant farmed animals, other than fur animals, with protein derived from animal. Official Journal of the European Union. 2021, 295: 1.

De Zutter, N., Audenaert, K., Arroyo-Manzanares, N., De Boevre, M., Van Poucke, C., De Saeger, S., Haesaert, G. and Smagghe, G. 2016. Aphids transform and detoxify the mycotoxin deoxynivalenol via a type II biotransformation mechanism yet unknown in animals. Scientific Reports, 6: 38640. https://doi.org/10.1038/srep38640.

DiGiacomo, K. and Leury, B.J., 2019. Review: Insect meal: a future source of protein feed for pigs? Animal 13: 3022-3030. https://doi.org/10.1017/S1751731119001873.

Dowd, P.F. 1990. Responses of representative midgut detoxifying enzymes from *Heliothis zea* and *Spodoptera frugiperda* to trichothecenes. Insect Biochemistry 20: 349–356. https://doi.org/10.1016/0020-1790(90)90054-X.

Dowd, P.F. 1992. Insect interactions with mycotoxin-producing fungi and their hosts. In:

Bhatnagar, D., Lillehoj, E.B. and Arora, D.K. (eds.) Handbook of Applied Mycology. Mycotoxins in Ecological Systems V. Marcel Dekker, New York, p. 137–155.

Eriksen, S. and Petterson, H., 2004. Toxicological evaluation of trichothecenes in animal feed. Animal Feed Science and Technology 114: 205-239. https://doi.org/10.1016/j.anifeedsci.2003.08.008.

European Commission Recommendation 2006/576/EC of 17 August 2006 on the presence of deoxynivalenol, zearalenone, ochratoxin A, T-2 and HT-2 and fumonisins in products intended for animal feeding. Official Journal of the European Union. 2006, 229: 7–9.

European Commission Recommendation 2015/2283/EC of 11 December 2015 on the novel foods, amending Regulation (EU) No 1169/2011 of the European Parliament and of the Council and repealing Regulation (EC) No 258/97 of the European Parliament and of the Council and Commission Regulation (EC) No 1852/2001. Official Journal of the European Union. 2015, 327: 1–22.

European Commission Recommendation 2017/893/EC of 25 May 2017 on the amending Annexes I and IV to Regulation (EC) No. 999/2001 of the European Parliament and of the Council and Annexes X, XIV and XV to Commission Regulation (EU) No 142/2011 as regards the provisions on processed animal protein. Official Journal of the European Union. 2017, 138: 92-116.

European Food Safety Authority 2015. Risk profile related to production and consumption of insects as food and feed. EFSA Scientific Committee. EFSA Journal 13: 4257. https://doi.org/10.2903/j.efsa.2015.4257.

Evans, N.M. and Shao, S. 2022. Mycotoxin Metabolism by Edible Insects. Toxins 14: 217. https://doi.org/10.3390/toxins14030217.

Gudauskas, R.T., Davis, N.D. and Diener, U.L. 1967. Sensitivity of *Heliothis virescens* larvae to aflatoxin in ad libitum feeding. Journal of Invertebrate Pathology 9: 132–133. https://doi.org/10.1016/0022-2011(67)90055-9.

Guo, Z., Döll, K., Dastjerdi, R., Karlovsky, P., Dehne, H.-W. and Altincicek, B., 2014. Effect of fungal colonization of wheat grains with *Fusarium* spp. on food choice, weight gain and mortality of meal beetle larvae (*Tenebrio molitor*). PLoS ONE 9: e100112. https://doi.org/10.1371/journal.pone.0100112.

Guodong, N., 2010. Toxicity of mycotoxins to insects and underlying molecular and biochemical mechanisms. Ph.D. Dissertation University of Illinois at Urbana-Champaign. http://hdl.handle.net/2142/17002.

Gützkow, K.L., Ebmeyer, J., Kröncke, N., Kampschulte, N., Böhmert, L., Schöne, C., Schebb, N.H., Benning, R., Braeuning, A. and Maul, R., 2021. Metabolic fate and toxicity reduction of aflatoxin B1 after uptake by edible *Tenebrio molitor* larvae. Food and Chemical Toxicology, 155:112375. http://doi.org/10.1016/j.fct.2021.112375.

Janković-Tomanić, M. Petković, B., Todorović, D., Vranković, J. and Perić-Mataruga, V., 2019. Physiological and behavioral effects of the mycotoxin deoxynivalenol in *Tenebrio molitor* larvae. Journal of Stored Products Research 83: 236-242. https://doi.org/10.1016/j.jspr.2019.07.008.

Jones, A.M. and Panaccione, D.G., 2023. Ergot alkaloids contribute to the pathogenic potential of the fungus *Aspergillus leporisi*. Applied and Environmental Microbiology 89: e00415-23. https://doi.org/10.1128/aem.00415-23.

Karabulut, S., Paytakov, G. and Leszczynski, J., 2014. Reduction of aflatoxin B1 to aflatoxicol: a comprehensive DFT study provides clues to its toxicity. Journal of the Science of Food and Agriculture 94: 3134–3140. https://doi.org/10.1002/jsfa.6663.

Khachatorians, G.G., 1996. Biochemistry and molecular biology of entomopathogenic fungi. In: Howard, D.H. and Miller, J.D. (eds.) The Mycota VI. Springer, Heidelberg p. 331–363.

Lange, K.W. and Nakamura, Y., 2021. Edible insects as future food: chances and challenges. Journal of Future Foods, 1: 38-46. https://doi.org/10.1016/j.jfutfo.2021.10.001.

Leadmon, C.E., Sampson, J.K., Maust, M.D., Macias, A.M., Rehner, S.A., Kasson, M.T. and Panaccione, D.G., 2020. Several *Metarhizium* species produce ergot alkaloids in a condition-specific manner. Applied and Environmental Microbiology. 86:e00373-20. https://doi.org/10.1128/AEM.00373-20.

Lee, S.E. and Campbell, B.C., 2000. In vitro metabolism of aflatoxin B1 by larvae of navel rangeworm, *Amyelois transitella* (Walker) (Insecta, Lepidoptera, Pyralidae) and codling moth, *Cydia pomonella* (L.) (Insecta, Lepidoptera, Tortricidae). Archives of Insect Biochemistry and Physiology 45: 166–174. https://doi.org/10.1002/1520-6327.

Leni, G., Cirlini, M., Jacobs, J., Depraetere, S., Gianotten, N., Sforza, S. and Dall'Asta, C., 2019. Impact of naturally contaminated substrates on *Alphitobius diaperinus* and *Hermetia illucens*: uptake and excretion of mycotoxins. Toxins, 11: 476. https://doi.org/10.3390/toxins11080476.

Li, S., Min, L., Wang, G., Li, D., Zheng, N. and Wang, J., 2018. Occurrence of aflatoxin M1 in raw milk from manufacturers of infant milk powder in China. International Journal of Environmental Research and Public Health 15: 879. https://doi:org/10.3390/ijerph15050879.

Mézes, M., 2018. Food safety aspects of insects: A review. Acta Alimentaria 47: 513–522. https://doi.org/10.1556/066.2018.47.4.15.

Mpuchane, S., Taligoola, H.K. and Gashe, B.A., 1996. Fungi associates with *Imbrasia belina*, an edible grasshopper. Botswana Notes and Records 28:193–197.

Mulè, G., D'Ambrosio, A., Logrieco, A. and Bottalico, A., 1992. Toxicity of mycotoxins of *Fusarium sambucinum* for feeding in *Galleria mellonella*. Entomologia Experimentalis et Applicata 62: 17–22. https://doi.org/10.1111/j.1570-7458.1992.tb00636.x.

Niermans, K., Meyer, A.M., Hoek-van den Hil, E.F., van Loon, J.J.A. and van der Fels-Klerx, H.J., 2021. A systematic literature review on the effects of mycotoxin exposure on insects and mycotoxin accumulation and biotransformation. Mycotoxin Research 37: 279-295. https://doi.org/10.1007/s12550-021-00441-z.

Niermans, K., Woyzichovski,, J., Kroncke, N., Benning, R. and Maul, R., 2019. Feeding study for the mycotoxin zearalenone in yellow mealworm (*Tenebrio molitor*) larvae-investigation of biological impact and metabolic conversion. Mycotoxin Research 35: 231–242. https://doi.org/10.1007/s12550-019-00346-y.

Ochoa-Sanabria, C.O., Hogan, N., Madder, K., Gillott, C., Blakley, B., Reaney, M., Beattie, A. and Buchanan, F., 2019. Yellow mealworm larvae (*Tenebrio molitor*) fed mycotoxin-contaminated wheat — A possible safe, sustainable protein source for animal feed? *Toxins* 11: 282. https://doi.org/10.3390/toxins11050282.

Piacenza, N., Kaltner, F., Maul, R., Gareis, M., Schwaiger, K. and Gottschalk, C., 2021. Distribution of T-2 toxin and HT-2 toxin during experimental feeding of yellow mealworm (*Tenebrio molitor*). Mycotoxin Research 37: 11-21. https://doi.org/10.1007/s12550-020-00411-x.

Purschke B., Scheibelberger, R., Axmann, S., Adler, A. and Jager, H., 2017. Impact of substrate contamination with mycotoxins, heavy metals and pesticides on growth performance and composition of black soldier fly larvae (*Hermetia illucens*) for use in the feed and food value chain. Food Additives and Contaminants Part A 34: 1410–1420. https://doi.org/10.1080/19440049.2017.1299946.

Qiongbo, H., Li, F. and Zhang, Y., 2016. Risks of mycotoxins from mycoinsecticides to humans. BioMed Research International 2016: 3194321. https://dx.doi.org/10.1155/2016/3194321.

Reeves, E.P., Messina, C.G.M., Doyle, S. and Kavanagh, K., 2004. Correlation between gliotoxin production and virulence of *Aspergillus fumigatus* in *Galleria mellonella*. Mycopathologia 158: 73-79. https://doi.org/10.1023/b:myco.0000038434.55764.16.

Reiss, J., 1975. Insecticidal and larvicidal activities of the mycotoxins, aflatoxin B1, rubratoxin B, patulin and diacetoxyscirpenol towards *Drosophila melanogaster*. Chemico-Biological Interactions 10: 339–342. https://doi.org/10.1016/0009-2797(75)90055-1.

Salimova, D., Dalinova, A., Dubovik, V., Senderskiy, I., Stepanycheva, E., Tomilova, O., Hu, Q. and Berestetskiy, A., 2021. Entomotoxic activity of the extracts from the fungus, *Alternaria tenuissima* and its major metabolite, tenuazonic acid. Jornal of Fungi 7: 774.

Schabel, H.G., 2010. Forest insects as food: A global review. In: Forest insects as food: humans bite back. FAO Regional Office for Asia and the Pacific, Bangkok, Thailand. Proceedings of a workshop on Asia-Pacific resources and their potential for development. 19–21 February 2008, Chiang Mai, Thailand, pp. 37–64.

Schrögel, P. and Wätjen, W. 2019. Insects for food and feed-safety aspects related to mycotoxins and metals. Foods 8: 288. https://doi.org/10.3390/foods8080288.

Suh, S.O., Nguyen, N.H. and Blackwell, M., 2008. Yeasts isolated from plant-associated beetles and other insects: seven novel Candida species near *Candida albicans*. FEMS Yeast Research 8: 88–102. https://doi.org/10.1111/j.1567-1364.2007.00320.x.

uptake and excretion of mycotoxins. Toxins 11: 476. https://doi.org/10.3390/toxins11080476.

Tournas, V., 1994. Heat-resistant fungi of importance to the food and beverage industry Critical Reviews in Microbiology 20: 243-263. https://doi.org/10.3109/10408419409113558.

Tucker, J.R., Badea, A., Blagden, R., Pleskach, K., Tittlemier, S.A. and Dilantha Fernando, W.G., 2019. Deoxynivalenol-3-glucoside content is highly associated with deoxynivalenol levels in two-row barley genotypes of importance to Canadian barley breeding programs. Toxins 11: 319. https://doi.org/10.3390/toxins11060319.

van Broekhoven, S., Gutierrez, J. M., De Rijk, T. C., De Nijs, W. C. M. and Van Loon, J. J. A., 2017. Degradation and excretion of the fusarium toxin deoxynivalenol by an edible insect, the yellow mealworm (*Tenebrio molitor L.*). World Mycotoxin Journal 10: 163–169. https://doi.org/10.3920/WMJ2016.2102.

van der Fels-Klerx, H.J., Camenzuli, L., Belluco, S., Meijer, N. and Ricci, A., 2018. Food safety issues related to uses of insects for feeds and foods. Comprehensive Reviews in Food Science and Food Safety 17: 1172-1183. https://doi.org/10.1111/1541-4337.12385.

van Huis, A., Van Itterbeck, J., Klunder, H., Mertens, E., Halloran, A., Muir, G. and Van Tomme, P., 2013. Edible insects: future prospects for food and feed security. FAO Forestry, Paper 171. FAO, Rome. http://www.fao.org/docrep/018/i3253e/i3253e00.htm.

Vega, F.E., Meyling, N.V., Luangsa-Ard, J.J. and Blackwell, M., 2012. Fungal entomopathogens. In: Vega, F.F. and Kaya, H.K. (eds.) Insect pathology. 2nd ed. Academic Press, San Diego p. 171–220.

Vilcinskas, A. and Götz, P., 1999. Parasitic fungi and their interactions with the insect immune system. Advances in Parasitology 43: 267–313. https://doi.org/10.1016/S0065-308X(08)60244-4.

Wang, Q. and Xu, L., 2012. Beauvericin, a bioactive compound produced by fungi: a short review. Molecules 17: 2367–2377. https://doi.org/10.3390/molecules17032367.

Wright, V.F., De Las Casas, E. and Harein, P.K., 1976. The response of confused flour beetle (*Tribolium confusum*) to the mycotoxins zearalenone (F-2) and T-2 toxin. Environmental Entomology 5: 371–374. https://doi.org/10.1093/ee/5.2.371.

CHAPTER 11

Sampling and analysis of feedstuffs and diets for mycotoxins

T. van Leeuwen[1], H. Janssens[1] and A. Gallo[2]*

[1]TLR International Laboratories, Ridderkerk, the Netherlands
[2]Università Cattolica del Sacro Cuore, Piacenza, Italy
**correspondence: tvleeuwen@tlr.nl*

1 Introduction

It is almost impossible to prepare a mycotoxin-free animal diet. In fact, co-contamination is the general rule. One of the crucial steps for minimizing animal exposure to mycotoxins consists of selecting feedstuffs marginally contaminated with mycotoxins. However, such a screening process is complicated because mycotoxins are not homogeneously distributed in feedstuffs and they can occur within a concentration range of μg/kg (ppb) to mg/kg (ppm). As a result, animal diets will certainly contain several mycotoxins that were already present in the ingredients during mixing. Furthermore, extra mycotoxins can be produced as a result of poor storage conditions, for example, when the moisture level is higher than 14%.

Decisions about the acceptance or rejection of a feedstuff batch, as well as quality control of feeds including silage are based on sample analysis. It is therefore assumed that the amount of mycotoxins detected in a sample represents the actual contamination in the whole lot. However, uncertainty is more of a reality than 100% certainty.

Heterogeneity is only one of the source of errors in mycotoxin determination. According to the Eurachem Uncertainty Guide it is necessary to consider uncertainty, which includes random (sampling variability-heterogeneity and analytical variability) and systematic (selection bias and analytical bias) errors that may influence a result (Ramsey et al., 2019). The Eurachem guide also indicates how to calculate uncertainty, which can be associated with sampling, sample preparation, and analysis. Uncertainty of measurement is defined as 'a parameter, associated with the result of a measurement, that characterizes the dispersion of the values that could reasonably be attributed to the measurement' (IUPAC, 2023). This measurement of uncertainty is often expressed relative to the reported value, that being the obtained percentage commonly applied. Considering that most errors related to mycotoxin detection and quantification occur during sampling, this chapter summarizes some important as-

 | DOI: 10.1163/9789004712720_12

pects for a proper sampling plan of feedstuffs, feeds, and silage to improve accuracy and precision when determining mycotoxin contamination.

2 Mycotoxin distribution

The uneven distribution of mycotoxins in feedstuffs is a challenge for their accurate determination. For instance, a maize batch may contain a highly contaminated maize kernel mixed with uncontaminated or marginally contaminated ones. Moreover, symptoms of infection in plants are not always clear. For example, *Fusarium* infection in wheat is characterized by visible infection in neighbouring spikelets and the damaged kernels lose up to 50% of their weight, whereas infected barley kernels may have neighbouring spikelets free of *Fusarium* contamination and the contaminated kernel loses only 20% of its weight (Janssen et al., 2018). The irregular distribution of mycotoxins also occurs after the loading of ships with feedstuffs that probably originated from several terminals and different sets of storage and silos (Bourgeois and Lyman, 2012). The contamination in feeds will be more homogeneously distributed due to the mixing of the ingredients during diet preparation, but that does not exclude the risk of concentrated contamination. This explains why just mixing highly contaminated with non-contaminated grains will not prevent exposure to critical levels of mycotoxins (Peng et al., 2018).

Cattle ration includes forage, which can be fed as silage because it preserves nutrients from harvest through storage, and facilitates feeding (Mahanna and Chase, 2003). However, depending on the forage quality and storage, mycotoxin presence before ensiling and production in silage may occur (see Chapter 2). Cattle is usually fed grass and/or maize silage. The latter is a common choice because of its high energy value, but it is also a source of mycotoxin contamination. It was demonstrated that maize silage contains a greater diversity of mycotoxin-producing fungi species than grass silage (Wambacq et al., 2016). However, this does not discard the possibility of higher OTA contamination in grass than in maize silages (Tangni et al., 2013).

Selecting only the mouldy areas during sampling will not result in accurate measurement of mycotoxins levels. These findings were previously summarized by Wambacq et al. (2016). In brief, in maize silages, gliotoxin was detected in 14% of non-mouldy samples (up to 1.5 mg/kg) and not detected in mouldy samples, whereas roquefortine C was present at low levels in non-mouldy samples (0.74 mg/kg; positive in 62% of the samples) and 100% of the mouldy samples were positive containing up to 15 mg/kg of roquefortine C. In grass silage, OTA was detected in 85% of non-mouldy samples (up to 0.31 mg/kg) and in 80% of the mouldy samples but at a much lower level (up to 0.09 mg/kg). Low levels of roquefortine C were detected in non-mouldy samples (1.11 mg/kg; positive in 45% of the samples) and 75% of the mouldy samples were positive for this mycotoxin containing up to 7.2 mg/kg (Wambacq et al., 2016).

3 Mycotoxin sampling plan

Mycotoxin detection, identification, and quantification depends on the following labour-intensive steps: accurate and precise sampling, proper sample preparation, and adequate analysis (i.e., extraction, purification, separation, detection, confirmation, and quantification) (Whitaker, 2006; Anfossi et al., 2016; Tittlemier and Whitaker, 2023) (Figure 11.1). Besides the diverse variety of mycotoxins, their analysis is notoriously difficult since they are often present in trace amounts in samples. A proper sampling plan will be important to decrease the heterogeneity of the sample and to decrease the risks of rejecting a 'marginally contaminated lot' (seller's risk) or accepting a 'highly contaminated lot' (buyer's risk).

The sampling step will be responsible for circa 80% of the total error (or total variance) in a mycotoxin sampling plan, because of the risk of the selected sample not representing the whole feed material (Whitaker, 2004; Miraglia et al., 2005). To achieve a high accuracy and precision, sampling errors should be decreased by increasing sample size and establishing sampling methods, hence decreasing variability. This accuracy should not be confounded with laboratorial accuracy, which is

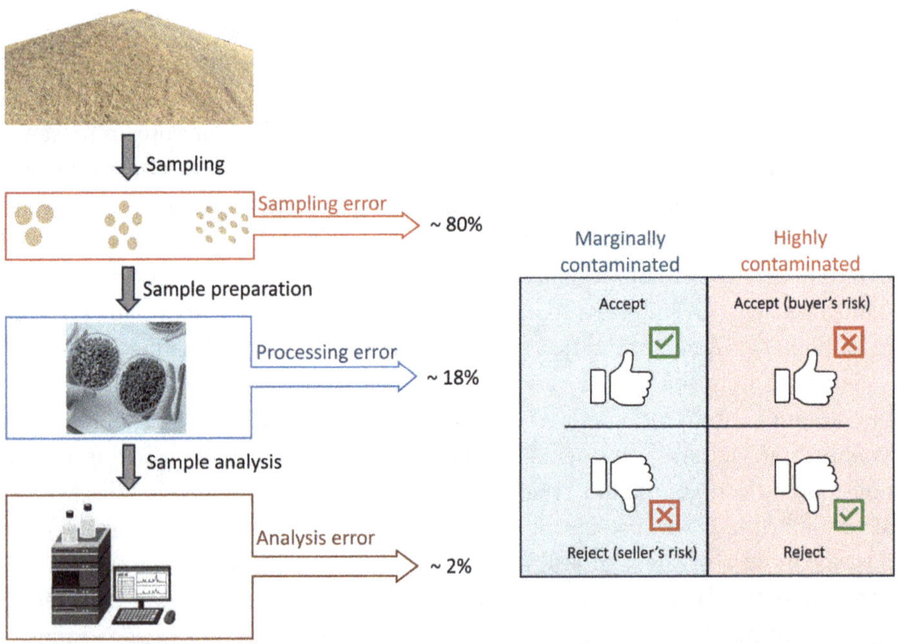

FIGURE 11.1 Schematic diagram showing the three main steps in a sampling plan: sampling, sample preparation, and sample analysis, which are associated with sources of variability and error, and the risks of an inappropriate sampling plan, i.e., the rejection of a good lot and the acceptance of a bad lot.
SOURCE: ADAPTED FROM WHITAKER AND JOHANSSON (2005)

related to the analysis compared with a reliable reference (Whitaker, 2006). The errors related to sampling, sample preparation, and analysis will depend on the contaminated material, as well as mycotoxin type and concentration. Whitaker (2004) analysed a sample of 0.91 kg of shelled maize containing 20 μg/kg AFB_1 using an immunoassay and observed error contribution of 75.6, 15.9, and 8.5% for the sampling, sample preparation and analysis, respectively. The error contribution of sampling, sampling preparation and analysis in wheat kernels was 53.2, 34.0. and 12.8 respectively for DON, and 65.4, 30.7, and 3.9, respectively for OTA (Biselli et al., 2008).

3.1 *Sampling*

Sampling methods need to follow a guideline to minimize the effects of heterogeneous distribution of mycotoxins and in this way prevent under- or overestimation of the actual contamination. Sample collection can be performed by applying dynamic or static methods. In the dynamic sampling the product is in motion (loading or unloading of the product), i.e. in processing lines, conveyors, elevators, or during packaging. It is important to guarantee an uninterrupted flow to allow the free fall of the sampled product, and it must pass at speed through the sampler to avoid blockage. In the case of cereal shipments, e.g. lots of more than 1000 tonnes, the cross-stream sampling appears to allow the selection of representative samples, but variability cannot be excluded as well as false negatives. In a study with cereal shipments (2000 tonnes) tested for OTA contamination, a 40-second sampling frequency to obtain circa 10 kg sample will have a 10% chance of no OTA detection if the level of this mycotoxin is 2 μg/kg, and this risk will decrease if the OTA level in the lot increases (Bourgeois and Lyman, 2012). These authors also showed that the larger the shipment, the lower the risk of false positives. In the static sampling the product is not in motion, i.e. in sacks, bags, transport containers, and retail packaging. The dynamic method allows an uninterrupted collection of a bigger number of incremental samples when compared to the static one, resulting in a more representative sample. The error associated with sampling is 3.9 to 5.6 times lower when a dynamic method is applied compared to a static one (Mallmann et al., 2014). These authors demonstrated that a maize lot contaminated with 20 μg/kg AFB_1, and a 15 μg/kg AFB_1 limit for acceptance/rejection, had 3% chance of being accepted after dynamic sampling and 21% chance of acceptance after static sampling.

The device used for sampling needs to be cleaned between each sampled lot to avoid cross-contamination. In the case of dynamic sampling, it is recommended to cut the flow in differing directions from one sample to the next in a frequency designed to distribute sampling points over the whole lot.

There are well-described and internationally accepted sampling plans for feedstuffs like those found in the Grain Inspection Handbook (www.gipsa.usda.gov/fgis/public_handbooks.aspx) in the USA, the Canadian Grain Commission in Canada (www.grainscanada.gc.ca), and in the European Union (Commission Regulation (EC) N° 401/2006). As an example, the EU guidelines for cereal sampling are summarized in Table 11.1. The lot represents the quantity of the batch delivered, the sublot is the designated part for the sampling, the incremental samples are the number of samples taken from the sublot, where each incremental sample weighs circa 100 g, and the aggregate sample is a combination of the mixed incremental samples taken from the sublot. The number of incremental samples and the weight of the aggregate sample will increase with the original weight of the lot.

Furthermore, the particle size must be considered during sampling, where maize batches require more incremental samples than wheat ones. Surveillance of DON and T-2 toxins in whole maize will require twice the number of incremental samples and sample weight when compared to wheat, barley, oats, and rye (Freese et al., 2000; Whitaker et al., 2000; IARC, 2012). For OTA, the number of incremental samples is similar, but the sample weight is 40% lower for wheat, barley, rice, and peas compared to maize (Coker et al., 2000; IARC, 2012).

If a sample is taken when filling the silage, a more representative sample is obtained. However, considering the fermentation during storage and the risks of mycotoxin production with time depending on the quality of the silage, it is more reliable to collect the samples during unloading and from the face of the silos after feeding (Allen et al., 2003).

Baled silage should be sampled in at least eight points with a steel corer able to reach at least 20 cm in depth during sampling (O'Brien et al., 2006). Each sub-sample should consist of at least 50 g and be thoroughly mixed to result in a composite sample of 400 g. However, the risk of underestimating contamination still exists and other authors suggest sampling the next-to-be-fed bale in 20 points, each of 25 g, resulting in a pool of 500 g aggregate sample (McElhinney et al., 2016) (Figure

TABLE 11.1 Division of cereal lots into sublots based on the feedstuff and batch weight

Lot weight (tonnes)	Sublots (weight or number)	Incremental samples (number)	Aggregate sample (kg)
≥ 1500	500 tonnes	100	30
> 300 and < 1500	3 sub lotes	100	30
≥ 50 and ≤ 300	100 tonnes	100	30
> 20 and ≤ 50	–	100	10
> 10 and ≤ 20	–	60	6
> 3 and ≤ 10	–	40	4
> 1 and ≤ 3	–	20	2
> 0.5 and ≤ 1	–	10	1
> 0.05 and ≤ 0.5	–	5	1
≤ 0.05	–	3	1

The weight of the incremental samples is of about 100 g, and the aggregate sample should be well mixed. In case of retail samples, the incremental sample weight will depend on the weight of the retail packing, but the aggregate sample must be of at least 1 kg.

SOURCE: COMMISSION REGULATION EC N° 401/2006

11.2). To decrease variation, the analysis from the feed-through is indicated specially when sampling baled silage.

Horizontal pit silage should be sampled in the next area ready to be fed (at its face; at least 60 cm deep), or vertically from 3 m behind the face (60 cm deep). This depth should also take into account the thickness of front removal for animal feeding (Dell'Orto et al., 2015). Vertical sampling requires a careful refill of all the holes made immediately after sampling and may bring extra risks of air getting in. It was previously shown that sampling the face is more practical and is as representative as sampling from the top surface of the silo (McElhinney et al., 2016). To obtain a more representative sample, it is recommended to collect at least nine subsamples following a W-shape from the silage face to form a pool of 500-1000 g aggregate sample (Figure 11.2).

The sampling of the total mixed ration (TMR) must be done up to a maximum of 20 minutes after distribution (Ishler, 2008). At least 10 incremental samples are indicated to produce the final aggregated sample. After being well mixed, a laboratorial sample of 500 g can be sent to a laboratory for mycotoxin analysis.

3.2 *Sample preparation, and transport*

All relevant information related to the sample must be recorded, such as sample code, type, place, and date of collection. When other additional information is available, it should also be recorded. This includes grain damage, moisture, storage period, presence of dust or insects, fungicide or pesticide treatment, etc.

Protect the samples from humidity and high temperatures. The samples need to be directly transferred to a clean plastic bag, and the air removed before clos-

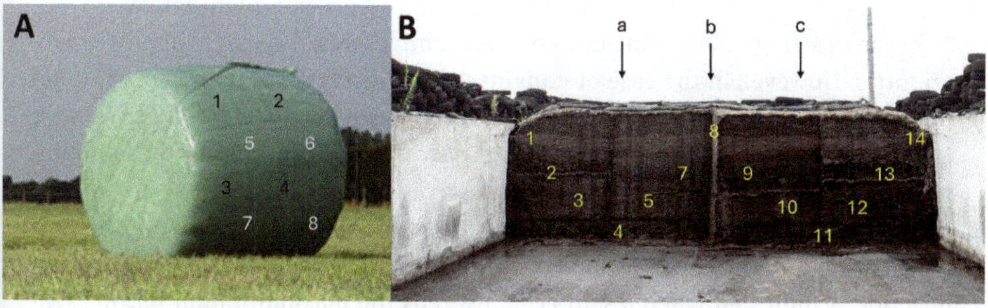

FIGURE 11.2 Pre-determined sampling positions on a bale (A) and pit (B) silage. A: A minimum of eight sampling points (1-4: front side; 5-8: back side) in a bale silage is indicated but decreasing the size (25 g instead of 50 g) and increasing the number of the subsamples (20 instead of 8) will increase accuracy. B: Pit silage can be sampled from its face (1 to 14) or vertically (a, b, and c) from the top surface. As no clear differences are observed between the positions, sampling in the next to fed face applying a W-shaped curve (9-14 subsamples) with at least nine samples is a practical method.

SOURCE: ADAPTED FROM O'BRIEN ET AL., 2006; DELL'ORTO ET AL., 2015; MCELHINNEY ET AL., 2016

ing them. Keep the samples cold (4 °C) until analysis. If mould is visible, samples should be stored frozen (-20 °C) until analysis. Avoid long-term storage of the samples. Samples stored cold must be analysed within three months, and those frozen at -20 °C should be analysed within six months.

Preferentially, the aggregate sample should be grinded/milled and well mixed before one portion of this sample is sent for analysis in the laboratory, being denominated as laboratory sample. If an aggregate sample contaminated by DON is ground to reach a particle size of 2 mm, a 16% coefficient of variance for the analysis will be obtained compared to a sample finely ground to a particle size of 0.5 mm. If the commodity is contaminated by OTA, this difference in particle size will result in a 94% coefficient of variance (Biselli et al., 2008). This is a result of a more uneven distribution of OTA compared to DON, as well as the fact that OTA levels are usually close to µg/kg, whereas DON can be found at much higher levels, i.e. mg/kg. Although sample preparation constitutes an important step before analysis, there is a scarcity of information about the applied instrument and method of grinding or milling the sample, optimal particle size of the aggregate and laboratorial sample, sample mass according to the analysed feedstuff, as observed in a meta-analysis focused on sample size and sample preparation for mycotoxin quantification in grain products (Kumphanda et al., 2021).

3.3 *Sample analysis*

Samples can be screened on site for major mycotoxins using rapid test kits or submitted to a laboratory for analysis. The rapid in-field test kits usually require sample grinding and are based on antibody technologies, providing results without the need to send samples to a specialized laboratory. This method is applicable, for example, when a rapid decision must be taken about the acceptance or not of a batch in a feed mill. However, the results can be qualitative or semi-quantitative.

In the case of marketing of grains or feed, samples must be sent to approved laboratories. However, in the case of crop insurance benefits, it is mandatory to send the sample to a laboratory approved by the crop insurance provider. Together with each analysed mycotoxin, a relative uncertainty must be provided by the laboratory responsible for the mycotoxin analysis. According to the Commission Directive 2003/121/EC, 'The analytical result has to be reported as $x \pm U$ whereby x is the analytical result and U is the expanded measured uncertainty'. Based on this directive, rejection is indicated when a sample '... exceeds the maximum limit beyond reasonable doubt taking into account the measurement uncertainty and correction for recovery'. The relative uncertainty of 50% is proposed for multi-mycotoxin analysis considering extraction, dilution, mass spectrophotometry analysis, and does not include sampling uncertainty (Stadler et al., 2018). When the total expanded uncertainty is considered, i.e. combining sampling, sample preparation, and analysis, this value is close to 12.8% for DON and 71.3% for OTA (Biselli et al., 2008). This is a result of a low coefficient of variation for DON during sampling (25%) and sample preparation (16%) compared to the high coefficient of variation during sampling (200%) and sample preparation (94%) of OTA.

The laboratory methods of analysis include immunoassays, thin layer chromatography, liquid chromatography or gas chromatography, fluorescence, or mass spectrometry, which are discussed in detail in Chapter 12. The selected methods of analysis will depend on the main aim, i.e. screening, quality control, marketing or insurance benefits.

4 Conclusions

Determining correct levels of mycotoxins in feedstuffs and animal diets is a laborious process because these contaminants are not homogeneously distributed and are found in very low concentrations ranging from µg to mg/kg. Moreover, the number of contaminated kernels can be low, but the level of contamination can be high. Therefore, the mycotoxin sampling plan is not valid unless representative sampling is performed. Applying an optimal sampling plan will reduce the errors considerably, and this can be done by increasing the number and pre-determining different collection positions of subsamples (incremental samples) without any unnecessary increase in the sample size/weight (aggregate or composite sample). To harmonize and improve sampling plans, the FAO has developed an online and free-to-use Mycotoxin Sampling Tool (http://tools.fstools.org/mycotoxins/). This system allows the user to adjust its sampling plan for several mycotoxins such as AFs (maize and cottonseed), DON (maize, wheat, and barley), FUM (maize), and OTA (wheat and oats) (Tittlemier and Whitaker, 2023). The plan can be edited based on the mycotoxin and commodity, regulatory limits, laboratory sample size, analytical variability, number of laboratory samples and replicates, as well as acceptance and rejection limits. Although this tool was developed for food commodities, it remains a practical tool to enhance the sampling plan of feedstuffs.

References

Allen, M., Coors, J.G. and Roth, G., 2003. Corn silage. In: Silage Science and Technology. Agronomy Monograph No. 42. ASA-CSSA-SSA, Madison, WI. 547–608.

Anfossi, L., Giovannoli, C. and Baggiani, C., 2016. Mycotoxin detection. Current Opinion in Biotechnology 37: 120–126. https://doi.org/10.1016/j.copbio.2015.11.005.

Biselli, S., Presin, C. and Syben, M., 2008. Investigation of the distribution of deoxynivalenol and ochratoxin A contamination within a 26 t truckload of wheat kernels. Mycotoxin Research 24: 98-104. https://doi.org/10.1007/BF02985287.

Bourgeois, F. and Lyman, G.J., 2012. Quantitative estimation of sampling uncertainties for mycotoxins in cereal shipments. Food Additives and Contaminants: Part A 29: 1141-1156. https://doi.org/10.1080/19440049.2012.675594.

Coker, R.D., Nagler, M.J., Defizem P.R., Derksen, G.B., Buchholz, H., Putzka, H.A., Hoogland, H.P., Roos, A.H. and Boenke, A., 2000. Sampling plans for the determination of aflatoxin B1 in large shipments of animal feedstuffs. Journal of AOAC International 83: 1252–1258.

Commission Directive 2003/121/EC, 2003. Official Journal of the European Union L 332: 38–40

Commission Regulation (EC) N° 401/2006. 2006. Official Journal of the European Union L70: 12-34.

Dell'Orto, V., Baldi, G. and Chelli, F., 2015. Mycotoxins in silage: checkpoints for effective management and control. World Mycotoxin Journal 8: 603-617. https://doi.org/10.3920/WMJ2014.1866.

Freese, L., Friedrich, R., Kendall, D. and Tanner, S., 2000. Variability of deoxynivalenol measurements in barley. Journal of AOAC International 83: 1259-1263.

IARC (International Agency for Research on Cancer), 2012. Sampling and sample preparation methods for determining concentrations of mycotoxins in foods and feeds. IARC Scientific Publications 158: 39-51.

Ishler, V., 2008. Resources for sampling and evaluating forages and total mixed rations (TMR). Penn State University, College of Agricultural Science, Pennsylvania, PA, USA. http://tinyurl.com/p79vlgy.

IUPAC (International Union of Applied and Pure Chemistry), 2023. Compendium of Terminology in Analytical Chemistry. Hibbert, B. (ed.) IUPAC Color Books. ISBN (978-1-78262-947-4). https://doi.org/10.1039/9781788012881.

Janssen, EM., Liu, C. and Van der Fels-Klerx, H.J., 2018. *Fusarium* detection and trichothecenes in barley and its comparison in wheat. World Mycotoxin Journal 11: 33-46. https://doi.org/10.3920/WMJ2017.2255.

Kumphanda, J., Matumba, L., Monjerezi, M., Whitaker, T.B., De Saeger, S. and Makun, H.A., 2021. Are sample size and sample preparation for mycotoxin quantitation in grain products getting trivialized? Food Control 130: 108400. https://doi.org/10.1039/c1an15124j.

Mahanna, W. and Chase, L.E., 2003. Practical applications and solutions to silage problems. In: Silage Science and Technology. Agronomy Monograph No. 42. ASA-CSSA-SSA, Madison, WI. 855–895.

Mallmann, A.O., Marchioro, A., Oliveira, M.S., Rauber, R.H., Dilkin, P. and Mallmann, C.A., 2014. Comparison of the efficiency between two sampling plans for aflatoxins analysis in maize. Brazilian Journal of Microbiology 45: 35-42. https://doi.org/10.1590/s1517-83822014000100006.

McElhinney, C., Danaher, M., Grant, J., Elliot, C.T. and O'Kiely, P., 2016. Variation associated with sampling bale or pit silage for mycotoxins and conventional chemical characteristics. World Mycotoxin Journal 9: 331-342. https://doi.org/10.3920/WMJ2015.1964.

Miraglia, M., De Santis, B., Minardi, V., Debegnach, F. and Brera, C., 2005. The role of sampling in mycotoxin contamination: an holistic view. Food Additives and Contaminants 22: 31-36. https://doi.org/10.1080/02652030500389055.

O'Brien, M.O., Kiely, P.O., Forristal, P.D. and Fuller, H.T., 2006. A note on sampling baled grass silage for fungal propagules. Journal of Animal and Feed Sciences 15: 305-311. https://doi.org/10.22358/jafs/66902/2006.

Peng, W.X., Marchal, L. and van der Poel, A.F.B., 2018. Strategies to prevent and reduce mycotoxins for compound feed manufacgturing. Animal Feed Science and Technology 237: 129-153. https://doi.org/10.1016/j.anifeedsci.2018.01.017.

Ramsey, M.H., Ellison, S.L.R. and Rostron, P. (eds.), 2019. Eurachem/EUROLAB/CITAC/Nordtest/AMC Guide: Measurement uncertainty arising from sampling: a guide to methods and approaches. Second Edition, Eurachem. ISBN (978-0-948926-35-8). http://www.eurachem.org.

Stadler, D., Sulyok, M., Schuhmacher, R., Berthiller, F. and Krska, R., 2018. The contribution of lot-to-lot variation to the measurement uncertainty of an LC-MS-based multi-mycotoxin assay. Analytical and Bioanalytical Chemistry 410: 4409-4418. https://doi.org/10.1007/s00216-018-1096-5.

Tangni, E.K., Pussemier, L., Bastiaanse, H., Haesaert, G., Foucart, G. and Van Hove, F., 2013. Presence of mycophenolic acid, roquefortine C, citrinin and ochratoxin A in maize and grass silages supplied to dairy cattle in Belgium. Journal of Animal Science Advances 3: 598–612.

Tittlemier, S.A. and Whitaker, T.B., 2023. Current sampling plans can introduce high variance in mycotoxin testing results as demonstrated by the online FAO Mycotoxin Sampling Tool. World Mycotoxin Journal 16: 115-126. https://doi.org/10.3920/WMJ2022.2804.

Wambacq, E., Vanhoutte, I., Audenaert, K., De Gelder, L. and Haesaert, G., 2016. Occurrence, prevention and remediation of toxigenic fungi and mycotoxins in silage: a review. Journal of Science Food and Agriculture 96: 2284-2302. https://doi.org/10.1002/jsfa.7565.

Whitaker, T.B. 2004. Mycotoxins in Foods: Detection and Control. Magan, N. and Olsen, M. (eds). Woodhead Publishing Limited, Cambridge. Sampling for Mycotoxins 69-87.

Whitaker, T.B. and Johansson, A.S., 2005. Sampling uncertainties for the detection of chemical agents in complex food matrices. Journal of Food Protection 68: 1306-1313. https://doi.org/10.4315/0362-028X-68.6.1306.

Whitaker, T.B., 2006. Sampling foods for mycotoxins. Food Additives and Contaminants 23: 50-61. https://doi.org/10.1080/02652030500241587.

Whitaker, T.B., Hagler, W.M. Jr., Giesbrecht, F.G. and Johansson, A.S., 2000. Sampling, sample preparation, and analytical variability associated with testing wheat for deoxynivalenol. Journal of AOAC International 83:1285–1292.

CHAPTER 12

Mycotoxin analysis in biological samples

L. Escriva[1]*, C. Luz*[1]*, J.M. Quiles*[1]*, V. D'Opazo*[1]*, J. Calpe*[1]*, A. Moreno*[1]*, R.R. Santos*[2] *and G. Meca*[1]*

[1]*Laboratory of Food Chemistry and Toxicology, Faculty of Pharmacy, University of Valencia, Spain*
[2]*Schothorst Feed Research, Lelystad, the Netherlands*
**correspondence: giuseppe.meca@uv.es*

1 Introduction

Identifying and quantifying mycotoxins in feed is essential for detoxifying or rejecting contaminated products, thus preserving and protecting animal and public health. However, in addition to their notorious toxicity, some mycotoxins have shown various levels of bioaccumulation in organs and tissues of animals, and sometimes when symptoms or suspicion of mycotoxicosis occurs, there is the risk that the complete contaminated diets have already been consumed because routine storage of dietary samples in the farm is not applicable.

The presence of elevated levels of mycotoxins in feed can lead to the presence of mycotoxins in animal products. For example, the degree of feed-to-milk carry-over of AFs was estimated to be approx. 6.2% in high-yielding cows consuming significantly high amounts of contaminated feed (Fink-Gremmels, 2008). Several studies report that some mycotoxins, such as DON, ZEN, and OTA, accumulate in living organisms and are able to contaminate all trophic levels of the food chain, from crop plants to animals and humans (Soto et al., 2015, Kwasniewska et al., 2015, Amuzie et al., 2008, Schneweis et al., 2005). Also, transplacental transfer of mycotoxins such as OTA and ZEN in pigs has been shown, where OTA concentrations in foetal serum double the maternal concentration (Minervini et al., 2013; Schoevers et al., 2012). In addition to their potential bioaccumulation in the organism, mycotoxins can be excreted to a greater or lesser extent by biological fluids (including colostrum and milk) or faeces (Arroyo-Manzanares et al., 2021; Benthem de Grave et al., 2021).

Exposure assessment of mycotoxins is usually performed by combining data on their presence in feedstuffs and diets with FI data. However, actual mycotoxin exposure is difficult to accurately determine using the classical indirect approach, which has certain limitations especially in farm conditions. Several factors influence individual exposure, such as the heterogeneous dietary distribution of mycotoxins, individual FI, and the presence of masked/conjugated mycotoxins (Huybrechts et al., 2015).

 | DOI: 10.1163/9789004712720_13

The analysis of mycotoxins in biological samples has attracted great interest in recent years since it allows wide and varied information to be gathered on the levels of mycotoxin exposure. The analysis of biological fluids such as blood and urine can be very useful in the evaluation of short- and long-term exposure to mycotoxins. In addition, the analysis of mycotoxins in these and other biological samples allows the study of toxicokinetic, ADME processes, and oral bioavailability of these compounds; these are extremely valuable data for evaluating the behaviour of these toxic substances in the organism of different animal species.

The most frequently analysed biological samples are blood (serum, plasma), urine, and faeces. These are considered integrative biofluids that incorporate the functions of different parts of the body into a single sample by containing a large number of metabolites, providing an overview of the metabolism of many areas of the body. However, the analysis of other fluids, tissues and organs such as bile, colostrum, and milk can also be of great interest (Arroyo-Manzanares et al., 2021; Benthem de Grave et al., 2021).

Analytical techniques for determining mycotoxins are essentially focused on their unique properties, such as solubility in polar organic solvents and the possibility that mycotoxins, either naturally or by derivatization, generate fluorescent compounds (Zhang et al., 2018). However, since mycotoxins have such a wide variety of chemical structures and physicochemical characteristics, the methods for analysing them vary considerably. On the other hand, owing to the diverse distribution of mycotoxins in contaminated samples and the interference of the matrix in which they are present, such analysis becomes even more challenging (Singh and Mehta, 2020), especially when talking about biological samples. The physicochemical properties of mycotoxins, the low concentrations involved in biological samples, and the complexity of such kinds of matrices lead to a constant analytical challenge. In recent years, extensive efforts have been made in the analytical field to achieve methodologies that allow effective, sensitive and accurate analyses of a large number of mycotoxins in various biological samples (Escrivá et al., 2017b). In this sense, sample treatment is a key step in any analytical process and its efficiency directly affects the reliability of the final data. The extraction and preparation process of a given sample will depend on the sample matrix and properties, as well as the expected analyte concentrations. The determination of mycotoxins in biological samples requires reliable and extremely sensitive analytical procedures due to the very low concentrations involved. In addition, these samples are complex, and extensive pre-treatment is usually necessary to eliminate possible interferences. Most of these procedures use a liquid-liquid extraction (LLE) or solid-liquid extraction (SLE) step, sometimes with additional energy to increase extraction efficiency (e.g. ultrasonic extraction, pressurized liquid extraction, microwave-assisted extraction, etc.), followed by a clean-up or purification step where immunoaffinity columns (IAC) or solid phase extraction (SPE) cartridges are often used. During the last decades, numerous studies have focused on the miniaturization and development of highly efficient and versatile extraction techniques based on liquid-phase microextraction (e.g. dispersive liquid-liquid microextraction; DLLME) or solid-phase microextraction methodologies (Arroyo-Manzanares et al., 2021, Escrivá et al., 2017a).

Regarding separation and detection methods, high-performance liquid chromatography (HPLC), and to a lesser extent gas chromatography (GC), is one of the most widespread techniques currently used for mycotoxin analysis, mainly due to its high performance and versatility to be coupled to multiple detection systems, such as fluorescence (FLD), ultraviolet (UV), or tandem mass spectrometry (MS/MS). Recently, high-resolution mass spectrometry (HRMS), together with time-of-flight (TOF) and Orbitrap spectrometers, have demonstrated their wide-ranging usefulness in the analysis of biological samples allowing applications such as the identification and detection of non-target compounds (e.g. identification of new mycotoxin metabolites), as well as for the retrospective analysis of large volumes of data (Arroyo-Manzanares et al., 2021, Escrivá et al., 2017a).

2 Mycotoxin analysis in animal samples

There are basically two different analytical techniques, depending on their intended use: qualitative and quantitative approaches.

Qualitative approaches are used to analyse mycotoxins and identify their presence or absence. Typically, they only offer a semi-quantitative level. Their use is more relevant in locations where raw materials are received and stored since it may enable preliminary screening of contaminated feed (Pleadin et al., 2019). However, it could be applied in biological matrices in order to determine mycotoxin presence (den Hollander et al., 2021).

Quantitative approaches are used to identify and quantify the presence of mycotoxins in a sample (den Hollander et al., 2021). They are, however, more expensive, time-intensive, and need specialized technical personnel. Mycotoxins are notoriously difficult to analyse in the laboratory since they are often present in trace amounts in samples. Furthermore, conventional analysis techniques for mycotoxin detection, identification, and quantification usually include the following laborious steps: sampling, sample preparation, extraction, purification, separation, detection, confirmation, and quantification (Anfossi et al., 2016). The basic procedures often entail mycotoxin detection and quantification methods, making them time-consuming, costly, and requiring experienced and competent workers to accomplish. With advancements in mycotoxin equipment and analysis, more efficient and sensitive procedures have been created. On the other hand, they enable a faster reaction since they are simpler and faster to execute.

Three approaches are primarily utilized to analyse and detect mycotoxins: HPLC or GC coupled to different detectors (FLD, diode array, UV, and MS), thin-layer chromatography (TLC), and immunological ELISA (Agriopoulou et al., 2020; Alshannaq and Yu, 2017). However, TLC has become an outdated and less conventional technique even though it is easy to apply. Table 1 shows the mycotoxin analysis studies applied to detect mycotoxins in several biological samples.

3 Methods of analyses

3.1 *Chromatographic methods*

These methods use liquid chromatography or GC with UV, FLD, or MS detection to quantify mycotoxins in biological samples. Using modern instrumentation and careful sample preparation allows for the most sensitive detection of mycotoxins. Chromatographic methods with UV and FLD detection are mainly used to confirm non-compliance with rules detected by a screening test (especially confirmatory analyses).

These methods are created for a single compound or a group of related mycotoxins. They are sometimes used to verify immunochemical testing. Determining the mycotoxins in all controlled samples often takes several processes. Otherwise, MS/MS detectors allow for the analytical method in mycotoxin identification to be altered and enable a response to most of the analytical problems. Several benefits of mass spectrometry make it the method of choice for multi-residue analyses, including accuracy, high sensitivity, precision, performance, and selectivity (Li et al., 2013; Streit et al., 2013). Additionally, due to the intrinsic selectivity of MS/MS detectors, extraction techniques requiring little or no sample clean-up might be designed effectively. In this context, the QuEChERS (Quick, Easy, Cheap, Effective, Rugged, and Safe) sample preparation method significantly simplified analytical procedures and allowed for the simultaneous extraction of a large number of mycotoxins, some of which belonged to different classes (Pereira et al., 2015). However, conventional QuEChERS methods, which are based on an exchange between optimum extraction conditions for various compounds, are intrinsically inefficient and limit the analytical method's sensitivity. As a result, unique methods involving pre-concentration stages must be used when ultrahigh sensitivity is required (Yan et al., 2018).

HRMS involves the use of mass spectrometers with high resolution and mass accuracy. These devices can be used to differentiate between compounds with identical nominal masses, ascertain their elemental compositions, and identify unknowns (Stock, 2017). HRMS and MS/MS detectors give structural information and the possibility of identifying undiscovered chemicals. Combining non-selective extraction techniques with mass screening by HRMS or MS/MS enabled the discovery of novel masked mycotoxins and group members (Nathanail et al., 2015). The use of GC–MS/MS is almost entirely limited to volatile substances such as *Fusarium* toxins and PAT (Gadzała-Kopciuch et al., 2020; Vidal et al., 2019).

TABLE 12.1 Mycotoxin analysis studies in animal samples, including the method, detector, mycotoxins, exposure, sam and detected mycotoxin level

Method	Detector	Animal species	Mycotoxins	Dietary level (exposure period)	Sample	Detected mycotoxin le
TLC		Bovine[1]	AFB$_1$	Not reported	Liver	0.10 ng/g
			AFB$_2$		Liver	0.03 ng/g
			AFM$_1$		Liver	0.08 ng/g
		Calf[2]	AFB$_1$	0.29 mg AFB$_1$/kg BW	Liver	2.7 – 5.8 ng/g
			AFM$_1$	(oral bolus for two days)	Liver	1.4 – 5.9 ng/g
		Pigs[2]	AFB$_1$	1.5 mg AFB$_1$/kg feed (35 days)	Liver	0.07 – 0.08 ng/g
		Pigs[3]	AFB$_1$	Up to 0.40 mg AFs/kg corn	Liver	0.01 – 0.24 ng/g
			AFM$_1$	(not reported)	Liver	0.03 – 0.44 ng/g
			AFB$_1$ and AFM$_1$		Muscle	Not detected
		Pregnant gilts[4]	OTA	0.38 mg OTA/kg BW	Serum	Not detected
				(7 days)	Kidney	0.43 - 0.70 μg/g
					Liver	0.30 - 0.34 μg/g
					Muscle	0.13 - 0.15 μg/g
					Placenta	0.04 - 0.06 μg/g
					Urine	0.04 - 0.33 μg/mg creati
					Faeces	1.30 - 8.10 μg/g dry matt
		Chicken[5]	T-2 toxin	5 mg T-2/kg BW	Liver	0.13 μg/g
				(single oral bolus)	Kidney	0.03 μg/g
					Muscle	0.04 μg/g
		Duck[5]	T-2 toxin	5 mg T-2/kg BW	Liver	0.09 μg/g
				(single oral bolus)	Kidney	0.03 μg/g
					Muscle	0.05 μg/g
Chromatography	HPLC-TOF/MS	Chickens, swine, goats, and cows[6]	OTA	Liver incubated with OTA	Liver	Not applicable
	HPLC-MS/MS	Pigs[7]	AFB$_1$	Unknown	Muscle from supermarket	0.46 – 0.74 ng/g
			AFB$_2$			0.52 ng/g
			AFG$_1$			< LOQ
			AFG$_2$			< LOQ
			AFM$_1$			< LOQ
			OTA			0.88 ng/g
			CIT			< LOQ
			PAT and sterigmatocystin			< LOQ
						0.76 – 1.23 ng/g
		Pigs[7]	AFB$_1$	Unknown	Liver from supermarket	0.68 ng/g
			AFB$_2$			0.52 ng/g
			AFG$_1$			< LOQ
			AFG$_2$			< LOQ
			AFM$_1$			0.63 ng/g
			OTA			< LOQ
			CIT			1.45 ng/g
			PAT and sterigmatocystin			< LOQ
						< LOQ

E 12.1 Mycotoxin analysis studies in animal samples, including the method, detector, mycotoxins, exposure, sample, and detected mycotoxin level (*cont.*)

od	Detector	Animal species	Mycotoxins	Dietary level (exposure period)	Sample	Detected mycotoxin level
		chickens[7]	AFB_1	Unknown	Muscle from supermarket	0.53 ng/g
			AFB_2			< LOQ
			AFG_1			< LOQ
			AFG_2			< LOQ
			AFM_1			0.87 ng/g
			OTA			< LOQ
			CIT			1.21 ng/g
			PAT and sterig-			0.69 ng/g
			matocystin			< LOQ
		chickens[7]	AFB_1	Unknown	Liver from supermarket	0.36 ng/g
			AFB_2			< LOQ
			AFG_1			< LOQ
			AFG_2			< LOQ
			AFM_1			< LOQ
			OTA			1.05 ng/g
			CIT			0.89 ng/g
			PAT and sterig-			< LOQ
			matocystin			< LOQ
	HPLC-FLD	Turkey[8]	CIT	Unknown	Muscle from supermarket	0.083 – 1.79 ng/g
	GC/MS	Bovine[9]	Zeranol	Unknown	Urine	1.4 – 40.9 ng/ml
			ZEN			< LOQ
			α-ZEL			< LOQ
			β-ZEL			< LOQ
	LC-MS/MS	Bovine[9]	Zeranol	Unknown	Urine	1.0 – 38.1 ng/ml
			ZEN			1.0 – 5.0 ng/ml
			α-ZEL			1.1 – 1.3 ng/ml
			β-ZEL			1.0 – 5.2 ng/ml
		Sows[10]	ZEN	300 µg ZEN + 250 µg DON /kg feed (33 days)	Milk	0.12 ng/ml
			α-ZEL			0.16 ng/ml
			β-ZEL			< LOQ
			DON			2.4 ng/ml
			DOM-1			< LOQ
			ZEN	300 µg ZEN + 250 µg DON /kg feed (33 days)	Colostrum	0.08 ng/ml
			α-ZEL			0.22 ng/ml
			β-ZEL			< LOQ
			DON			< LOQ
			DOM-1			< LOQ
			ZEN	300 µg ZEN + 250 µg DON /kg feed (33 days)	Serum	1.1 ng/ml
			α-ZEL			3.4 ng/ml
			β-ZEL			0.5 ng/ml
			DON			2.1 ng/ml
			DOM-1			0.2 ng/ml

TABLE 12.1 Mycotoxin analysis studies in animal samples, including the method, detector, mycotoxins, exposure, sa and detected mycotoxin level (*cont.*)

Method	Detector	Animal species	Mycotoxins	Dietary level (exposure period)	Sample	Detected mycotoxin l
		Piglets[10]	ZEN α-ZEL β-ZEL DON DOM-1	Lactation	Serum	0.02 ng/ml 0.04 ng/ml 0.04 ng/ml 0.10 ng/ml 0.06 ng/ml
	UHPLC-MS/MS	Dairy cows[11]	AFM_1 AFM_2	Unknown	Commercial milk samples	0.005 – 5.0 µg/L 0.009 – 0.08 µg/L
	HPLC-FLD	Donkey[12]	AFM_1 AFM_2	0.20 µg AFB_1 + 0.11 µg AFB_2 /kg feed (12 days)	Milk	32.3 ng/l 23.3 ng/l
	HPLC-TOF/MS	Chickens[13]	ZEN	5 mg ZEN/kg BW (single oral bolus)	Liver	Data given as absolute
	HPLC-MS/MS	Pigs[14]	ZEN α-ZEL ZEN-14G	10 µg ZEN/kg BW (single oral bolus)	Urine	33 ± 12 nmol ND 53 ± 44 nmol
	HPLC-MS/MS	Pigs[14]	ZEN α-ZEL ZEN-14G	10 µg ZEN/kg BW (single oral bolus)	Faeces	31 ± 9 nmol 14 ± 6 nmol ND
ELISA		Pigs[15]	DON ZEN	1.1 – 5.0 mg DON/kg diet + 0.4 – 2.4 mg ZEN/kg diet	Urine	Numerical data not available
		Dairy cows[16]	AFM_1	Unknown	Milk (farm bulk tanks)	Numerical data not available
		Pigs[17]	OTA	250 µg OTA/kg feed (28 days)	Kidney Liver Lungs Heart Spleen Brain Muscle Fatty tissue Urine Serum	13.9 – 11.9 ng/g 4.9 – 9.8 ng/g 8.0 – 14.0 ng/g 2.5 – 5.3 ng/g 2.9 – 6.5 ng/g 2.6 – 3.2 ng/g 2.8 – 5.4 ng/g 3.0 – 5.3 ng/g 13.9 – 18.2 ng/ml 3.7 – 6.6 ng/ml

a van Egmond et al., 1990; b Stoloff and Trucksess, 1979; c Stubblefield et al., 1991; d Patterson et al., 1976; e Giroir et al., 1991; f Yang et al., 2015; g Cao et al., 2018; h Sari et al., 2020; i Echarte et al., 2014; j Benthem de Grave et al., 2021; k Sartori et al., 2015; l Tozzi et al., 2016; m Yang et al., 2017; n Binder et al., 2017; o Thanner et al., 2016; p Pecorelli et al., 2020; q Pleadin et al., 2016.
ELISA: enzyme-linked immunosorbent assay; GC/MS: gas chromatography/mass spectrometry; HPLC-FLD: high performance liquid matography with fluorescence detection; HPLC-MS/MS: high performance liquid chromatography tandem mass spectrometry; HPLC MS: high performance liquid chromatography time-of-flight/mass spectrometry; LC-MS/MS: liquid chromatography-tandem mass sp trometry; LOQ: Limit of quantification; TLC: thin-layer chromatography; UHPLC-MS/MS: ultra-high performance liquid chromatogra tandem mass spectrometry.

Cao et al. (2018) used liquid chromatography-tandem mass spectrometry (LC-MS/MS) to create a multi-mycotoxin-based approach for identifying and quantifying numerous carcinogenic mycotoxins in edible animal tissues, including muscle and liver tissue from pigs and chickens. To investigate the toxicokinetics of particular mycotoxins, extremely sensitive and selective LC-MS/MS techniques for rat plasma and urine were established. The authors detected six types of mycotoxins in swine and chicken muscle and livers, including AFs, PAT, sterigmatocystin, CIT, and ochratoxins.

Sari et al. (2020) extracted CIT from meat products using a variety of extraction techniques, including LLE, QuEChERS, and immunoaffinity column extraction, and then analysed CIT using HPLC-FLD. CIT was detected in six out of 19 samples, with concentrations ranging from 0.28 to 1.79 ng/g (36.8%). On the other hand, CIT was not discovered in minced meat samples.

Echarte et al. (2014) confirmed zeranol, ZEN and α- and β-ZEL in bovine urine samples using a validated GC/MS technique. Further, they created a more sensitive HPLC MS/MS method for these four analytes. Their performances were compared, and conclusions were taken from their findings. Between 2005 and 2011, around 1262 bovine urine samples were analysed. This survey showed that when the quantities of zeranol and taleranol in urine surpassed 20 μg/L, the LC-MS/MS approach consistently delivered lower amounts than the GC/MS method. The LC-MS/MS approach found meagre amounts of *Fusarium* metabolites ranging from 1 to 5 μg/L in the same batch of samples, which were not detected by the GC/MS method. The comparison investigation established that the LC-MS/MS technique was more sensitive than the GC/MS technique.

3.2 *Immunoassay methods*

Immunochemical techniques are based on antibodies attaching to antigens selectively, reversibly, and non-covalently. Due to their simplicity and low cost, immunoassays are used for initial screening and survey investigations on mycotoxin contamination. Commercially accessible ELISA-based kits are available for all regulated mycotoxins and are the most often utilized analytical instrument for ensuring animal origin food safety (Matabaro et al., 2017). In particular, immunoassays may be used to quickly and easily evaluate biological samples such as serum and urine (Han et al., 2016). In addition, immunochemical assays in a variety of forms are constantly being developed with the goal of obtaining systems that are quick, portable, and simple to handle (Beloglazova and Eremin, 2015; Han et al., 2016; Zhang et al., 2015). Among them, the immunochromatographic technique is the most extensively used and often provides qualitative responses of yes or no, indicating the presence or absence of analytes at concentrations greater than the predetermined LOD level (Di Nardo et al., 2016; Wu et al., 2021).

In contrast, immunochemistry-based methods appear to have a potential limitation in the new scenario of mycotoxin research due to the analyte recognition mechanism's extreme selectivity, which precludes the simultaneous determination of multiple compounds and the detection of unknown toxins as well as modified structures produced by animal metabolism. The development of immunochemical methods capable of detecting single to multiple analytes remains a scientific challenge (Goryacheva et al., 2007). Nonetheless, immunochemical approaches using conventional ELISA-based formats enable large-scale and regular surveys, which make them applicable for monitoring the occurrence of mycotoxins and ensuring the safety of products of animal origin (Jia et al., 2021).

For instance, Thanner et al. (2016) investigated the urine of pigs fed diets containing DON and ZEN. Additionally, urinary ZEN equivalents were determined using an ELISA kit after glucuronide hydrolysis. The quantities of ZEN in the urine as determined by LC-MS/MS and ELISA were substantially related. The strong correlation between the urine ZEN levels determined by the two analytical procedures demonstrated that ELISA is valid for determining urinary ZEN concentration. Pecorelli et al. (2020) evaluated the analytical performance of two commonly used commercial immunoassays (ELISA) for AFM_1 detection in milk. Both tests demonstrated acceptable accuracy, recovery rates, and false-positive and -negative rates. By comparing the ELISA to the HPLC-FLD, the scientists determined that the ELISA showed a high correlation with the findings obtained from naturally contaminated samples. Then, because ELISA tests allowed for the simultaneous analysis of up to 48 samples, the authors concluded that they could be more efficient when a high volume of samples has to be consistently evaluated by skilled technicians, as strip tests may be stacked to process many samples in a very short period of time.

Pleadin et al. (2016) also performed a similar investigation in which they examined the levels of OTA in pig tissues and biological fluids after exposure to a feed contaminated with 250 µg of OTA/kg. The amounts of OTA were determined using a validated immunoassay and HPLC-FLD. The kidneys of exposed pigs had the greatest mean OTA content, followed by the lungs, liver, spleen, muscle tissue, fat tissue, heart, and brain. Additionally, on the last day of exposure, urine had considerably higher mean OTA levels than serum, indicating that urine analysis of OTA may be a useful biomarker for detecting excessive levels of this toxin in swine tissues. To summarize, the authors revealed recommendations for more effective management of OTA in biological materials obtained from swine that may be implemented in abattoirs. Considering that these tissues are employed as raw materials in the meat industry, the authors also proposed that extra attention must be paid to OTA determination.

4 Biomarkers

The use of biomarkers is a method that establishes the exposure of xenobiotics by determining their metabolic forms in biological fluids, such as urine, blood, milk, and faeces. Biomarker analysis has gained popularity in recent years as a diagnostic tool to determine the individual mycotoxin exposure in humans, since the approach of analysing mycotoxins supplied by the diet has clear limitations (individual differences in consumption, absorption, metabolism, and excretion) (Tkaczyk and Jedziniak, 2021). The analysis of the main mycotoxins can be performed in animal feed, but conjugated mycotoxins such as DON-3G and its hydrolysis in the animal resulting in extra DON exposure is not measured (Nagl et al., 2014). Furthermore, FI and subsequent exposure is measured per group. However, the use of biomarkers to detect exposure appears as a tool for the biomonitoring of animals affected by mycotoxins. In the case of registration of mycotoxin decontaminants, the determination of endpoint biomarkers is mandatory. Progress in chromatographic, particularly MS techniques, has been fundamental for the determination of biomarkers. The biomarkers described for the exposure of the main mycotoxins (AFs, OTA, DON, and ZEN) are summarized below.

The main analysed biomarkers for AFB_1 exposure in animals are AFM_1, AFQ_1, AFL, AFBO and AFB_1 conjugated with glutathione. However, the concentration of these metabolites in biological samples can vary considerably depending on the animal species (Vidal et al., 2018). In milk, Sartori et al., (2015) developed a methodology for analysing AFB_1, AFB_2, AFG_1, AFG_2, AFM_1, and AFM_2 with the QuEChERS (quick, easy, cheap, effective, rugged, and safe) method and subsequent determination with ultra-high-performance liquid chromatography-tandem mass spectrometry (UHPLC-MS/MS). Elsewhere, Tozzi et al. (2016) analysed donkey milk after exposure to feed contaminated with AFB_1 and AFB_2. The authors demonstrated the conversion to AFM_1 and AFM_2 by 0.02 and 0.31%, respectively.

Absorbed OTA can be metabolised into hydroxylated forms and OTB, the dechlorinated form. The amount of OTA absorbed and its metabolism depends on the animal species. Yang et al. (2015) have determined as biomarkers of exposure to OTA in urine and faeces of rats and chickens the following compounds: (4R)-hydroxyochratoxin A, (4S)-hydroxyochratoxin A, 5-hydroxyochratoxin A, 7-hydroxyochratoxin A, and 10-hydroxyochratoxin A. The analytic determination was performed with ultra-performance liquid chromatography-quadrupole/time-of-flight hybrid mass spectrometry (UPLC-Q/TOF-MS). Also, in rats, Abbas et al. (2013) verified that feeding with OTA resulted in urinary and faecal excretion of OTα.

In pigs, DOM-1 is the major phase I metabolite, whereas DON-3-glucuronide is the major phase II metabolite of DON (Nagl et al., 2014). In broiler chickens the sulfation to DON-3α-sulphate (DON-3S) occurs, representing more than 85% of the ingested DON (Devreese et al., 2015). In ruminants, a substantial glucuronidation of DON (Maresca, 2013) is observed, as well as its de-epoxidation into DOM-1 by the rumen microorganisms (Pestka, 2007).

Regarding ZEN, the main metabolites studied *in vivo* in animals have been α- and β-ZEL. The study conducted by Yang et al. (2017) described the following biomarkers after exposure to ZEN in rats: α-ZEL, β-ZEL and hydroxylated metabolites of ZEN. In pigs, Binder et al. (2017) found that a large amount of ZEN is recovered in faeces, in its free form and as ZEN-14G and α-ZEL.

4.1 *Biomarkers for the registration of feed additives as mycotoxin decontaminants*

In the EU, in the registration procedure of technological feed additives, a special functional group named mycotoxin decontaminants (EC 1831/2003) requires the quantification of endpoint biomarkers for the target mycotoxins in serum/plasma, milk, eggs, or tissues to demonstrate their efficacy (Table 2). It is required that the experimental animals are exposed to mycotoxins within the maximum regulated/recommended mycotoxin level per animal species. When a mycotoxin has no guidance for maximum inclusion level in the diet, the chosen level must ensure the absence of adverse effects in the experimental animal. It is important to bear in mind that the endpoint mycotoxin metabolite may vary among species. For instance, usually DOM-1 is a marker used for pigs, whereas for poultry DON-3S is the most indicated. Unfortunately, it is useful to evaluate these biomarkers in a supervised experiment, where control group data is compared with those from exposed animals. Moreover, there is no threshold level to be applied in biological samples when working under farm conditions.

TABLE 12.2 Relevant biomarker for feed additives intended to be registered as mycotoxin decontaminant in the EU

Target mycotoxin	Biomarkers
AFB_1	AFM_1 in milk or egg yolk
DON	DON/metabolites in blood serum
ZEN	ZEN + α- and β-ZEL in plasma Excretion of ZEN/metabolites
OTA	OTA in kidney (or blood serum)
$FB_1 + FB_2$	Sa/So ration in blood, plasma or tissues

SOURCE: EC 1831/2003

4.2 *Biomarkers for animal biomonitoring*

4.2.1 Blood, plasma, and serum

Biomonitoring of blood and its harvested plasma and serum is claimed to be a useful method for determining mycotoxin exposure. However, there is the risk of an underestimation or miscalculation of the actual mycotoxin exposure. The measured levels of mycotoxins in blood will depend on oral absorption, distribution to other organs declining plasma levels, elimination, and time of sampling, which will vary according to the animal species (Muñoz-Solano and González-Peñas, 2023). Most toxicokinetic studies have been carried out in pigs and chickens (Devreese et al., 2018; Lauwers et al., 2019; Tkaczyk and Jedziniak, 2021). When comparing feed and plasma levels of 19 mycotoxins in pigs, poultry, cattle and sheep, it was observed that plasma samples were negative for mycotoxin contamination if they were not previously enzymatically treated, and the high plasma prevalence of certain mycotoxins (sterigmatocystin) differing from low feed levels indicates that plasma analysis will actually indicate contamination within the farm environment (Muñoz-Solano and González-Peñas, 2023). No correlation was found between feed and serum levels of 24 mycotoxins in poultry (den Hollander et al., 2021), whereas in pig a positive correlation ($R^2 = 0.89$) between dietary and serum levels of DON was detected (Brezina et al., 2014a). However, DON serum levels cannot be used to estimate or predict health or growth performance in pigs (Dänicke et al., 2004). Dänicke et al. (2023) established equations to predict dietary exposure of dairy cows to DON. Based on that, DON residues (ng/ml) in blood, urine and milk DON multiplied by 2.52, 0.022, and 2.47, respectively, can be used to predict exposure as µg/kg BW/day. However, blood remains the most indicated biological sample to predict DON exposure, and the equations should be applied on a herd and not individual basis. A correlation between FI and serum levels of ZEN is observed only when the levels of this mycotoxin in the feed are much higher than the European guidance level (Lauwers et al., 2019). It is indicated to evaluate the levels of its main metabolites α- and β-ZEL but, still, the sampling moment after FI will interfere with the results and may vary for metabolites. For instance, the optimal moment for determining the maximum serum levels of ZEN, α- and β-ZEL in the serum of pigs orally exposed to 1 mg/kg ZEN has a wide range, i.e. 0.25-2.0, 0.25-4.0, and 0.5-9h, respectively, due to animal individual differences (Fleck et al., 2017). Serum levels of OTA in pigs reaches a plateau after 10-13 days (Hult et al., 1979). When combining several studies where pigs were subjected to an OTA dietary exposure for at least 14 days, a positive correlation ($R^2 = 0.85$) between feed (OTA concentration ranging from 0.032 to 3.3 mg/kg) and serum levels was demonstrated (Battacone et al., 2010). Regarding long-term exposure, it was demonstrated that pigs exposed to OTA-contaminated diet for six months may present higher serum levels of OTA than those fed the same diet for three months (Stoev et al., 2002). Broiler chickens fed a diet contaminated with 0.5 or 1.0 mg/kg OTA had a recovery in serum of this mycotoxin of approx. 5% at day 28 and 7-9% at day 42 of age, whereas no mycotoxin was detected after 14 days' dietary exposure (Hanif et al., 2012). When ruminant calves were orally exposed to OTA (2.0 mg/kg BW), this mycotoxin was rapidly absorbed, and serum levels presented peaks at two very distinct time-points (1988). These authors observed a first peak after 2-4 h and a second one 12.5 h after OTA intake, as a result of biliary excretion and intestine reabsorption of OTA. A similar biphasic disappearance was observed in sheep treated with OTA injection (Xiao et al., 1991). In sheep, OTA serum levels reach a plateau in a dose-dependent manner, where highest values are reached after two weeks' exposure to 2 mg/kg of concentrate (22 mg OTA/kg BW/day), but it takes four weeks when the animals are exposed to 5 mg/kg of concentrate (55 mg OTA/kg BW/day) (Höhler et al., 1999). It is important to mention that serum levels of OTA may decrease when the dietary exposure levels are much higher, due to feed refusal. All these previously mentioned studies were performed with extremely high OTA dietary levels. In health ruminants, OTA is hydrolyzed into a far less toxic compound, the OTα, by the rumen microbes (Sreemannarayana et al., 1988).

4.2.2 Urine and faeces

Urine samples can contain biomarkers of different mycotoxins. However, urine biomarkers are susceptible to daily variations in mycotoxin intake, and the urine volume is individual, resulting in changes in the concentration of excreted compounds in the samples. This discrepancy can be reduced by applying a common method (which is still however questioned by the scientific community) for adjusting mycotoxin levels in creatinine concentrations. In humans, it is still unclear if the mycotoxin/creatinine ratio is applicable for comparison between individuals since creatinine is influenced by many factors, such as muscle mass, sex, age, season, diet, etc. (Arce-López et al., 2020). However, when collecting samples from farm animals, muscle mass, age, season, and diet will be very similar within all individuals, leaving only sex as a factor. Urinary studies often encompassed a large number of mycotoxins and metabolites, where various extraction techniques were combined to achieve the highest variety of studied compounds. High method sensitivity is of the utmost importance since the concentration of these analytes in urine samples is often in a very low concentration range. The urine volume used for mycotoxin analysis is generally considerably higher than in other biological fluids, ranging from 100 μl to 20 ml. From the relatively few studies performed on one single mycotoxin in urine, AFB_1, DON, ZEN, and OTA were the most commonly analysed. Although urine sampling has the advantage of being a non-invasive method for biomonitoring, collection will often require animal handling, sometimes including individual housing. The analysis of mycotoxins in faeces is one of the most interesting fields of study not only because it is a matrix with which the biomonitoring of these toxins can be carried out, but also because these samples can be obtained in a non-invasive and relatively simple way.

In pigs subjected to dietary exposure to AFB_1, the excretion of AFB_1 and its metabolite AFM_1 via urine accounted for 30% of the oral dose, and the ratio of AFB_1

to AFM_1 was 23% (Thieu and Pettersson, 2009). Ileum content (faecal sample) and excreta (faecal and urine sample) of broiler chickens exposed to AFB_1 were analysed for the presence of this mycotoxin and metabolites (Jurišić et al., 2019). These authors showed that the levels of AFM_1 were detected in excreta at a level five times higher than in the ileum. This not only indicates that primary excretion occurs via urine but also that biomonitoring can be easily performed in a non-invasive manner. Most of the DON that pigs excrete (54–85%) is found in their urine (Prelusky et al., 1988). About half of this DON is glucuronide conjugated, while only about five percent of the DON they ingest is excreted as DOM-1 (Goyarts and Dänicke, 2006). A positive correlation between ingested and excreted DON via urine has been reported, whereas the concentration of DON excreted via faeces is 50 times lower than in urine (Dänicke et al., 2004), and this type of sample should not be used as a biomarker for exposure. About 15% of the ZEN that is eaten is passed out of the body as α-ZEL in the urine 8 hours after feeding (Olsen et al., 1985; Zöllner et al., 2002). There was a positive correlation between the amount of ZEN (R2 = 0.69) and α-ZEL (R2 = 0.89) in the urine and the amount of ZEN that was ingested (Dänicke et al., 2005; Tkaczyk and Jedziniak, 2021). Approx. 10–18% of ZEN and α-ZEL can be detected in faeces between 24 and 48 hours after dietary exposure to ZEN (Binder et al., 2017). The highest levels of these mycotoxins are obtained from the excreta, which is collected between 12 and 24 hours after oral exposure (Brezina et al., 2014b). In cows, most ingested ZEN is excreted as β-ZEL in urine (51%) and faeces (58%), and α-ZEL represents 20% of the excreted mycotoxin in urine and 12% in faeces (Mirocha et al., 1981). Urine and faecal analysis of FB_1 was performed after a single oral exposure to 5 mg/kg of this mycotoxin, resulting in a low recovery in urine (< 1%) after monitoring up to 41 hours after oral administration, whereas 76.5% of the FB_1 was recovered in faeces between 8 and 24 hours after oral administration (Dilkin et al., 2010). The excreta from turkeys fed a diet contaminated with 15 mg/kg FB_1 + FB_2 contained 2.6 mg/kg FB_1 and 0.04 mg/kg of the less toxic metabolite hydrolyzed FB_1 (HFB_1) after 7 days of exposure, and 5.2 mg/kg FB_1 and no HFB_1 after 14 days of exposure (Masching et al., 2016). Although a positive correlation (R^2 = 0.68) is observed between ingested and OTA and excretion via urine in pigs, less than 3% of the dietary amount is recovered via urine (Gambacorta et al., 2013), and around a 2-fold lower level of OTα can be detected. The biomonitoring for this mycotoxin, however, has the advantage of slow elimination (72–120 hours) due to its capacity to bind serum albumin (Tkaczyk et al., 2021). Although after oral administration OTA undergoes extensive metabolism, 56% of this mycotoxin is recovered in the excreta from chickens, and 20% is recovered as the hydroxyochratoxin metabolite 7'-OH-OTA (Yang et al., 2015). In sheep, more than 90% of ingested OTA is excreted as OTα, whereas only 0.5–3.0% is excreted as unchanged OTA via urine (Xiao et al., 1991). In 16–21-day-old calves receiving OTA (500 ml/kg BW/day) directly in the stomach, it was observed that there was an 88% recovery of OTα, 3.4% OTA in urine, and 9.2% OTA in faeces (Sreemannarayana et al., 1988). The biomonitoring of OTA exposure in ruminants, therefore, seems not to be necessary, as these animals are much less sensitive to OTA (Mobashar et al., 2010).

4.2.3 Colostrum/milk

Like most biological fluids, breast milk is a matrix containing compounds that can interfere with the signal of mycotoxins during their analysis by LC-MS, so prior purification is recommended to increase the sensitivity of the analysis, and direct ELISA methods may underestimate or overestimate the levels of the analysed mycotoxins. Due to human health concerns, especially in infants who consume a large amount of milk, the analysis for the presence of AFM_1 is regulated in many countries. Several factors will affect the amount of AFM_1 transferred to milk, such as the concentration of the mycotoxin in the diet, the period of exposure, and milk yield. For instance, an estimation of 1-2% of transfer to milk has been calculated, but this transfer increases three times for high-yielding cows (> 6%). *Fusarium* mycotoxins have been detected in the milk and colostrum of pigs. Feeding sows with 250 μg/kg DON resulted in a transfer of 0.01–0.04% DON to milk, whereas dietary contamination with 300 μg/kg ZEN resulted in a transfer of 0.002% ZEN to both colostrum and milk (Benthem de Grave et al., 2021). In ruminants, the transfer of DON into milk is much lower than in pigs, and its metabolite DOM-1 is recovered at a very low concentration (0.0001-0.0011%) (Keese et al., 2008). Cows fed a diet contaminated with 25 mg/kg ZEN for 7 days had 210 ng/ml ZEN, 186 ng/ml α-ZEL, and 189 ng/ml β-ZEL in milk, besides other glucuronide (322 ng/ml α-ZEL conjugated) and sulphate conjugates in a smaller amount (Mirocha et al., 1981). Transfer of OTA into the milk of ruminants is extremely low. Cows fed an OTA-contaminated diet (0.32 – 1.13 mg/kg diet) for 77 days presented no OTA in their milk (Shreeve et al., 1979). After a single dose of 13.3 mg OTA/kg diet, cows presented OTA in milk (650 mg), but on the following day only Otα (4500 mg) was detected in the milk (Ribelin et al., 1978). Unfortunately the former authors do not report the volume of the milk sample. In surveys on 190 milk samples of cows from Germany and 100 dairy cows from the United Kingdom, OTA was absent and only traces of Otα could be detected, whereas OTA was detected in the milk of cows in Scandinavian countries, ranging from 0.01 to 0.04 ng/ml in Sweden, and 0.01 to 0.06 ng/ml in Norway (Mobashar et al., 2010).

4.2.4 Tissues

Broiler chickens and laying hens exposed to diets containing 50 μg/kg AFB_1 showed residual levels of AFB_1, AFL, AFM_1, and kg AFB_2, as detected by HPLC in the liver and kidneys (Micco et al., 1988). These authors showed that after feed replacement by a non-contaminated diet, no AFB_1 or metabolites were detected by HPLC after 14 and 33 days for broiler chickens and laying hens, respectively. The levels of AFB_1 and its metabolites in the tissues will also depend on the poultry species (duck > turkey > quail > chicken) and sex (higher levels of the highly toxic AFBO in males than in females) (Lozano and Diaz, 2006; Diaz and Murcia, 2011). On the other hand, the reduction of AFB_1 to AFL is inversely related to the susceptibility to AFB_1 (Murcia and Diaz, 2020). Due to a strong positive correlation, it is possible to estimate the amount of oral exposure to OTA based on the residues of this mycotoxin found in the kidneys, liver, and muscle tissue of pigs (Battacone et al., 2010). Pigs naturally exposed to OTA via a diet containing contaminated barley (4.3 mg/kg) presented high OTA levels in the kidneys during an 18-month period, where after 12 months the levels of OTA in the kidneys (> 70 µg/kg) were higher than in the liver (> 60 µg/kg), and the levels in plasma were higher (> 330 µg/kg) than in the kidneys (Rutqvist et al., 1978). The correlation between OTA in plasma and kidney is higher ($R^2 = 0.81$) than with liver ($R^2 = 0.61$) (Aoudia et al., 2009). In this trial, it seems that barley grown at the farm was not properly stored, favouring the growth of *Penicillium verrucosum*, and the contaminated feedstuff was used as a feed ingredient for two years, which explains the long-term presence of OTA in the tissues of the pigs. Moreover, depending on the period of dietary exposure, the OTA levels in kidneys can be higher than those measured in the diet of pigs because of a high blood flow in this organ, together with the kidney reabsorption, resulting in a delay in OTA elimination (Pfohl-Leszkowicz et al., 2007). As observed in pigs, OTA levels in the kidneys are higher than in the liver. Broiler chickens fed a diet contaminated with 0.5 or 1.0 mg/kg OTA presented a 10-fold higher level of OTA in the kidneys than in the liver, and recovery in the kidney after 42 days of exposure was approx. 2.0% regardless of the OTA concentration in the diet (Hanif et al., 2012).

5 Conclusions

In conclusion, it should be noted that the analysis of mycotoxins from biological samples is a growing field of study. Multi-mycotoxin methodologies capable of achieving lower values with increasingly homogeneous and simple extraction procedures, generally by injection in LC-MS/MS systems, are being published. These results demonstrate the importance of the analysis of biological samples as a useful tool for the quantification of mycotoxins and metabolites in animal samples as a means of mycotoxin decontaminant registration. However, there are some important limitations when biomonitoring groups of animals. Due to the lack of comparative data, the type of sample will limit the type of detected mycotoxins, and the levels of mycotoxin detection cannot always be used as a marker for disease prevention.

References

Abbas, Z., Blank, R., Wein, S. and Wolffram, S., 2013. Effect of quercetin on the toxicokinetics of ochratoxin A in rats. Food Additives and Contaminants. Part A, Chemistry, Analysis, Control, Exposure and Risk Assessment 30: 861–866. https://doi.org/10.1080/19440049.2013.793823.

Agriopoulou, S., Stamatelopoulou, E. and Varzakas, T., 2020. Advances in analysis and detection of major mycotoxins in foods. Foods 9: 518. https://doi.org/10.3390/foods9040518.

Alshannaq, A. and Yu, J.H., 2017. Occurrence, toxicity, and analysis of major mycotoxins in food. International Journal of Environmental Research and Public Health 14: 632. https://doi.org/10.3390/ijerph14060632.

Amuzie, C.J., Harkema, J.R. and Pestka, J.J. 2008. Tissue distribution and proinflammatory cytokine induction by the trichothecene deoxynivalenol in the mouse: Comparison of nasal vs. oral exposure. Toxicology 248: 39–44. https://doi.org/10.1016/j.tox.2008.03.005.

Anfossi, L., Giovannoli, C. and Baggiani, C., 2016. Mycotoxin detection. Current Opinion in Biotechnology 37: 120–126. https://doi.org/10.1016/j.copbio.2015.11.005.

Aoudia, N., Callu, P., Grosjean, F., Larondelle Y., 2009. Effectiveness of mycotoxin sequestration activity of micronized wheat fibres on distribution of ochratoxin A in plasma, liver and kidney of piglets fed a naturally contaminated diet. Food and Chemical Toxicology 47: 1485–1489. https://doi.org/10.1016/j.fct.2009.03.033.

Arce-López, B., Lizarraga, E., Vettorazzi, A. and González-Peñas, E., 2020. Human biomonitoring of mycotoxins in blood, plasma and serum in recent years: a review. Toxins 12: 147. https://doi.org/10.3390/toxins12030147.

Arroyo-Manzanares, N., Campillo, N., López-García, I., Hernández-Córdoba, M. and Viñas P., 2021. High-resolution mass spectrometry for the determination of mycotoxins in biological samples. A review. Microchemical Journal. 166: 106197. https://doi.org/10.1016/j.microc.2021.106197.

Battacone, G., Nudda, A. and Pulina, G., 2010. Effects of Ochratoxin A on livestock production. Toxins 2: 1796-1824. https://doi.org/10.3390/toxins2071796.

Beloglazova, N.V. and Eremin, S.A., 2015. Rapid screening of aflatoxin B1 in beer by fluorescence polarization immunoassay. Talanta 142: 170–175. https://doi.org/10.1016/j.talanta.2015.04.027.

Benthem de Grave, X., Saltzmann, J., Laurain, J., Rodriguez, M.A., Molist, F., Dänicke, S. and Santos, R.R., 2021. Transmission of zearalenone, deoxynivalenol, and their derivatives from sows to piglets during lactation. Toxins 13: 37. https://doi.org/10.3390/toxins9080251.

Binder, S.B., Schwartz-Zimmermann, H.E., Varga, E., Bichl, G., Michlmayr, H., Adam, G. and Berthiller, F., 2017. Metabolism of zearalenone and its major modified forms in pigs. Toxins 9: 56. https://doi.org/10.3390/toxins9020056.

Brezina, U., Valenta, H., Rempe, I., Kersten, S., Humpf, H.U. and Dänicke, S. 2014a. Development of a liquid chromatography tandem mass spectrometry method for the simultaneous determination of zearalenone, deoxynivalenol and their metabolites in pig serum. Mycotoxin Research 30: 171–186. https://doi.org/10.1007/s12550-014-0200-8.

Brezina, U., Rempe, I., Kersten, S., Valenta, H., Humpf, H.U. and Dänicke, S. 2014b. Diagnosis of intoxications of piglets fed with *Fusarium* toxin-contaminated maize by the analysis of mycotoxin residues in serum, liquor and urine with LC-MS/MS. Archives of Animal Nutrition 68: 425–447. https://doi.org/10.1080/1745039X.2014.973227.

Cao, X., Li, X., Li, J., Niu, Y., Shi, L., Fang, Z., Zhang, T., Ding, H., 2018. Quantitative determination of carcinogenic mycotoxins in human and animal biological matrices and animal-derived foods using multi-mycotoxin and analyte-specific high performance liquid chromatography-tandem mass spectrometric methods. Journal of Chromatography B 1073: 191–200. https://doi.org/10.1016/j.jchromb.2017.10.006.

Dänicke, S., Kersten, S., Billenkamp, F., Spilke, J., Starke, A. and Saltzmann, J., 2023. Estimation of oral exposure of dairy cows to the mycotoxin deoxynivalenol (DON) through toxin residues in blood and other physiological matrices with a special focus on sampling size for future predictions. Dairy 4: 360-391. https://doi.org/10.3390/dairy4020024.

Dänicke, S., Valenta, H., Klobasa, F., Döll, S., Ganter, M. and Flachowsky, G., 2004. Effects of graded levels of *Fusarium* toxin contaminated wheat in diets for fattening pigs on growth performance, nutrient digestibility, deoxynivalenol balance and clinical serum characteristics. Archives of Animal Nutrition 58:1–17. https://doi.org/10.1080/00039420 31000161045.

Dänicke, S., Brüssow, K.P., Valenta, H., Ueberschär, K.H., Tiemann, U. and Schollenberger, M. 2005. On the effects of graded levels of *Fusarium* toxin contaminated wheat in diets for gilts on feed intake, growth performance and metabolism of deoxynivalenol and zearalenone. Molecular Nutrition and Food Research 49: 932-943. https://doi.org/10.1002/mnfr.200500050.

den Hollander, D., Croubels, S., Lauwers, M., Caekebeke, N., Ringenier, M., De Meyer, F., Reisinger, N., Van Immerseel, F., Dewulf, J. and Antonissen, G., 2021. Applied Research Note: Biomonitoring of mycotoxins in blood serum and feed to assess exposure of broiler chickens. Journal of Applied Poultry Research 30: 100111. https://doi.org/10.1016/j.japr.2020.10.010.

Devreese, M., Antonissen, G., Broekaert, N., De Mil, T., De Baere, S., Vanhaecke, L., De Backer, P. and Croubels, S., 2015. Toxicokinetic study and oral bioavailability of deoxynivalenol in Turkey poults, and comparative biotransformation between broilers and turkeys. World Mycotoxin Journal 8: 533-539. https://doi.org/10.103920/WMJ2014.1843.

Devreese, M., Croubels, S., De Baere, S., Gehring, R. and Antonissen, G., 2018. Comparative toxicokinetics and plasma protein binding of ochratoxin a in four avian species. Journal of Agricultural and Food Chemistry 66: 2129-2135. https://doi.org/10.1021/acs.jafc.7b06048.

Di Nardo, F., Anfossi, L., Ozella, L., Saccani, A., Giovannoli, C., Spano, G. and Baggiani, C., 2016. Validation of a qualitative immunochromatographic test for the noninvasive assessment of stress in dogs. Journal of Chromatography B: Analytical Technologies in the Biomedical and Life Sciences 1028: 192–198. https://doi.org/10.1016/j.jchromb.2016.06.019.

Diaz, G.J. and Murcia, H.W., 2011. Biotransformation of aflatoxin B1 and its relationship with the differential toxicological response to aflatoxin in commercial poultry species. Aflatoxins-Biochemistry and Molecular Biology 1: 3–20. https://doi.org/10.5772/22109.

Dilkin, P., Direito, G., Simas, M.M.S., Mallmann, C.A. and Correa, B., 2010. Toxicokinetics and toxicological effects of single oral dose of fumonisin B1 containing *Fusarium verticillioides* culture material in weaned piglets. Chemico-Biological Interactions 185: 157-162. https://doi.org/10.1016/j.cbi.2010.03.025.

Echarte, J. M., Fernández, D. C., Chiacchio, C. A. and Torres Leedham, V. M., 2014. Comparison of a validated LC-MS/MS method with a validated GC/MS method for the analysis of zeranol and its related mycotoxin residues in bovine urine samples collected during Argentina's residue monitoring control program (2005-2012). Journal of AOAC International 97: 1470–1475. https://doi.org/10.5740/jaoacint.13-283.

Escrivá, L., Font, G., Manyes and L., Berrada, H., 2017a. Studies on the presence of mycotoxins in biological samples: an overview. Toxins 9: 251. https://doi.org/10.3390/toxins5010084.

Escrivá, L., Manyes, L., Font, G. and Berrada, H., 2017b. Mycotoxin analysis of human urine by LC-MS/MS: a comparative extraction study. Toxins 9: 330. https://doi.org/10.3390/toxins9100330.

European Union Register Of Feed Additives Pursuant To Regulation (EC) no 1831/2003 Appendixes 3e and 4(I). Annex I: List of additives. Available online at: http://ec.europa.eu/food/food/animalnutrition/feedadditives/comm_register_feed_additives_1831-03.pdf

Fink-Gremmels, J., 2008. Mycotoxins in cattle feeds and carry-over to dairy milk: A review. Food Additives and Contaminants – Part A Chemistry, Analysis, Control, Exposure and Risk Assessment 25: 172–180. https://doi.org/10.1080/02652030701823142.

Fleck, S.C., Churchwell, M.I. and Doerge, D.R., 2017. Metabolism and pharmacokinetics of zearalenone following oral and intravenous administration in juvenile female pigs. Food and Chemical Toxicology 106: 193–201. https://doi.org/10.1016/j.fct.2017.05.048.

Gadzała-Kopciuch, R., Kuźniewska, A. and Buszewski, B., 2020. Analytical approaches and preparation of biological, food and environmental samples for analyses of zearalenone and its metabolites. Reviews in Analytical Chemistry 39: 157–167. https://doi.org/10.1515/revac-2020-0105.

Gambacorta, S., Solfrizzo, H., Visconti, A., Powers, S., Cossalter, A.M., Pinton, P. and Oswald, I.P., 2013. Validation study on urinary biomarkers of exposure for aflatoxin B1, ochratoxin A, fumonisin B1, deoxynivalenol and zearalenone in piglets. World Mycotoxin Journal. 6: 299–308. https://doi.org/10.3920/WMJ2013.1549.

Giroir, L.E., Ivie, G.W. and Huff, W.E., 1991. Comparative fate of the tritiated trichothecene mycotoxin, T-2 toxin, in chickens and ducks. Poultry Science 70: 1138-1143. https://doi.org/10.3382/ps.0701138.

Goryacheva, I. Y., de Saeger, S., Eremin, S. A. and van Peteghem, C., 2007. Immunochemical methods for rapid mycotoxin detection: Evolution from single to multiple analyte screening: A review. Food Additives and Contaminants 24: 1169–1183. https://doi.org/10.1080/02652030701557179.

Goyarts, T. and Dänicke, S., 2006. Bioavailability of the *Fusarium* toxin deoxynivalenol (DON) from naturally contaminated wheat for the pig. Toxicology Letter 163: 171–182. https://doi.org/10.1016/j.toxlet.2005.10.007.

Han, S., Zhou, T., Yin, B. and He, P., 2016. A sensitive and semi-quantitative method for determination of multi-drug residues in animal body fluids using multiplex dipstick immunoassay. Analytica Chimica Acta 927: 64–71. https://doi.org/10.1016/j.aca.2016.05.004.

Hanif, N.Q., Muhammad, G., Muhammad, K., Tahira, I. and Raja, G.K., 2012. Reduction of ochratoxin A in broiler serum and tissues by *Trichosporon mycotoxinivorans*. Research in Veterinary Science 93: 795–797. https://doi.org/10.1016/j.rvsc.2011.09.013.

Höhler D., Sudekum K.H., Wolffram S., Frohlich A.A. and Marquardt R.R., 1999. Metabolism and excretion of ochratoxin A fed to sheep. Journal of Animal Science 77: 1217–1223. https://doi.org/10.2527/1999.7751217x.

Hult, K., Hokby, E., Hagglund, U., Gatenbeck, S., Rutquist, L. and Sellyey, G., 1979. Ochratoxin A inpig blood: Method of analysis and use of a tool for feed studies. Applied and Environmental Microbiology 38: 772–776. https://doi.org/10.1128/aem.38.53772-776.1979.

Huybrechts, B., Martins, J. C., Debongnie, Ph., Uhlig, S. and Callebaut, A., 2015. Fast and sensitive LC-MS/MS method measuring human mycotoxin exposure using biomarkers in urine. Archives of Toxicology 89: 1993-2005. https://doi.org/10.1007/s00204-014-1358-8.

Jia, M., Liao, X., Fang, L., Jia, B., Liu, M., Li, D., Zhou, L. and Kong, W., 2021. Recent advances on immunosensors for mycotoxins in foods and other commodities. TrAC – Trends in Analytical Chemistry 136: 116193. https://doi.org/10.1016/J.TRAC.2021.116193.

Jurišić, N., Schwartz-Zimmermann, H.E., Kunz-Vekiru, E., Moll, W.D., Schweiger, W., Fowler, J, and Berthiller, F., 2019. Determination of aflatoxin biomarkers in excreta and ileal content of chickens. Poultry Science 98: 5551-5561. https://doi.org/10.3382/ps/pez308.

Keese, C., Meyer, U., Valenta, H., Schollenberger, M., Starke, A., Weber, I.-A., Rehage, J., Breves, G. and Dänicke, S., 2008. No carry over of unmetabolised deoxynivalenol in milk of dairy cows fed high concentrate proportions. Molecular Nutrition and Food Research 52: 1514-1529. https://doi.org/10.1002/mnfr.200800077.

Kwasniewska, K., Gadzała-Kopciuch, R. and Cendrowski, K., 2015. Analytical procedure for the determination of zearalenone in environmental and biological. Critical Reviews in Analytical Chemistry 45: 119–130. https://doi.org/10.1080/10408347.2014.896731.

Lauwers, M., De Baere, S., Letor, B., Rychlik, M., Croubels, S. and Devreese, M., 2019. Multi LC-MS/MS and LC-HRMS methods for determination of 24 mycotoxins including major phase i and ii biomarker metabolites in biological matrices from pigs and broiler chickens. Toxins 11: 171. https://doi.org/10.3390/toxins11030171.

Li, P., Zhang, Z., Hu, X. and Zhang, Q., 2013. Advanced hyphenated chromatographic-mass spectrometry in mycotoxin determination: Current status and prospects. Mass Spectrometry Reviews 32: 420–452. https://doi.org/10.1002/mas.21377.

Lozano, M.C. and Diaz, G.J., 2006. Microsomal and cytosolic biotransformation of aflatoxin B1 in four poultry species. British Poultry Nutrition 47: 734-741. https://doi.org/10.1080/00071660601084390.

Maresca, M., 2013. From the gut to the brain: journey and pathophysiological effects of the food-associated trichothecene mycotoxin deoxynivalenol. Toxins 5: 784-820. https://doi.org/10.3390/toxins5040784.

Masching, S., Naehrer, K., Schwartz-Zimmermann, H.-E., Sarandan, M., Schaumberger, S., Dohnal, I., Nagl, V. and Schatzmayr, D., 2016. Gastrointestinal degradation of fumonisin

B1 by carboxylesterase FumD prevents fumonisin induced alteration of sphingolipid metabolism in turkey and swine. Toxins 8: 84. https://doi.org/10.3390/toxins8030084.

Matabaro, E., Ishimwe, N., Uwimbabazi, E. and Lee, B.H., 2017. Current immunoassay methods for the rapid detection of aflatoxin in milk and dairy products. Comprehensive Reviews in Food Science and Food Safety 16: 808–820. https://doi.org/10.1111/1541.4337.12287.

Micco, C., Miraglia, M., Onori, R., Brera, C., Mantovani, A., Ioppolo, A. and Stasolla, D., 1988. Long-term administration of low doses of mycotoxins to poultry. 1. Residues of aflatoxin B_1 and its metabolites in broilers and laying hens. Food Additives and Contaminants 5: 303–308. https://doi.org/10.1080/02652038809373708.

Minervini, F., Giannoccaro, A., Nicassio, M., Panzarini, G. and Lacalandra, J.M., 2013. First evidence of placental transfer of ochratoxin A in horses. Toxins 5: 84–92. https://doi.org/10.3390/toxins5010084.

Mirocha, C. J., Pathre, S.V. and Robison, T.S., 1981. Comparative metabolism of zearalenone and transmission into bovine milk. Food and Cosmetics Toxicology 19: 25–30. https://doi.org/10.1016/0015-6264(81)90299-6.

Mobashar, M., Hummel, J., Blank, R. and Südekum, K.H., 2010. Ochratoxin A in ruminants–A review on its degradation by gut microbes and effects on animals. Toxins 2: 809-839. https://doi.org/10.3390/toxins204809.

Muñoz-Solano, B. and González-Peñas, E. 2023. Biomonitoring of 19 mycotoxins in plasma from food-producing animals (cattle, poultry, pigs, and sheep). Toxins 15: 295. https://doi.org/10.3390/toxins15040295.

Murcia, H.W. and Diaz, G.J., 2020. *In vitro* hepatic aflatoxicol production is related to a higher resistance to aflatoxin B_1 in poultry. Nature Scientific Reports 10: 5508. https://doi.org/10.1038/s41598-020-62415-y.

Nagl, V., Woechtl, B., Schwartz-Zimmermann, H. E., Hennig-Pauka, I., Moll, W. D., Adam, G. and Berthiller, F., 2014. Metabolism of the masked mycotoxin deoxynivalenol-3-glucoside in pigs. Toxicology Letters 229: 190–197. https://doi.org/10.1016/j.toxlet.2014.06.032.

Nathanail, A. V., Syvähuoko, J., Malachová, A., Jestoi, M., Varga, E., Michlmayr, H., Adam, G., Sieviläinen, E., Berthiller, F. and Peltonen, K., 2015. Simultaneous determination of major type A and B trichothecenes, zearalenone and certain modified metabolites in Finnish cereal grains with a novel liquid chromatography-tandem mass spectrometric method. Analytical and Bioanalytical Chemistry 407: 4745–4755. https://doi.org/10.1007/s00216-015-8676-4.

Olsen, M., Malmlof, K., Pettersson, H., Sandholm, K. and Kiessling, K.H., 1985. Plasma and urinary levels of zearalenone and alphazearalenol in a prepubertal gilt fed zearalenone. Acta Pharmacologica et Toxicologica 56: 239–243. https://doi.org/10.1111/j.1600-0773.1985.tb01282.x.

Patterson, D.S., Roberts, B.A. and Small, B.J., 1976. Metabolism of ochratoxins A and B in the pig during early pregnancy and the accumulation in body tissues of ochratoxin A only. 14: 439-442.

Pecorelli, I., Guarducci, N., von Holst, C., Bibi, R., Pascale, M., Ciasca, B., Logrieco, A. F. and Lattanzio, V.M.T., 2020. Critical comparison of analytical performances of two immunoassay methods for rapid detection of aflatoxin M1 in milk. Toxins 12: 270. https://doi.org/10.3390/toxins12040270.

Pereira, V. L., Fernandes, J. O. and Cunha, S. C., 2015. Comparative assessment of three clean-up procedures after QuEChERS extraction for determination of trichothecenes (type A and type B) in processed cereal-based baby foods by GC-MS. Food Chemistry 182: 143–149. https://doi.org/10.1016/j.foochem.2015.01.047.

Pestka, J.J., 2007. Deoxynivalenol: toxicity, mechanisms and animal health risks. Animal Feed Science and Technology 137: 283-298. https://doi.org/10.1016/j.anifeedsci.2007.06.006.

Pfohl-Leszkowicz, A., Tozlovanu, M., Manderville, R., Peraica, M., Castegnaro, M. and Stefanovic, V., 2007. New molecular and field evidences for the implication of mycotoxins but not aristolochic acid in human nephropathy and urinary tract tumor. Molecular Nutrition Food Research 51: 131–146. https://doi.org/10.1002/mnfr.200700045.

Pleadin, J., Frece, J. and Markov, K., 2019. Mycotoxins in food and feed. Advances in Food and Nutrition Research 2019: 297-345. https://doi.org/10.1016/bs.afnr.2019.02.007.

Pleadin, J., Kudumija, N., Kovačević, D., Scortichini, G., Milone, S. and Kmetič, I., 2016. Comparison of ochratoxin A levels in edible pig tissues and in biological fluids after exposure to a contaminated diet. Mycotoxin Research 32: 145–151. https://doi.org/10.1007/s12550-016-0249-7.

Prelusky, D.B., Hartin, K.E., Trenholm, H.L. and Miller, J.D., 1988. Pharmacokinetic fate of 14C-labeled deoxynivalenol in swine. Fundamental and Applied Toxicology 10: 276–286. https://doi.org/10.1016/0272-0590(88)90312-0.

Ribelin, W.E., Fukushima, K. and Still, P.E., 1978. The toxicity of ochratoxin to ruminants. Canadian Journal of Comparative Medicine 42: 172–176.

Rutqvist, L., Bjorklund, N.E., Hult, K., Hokby, E., Carlsson, B., 1978. Ochratoxin A as the cause of spontaneous nephropathy in fattening pigs. Applied and Environmental Microbiology 36: 920–925. https://doi.org/10.1128/aem.36.6.920-925.1978.

Sari, F.M., Öztas, E., Özden, S. and Özhan, G., 2020. Liquid chromatographic determination of citrinin residues in various meat products: A pioneer survey in Turkey. Intanbul J. Pharm 50: 195-201. https://doi.org/10.26650/IstanbulJPharm.2020.0026.

Sartori, A.V., Swensson de Mattos, J., de Moraes, M.H.P. and da Nóbrega, A.W., 2015. Determination of aflatoxins M1, M2, B1, B2, G1, and G2 and ochratoxin A in UHT and powdered milk by modified QuEChERS method and ultra-high-performance liquid chromatography tandem mass spectrometry. Food Analytical Methods 8: 2321–2330. https://doi.org/10.1007/s12161-015-0128-4.

Schneweis, I., Meyer, K., Ritzmann, M., Hoffmann, P., Dempfle, L. and Bauer, J., 2005. Influence of organically or conventionally produced wheat on health, performance and mycotoxin residues in tissues and bile of growing pigs. Archives of Animal Nutrition 59: 155–163. https://doi.org/10.1080/17450390500147594.

Schoevers, E.J., Santos, R.R., Colenbrander, B., Fink-Gremmels, J. and Roelen, B.A.J., 2012. Transgenerational toxicity of zearalenone in pigs. Reproductive Toxicology 34: 110-119. https://doi.org/10.1016/j.reprotox.2012.03.004.

Shreeve, B.J., Patterson, D.S.P. and Roberts, B.A., 1979. The carryover of aflatoxin, ochratoxin and zearalenone from naturally contaminated feed to tissues, urine and milk of dairy cows. Food and Cosmetics Toxicology 17: 151–152. https://doi.org/10.1016/0015-6264(79)90215-3.

Singh, J. and Mehta, A., 2020. Rapid and sensitive detection of mycotoxins by advanced and emerging analytical methods: A review. Food Science and Nutrition 8: 2183–2204. https://doi.org/10.1002/fsn3.1474.

Soto, J.B., Ruiz, M.J., Manyes, L. and Juan-García, A., 2015. Blood, breast milk and urine: Potential biomarkers of exposure and estimated daily intake of ochratoxin A: A review. Food Additives and Contaminants – Part A Chemistry, Analysis, Control, Exposure and Risk Assessment 33; 313–328. https://doi.org/10.1080/19440049.2015.1118160.

Sreemannarayana, O., Frohlich, A.A., Vitti, T.G., Marquardt, R.R. and Abramson D., 1988. Studies of tolerance and disposition of ochratoxin A in young calves. Journal of Animal Science 66: 1703–1711. https://doi.org/10.2527/jas1988.6671703x.

Stock, N. L., 2017. Introducing graduate students to high-resolution mass spectrometry (HRMS) using a hands-on approach. Journal of Chemical Education 94: 1978–1982. https://doi.org/10.1021/acs.jchemed.7b00569.

Stoev, S.D., Paskalev, M., MacDonald, S. and Mantle, P.G., 2002. Experimental one year ochratoxin A toxicosis in pigs. Experimental and Toxicologic Pathology 53: 481–487. https://doi.org/10.1078/0940-2993-00213.

Stoloff, L. and Trucksess, M.W., 1979. Distribution of aflatoxins B1 and M1 in contaminated calf and pig livers. Journal of the Association of the Official Analytical Chemists 62: 1361-1362.

Streit, E., Schwab, C., Sulyok, M., Naehrer, K., Krska, R. and Schatzmayr, G., 2013. Multi-mycotoxin screening reveals the occurrence of 139 different secondary metabolites in feed and feed ingredients. Toxins 5: 504–523. https://doi.org/10.3390/toxins5030504.

Stubblefield, R.D., Honstead, J.P. and Shotwell, O.L., 1991. An analytical survey of aflatoxins in tissues from swine grown in regions reporting 1988 aflatoxin-contaminated corn. Food and Cosmetics Toxicology 74: 897-899.

Thanner, S., Czeglédi, L., Schwartz-Zimmermann, H. E., Berthiller, F. and Gutzwiller, A., 2016. Urinary deoxynivalenol (DON) and zearalenone (ZEA) as biomarkers of DON and ZEA exposure of pigs. Mycotoxin Research 32: 69–75. https://doi.org/10.1007/s12550-016-0241-2.

Thieu, N.Q. and Pettersson, H., 2009. Zearalenone, deoxynivalenol and aflatoxin B1 and their metabolites in pig urine as biomarkers for mycotoxin exposure. Mycotoxin Research 25: 59–66. https://doi.org/10.1007/s12550-009-0009-z.

Tkaczyk, A. and Jedziniak, P., 2021. Mycotoxin biomarkers in pigs current state of knowledge and analytics. Toxins 13: 586. https://doi.org/10.3390/toxins13080586.

Tkaczyk, A., Jedziniak, P., Zielonka, Ł., Dabrowski, M., Ochodzki, P. and Rudawska, A., 2021. Biomarkers of deoxynivalenol, citrinin, ochratoxin a and zearalenone in pigs after exposure of naturally contaminated feed close to guidance values. Toxins 13: 750. https://doi.org/10.3390/toxins13110750.

Tozzi, B., Liponi, G. B., Meucci, V., Casini, L., Dall'Asta, C., Intorre, L. and Gatta, D., 2016.

Aflatoxins M1 and M2 in the milk of donkeys fed with naturally contaminated diet. Dairy Science and Technology 96: 513–523. https://doi.org/10.1007/s13594-016-0285-2.

Van Egmond, H.P., Paulsch, W.E., Sizoo, E.A. and Schuller, P.L., 1990. Two-dimensional chromatographic method for determination of aflatoxins in liver; occurrence of aflatoxins in animal liver. Journal of Environmental Pathology, Toxicology, and Oncology 10: 120-123.

Vidal, A., Mengelers, M., Yang, S., De Saeger, S. and De Boevre, M., 2018. Mycotoxin Biomarkers of Exposure: A Comprehensive Review. Comprehensive Reviews in Food Science and Food Safety 17: 1127–1155. https://doi.org/10.1111/1541-4337.12367.

Vidal, A., Ouhibi, S., Ghali, R., Hedhili, A., De Saeger, S. and De Boevre, M., 2019. The mycotoxin patulin: An updated short review on occurrence, toxicity and analytical challenges. Food and Chemical Toxicology 129: 249–256. https://doi.org/10.1016/j.fct.2019.04.048.

Wu, Y., Sun, J., Huang, X., Lai, W. and Xiong, Y., 2021. Ensuring food safety using fluorescent nanoparticles-based immunochromatographic test strips. Trends in Food Science and Technology 118: 658–678. https://doi.org/10.1016/j.tifs.2021.10.025.

Xiao, H., Marquardt, R.R., Frohlich, A.A., Phillips, G.D. and Vitti, T.G., 1991. Effect of a hay and a grain diet on the bioavailability of ochratoxin A in the rumen of sheep. Journal of Animal Science 69: 3715–3723. https://doi.org/10.2527/1991.6993715x.

Yan, Z., Wang, L., Wang, J., Tan, Y., Yu, D., Chang, X., Fan, Y., Zhao, D., Wang, C., De Boevre, M., De Saeger, S., Sun, C. and Wu, A., 2018. A quechers-based liquid chromatography-tandem mass spectrometry method for the simultaneous determination of nine zearalenone-like mycotoxins in pigs. Toxins 10: 129. https://doi.org/10.3390/toxins10030129.

Yang, S., Zhang, H., De Saeger, S., De Boevre, M., Sun, F., Zhang, S., Cao, X. and Wang, Z., 2015. *In vitro* and *in vivo* metabolism of ochratoxin A: a comparative study using ultra-performance liquid chromatography-quadrupole/time-of-flight hybrid mass spectrometry. Analytical and Bioanalytical Chemistry 407: 3579-3589. https://doi.org/10.1007/s00216-015-8570-0.

Yang, S., Zhang, H., Sun, F., De Ruyck, K., Zhang, J., Jin, Y., Li, Y., Wang, Z., Zhang, S., De Saeger, S., Zhou, J., Li, Y. and De Boevre, M., 2017. Metabolic profile of zearalenone in liver microsomes from different species and its in vivo metabolism in rats and chickens using ultra high-pressure liquid chromatography-quadrupole/time-of-flight mass spectrometry. Journal of Agricultural and Food Chemistry 65: 11292–11303. https://doi.org/10.1021/acs.jafc.7b04663.

Zhang, L., Dou, X.W., Zhang, C., Logrieco, A.F. and Yang. M.H., 2018. A review of current methods for analysis of mycotoxins in herbal medicines. Toxins 10: 65. https://doi.org/10.3390/toxins10020065.

Zhang, Z., Tang, X., Wang, D., Zhang, Q., Li, P. and Ding, X., 2015. Rapid on-site sensing aflatoxin B1 in food and feed via a chromatographic time-resolved fluoroimmunoassay. PLoS ONE 10: e0123266. https://doi.org/10.1371/j.aca.2016.05.004.

Zöllner, P., Jodlbauer, J., Kleinova, M., Kahlbacher, H., Kuhn, T., Hochsteiner, W. and Lindner, W. 2002. Concentration levels of zearalenone and its metabolites in urine, muscle tissue, and liver samples of pigs fed with mycotoxin-contaminated oats. Journal of Agricultural and Food Chemistry 50: 2494–2501. https://doi.org/10.1021/jf0113631.

CHAPTER 13

In vitro models to evaluate the effects of mycotoxins on the gut and other organs

J. Jin[1], F. Xing[1] and X. Wu[2]

[1]Institute of Food Science and Technology, Chinese Academy of Agricultural Sciences, 2 Yuanmingyuan West Road, Haidian District, Beijing 100193, P.R. China; [2]Department of Infectious Diseases, Sir Run Run Shaw Hospital, Zhejiang University School of Medicine, Hangzhou 310016, P.R. China
**correspondence: jinjing@caas.cn; xingfuguo@caas.cn; x.wu@outlook.com*

1 Introduction

Feed is considered to be the main route of exposure to mycotoxins, although cases of skin contact and inhalation exposure have also been reported (Bush et al., 2006). Among the many other organs exposed to mycotoxins, the intestinal epithelium is exposed to high concentrations. The GIT is the initial site for the interaction of ingested mycotoxins and living organisms and it plays a crucial role in mycotoxin absorption and oral bioavailability. The toxic effects of mycotoxins include two types of pathway, toxicodynamic (effects of mycotoxins on the gut and other organs) and toxicokinetic (metabolism of the mycotoxins by the gut and other organs) (Jin et al., 2021a). Mycotoxin effects on animals can be categorized as acute or chronic. Acute toxicity causes a rapid response. In contrast, chronic effects appear over a relatively long time period. The issues caused by mycotoxin contamination in feed do not normally result in acute disease, but low-level, long-term exposure can cause a series of metabolic disturbances (Bryden, 2012). *In vivo* and *in vitro* studies have been conducted considering the complex impacts of mycotoxins on health. The *in vivo* studies provide information on how mycotoxins act upon the whole organism. The *in vitro* studies, such as cell-based models, provide information on toxicity, mode of action, and oral bioavailability of mycotoxins. More *in vitro* models have been developed given the request for reduction of *in vivo* animal studies within the 3R principles (replacement, reduction, and refinement) (Flecknell, 2002). The aim of this chapter is to introduce the various *in vitro* models used for studying the effects of mycotoxins on the gut and other organs.

 | DOI: 10.1163/9789004712720_14

2 Cell-based models

Isolated cells have been extensively used in the toxicological study of mycotoxins. Cultured cells are a useful system for studying the normal physiological and biochemical processes of cells, and can even be used for diagnostic purposes. The main advantage of using the cell culture system is the consistency and reproducibility obtained (Verhoeckx et al., 2015). *In vitro* cell-based models are a highly defined tool, in which the cultural conditions and parameters can be easily changed. They are useful tools to assess the toxicity of mycotoxins on gut and other organs (Allen et al., 2005). However, *in vitro* models lack the signalling and interaction between organs and are thus not able to reflect the *in vivo* situation of cells and tissues in the organisms. Typical approaches are classified as primary cell cultures, cells derived from healthy organs and cell lines from tumour origins. In this chapter, the cell models are sorted into two categories: from the gut and from other organs.

2.2 *Intestinal epithelial cell models*

Intestinal epithelial cell models can be used to assess intestinal barrier function, metabolism and cell proliferation. The gut is composed of four functional layers: the most inner mucosal or epithelial layer, the submucosa, the muscularis propria, and the outer layer (Bagnoli and Rappuoli, 2021). Although all intestinal layers may be exposed to and affected by ingested mycotoxins, the mucosal or epithelial layer is directly affected and so the exposure is the most severe. The cells that constitute the mucosa or epithelium play an important role in the effect of mycotoxins on the gut (Robert et al., 2017). Intestinal epithelial cell lines such as Caco-2, HT29, and IPEC-J2 are most commonly used to study the effects of xenobiotics on the gut epithelium.

The cell line of Caco-2 consists of epithelial cells isolated from colon tissue, derived from a male patient with adenocarcinoma, and shows many properties related to the cells of the small intestine. When growing on filter scaffolds, Caco-2 cells spontaneously begin to differentiate into a polarized cell layer after confluence. They express many of the enzyme markers and transport systems of primary epithelial cells. However, Caco-2 cells do not produce mucins. The HT29 cell line consists of cells from a female patient with colorectal adenocarcinoma (Verhoeckx et al., 2015). The HT29-MTX cell line is a cell population isolated from the HT29 cell line, but adapted with methotrexate (MTX). This cell line is able to secret mucus and develops confluent monolayers forming functional tight junction complexes. The Caco-2 cell line is also able to form tight junction complexes, but the extent differs from the HT29-MTX cell line. For this reason, the Caco-2 cell line is more often adopted to study the epithelial barrier function and chemical transport. For example, DON inhibits the absorption of conjugated bile acids and the expression of transporters in Caco-2 cells (Wang et al., 2022). The Caco-2 and the HT29-MTX cell lines are often co-cultured in one cell system. This developed method can compensate for the lack of mucus production by the Caco-2 cell line. The intestinal epithelial cell models are a powerful tool for screening and predicting the effects of mycotoxins in contaminated food and feed in human and animal gut. Differentiated monolayers of Caco-2 cells alone and co-culture with HT29-MTX cells have been used to study mycotoxin absorption and transport (Cheli et al., 2015). *In vitro* cell-based models offer a powerful approach for studying oral bioavailability, compound modifications in the GIT of masked mycotoxins. However, it is important to remark that these are human cancer cell lines and are not derived from animals.

The intestinal porcine epithelial cells 1 (IPEC-1) and J2 (IPEC-J2) are cell lines derived from neonatal unsuckled piglets younger than 12 hours old (Berschneider, 1989). These two cell lines differ in metabolism and morphology (Nossol et al., 2015). For instance, the absorption area of IPEC-J2 cells is the higher because the number of microvilli in IPEC-J2 cells is twice that in IPEC-1 cells and IPEC-J2 cells have a higher ATP-content and thus a higher oxygen consumption than IPEC-1 cells (Nossol et al., 2015). For these reasons, most studies use IPEC-J2 cells when evaluating the mycotoxin effect on animal intestinal cells. The differences between these cell types is reflected in the results of *in vitro* studies of mycotoxins. The IC_{50} values of DON for IPEC-1 cells is 1.3 μM, but 3 times higher for IPEC-J2 cells, i.e., 3 μM (Dänicke et al., 2010). The cytotoxic concentrations of T-2 are 0.093 μM for IPEC-1 cells (Kolf-Clauw et al., 2013) and 0.005 μM for IPEC-J2 (Pomothy et al., 2020). For intestinal explants, the T-2 toxin cytotoxic concentration is 0.015 μM (Kolf-Clauw et al., 2013). There are several studies demonstrating that DON promotes the translocation of *Salmonella typhimurium* (exposure of IPEC-J2 cells to 148 mM DON for 24 hours followed by 1 hour incubation with *S. typhimurium*; Vandenbroucke et al., 2011) This effect was confirmed by *in vivo* studies where broiler chickens diets were contaminated with DON and exposed to *Escherichia coli* (Awad et al., 2019) and *Campylobacter jejuni* (Ruhnau et al., 2020). T-2 also leads to the translocation of *S. typhimurium* (exposure of IPEC-J2 cells to 460 μM T-2 for 24 hours followed by 1 hour incubation with *S. typhiurium*; Verbrugghe et al., 2012). IPEC-J2 cells were successfully used to study the intestinal epithelial barrier function, demonstrating the risks of the passage of antibiotics during exposure to *Fusarium* mycotoxins such as DON and T-2 toxin, whereas this effect was not observed with FB_1 and ZEN (Goossens et al., 2012).

Epithelial integrity is essential for maintaining a physical but selective barrier between external and internal environments. The barrier function is maintained by well-organized intercellular structures (Cheli et al., 2015). When intestinal cells are differentiated into polarized monolayers, they become useful *in vitro* models. Several cell lines such as Caco-2, IPEC-1, IPEC-J2 cells, or intestinal organoids have been used in three-dimensional models. Transepithelial electrical resistance (TEER) is a widely accepted quantitative technique for measuring epithelial integrity and the tight junctions of the cell monolayer. TEER measurements are in real time and without cell damage. Previous studies have investigated the effects of mycotoxins on intestinal barrier integrity. Of all the tested mycotoxins, DON has clearly been proved to affect epithelial integrity and to significantly reduce the measured TEER value. The mechanism involved may be one where the mycotoxins affect the expression of tight junction proteins (Cheli et al., 2015).

2.3 *Other cell models*

Mycotoxins may affect different targeted organs, such as stomach, liver, kidney, immune system, nervous system and so on (Hussein and Brasel, 2001). The most adequate cell lines are chosen considering the different target organs of interest. The sensitivities of various cell lines to mycotoxins were demonstrated (Cheli et al., 2014). A range of cell lines often used for the development of cell-based models for OTA; cell lines of GES-1 (normal human gastric epithelial cells), HK-2 (proximal tubular cell line derived from normal adult male kidney), J774A.1 (cells from the ascites of an adult female mouse with reticulum cell sarcoma), SK-N-MC (neuroblastoma cell line), MDCK (Madin-Darby canine kidney cells), AML-12 (hepatocytes from the normal liver of a 3-month old mouse), LLC-PK1 (epithelial kidney cells from the kidney of a male 3-4 week-old pig), and BME-UV1 (bovine mammary epithelial cells). Whereas cell lines HepG2 (liver cells from a hepatocellular carcinoma of a 15-year-old human male), HCT116 (cells from the colon of an adult male with colon cancer), and PBMC (human peripheral blood mononuclear cells) are normally used for building the *in vitro* cell models for DON (Cheli et al., 2015).

The nephrotoxic effect of OTA was studied with MDCK cells, which were exposed to 0.25, 0.5, 1.0, 2.0, or 4.0 μg OTA/ml medium for 24 hours (Li et al., 2021). These authors showed that 2.0, or 4.0 μg OTA/ml induced inflammation in MDCK cell fibrosis related markers (α-smooth muscle actin, vimentin, transforming growth factor-β), which was confirmed by kidney fibrosis in mice exposed to 2 mg/kg BW OTA. The MDCK cells are also used to test compounds able to counteract the negative effects of OTA. Fusi et al. (2018) showed that 1 nM α-tocopherol improves cell viability and protects tight junctions, maintaining cell-cell interactions during a 24-hours exposure of MDCK cells to 0.3 and 0.6 μg OTA/ml. This is probably due to its antioxidant capacity. However, no positive effect was observed when cells were exposed to 1.2 μg OTA/ml.

Using HepG2 cells as *in vitro* model, Juan-Garcia et al. (2019) studied the effect of DON and its two acetylated forms (3-ADON and 15-ADON) on oxidative stress status. This study proved that HepG2 cells, exposed to DON, 3-ADON and 15-ADON, either individually or in combination, decreased the production of ROS, but increased the production of lipid peroxidation (LPO). Combinations of two mycotoxins containing 15-ADON produced the highest values, however, with two mycotoxins containing 3-ADON, the values were minor. A combination of all three mycotoxins alleviated ROS production and increased the levels of LPO.

The main disadvantages of the models are the lack of interactions with other cells, extracellular factors, hormonal or immunological influences and the expression of genes involved in the overall response to mycotoxins. Despite the limitations, *in vitro* cell-based models represent an effective method to assess the combined effects in terms of additive, antagonistic or synergistic toxicity of mycotoxins. *in vitro* cell-based models have also proved to be powerful tools when screening for and predicting the toxicities of food borne mycotoxins and in supporting the risk assessment of these toxic compounds.

2.3.1 Microsome/supersomes/S9 fraction models

Biotransformation occurs in many organs, such as liver, kidney, intestine, nasal epithelium, brain, lung and skin. Among these tissues, liver is the most important organ because it is rich in CYP enzymes, which play a major role in phase I oxidative reactions (Asha and Vidyavathi, 2010). Phase II enzymes, such as GST, N-acetyl transferase (NAT), sulfotransferase (ST) and UDP-glucuronosyltransferases, also play a vital role in the detoxification and excretion of xenobiotics (Lu et al., 1996). Several *in vitro* liver models, such as microsomes, supersomes, cytosol, and S9 fractions have been developed in the past decades. These represent effective *in vitro* models for estimating the xenobiotic metabolic fates *in vivo* (Lu, 1996).

Metabolism of mycotoxins by hepatic microsomes has been reported in the literature. Pfeiffer et al. (2009) reported that ZEN can be monohydroxylated by human liver microsomes *in vitro*. Two of the major metabolites arise through aromatic hydroxylation and are catechols, but the chemical structures are not clear. The author proved that only hCYP1A2 was able to hydroxylate ZEN to its catechols. Another study conducted by Maul et al. (2012) investigated the metabolism of DON using liver microsomes from human and different species of animals. The results showed that metabolism by all animal and human liver microsomes led to the formation of mono-O-glucuronides, but that the chemical structures and activity varied from species to species. The activity of human liver microsomes was lower than bovine liver and rat liver.

2.3.2 Batch fermentation models

Commensal microbes occur in the GIT at densities exceeding 10^8 cells/ml in the small intestine and 10^{11} cells/ml in the large intestine. The estimated 10^{13} microorganisms inhabiting the human gut play a vital role in many processes, such as colonization resistance, immune system modulation, synthesis of essential vitamins and digestion of feed constituents (Koppel et al., 2017). Gut microbes co-encode 150 times more genes than the human genome, including a richer library of enzymes with metabolic potential for xenobiotics, thereby affecting their efficiency and toxicity (Zimmermann et al., 2019). The intestinal microbiota is now considered an extra organ because it plays a key role in almost all intestinal functions, including, but not limited to, the degradation or metabolism of ingested mycotoxins. Considering the important role that the gut microbiome plays in host metabolism, it is necessary to develop suitable *in vitro* batch incubation models with a view to achieving a deeper understanding of it.

The ideal *in vitro* models should present similar functionality, community structure or ability to the host microbiome to degrade certain xenobiotics. Many efforts have been put into developing different strategies to build *in vitro* models, such as static batch fermentation and continuous systems. Batch fermentation is generally carried out in test tubes, and enables many samples to be studied simultaneously and within a short period of time, usually 24-48 h. The disadvantage of batch fermentation is its lower physiological relevance, and the fact that bacterial waste products will eventually accumulate, hence the need for a shorter experimental duration. Incubation conditions, such as faecal material origin, faecal slurry concentration, incubation time and culture composition affect microbial communities and their metabolism. In recent studies (Jin et al., 2021b; Mendez-Catala et al., 2020), targeted mycotoxins were incubated with faecal or intestinal content samples in deoxygenated PBS under anaerobic conditions. The PBS used in this thesis represents a minimum medium. For batch incubation, a minimum medium can help to highlight the metabolism of the compound under study and quickly identify the metabolites produced or the bacteria involved, with results that are easier to interpret than when using a rich nutrient culture medium. Faecal or intestinal content source and slurry concentration may vary depending on the experimental goals. When calculating the conversion rate by the gut microbiome and quantifying kinetic parameters, source and concentration should be carefully optimized to ascertain linearity in time and relative to the amount of faecal slurry. Obviously, this is more easily done in batch cultures than in long-term continuous cultures. Batch fermentation models allow the assessment of mycotoxin conversion by the microbiome of different species, different intestinal segments, and different individuals. Using the *in vitro* anaerobic batch incubation method, metabolic pathways and formation and subsequent conversion of intermediate metabolites over time become visible (Jin et al., 2021b). Furthermore, using the batch *in vitro* incubation method is less time consuming and less costly than *in vitro* continuous systems, animal studies or human studies. The main limitations of this method relate to the fact that they are distanced from physiological conditions.

Several research studies have revealed the kinetic studies using *in vitro* batch fermentation models. Berthiller et al. (2011) demonstrated the possible toxicological relevance showing that several lactic acid bacteria hydrolyse DON-3G during digestion *in vitro*. Jin et al. (2021b) utilize this model, quantifying kinetics and interspecies differences in intestinal microbiota mediated deconjugation of 3-ADON and 15-ADON and evaluating the results within the framework of a risk assessment of acetyl-DONs. The *in vitro* model system has also been used to quantify intestinal microbial metabolism in the bioactivation and detoxification of ZEN (Mendez-Catala et al., 2020). The study reveals the V_{max}, K_m, and catalytic efficiency (k_{cat}) values for the formation of metabolites, α-ZEL and β-ZEL. The authors estimated that the metabolic activity of colon microbiota accounts for 36% of liver activity.

3 Intestinal organoids

Intestinal organoids (or miniguts) can be used as promising models to assess the effects of mycotoxins *in vitro and as an alternative to* cell lines. These organoids are differentiated intestinal stem cells that can be cultured in 2D or 3D systems. With the exception of immune and glial cells, the intestinal organoids contain the main cell types present in the intestinal structure, i.e., enterocytes, goblet cells, enteroendocrine cells, and Paneth-like cells (Angus et al., 2019). This means that it is possible to evaluate villus and crypt development, mucus production, hormonal activity, the synthesis of defensins, allowing a close mimicking of *in vivo* conditions. These organoids can be culture dissociated in a 2D system or kept as a 3D miniguts structure. Exposure of porcine ileal organoid monolayers (2D system) to a mixture of 0.1 μM DON and 0.01 μM T-2 toxins together with *Streptococcus suis* resulted in increased permeability, favouring the translocation of this bacterium (Guan et al., 2023).

In a 3D system, murine jejunum organoids were not dissociated and the obtained miniguts received a luminal microinjection with 1 μM DON or were just exposed to 1 μM DON via culture medium (basolateral) (Hanyu et al., 2020). These authors observed no differences in cell permeability, but disruption of tight junctions were observed only when basolateral exposure was performed. Chicken duodenum, jejunum, and ileum organoids were developed and exposed to 2 μg DON/ml (~ 6.5 μM DON), resulting in increased intestinal permeability, and decreased uptake of nutrient-like fatty acids and glucose (Kang and Lee, et al., 2024). In an *in vivo* study performed with broiler chickens being fed a DON-contaminated diet, it was observed that markers for the uptake of peptides decreased, but that glucose uptake increased, probably as a compensatory mechanism (Santos et al., 2021).

4 Continuous system models

The determination of mycotoxin content in feed or *in vitro* cytotoxicity tests on cell lines are insufficient to predict the effects of mycotoxins in a host. *In vivo* models have clearly played an indisputable role by providing a real gut environment for finally understanding and resolving the issue of mycotoxins. However, *in vivo* models have some drawbacks such as differences in biokinetic parameters in mice, difficulty in extrapolation of results to other animals and difficulty in maintaining a stable gut environment in different trials. In the meantime, in order to overcome the disadvantages of a simple cell culture, or an organotypic model, continuous system models offer a better simulation of the *in vivo* gut environment when studying the effects of mycotoxins.

4.1 *Ussing chamber system*

An Ussing chamber system is an apparatus that was invented by the Danish zoologist and physiologist Hans Henriksen Ussing in 1946 (Ussing). It consists of two halves separated by a sheet of mucosa or a monolayer of epithelial cells grown on a permeable support, e.g. snapwells (Corning, USA), and attached to a continuous liquid system. The Ussing chamber system would be utilized to detect and quantify the transport and barrier functions of epithelia or of living tissue. This system is the most commonly used continuous *ex vivo* model in mycotoxin research. For example, a study using the Ussing chamber system to co-expose intestine to DON and *C. jejuni*, found that DON promoted *C. jejuni* multiplication in the intestine of broiler chickens (Ruhnau et al., 2020). Using the same method, the same group reported that the detoxification product of DON, DOM-1, does not have negative effects on the GIT (Ruhnau et al., 2021). Yiannikouris et al. (2021) reported that the carcinogenic mycotoxin AFB_1's physiological bioavailability was limited by yeast cell wall extract in the dietary ration. Such a result was obtained by effective measurement of AFB_1 transport across live intestinal tissue in an Ussing system that was able to 'complete the journey' of AFB_1 through the digestive tract (Yiannikouris et al., 2021). Another group used Ussing chambers to test the putative direct effects of AFB_1 on the anterior intestine of fish (Ouethrani et al., 2013). Overall, the Ussing chamber is a useful model to investigate mycotoxin toxicity on the intestinal mucosa of diverse species.

4.2 *InTESTine™ system*

The InTESTine™ system is a continuous intestinal model using healthy porcine intestine and was developed by the Netherlands Organization for Applied Scientific Research (TNO, Zeist, the Netherlands). Using the InTESTine™ system allows researchers to study multiple intestinal segments in parallel and under controlled conditions and to observe regional differences in intestine absorption (Westerhout et al., 2014). Comparison of InTESTine™ and Ussing Chamber systems indicates that both models would provide stable testing parameters, including small molecule absorption. The major limitations of the Ussing chamber system are the low throughput and the complexity of the operation. The InTESTine™ model is easy to operate with region-specific measurements (duodenum, jejunum, ileum, and colon) that can be studied with a higher throughput (Stevens et al., 2019). So far, no mycotoxin-related study has been found to utilize the InTESTine™ system, but this system may provide a good *ex vivo* model to study the effects of different mycotoxins in the host intestine system (Pearce et al., 2018). The same research group from TNO took one further step and developed a microfluidic chip, the Intestinal Explant Barrier Chip (IEBC), which incorporates intestinal tissue explants in a two microchannel setup that simulates an *in vivo* microenvironment (Eslami Amirabadi et al., 2022). IEBC prolongs tissue viability, maintaining stability of the tested intestinal tissue and properly ranking high to low permeability molecules, including mycotoxins.

4.3 *SHIME model*

A Simulator of the Human Intestinal Microbial Ecosystem (SHIME) system was developed in 1993. It is a multi-compartment dynamic simulator of the human gut and can be technically modified to target digestive conditions of interest (Van de Wiele et al., 2015). The advantages of the SHIME system include the ability to mimic the entire GIT, study microbiome inoculation from different target groups (Vermeiren et al., 2012), maintain culture stability over a long timeframe, control multiple parameters in mechanistic research and differentiate between mucosal and luminal microbiome (Van den Abbeele et al., 2013). Bondue et al. developed a toddler SHIME model that was well adapted to evaluate prebiotic and probiotic impact on the specific microbiota of toddlers. They inoculated the system with faeces from young children in ascending, transverse, and descending colons to determine if this toddler SHIME system was also applicable to mycotoxin studies. For example, the efficacy of an aflatoxin bentonite clay binder and a FUM esterase were investigated in a SHIME model. Meanwhile, the researchers also investigated the detoxification effect on gut microbiota in this system (Neckermann et al., 2021). The capability of the human gut microbiota to degrade coffee mycotoxin contaminates OTA in a SHIME system. The OTA concentrations were measured during the SHIME run, revealing that the descending colon is the predominant site for microbial breakdown of OTA (Ouethrani et al., 2013). This three-stage continuous culture system can be extended to M-SHIME (including porcine mucin gels in small glass tubes) or to twin-SHIME (two parallel running reactors) that may adapt to different investigation purposes, allowing broader application of SHIME in mycotoxin studies.

4.4 *Organ-on-chip*

Although the above-mentioned *in vitro* models are frequently used to test the effect of mycotoxins on organisms, there is still a significant gap between these *in vitro* assays (3D cell culture) and animal tests. Organ-on-chip (OOC) has been developed in recent years, and is considered a promising platform to minimize such gaps (Akarapipad et al., 2021). OOCs are composed of microfluidic-based systems and 3D cell culture. OOCs are gaining more attention because microenvironments, physical and biochemical stimuli and sensing and biosensing systems can be incorporated in OOC devices. Microfluidic chips can be engineered to mimic the microstructure of different organs and the surface of the chip can allow cells to approximate to their native behaviours and functions. In OOC systems, the applied flow can stimulate the fluid shear stress and the microenvironments in the body. This allows the development of biochemical molecules' concentration gradients and influences cell migration, proliferation, and differentiation (Akarapipad et al., 2021). These properties to enable OOCs to mimic *in vivo* tissue and organ behaviour and metabolism. OOCs are suitable as disease modelling, biosensing, and toxicity screening platforms. The advantages of OOCs include miniaturization of the system, easy chemical loading, real-time monitoring, minimizing time and sample size, and reduction in cost. These advantages give OOCs high potential as toxicity screening platforms. OOCs can be utilized for toxicology studies, such as pharmacokinetic, pharmacodynamic and ADME analyses, TEER analyses, and biomarker assays. A number of studies have used OOCs for drug toxicity and environmental toxicity testing (Akarapipad et al., 2021). However, their use in evaluation of mycotoxin toxicology effects has been limited. OOCs will be a promising tool in future toxicology studies of mycotoxins.

4.5 *HuMiX model*

The Human-microbial crosstalk model (HuMiX) is a microfluidics-based model, which allows researchers to recapitulate *in vivo,* transcriptional, metabolic, and immunological responses to the microbiome or other chemicals in the gut environment (Ouethrani et al., 2013). The HuMiX model consists of a medium perfusion microchamber, a human epithelial cell culture microchamber, and a microbial culture microchamber. Each has an inlet and outlet for the inoculation of cells and for the control of physicochemical parameters through the perfusion of laminar streams of culture media. Moreover, it is also combined with measuring TEER on a chip. Furthermore, the HuMiX model also integrates oxygen sensors for the real-time monitoring of the dissolved oxygen concentrations within the device. The HuMiX model was initially designed to study the interaction between microbes and host. We believe that it is also a good model for the study of mycotoxins using the above unique features. No reported investigation was yet found to utilize HuMiX model, yet it might be a potential alternative when other continuous models are not available.

5 Zebrafish model

The zebrafish model was proposed by the Organization for Economic Co-operation and Development (OECD) guidelines (OECD 236, 2013) as a good model system for the study of ecotoxicology and neurodevelopmental toxicity. The use of the zebrafish model is within the principle of 3Rs. The zebrafish is easy to keep and readily produced in the laboratory environment. Their embryos are transparent, which makes them suitable for examining developmental and neuro behavioural toxicity. Zebrafish embryos are considered to be non-protected life stage within five days after fertilization (dpf) by EU directive 2010/63/EU and by the Institutional Animal Care and Use Committee (IACUC) in the USA. Zebrafish embryos, therefore, present a replacement method for testing adult animals, and provide new insights into the toxicological effects of mycotoxins (Juan-Garcia et al., 2020).

Several mycotoxins have been studied using the larval zebrafish model (Juan-Garcia et al., 2020). In order to assess the effects of mycotoxin exposure in the zebrafish embryos, chemicals are dissolved in water at certain concentrations. ZEN is the most widely studied mycotoxin using zebrafish. As reported by previous studies, ZEN can induce the yolk precursor protein vitellogenin (VTG), can have an effect on gonad development, inhibit spermatogenesis, induce atresia of oocytes and can affect egg production, viability, fertilization, and sexual differentiation (Juan-Garcia et al., 2020). Muthulakshmi et al. (2018) and Bakos et al. (2013) studied the fish embryo toxicity (FET) of ZEN in AB strain zebrafish embryos. These tests showed that the ZEN LC_{50} value was 950 μg/l and that the induction of developmental defects occurred above 550 μg/L. For longer exposure time, the LC_{50} value was slightly lower (893 μg/l) (Bakos et al., 2013). Bakos' study also observed morphological and developmental effects in heart and eyes. These data supported the evidence that ZEN might have an effect on some ontogenic pathways. The toxicity of ZEN was coupled with oxidative stress by inducing developmental alterations, neurotoxicity and genotoxicity (Muthulakshmi et al., 2018). The oxidative stress and reduction of antioxidant defence were valued by LPO, ROS, NO and LDH, SOD, GSH, GPx, GST, CAT in zebrafish embryo AB strain. Neurotoxicity was assessed by testing acetyl cholinesterase (AchE) activity. The highest ZEN concentration levels were 750 and 950 μg/L. In brief, the toxicological impact of ZEN in zebrafish larvae shows noticeable effects when tested with high concentrations (> 700 μg/L). When using the zebrafish larval essay for studying the major effect, oestrogenicity of ZEN, effects were detected at very low concentrations (about 0.5 μg/L).

In addition to ZEN, zebrafish larval assays were also used in evaluating the toxicological effects of OTA, AFB_1, PAT, DON and other mycotoxins. These studies support the proof of toxicological effects of mycotoxins in the early developmental phases of zebrafish. Because of the similarities between zebrafish and humans, these toxicological effects can be assumed to occur in humans. Further studies of the biological interactions of all organs and systems with mycotoxins will still need to use the zebrafish model. The model provides a wide range of options for studying complex organisms, avoiding the use of animals. The zebrafish model facilitates the assessment of mycotoxins and translation of the results to humans.

The zebrafish model can also be used to evaluate the effects of multi-mycotoxin exposure, as well as the study of mycotoxin interactions with organs and systems. It was demonstrated that zebrafish liver has a decreasing order of sensitivity to mycotoxins as AFB_1 > DON > ZEN, a synergistic affect when combining AFB_1 + DON or AFB_1 + ZEN, but an antagonistic effect when exposing zebrafish to DON + ZEN (Zhou et al., 2017). This antagonism was previously reported in *in vivo* studies (Grenier and Olswald, 2011; Ji et al., 2016), which may be explained by competition for similar target sites. However, depending on the level of the mycotoxins and the exposure time, DON and ZEN may have a synergistic effect (Wan et al., 2013). Interestingly, Zhou et al. (2017) also demonstrated that the lower the mycotoxin dose, the higher the synergistic interaction, and that, at very high levels of the mycotoxins, this interaction can become antagonistic.

6 Conclusions

As described in this chapter, cell-based models, microsome/supersomes/S9 fraction models, batch fermentation, complex cell systems, continuous system models, and zebrafish are used in studies evaluating the effects of mycotoxins on the gut and other organs. These in *vitro* models have different properties as well as different advantages and disadvantages. Some *in vitro* models, such as organ-on-chip, are promising tools for studying the toxicological effects of mycotoxins. However, currently, this model is rarely used for this purpose. The availability of *in vitro* tools, ranging from simple to sophisticated, will allow timely determination of mycotoxin risks and the testing of promising compounds without the use of a great number of experimental animals in long-term field trials.

References

Akarapipad, P., Kaarj, K., Liang, Y. and Yoon, J. Y., 2021. Environmental toxicology assays using organ-on-chip. Annual Review of Analytical Chemistry 14: 155-183. https://doi.org/10.1146/annurev-anchem-091620-091335.

Allen, D. D., Caviedes, R., Cárdenas, A. M., Shimahara, T., Segura-Aguilar, J. and Caviedes, P. A., 2005. Cell lines as *in vitro* models for drug screening and toxicity studies. Drug Development and Industrial Pharmacy 31: 757-768. https://doi.org/10.1080/03639040500216246.

Angus, H.C.K., Butt, A.G., Schultz, M. and Kemp, R.A., 2020. Intestinal organoids as a tool for inflammatory bowel disease research. Frontiers in Medicine 6: 334. https://doi.org/10.3389/fmed.2019.00334.

Asha, S. and Vidyavathi, M., 2010. Role of human liver microsomes in *in vitro* metabolism of drugs – a review. Applied Biochemistry and Biotechnology, 160: 1699-1722. https://doi.org/10.1007/s12010-009-8689-6.

Awad, W.A., Ruhnau, D., Hess, C., Doupovec, B., Schatzmayr, D. and Hess, M., 2019. Feeding of deoxynivalenol increases the intestinal paracellular permeability of broiler chickens. Archives in Toxicology 93: 2057-2064. https://doi.org/10.1007/s00204-019-02460-3.

Bagnoli, F. and Rappuoli, R., 2021. Three dimensional human organotypic models for biomedical research. springer. part ii: chapter: overview and comparison of intestinal organotypic models, intestinal cells, and intestinal explants used for toxicity studies. https://doi.org/10.1007/978-3-030-62452-1.

Bakos, K., Kovács, R., Staszny, Á., Sipos, D.K., Urbányi, B., Müller, F., Csenki, Z. and Kovács, B.,

2013. Developmental toxicity and estrogenic potency of zearalenone in zebrafish (*Danio rerio*). Aquatic toxicology 136: 13-21. https://doi.org/10.1016/j.aquatox.2013.03.004.

Berschneider H. Development of normal cultured small intestinal epithelial cell lines which transport Na and Cl. Abstract of the Annual Meeting of the American Gastroenterological Association 1989.

Berthiller, F., Krska, R., Domig, K.J., Kneifel, W., Juge, N., Schuhmacher, R. and Adam, G., 2011. Hydrolytic fate of deoxynivalenol-3-glucoside during digestion. Toxicology Letters 206: 264-267. https://doi.org/10.1016/j.toxlet.2011.08.006.

Bryden, W.L., 2012. Mycotoxin contamination of the feed supply chain: implications for animal productivity and feed security. Animal Feed Science and Technology 173: 134-158. https://doi.org/10.1016/j.anifeedsci.2011.12.014.

Bush, R.K., Portnoy, J.M., Saxon, A., Terr, A.I. and Wood, R.A., 2006. The medical effects of mold exposure. Journal of Allergy and Clinical Immunology, 117: 326-333. https://doi.org/10.1016/j.jaci.2005.12.001.

Cheli, F., Fusi, E. and Baldi, A., 2014. Cell-based models for mycotoxin screening and toxicity evaluation: an update. World Mycotoxin Journal 7: 153-166. https://doi.org/10.3920/WMJ2013.1639.

Cheli, F., Giromini, C. and Baldi, A., 2015. Mycotoxin mechanisms of action and health impact: "*in vitro*" or "*in vivo*" tests, that is the question. World Mycotoxin Journal 8: 573-589. https://doi.org/10.3920/WMJ2014.1864.

Dänicke, S., Hegewald, A.K., Kahlert, S., Kluess, J., Rothkötter, H.J., Breves, G. and Döll, S., 2010. Studies on the toxicity of deoxynivalenol (DON), sodium metabisulfite, DON-sulfonate (DONS) and de-epoxy-DON for porcine peripheral blood mononuclear cells and the Intestinal porcine epithelial cell lines IPEC-1 and IPEC-J2, and on effects of DON and DONS on piglets. Food and Chemical Toxicol*ogy* 48: 2154-2162. https://doi.org/10.1016/j.fct.2010.05.022.

Eslami Amirabadi, H., Donkers, J. M., Wierenga, E., Ingenhut, B., Pieters, L., Stevens, L., Donkers, T., Westerhout, J., Masereeuw, R., Bobeldijk-Pastorova, I., Nooijen, I. and van de Steeg, E., 2022. Intestinal explant barrier chip: long-term intestinal absorption screening in a novel microphysiological system using tissue explants. Laboratory Chip 22: 326-342. https://doi.org/10.1039/d1lc00669j.

Flecknell, P., 2002. Replacement, reduction, refinement. ALTEX-alternatives to Animal Experimentation 19: 73-78. https://www.altex.org/index.php/altex/article/view/1106.

Fusi, E., Giromini, C., Rebucci, R., Pinotti, L., Caprarulo, V., Cheli, F., Vitari, F., Domeneghini, C. and Baldi, A., 2018. Ochratoxin A cytotoxicity on Madin-Darby canine kidney cells in the presence of alpha-tocopherol: effects on cell viability and tight junctions. Journal of Animal Physiology and Animal Nutrition 102: 350-355. https://doi.org/10.1111/jpn.12682.

Goossens, J., Pasmans, F., Verbrugghe, E., Vandenbroeckel, V., De Baerel, S., Meyer, E., Haesebrouck, F., De Backer, P. and Croubels, S., 2012. Porcine intestinal epithelial barrier disruption by the Fusarium mycotoxins deoxynivalenol and T-2 toxin promotes transepithelial passage of doxycycline and paromomycin. Veterinay Research 8: 245. https://doi.org/10.1186/1746-6148-8-245.

Grenier, B. and Oswald, I., 2011. Mycotoxin co-contamination of food and feed: meta-analysis of publications describing toxicological interactions. World Mycotoxin Journal 4: 285-313. https://doi.org/10.3920/WMJ2011.1281.

Guan, X., Martinez, A.R., Fernandez, M., Molist, F., Wells, J.M. and Santos, R.R., 2024. The mycotoxins T-2 and deoxynivalenol facilitate the translocation of *streptococcus suis* across porcine ileal organoid monolayers. Toxins 16: 382. https://doi.org/10.3390/toxins16090382.

Hanyu, H., Yokoi, Y., Nakamura, K., Ayabe, T., Tanaka, K., Uno, K., Miyajima, K., Saito, Y., Iwatsuki, K., Shimizu, M., Tadaishi, M. and Kobayashi-Hattori, K., 2020. Mycotoxin Deoxynivalenol Has Different Impacts on Intestinal Barrier and Stem Cells by Its Route of Exposure. Toxins 12: 610. https://doi.org/10.3390/toxins12100610.

Hussein, H.S. and Brasel, J.M., 2001. Toxicity, metabolism, and impact of mycotoxins on human and animals. Toxicology 167: 101-134. https://doi.org/10.1016/s0300-483x(01)00471-1.

Ji, J., Gu, W., Sun, C., Sun, J., Jiang, H., Zhang, Y., Sun, X., 2016. A novel recombinant cell fluorescence biosensor based on toxicity of pathway for rapid and simple evaluation of DON and ZEN. Scientific Reports 6: 31270. https://doi.org/10.1038/srep31270.

Jin, J., Beekmann, K., Ringø, E., Rietjens, I. M. and Xing, F., 2021a. Interaction between food-borne mycotoxins and gut microbiota: A review. Food Control 126: 107998. https://doi.org/10.1016/j.foodcont.2021.107998.

Jin, J., Spenkelink, A., Beekmann, K., Baccaro, M., Xing, F. and Rietjens, I.M., 2021b. Species differences in *in vitro* and estimated *in vivo* kinetics for intestinal microbiota mediated metabolism of acetyl-deoxynivalenols. Molecular Nutrition and Food Research 65: 2001085. https://doi.org/10.1002/mnfr.202001085.

Juan-García, A., Juan, C., Tolosa, J. and Ruiz, M.J., 2019. Effects of deoxynivalenol, 3-acetyl-deoxynivalenol and 15-acetyl-deoxynivalenol on parameters associated with oxidative stress in HepG2 cells. Mycotoxin Research 35: 197-205. https://doi.org/10.1007/s12550-019-00344-0.

Juan-García, A., Bind, M. A. and Engert, F., 2020. Larval zebrafish as an in vitro model for evaluating toxicological effects of mycotoxins. Ecotoxicology and Environmental Safety 202: 110909. https://doi.org/10.1016/j.ecoenv.2020.110909.

Kang, T.H. and Lee, A.I., 2024. Establishment of a chicken intestinal organoid culture system to assess deoxynivalenol-induced damage of the intestinal barrier function. Journal of Animal Science and Biotechnology 15: 30. https://doi.org/10.1186/s40104-023-00976-4.

Kolf-Clauw, M., Sassahara, M., Lucioli, J., Rubira-Gerez, J., Alassane-Kpembi, I., Lyazhri, F., Borin, C. and Oswald, I.P., 2013. Organ toxicity and mechanisms the emerging mycotoxin, enniatin B1, downmodulates the gastrointestinal toxicity of T2 toxin in vitro on intestinal epithelial cells and ex vivo on intestinal explants. Archives in Toxicology 87: 2233-2241. https://doi.org/10.1007/s00204-013-1067-8.

Koppel, N., Maini Rekdal, V., Balskus, E.P., 2017. Chemical transformation of xenobiotics by the human gut microbiota. Science 356: eaag2770. https://doi.org/10.1126/science.aag2770.

Li, H., Mao, X., Liu, K., Sun, J., Li, B., Malyar, R.M., Liu, D., Pan, C., Gan, F., Liu, Y., Huang, K.

and Chen, X., 2021. Ochratoxin A induces nephrotoxicity *in vitro* and *in vivo* via pytoptosis. Archives in Toxicology 95: 1489-1502. https://doi.org/10.1007/s000204-021-02993-6.

Lu, F.C., 1996. Basic toxicology: Fundamentals, target organs and risk assessment (3rd ed., pp. 27-39). Washington: Taylor and Francis.

Maul, R., Warth, B., Kant, J.S., Schebb, N.H., Krska, R. and Koch, M., 2012. Investigation of the hepatic glucuronidation pattern of the fusarium mycotoxin deoxynivalenol in various species. Chemical Research in Toxicology 25: 2715-2717. https://doi.org/10.1021/tx300348x.

Mendez-Catala, D.M., Spenkelink, A., Rietjens, I. M. and Beekmann, K., 2020. An *in vitro* model to quantify interspecies differences in kinetics for intestinal microbial bioactivation and detoxification of zearalenone. Toxicology Reports 7: 938-946. https://doi.org/10.1016/j.toxrep.2020.07.010.

Muthulakshmi, S., Maharajan, K., Habibi, H. R., Kadirvelu, K. and Venkataramana, M., 2018. Zearalenone induced embryo and neurotoxicity in zebrafish model (Danio rerio): role of oxidative stress revealed by a multi biomarker study. Chemosphere 198: 111-121. https://doi.org/10.1016/j.chemosphere.2018.01.141.

Neckermann, K., Claus, G., De Baere, S., Antonissen, G., Lebrun, S., Gemmi, C., Taminiau, B., Douny, C., Scippo, M.L., Schatzmayr, D., Gathumbi, J., Croubels, S. and Delcenserie, V., 2021. The efficacy and effect on gut microbiota of an aflatoxin binder and a fumonisin esterase using an in vitro simulator of the human intestinal microbial ecosystem (SHIME(R)). Food Research International 145: 110395. https://doi.org/10.1016/j.foodres.2021.110395.

Nossol, C., Barta-Böszörmenyi, A., Kahlert, S., Zuschratter, W., Faber-Zuschratter, H., Reinhardt, N., Ponsuksili, S., Wimmers, K., Diesing, A.K. and Rothkötter, H.J., 2015. Comparing two intestinal porcine epithelial cell lines (ipecs): morphological differentiation, function and metabolism. PLoS ONE 10: e0132323. https://doi.org/10.1371/journal.pone.0132323.

OECD, 2013. Test No. 236: Fish Embryo Acute Toxicity (FET) Test OECD Publishing. https://doi.org/10.1787/9789264203709-en.

Ouethrani, M., Van de Wiele, T., Verbeke, E., Bruneau, A., Carvalho, M., Rabot, S. andCamel, V., 2013. Metabolic fate of ochratoxin A as a coffee contaminant in a dynamic simulator of the human colon. Food Chemistry 141: 3291-3300. https://doi.org/10.1016/j.foodchem.2013.05.157.

Pearce, S.C., Coia, H.G., Karl, J.P., Pantoja-Feliciano, I.G., Zachos, N.C. and Racicot, K., 2018. Intestinal in vitro and ex vivo models to study host-microbiome interactions and acute stressors. Frontiers in Physiology 9: 1584. https://doi.org/10.3389/fphys.2018.01584.

Pfeiffer, E., Hildebrand, A., Damm, G., Rapp, A., Cramer, B., Humpf, H. U. and Metzler, M., 2009. Aromatic hydroxylation is a major metabolic pathway of the mycotoxin zearalenone *in vitro*. Molecular Nutrition and Food Research 53: 1123-1133. https://doi.org/10.1002/mnfr.200800584.

Pomothy, J.M., Paszti-Gere, E., Barna, R.F., Prokoly, D. and Jerzsele, A., 2020. The impact of fermented wheat germ extract on porcine epithelial cell line exposed to deoxynivalenol and T-2 mycotoxins. Oxidative Medicine and Cellular Longevity 2020: 3854247. https://doi.org/10.1155/2020/3854247.

Robert, H., Payros, D., Pinton, P., Théodorou, V., Mercier-Bonin, M. and Oswald, I.P., 2017. Impact of mycotoxins on the intestine: are mucus and microbiota new targets? Journal of Toxicology Environmental Health: Part B: Critical Review 20: 249-275. https://doi.org/10.1080/10937404.2017.1326071.

Ruhnau, D., Hess, C., Grenier, B., Doupovec, B., Schatzmayr, D., Hess, M. and Awad, W.A., 2020. The mycotoxin deoxynivalenol (DON) promotes *Campylobacter jejuni* multiplication in the intestine of broiler chickens with consequences on bacterial translocation and gut integrity. Frontiers in Veterinary Science 7: 573894. https://doi.org/10.3389/fvets.2020.573894.

Ruhnau, D., Hess, C., Doupovec, B., Grenier, B., Schatzmayr, D., Hess, M. and Awad, W., 2021. Deepoxy-deoxynivalenol (DOM-1), a derivate of deoxynivalenol (DON), exhibits less toxicity on intestinal barrier function, *Campylobacter jejuni* colonization and translocation in broiler chickens. Gut Pathogen 13: 44. https://doi.org/10.1186/s13099-021-00440-6.

Santos, R.R., Oosterveer-van der Doelen, M.A.M., Tersteeg-Zijderveld, M.H.G., Molist, F., Mézes, M. and Gehring, R., 2021. Susceptibility of broiler chickens to deoxynivalenol exposure via artificial or natural dietary contamination. Animals 11: 989. https://doi.org/10.3390/ani11040989.

Stevens, L.J., van Lipzig, M.M.H., Erpelinck, S.L.A., Pronk, A., van Gorp, J., Wortelboer, H.M. and van de Steeg, E., 2019. A higher throughput and physiologically relevant two-compartmental human ex vivo intestinal tissue system for studying gastrointestinal processes. European Journal of Pharmaceutical Sciences 137: 104989. https://doi.org/10.1016/j.ejps.2019.104989.

Ussing, H.H. Retrieved from https://royalsocietypublishing.org/doi/pdf/10.1098/rsbm.2009.0002.

van de Wiele, T., van den Abbeele, P., Ossieur, W., Possemiers, S. and Marzorati, M., 2015. The simulator of the human intestinal microbial ecosystem (SHIME((R))). In K. Verhoeckx, P. Cotter, I. Lopez-Exposito, C. Kleiveland, T. Lea, A. Mackie, T. Requena, D. Swiatecka and H. Wichers (Eds.), The Impact of Food Bioactives on Health: *in vitro* and *ex vivo* models (305-317).

Van den Abbeele, P., Belzer, C., Goossens, M., Kleerebezem, M., De Vos, W.M., Thas, O., de Weirdt, R., Kerckhof, F.M. and van de Wiele, T., 2013. Butyrate-producing *Clostridium* cluster XIVa species specifically colonize mucins in an *in vitro* gut model. ISME Journal 7: 949-961. https://doi.org/10.1038/ismej.2012.158.

Vandenbroucke, V., Croubels, S., Martel, A., Verbrugghe, E., Goossens, J., van Deun, K., Boyen, F., Thompson, A., Shearer, N., de Backer, P., Haesebrouck, F. and Pasmans, F., 2011. The mycotoxin deoxynivalenol potentiates intestinal inflammation by *Salmonella typhimurium* in porcine ileal loops. PLos ONE 6: e23871. https://doi.org/10.1371/journal.pone.0023871.

Verhoeckx, K., Cotter, P., López-Expósito, I., Kleiveland, C., Lea, T., Mackie, A., Requena, T., Swiatecka, D. and Wichers, H., 2015. The impact of food bioactives on health: *in vitro* and *ex vivo* models. Cham (CH). https://doi.org/10.1007/978-3-319-16104-4.

Vermeiren, J., Van den Abbeele, P., Laukens, D., Vigsnaes, L. K., De Vos, M., Boon, N. and van de Wiele, T., 2012. Decreased colonization of fecal *Clostridium coccoides*/*Eubacterium rectale* species from ulcerative colitis patients in an in vitro dynamic gut model with mucin environment. FEMS Microbiology Ecology 79: 685-696. https://doi.org/10.1111/j.1574-6941.2011.01252.x.

Wan, L.Y.M., Turner, P.C., El-Nezami, H., 2013. Individual and combined cytotoxic effects of *Fusarium* toxins (Deoxynivalenol, Nivalenol, Zearalenone and Fumonisins B1) on swine jejunal epithelial cells. Food and Chemical Toxicology 57: 276-283. https://doi.org/10.1016/j.fct.2013.03.034.

Wang, J., Bakker, W., Zheng, W., de Haan, L., Rietjens, I.M. and Bouwmeester, H., 2022. Exposure to the mycotoxin deoxynivalenol reduces the transport of conjugated bile acids by intestinal Caco-2 cells. Archives of Toxicology 96: 1473-1782. https://doi.org/10.1007/s00204-022-03256-8.

Westerhout, J., van de Steeg, E., Grossouw, D., Zeijdner, E.E., Krul, C.A., Verwei, M. and-Wortelboer, H. M., 2014. A new approach to predict human intestinal absorption using porcine intestinal tissue and biorelevant matrices. European Journal of Pharmaceutical Sciences 63: 167-177. https://doi.org/10.1016/j.ejps.2014.07.003.

Yiannikouris, A., Apajalahti, J., Kettunen, H., Ojanpera, S., Bell, A.N.W., Keegan, J. D. and Moran, C.A., 2021. Efficient Aflatoxin B_1 sequestration by yeast cell wall extract and hydrated sodium calcium aluminosilicate evaluated using a multimodal *in-vitro* and *ex-vivo* methodology. Toxins 13: 24. https://doi.org/10.3390/toxins13010024.

Zhou, H., George, S., Li, C., Gurusamy, S., Sun, X., Gong, Z. and Qian, H., 2017. Combined toxicity of prevalent mycotoxins studied in fish cell line and zebrafish larvae revealed that type of interactions is dose-dependent. Aquatic Toxicology 193: 60-71. https://doi.org/10.1016/j.aquatox.2017.09.030.

Zimmermann, M., Zimmermann-Kogadeeva, M., Wegmann, R. and Goodman, A.L., 2019. Separating host and microbiome contributions to drug pharmacokinetics and toxicity. Science 363: eaat9931. https://doi.org/10.1126/science.aat9931.

CHAPTER 14

In vitro models to evaluate the effects of mycotoxins on reproductive function

A. Mastrorocco[1], M.E. Dell'Aquila[1] and B.A.J. Roelen[2]*

[1]Department of Biosciences, Biotechnologies & Biopharmaceutics, University of Bari Aldo Moro, Bari, Italy
[2]Embryology, Anatomy and Physiology, Department Clinical Sciences, Faculty of Veterinary Medicine, Utrecht University, Utrecht, the Netherlands
**correspondence: B.A.J.Roelen@uu.nl*

1 Introduction

Reproduction in vertebrates includes a complex series of events that involves a large number of cell types, tissues and organs and spans a considerable time interval, which can last several decades, depending on the species. Understanding if and how the exposure to adverse environmental factors, such as mycotoxins, may affect the reproductive success is complex and difficult due to the multiple organs and tissues involved, different modes of toxic action and dependence on the endocrine system (Lorenzetti et al., 2011). Most importantly, reproductive toxicity testing, using laboratory animals, can require up to 3,200 animals per substance in 2-generation studies. Alternatives, such as *in vitro* studies, are therefore needed according to the 3Rs principle of reduction, refinement and replacement of animals used in scientific research, saving considerable time and resources, while preserving the ability to detect relevant reproductive risks (Jain et al., 2018). *In vitro* studies allow the dissection of and focusing on specific aspects of the reproductive system.

Understanding which combinations of different mycotoxins and at which concentrations they can disturb reproduction, in particular germ cell maturation and early embryonic development, can be even more complicated. Obviously, in humans, the effects of mycotoxins on reproductive function are mostly determined retrospectively from large cohort studies where the challenges are establishing a mycotoxin-poor or -free control cohort and ruling out confounding factors (Batista da Silva et al., 2021, and references therein). Efforts are also aimed at determining the concentrations available in blood and reproductive organs. In addition, and this also accounts for most livestock animals, the time between conception and germ cell maturation is several years or decades. Indeed, when a pregnant mother is exposed to (myco)toxins, possible effects can be present at least up to the second generation, since germ cells are already present in the developing foetus.

 | DOI: 10.1163/9789004712720_15

Alternative strategies for reproductive toxicology studies are therefore urgently needed. To date, animal models with a short reproductive timespan, such as mice and rats, have mostly been used. However, the use of these models has some disadvantages. Firstly, inter-species variances reduce the translatability of the results from one species to another. Indeed, there are no animal models that can perfectly predict reproductive toxicity in humans. Secondly, the use of laboratory animals is discouraged, particularly when alternatives exist or can be developed. However, it has been estimated that around 70% of all laboratory animals used in toxicity testing have been used to determine effects on reproduction (Spielmann, 2009). Most studies using laboratory animals have been focused on teratological effects of toxins, where timing of foetal development and exposure plays a crucial role, while reduced fertility has been less well studied (Spielmann, 2009).

2 Mammalian reproduction

The reproductive process can be divided into sequential events. During early development, around the time of gastrulation, some cells are allocated by inductive signals from other cells to become germ cells (Lawson et al., 1999). These primordial germ cells (PGCs) migrate to the developing gonads and, after reaching them, differentiate to either oocytes or sperm cells (McLaren, 1998; Saitou, 2021). After differentiation, the germ cells need to undergo maturation, through meiosis, to generate haploid germ cells (Hyttel, 2011; Sanchez and Smitz, 2012). The somatic tissue of the gonads is needed to support germ cell differentiation and maturation. After gamete release, fertilisation occurs by cell fusion and a diploid zygote is formed. The embryo undergoes a series of cleavage divisions by which it grows in cell number but not in size. At this stage, with species-specific timing, the zygotic genome is switched on and the embryo is no longer dependent on maternal mRNA. In mouse embryos, zygotic genome activation (ZGA) initiates at the 2-cell stage while in bovine embryos, for instance, this occurs at the 8-cell stage (Madeja et al., 2019). While undergoing cleavage divisions, the embryo passively travels through the oviduct in the direction of the uterus. The first lineage segregations occur and, when the embryo has reached the uterus, it is at the blastocyst stage with a pluripotent inner cell mass (ICM) and an outer layer of trophectoderm cells that are needed for implantation and placenta formation in eutherians.

3 *In vitro* models

In vitro models have been developed to study reproductive toxicity. Most commonly used *in vitro* systems make use of cell lines, such as granulosa cell (GC) lines, *in vitro* fertilisation (IVF) and *in vitro* embryo culture, but embryo models from stem cells and on-chip technologies are now being developed. In combination with -omics technology, these *in vitro* models may become very powerful.

3.1 In vitro *fertilisation models*

In rodents, particularly mice and rats, *in vitro* oocyte development, IVF and *in vitro* embryo culture are usually performed after superovulation of the female animals using hormone cocktails. The animals are subsequently euthanised and the oviducts removed to retrieve the ovulated oocytes, which are then already at the metaphase II (MII) stage. Oocytes can be *in vitro* fertilised and embryos cultured *in vitro* up to the blastocyst stage. Alternatively, the females are allowed to mate with males followed by removal of the oviduct and collection of the embryos at the desired stages. Particularly with mouse models, a combination of *in vitro* and *in vivo* embryo development has added advantages. Typically, mouse embryos can be derived from fertilised dams after euthanasia, and preimplantation embryos cultured *in vitro* for distinct periods of time and developmental stages up until the blastocyst stage in the presence of mycotoxins. The use of superovulation techniques by injecting the female animals before fertilisation with pregnant mare serum gonadotropin (PMSG) and human chorionic gonadotropin (hCG) allows for the ovulation of up to 40 oocytes per mouse depending on the animal strain (Luo et al., 2011). *In vitro* culture as opposed to *in vivo* development permits exposure at exact concentrations of mycotoxins, which would be difficult to achieve *in vivo*. This is only useful, however, when *in vivo* concentrations in the ovary or oviduct are known or can by calculated after ingestions of mycotoxin-contaminated food. In mice *in vitro* cultured preimplantation embryos can also be transferred to pseudo-pregnant recipient female mice to establish implantation capacity and, for instance, placental and foetal growth. These types of hybrid *in vitro*/*in vivo* experiments help to unravel if and how mycotoxins affect specific aspects of early development.

Oocytes can also be retrieved from large mammals, such as the cow, by ultrasound-guided transvaginal aspiration, without the need to euthanise the animal (Merton et al., 2013). This procedure is relatively time consuming and is subject to ethical approval. Large quantities of oocytes can however be obtained from animals that are being commercially culled. The ovaries of farm animals that are being reared for meat and milk production (in particular cattle, pigs and sheep) are usually discarded after slaughter. However, these ovaries still contain numerous healthy immature oocytes within their follicles, and when removed within a few hours after slaughter, these oocytes, as cumulus-oocyte complexes (COCs), can be *in vitro* matured to the MII stage, *in vitro* fertilised using fresh (pig) or frozen/thawed (cattle, sheep) sperm and the embryos can be cultured *in vitro* until the blastocyst stage. The maturation of oocytes to the MII stage can be performed by *in vitro* culture of the COCs for around 24 hours (bovine and ovine) or 48 hours (porcine) in specific culture media (Gandolfi and Brevini, 2010; Roelen, 2019) (Figure 14.1).

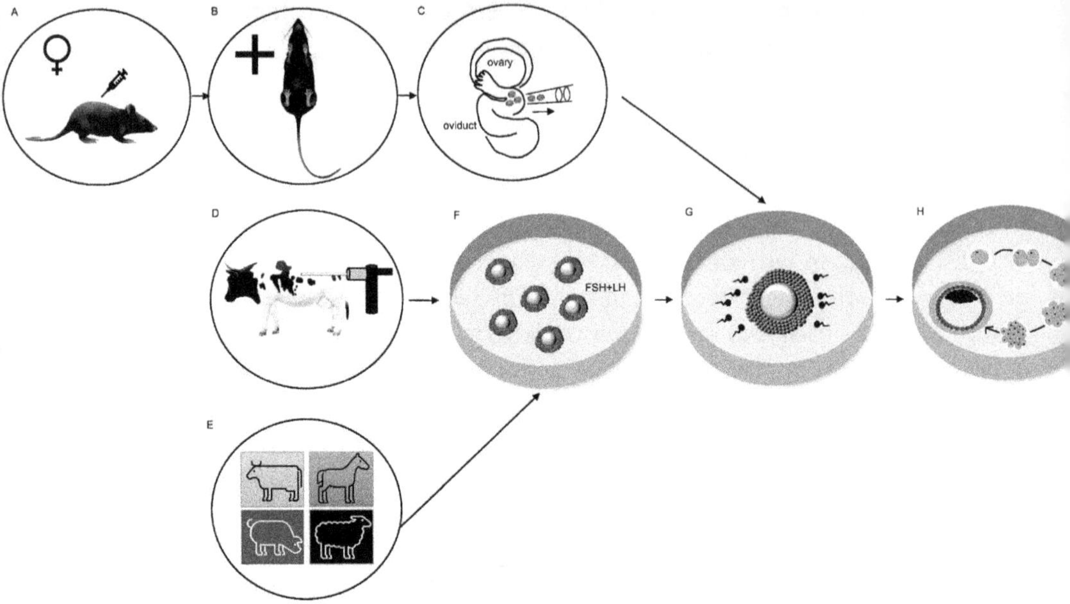

FIGURE 14.1 Flow chart showing the steps of the *in vitro* embryo production process. Metaphase II stage (MII) oocytes be obtained either following superovulation treatment (A) and euthanasia of the animal (B-C) or by u sound-guided transvaginal aspiration (D) or from the ovaries of slaughtered animals (E), subsequent *in* maturation (IVM; F), *in vitro* fertilisation (IVF; G) and *in vitro* embryo culture (H).

For bovine oocytes, the cumulus oophorus is essential at least for the first 8 hours of *in vitro* culture. Porcine oocytes derived from sow ovaries are most competent to mature and be *in vitro* fertilised, while oocytes retrieved from gilt ovaries are less competent and are more prone to polyspermic fertilisation. Porcine *in vitro* oocyte maturation (IVM) of gilt oocytes can be enhanced however when sow follicular fluid is included in the IVM medium (Schoevers et al., 2020). Since prepubertal large animals (e.g. gilts, lambs) are culled in relatively high numbers for their meat, ovaries from these animals can be available for research on impact of mycotoxin exposure. By definition, these oocytes have been exposed to much lower concentrations of reproductive hormones allowing for an investigation into the effects of exposure to immature oocytes. There are multiple reasons for the interest in using prepubertal animals as oocyte donors with applications for human and veterinary aspects. First, the ovarian pool of available oocytes is vast; prepubertal animals consistently yield large numbers of COCs compared to their adult counterparts (Tervit, 1996). Moreover, in some regions, the ovaries derived from prepubertal animals represent a more widely available biological material. Young animals and children are more sensitive to the adverse effects of mycotoxins than adults because of higher metabolic rate, underdeveloped organ functions, detoxication mechanisms, their smaller size relative to the amount of food consumed and due to the amounts of oat-, wheat-, and other cereal-based foods, an important contributor to dietary mycotoxins exposure, relative to other age groups (Palma et al., 1993; Peraica et al., 2014; Leoni et al., 2015). Moreover, prepubertal lamb oocytes are of relevant scientific interest as models of oocytes recovered at paediatric and adolescent age in humans (Torres Rovira et al., 2016). Unlike in the mouse, hybrid *in vitro/in vivo* experiments are less feasible in large animals.

3.2 *Primary cells and cell lines*

Primary cells derived from ovarian follicles, such as GC, but also endometrial cells can be cultured *in vitro* for several passages before they become senescent. Exposure of these cells to mycotoxins can provide useful information on reproductive toxicity, since GC and cumulus cells (CCs) that surround the oocyte are crucial for the oocyte to obtain developmental capacity. In particular, the manner in which hormone production and viability of GC are affected by mycotoxin exposure can exemplify mechanisms of reproductive toxicity. Exposure of endometrial cells may provide clues to the effects of mycotoxin exposure on embryo receptivity.

Embryonic stem cells (ESCs) are representative for the ICM of blastocyst stage embryos. ESC lines were already derived from mouse and human embryos decades ago, and cultured under the right conditions, these cells do not differentiate but maintain pluripotent and can self-renew (Pereira Daoud, 2020 and references therein). More recently well-characterised ESC lines have been derived from pig and cattle (Bogliotti et al., 2018; Kinoshita et al., 2021). Induced pluripotent stem cells (iPSCs) are generated from differentiated somatic cells by *in vitro* nuclear reprogramming using forced expression of a specific set of pluripotency genes (Takahashi et al., 2007). These iPSCs are very similar to ESCs and can therefore also be used as a model for embryonic cells (Figure 14.2).

For reproductive toxicity studies, ESCs and iPSCs cells can be useful to determine toxicity at early embryonic stages. Indeed, murine ESCs can be used in a so-called embryonic stem cell test (EST) to determine reproductive and developmental toxicity, and this test has been approved by the European Centre for the Validation of Alternative Methods (ECVAM) (Genschow et al., 2004). Because of their pluripotent character, after differentiation to form specialised tissues or more complex structures that resemble organs, so-called *organoids*, ESCs and iPSCs can be used to study toxicity for different tissues separately. Particularly iPSCs are useful as these can be derived from basically any differentiated cell type (Takahashi et al., 2007). Nuclear reprogramming of differentiated cells to iPSCs is common practice using rodent and human primary cells, and in the past few years bona fide iPSCs have been described for other species including livestock species (Yoshimatsu et al., 2021). In the mouse, ESCs and iPSCs have been differentiated *in vitro* to functional germ cells, both oocytes and sperm cells (Hayashi, 2019). This would allow an examination of how mycotoxins affect germ cell development. Functional germ cells have however only been developed from mouse pluripotent stem cells, although

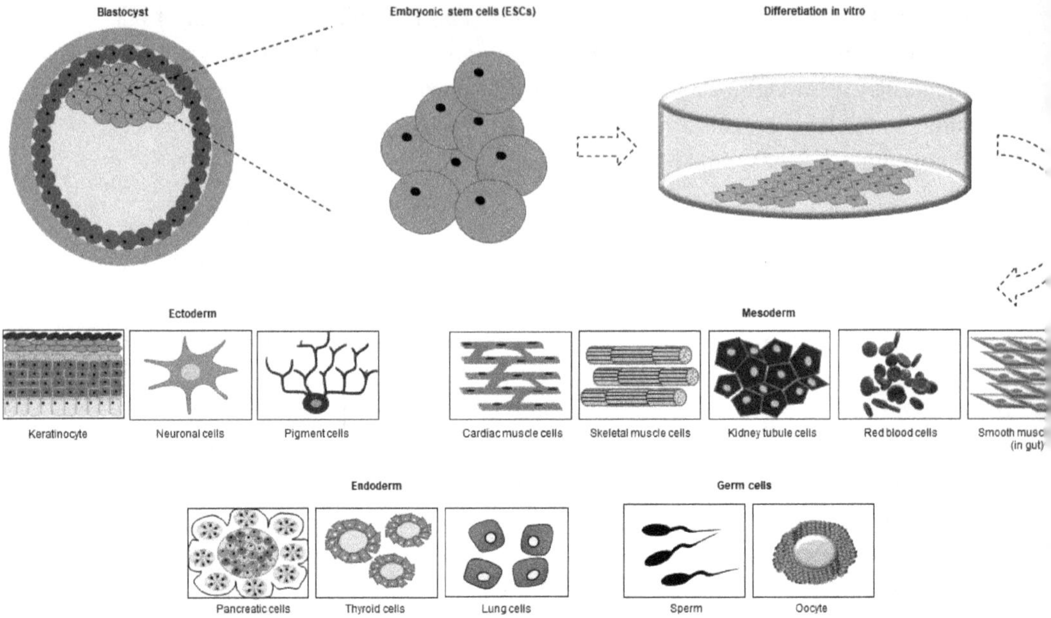

FIGURE 14.2 Representative images of embryonic stem cell differentiation *in vitro*. Isolated embryonic stem cells from inner cell mass of the blastocyst are cultured in a pluripotent state. The pluripotent stem cells can be diff- tiated into various cell lineages, such as cardiomyocytes, neurons, or liver cells.

PGC-like cells have been developed from human iPSCs (Saitou and Hayashi, 2021), so using this system to understand mycotoxin sensitivity for germ cell development is currently not practical. Validation to establish whether ESCs and iPSCs can replace animal models is important. Because of the accessibility and relative ease of culture, (stem) cell lines seem to be particularly promising for unravelling the modes of action of mycotoxins.

3.3 *Embryo models*

Although the development of mammalian embryos appears morphologically very similar, there are many important differences between different species, particularly in timing and activated signalling pathways (Rossant and Tam, 2018). For instance, the Hippo signalling pathway is important for specification of the trophectoderm in preimplantation embryos of different species (Gerri et al., 2020), but there are important differences (Yu et al., 2021). While FGF signalling is important for the formation of primitive endoderm in the mouse (Yamanaka et al., 2010,) it does not seem to be important in human embryos (Kuijk et al., 2012; Roode et al., 2012). Pluripotent stem cells such as ESCs and iPSCs can model the ICM of blastocyst stage embryos, and possibly the reprogramming of somatic cells can facilitate the differentiation towards trophectoderm stem cells and extraembryonic endoderm cells as well (Benchetrit et al., 2019). Trophectoderm stem cells can be assembled with pluripotent stem cells into 3D aggregates that will lead to the formation of blastocyst-like structures, termed blastoids (Rivron et al., 2019; Kagawa et al., 2022). Importantly, such embryo-like structures would provide a platform for reproductive toxicity testing without the need to use laboratory animals, and, when appropriate stem cells are available, in principle species-specific blastoids can be assembled. For human blastoids, depending on their developmental potential, legal, regulatory and ethical concerns may arise (Pereira Daoud et al., 2020).

Although promising, it has to be established how far these blastoids resemble embryos in terms of signalling pathways, developmental capacity, and response to toxic agents, such as mycotoxins.

4 *In vitro* models and mycotoxin exposure

4.2 *Deoxynivalenol*

DON has been found to reduce porcine oocyte maturation already at levels as low as 0.2 μmol/l. Although this inhibition of maturation was observed in oocytes from both prepubertal gilts and sows, the inhibition was more severe with oocytes from gilts (Schoevers et al., 2010; 2021) and may be dependent on the porcine breed (Alm et al., 2002). Of the DON-exposed oocytes that reached the MII stage and were fertilised, a higher percentage showed polyspermic fertilisation, and a lower percentage developed to blastocyst stage. The observation that a higher percentage of blastomeres was aneuploid after DON exposure is possibly related to polyspermic fertilisation, or due to abnormal meiotic spindle formation (Schoevers et al., 2010). Expansion of CCs surrounding the oocyte is important for oocytes to mature. In the presence of 0.02 μmol/l DON, cumulus expansion was severely compromised. Expansion was even absent in the presence of 2 μmol/l DON, partly because of CC apoptosis (Schoevers et al., 2010). In cultured porcine GCs, exposure to 3.4 μmol/l DON led to reduced insulin-like growth factor 1 (IGF-1) secretion by these cells. Interestingly, progesterone secretion was increased by DON at 3.4 μmol/l (Medvedova, et al., 2011). Cultured porcine endometrial cells were growth-inhibited by 3.8 μmol/l DON, but not significantly at 1.88 μmol/l, while at this concentration a significant decrease in cells at the S-phase was noted (Tiemann et al., 2003).

Mouse GCs exhibited reduced proliferation after 24 hours exposure to 2 μmol/l DON. An increased inflammatory response was also noted in these cells, due to activation of the MAP kinase and NF-κB signalling pathways (Fan et al., 2021). The only study available to date on the effects of combined exposure of DON and FB_1, the most common fusariotoxin, showed a greater inhibitory effect of DON and FB_1 on the proliferation of porcine GCs compared with the single effect of each individual mycotoxin (Gerez et al., 2021).

Melatonin supplementation *in vitro* was shown to alleviate the mouse and bovine oocyte meiotic arrest caused by DON (Tian et al., 2014; Lan et al., 2018) and protect ovarian GCs from the adverse effects of DON by ameliorating oxidative stress-induced apoptosis and the inflammatory response (Fan et al., 2021; Xue et al., 2021).

4.3 *T-2 toxin*

It has been reported that 100 nmol/l T-2 toxin affects the maturation of porcine oocytes and embryo development by altering the epigenetic modifications of oocytes (Zhang et al., 2016, 2017). T-2 toxin affected maturation of mouse oocytes from 20 nmol/l onwards (Zhu et al., 2016) and caused developmental slowdown or arrest in embryo development already at 0.5 ng/mL, resulting in decreased blastomere numbers and late blastocoel formation (Somoskoi et al., 2016a; 2016b). Co-contamination of T-2 with FB_1 resulted in a significantly lower blastocyst formation rate and disrupted blastocoel expansion of *in vitro* mouse embryo (Somoskoi et al., 2018).

T-2 toxin exposure induces apoptosis through oxidative stress and inhibition of steroidogenesis in rat ovarian GCs (Wu et al., 2013, 2015). T-2 toxin inhibited mouse ESC proliferation already from 0.21 μmol/l onwards (Fang et al., 2012). Interestingly, differentiation to cardiac muscle cells was also inhibited in the presence of T-2 toxin, demonstrating the versatility of ESCs in studying toxicity to specific differentiated cells (Fang et al., 2012).

4.4 *Zearalenone*

The mycotoxin ZEN, and its metabolites α-ZEL and β-ZEL, can cause reproductive disorders that have in the first instance been related to reproductive tract and mammary gland disorders (Haschek and Haliburton, 1986; Minervini et al., 2001; Minervini and Dell'Aquila, 2008). Pigs appear to be particularly sensitive to ZEN exposure exhibiting abnormal vulva, ovaries, uterus, testes and pregnancy loss depending on the sex and time of exposure (Haschek and Haliburton, 1986; Osweiler, 1986), while ruminants are less sensitive to ZEN toxicity (Osweiler, 1986).

Exposure of porcine oocytes to ZEN already inhibited *in vitro* oocyte maturation at 1.88 μmol/l, while *in vitro* embryo development of *in vivo*-derived zygotes was reduced after exposure to α-ZEL at concentrations of 7.5 μmol/l and higher (Alm et al., 2002). *In vitro* porcine oocyte maturation in the presence of ZEN is characterised by disruption of oocyte meiosis and failed polar body extrusion, and impaired functions of oocyte mitochondria, Golgi apparatus, endoplasmic reticulum (ER), cytoskeletal dynamics and lysosomes (Sambuu et al., 2011; Hou et al., 2015; Xu et al., 2019; Wang et al., 2022). Moreover, ZEN effects on oocyte maturation also appear to be related to ZEN-induced arrest of cell cycle progression and apoptosis of GCs (Liu et al., 2018; Zhang et al., 2018, 2019; Li et al., 2020). There are species-specific differences in the ZEN-induced GC toxicity and porcine GCs are more sensitive to ZEN than murine GCs (Zhang et al., 2018). In bovine oocytes, *in vitro* meiotic maturation was inhibited by ZEN at concentrations of 9.4 μmol/l and higher (Minervini et al., 2001), indicating that in general mammalian oocyte meiotic maturation is disturbed by ZEN exposure. Moreover, bovine oocytes supplemented with 1000 µg/l ZEN showed higher MI and lower MII rates after 24 hours, without affecting IVF and embryo cleavage rates, indicating a slowing down of meiotic progression (Takagi et al., 2008). However, it has been reported that porcine blastocysts derived from ZEN-exposed oocytes contained significantly more aneuploid blastomeres (Malekinejad et al., 2007). In addition, these results indicate a higher sensitivity of porcine oocytes. Because ZEN and its metabolites can bind and activate oestrogen receptors, ZEN-induced inhibition of oocyte maturation is likely to result from the estradiol agonistic activity of ZEN (Kuiper et al., 1998).

Since ZEN is produced by *Fusarium* moulds, it is likely that food and feed contamination by *Fusarium* species leads to exposure to a combination of mycotoxins, for instance ZEN and DON. Combined exposure can be relatively easily examined using IVF models. Although in cultured porcine oocytes both ZEN and DON exposure led to inhibition of oocyte maturation, a combined exposure did not lead to additional abnormalities (Malekinejad et al., 2007).

Mouse and human ESC lines have been examined for toxicity to ZEN and the results indicate that ESCs are proper alternatives for embryonic toxicity testing (Cao et al. 2019). ZEN exposure of 7.85 µmol/l led to decreased viability of both mouse and human ESCs. Although in oocytes ZEN toxicity is mainly due to its oestrogenic activities, in ESCs, ZEN exposure led to increased ROS accumulation. Indeed, simultaneous exposure of human ESCs to the antioxidant Trolox, a water-soluble vitamin E derivative, could mitigate ZEN-induced apoptosis (Cao et al., 2019).

4.5 *Ochratoxin A*

Among ochratoxins, the mycotoxin OTA is the most prevalent one and generally more toxic than other ochratoxins because of its late metabolism (Gautier et al., 2001). *In vitro* exposure studies helped explain a direct and marked damage caused by OTA to the female gamete which affects its developmental competence. At 10 µmol/l, OTA significantly decreased the *in vitro* mouse oocyte maturation rate and ability to be fertilised. In addition, the embryo cleavage rate to the 2-cell stage and development to the blastocyst stage were inhibited after OTA exposure. Moreover, blastocysts derived from OTA-exposed oocytes showed decreased numbers of ICM and trophoblast cells with an increase in the concentration of apoptotic cells (Huang and Chan, 2016; Huang et al., 2019). A comparative study between mouse and porcine oocytes confirmed reduced *in vitro* oocyte maturation rates in both species. It has been suggested that porcine oocytes are more sensitive to OTA than mouse oocytes; porcine oocytes were already affected after 5 µmol/l OTA while the first effects in mouse oocytes were observed at an OTA concentration of 300 µmol/l (Lu et al., 2018; Lan et al., 2020). Other authors however demonstrated impaired IVM rate of mouse oocytes and reduced cleavage rates after IVF already at 5 µmol/l OTA. The reduced quality of the oocytes was predominantly caused by disrupted meiotic spindle formation, oxidative damage and epigenetic modifications (Jia et al., 2020). The inhibition of oocyte meiosis by OTA is partly due to a reduced cumulus expansion in the presence of OTA (Lan et al., 2020; Park et al., 2022). Indeed, when porcine GCs were cultured as a monolayer, cells were arrested at the G2/M phase of the cell cycle and as a result the proliferation rate decreased with 20 μmol/l OTA (Zhang et al., 2021). Interestingly, melatonin, a known antioxidant, can partly ameliorate the increased oxidative stress induced by OTA in porcine oocytes (Lan et al., 2020). Although *in vivo* combined exposure to compounds will be more complex, these types of *in vitro* experiments can help to clarify the mechanisms of reproductive toxicity. In juvenile sheep, OTA has also been demonstrated to reduce cumulus viability and impair oocyte maturation at micromolar concentrations. Interestingly, at nanomolar concentrations, although cumulus expansion and polar body extrusion rates of *in vitro* cultured COCs were not affected, a higher percentage of apoptotic cells was observed after fertilisation in the resulting blastocysts (Dell'Aquila et al., 2021).

OTA exposure to human ESCs, from 12.35 μmol/l onwards, led to decreased survival, increased Caspase-mediated apoptosis, increased levels of necrosis and increased expression of oxidative stress markers (Erceg et al., 2019). When mouse ESCs were cultured in the presence of OTA, differentially expressed proteins were classified as predominantly associated with embryotoxicity (Groebe et al., 2010).

4.6 *Citrinin*

CIT exposure has been particularly associated with reproductive defects (de Oliveira Filho et al., 2017). Few studies have been directed to the *in vitro* culture of oocytes or embryos in the presence of CIT. Mouse oocytes appear more sensitive to CIT exposure than porcine embryos, as concentrations as low 30 μmol/l already halved the percentage of mouse oocytes that extruded the first polar body (Sun et al., 2020), while in porcine oocytes similar effects only became evident at 80 μmol/l (Jiang et al., 2022). The use of mouse oocytes has been helpful to determine which biological mechanisms are vulnerable to CIT, and particularly the distribution of mitochondria, ER and the Golgi apparatus seem disturbed after CIT exposure (Sun et al., 2020). Indeed, abnormal mitochondrial functioning and ER stress caused by CIT reduced oocyte quality. Pig oocytes appear less sensitive to CIT and in these animals, the cytoskeleton was particularly affected by CIT exposure leading to meiotic spindle defects (Jian et al., 2022). Mouse embryos exposed to CIT during *in vitro* culture showed reduced blastomere numbers, particularly in the ICM of blastocysts. These reduced cell numbers were mainly due to apoptosis, not to reduced cell division rates or timing (Chan and Shiao, 2007). Even though the trophectoderm seemed less affected, CIT exposed blastocysts were less efficient in uterine implantation compared with control embryos after transfer to control mice (Jiang and Shiao, 2007). Importantly, these experiments can discriminate between maternal and embryonic effects, and will be helpful in determining the most sensitive time period in the reproductive cycle.

4.7 *Beauvericin*

BEA is an emerging mycotoxin whose effects on reproduction have only recently been described (Santos et al., 2015). Indeed, little information is available on its potential reproductive and developmental toxicity despite the fact that BEA is frequently present in food and feed (Caloni et al., 2020). *In vitro* studies have indicated that BEA may directly affect reproductive cell viability and function. In porcine GCs, BEA reduced viability after exposure to 2.5 µmol/L BEA and inhibited steroidogenesis at 0.6 µmol/l (Santos at al., 2015). It seems likely that inhibition of GC steroidogenesis is a reflection of the quality of the oocytes enclosed by these cells in pig. Indeed, CC viability was reduced after 44 hours exposure to 5 and 10 µmol/L of BEA. In addition, BEA at a concentration of 10 µmol/l reduced cumulus expansion and plasma membrane integrity. In line with the decreased viability, 10 µmol/L BEA interfered with progesterone synthesis in CCs. Abnormal oocyte meiotic progression and impaired oocyte developmental competence were observed at concentrations exceeding 2.5 µmol/l (Schoevers et al., 2016). When maturing oocytes and developing embryos were continuously exposed to BEA, lower percentages of blastocyst stage embryos were obtained at concentrations of 1 µmol/l. When oocytes were matured *in vitro* in regular medium and embryos exposed to 1 µmol/l BEA during cleavage division, blastocyst formation was similarly reduced (Schoevers et al., 2016). These results indicate that maturing porcine oocytes are less sensitive to BEA than developing embryos. BEA sensibility in pigs also depends on donor age. Oocytes from gilts are more sensitive than those from sows when exposed to BEA during IVM. Indeed, gilt oocytes showed a decreased rate of nuclear maturation at concentrations as low as 0.5 µmol/l, while at this concentration, BEA did not affect the maturation of sow oocytes (Schoevers et al., 2021). BEA *in vitro* exposure also inhibited bovine GC proliferation at 10 µmol/l and a reduction in their oestrogenic activity was observed from 3 µmol/l onwards (Albonico et al., 2017). *In vitro* maturing oocytes from sheep exposed to BEA showed various toxic effects at 0.5 to 5 µmol/l. At the highest tested concentration (5 µmol/l), BEA showed a rapid effect on CCs and oocytes, with increased CC apoptosis and DNA fragmentation, impairment of oocyte maturation and abnormal mitochondrial distribution patterns in matured oocytes. At lower tested concentrations (3–0.5 µmol/l), BEA did not affect oocyte maturation, but matured oocytes showed reduced mitochondrial activity with long-term carryover effects on embryo developmental competence and blastocyst quality after IVF and *in vitro* embryo culture (Mastrorocco et al., 2019). The involvement of mitochondria, as a possible BEA target in ovine oocytes was demonstrated by reduced expression of nuclear and mitochondrial genes coding for key proteins of the mitochondrial chain in ovine COCs, with a greater impact on CCs rather than on the oocyte (Mastrorocco et al., 2020). Overall, these studies show a different sensitivity to BEA of CCs, oocytes and embryos probably linked to different mechanisms of action that contribute to the toxic effects. Moreover, a species-specific effect of BEA *in vitro* with greater sensitivity among pigs and sheep compared to cattle could be noted.

The results described above reflect exposure to a single mycotoxin species, but given the production of BEA by *Fusarium* fungi, combinatorial effects of BEA with other mycotoxins produced by these fungi are likely to occur. The only combinatorial study currently available analysed the effects of BEA in combination with FB_1. BEA was found to inhibit oestradiol production by GCs more potently with respect to FB_1 that did not influence BEA induced effects (Albonico et al., 2017).

5 Conclusions

Although *in vitro* studies can be helpful in determining possible reproductive toxicity of mycotoxins, there are clear limitations that should be taken into account when interpreting data (Figure 14.3). *In vivo*, mycotoxins are mainly introduced to the animal via feed. Depending on the animal species, the toxins may be metabolised in the digestive system or secreted via the kidneys without coming into contact with the reproductive system. *In vitro* exposures are obviously very different from *in vivo* exposures since ADME are largely absent in the *in vitro* tests. In addition, some adverse effects can only be observed in the generation next after parental crossing (Spielmann, 2009). The effects on the developing organism depend on the compound and/or its active metabolite(s) concentration and time of exposure in the target cells (Andrews et al., 1995). None of the tests available to date represent the entire spectrum of developmental events and embryotoxic mechanisms, and the metabolising capacity of embryonic cells, tissues, and organs is limited (Spielman, 2005). Sensitivity of the foetus to mycotoxins after implantation into the womb can currently not be studied *in vitro*, although artificial wombs are being developed (Partridge et al., 2017).

By definition, exposing germ cells, embryos or stem cells to mycotoxins will lead to cellular toxicity. What is critical is to determine the concentration at which toxic events occur, and to establish whether *in vivo* reproductive cells or embryos can be exposed to these concentrations. Therefore, with *in vitro* models it is essential that mycotoxin levels in organs and serum are established, or to know how these levels change after consumption of mycotoxin-contaminated food or feed. Moreover, it is important to understand how to interpret the findings in terms of toxicity and so at what level of response the observed effects reach the level of adversity. Many beneficial responses intervene physiologically to counteract the hazard and therefore not all detectable *in vitro* responses should be characterised as toxic or adverse. Furthermore, exposure levels over time may be largely stable *in vitro* whereas *in vivo* peak exposures may occur, complicating comparison (Piersma, 2013).

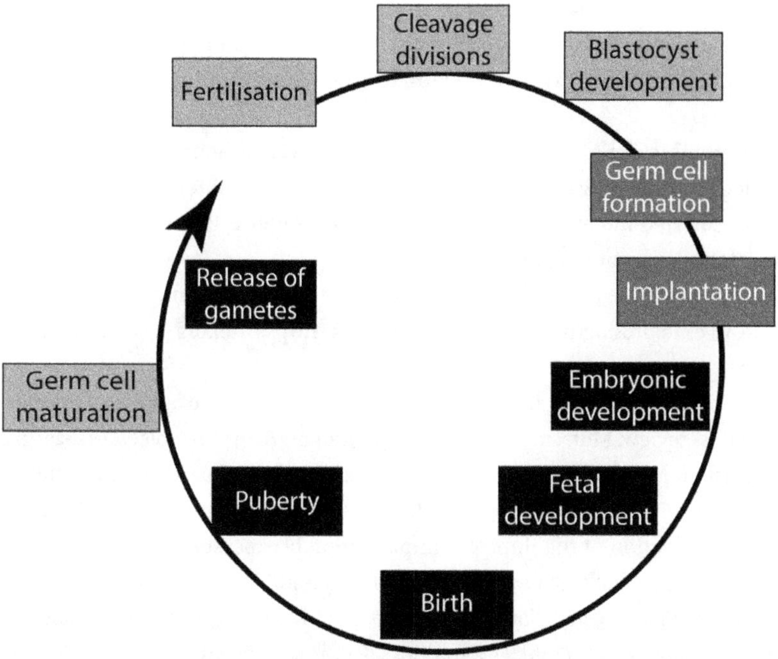

FIGURE 14.3 Schematic representation of the germ-cell production cycle from fertilisation until release of gametes. Events in light grey boxes with black letters can be reproduced *in vitro*, in the presence or absence of mycotoxins. Events in darker grey boxes with white letters may in the near future be studied *in vitro*, also depending on the animal species. Events in black boxes with white letters are difficult to recreate, and unlikely to be modelled *in vitro*.

Since different mycotoxins can be produced by the same fungal species, it is most likely that there will be exposure to a cocktail of different mycotoxins. While *in vitro* studies can decipher specific effects of individual toxins, indicating reproductive toxicity by a combination of mycotoxins can only be realistically done when the serum can be measured or approximated. Lastly, toxic effects on other organs can have a subsequent impact on the reproductive system. Although currently it does not appear possible to study this *in vitro*, organ-on-chip technology is potentially useful to study interactions between toxic exposed organs *in vitro* (Nitsche et al., 2022).

References

Albonico, M., Schutz, L. F., Caloni, F., Cortinovis, C. and Spicer, L. J., 2017. *In vitro* effects of the *Fusarium* mycotoxins fumonisin B1 and beauvericin on bovine granulosa cell proliferation and steroid production. Toxicon 128: 38–45. https://doi.org/10.1016/j.toxicon.2017.01.019.

Alm, H., Greising T., Brüssow, K.P., Torner, H. and Tiemann, U., 2002. The influence of the mycotoxins deoxynivalenol and zearalenol on *in vitro* maturation of pig oocytes and *in vitro* culture of pig zygotes. Toxicology *In Vitro* 16: 643-648. https://doi.org/10.1016/s0887-2333(02)00059-0.

Andrews, J.E., Ebron-McCoy, M., Bojic, U., Nau, H. and Kavlock, R.J., 1995. Validation of an *in vitro* teratology system using chiral substances: stereoselective teratogenicity of 4-yn-valproic acid in cultured mouse embryos. Toxicology and Applied Pharmacology 132: 310-316. https://doi.org/10.1006/taap.1995.1112.

Chan, W.H. and Shiao, N.H., 2007. Effect of citrinin on mouse embryonic development *in vitro* and *in vivo*. Reproductive Toxicology 2: 120-125. https://doi.org/10.1016/j.reprotox.2007.04.070.

da Silva, J.V.B., de Oliveira, C.A.F. and Ramalho, L.N.Z., 2021. Effects of prenatal exposure to Aflatoxin B1: a review. Molecules 26: 7312. https://doi.org/10.3390/molecules26237312.

Benchetrit, H., Jaber, M., Zayat, V., Sebban, S., Pushett, A., Makedonski, K., Zakheim, Z., Radwan, A., Maoz, N., Lasry, R., Renous, N., Inbar, M., Ram, O., Kaplan, T. and Buganim, Y., 2019. Direct induction of the three pre-implantation blastocyst cell types from fibroblasts. Cell Stem Cell 24: 983-994.e7. https://doi.org/10.1016/j.stem.2019.03.018.

Bogliotti, Y.S., Wu, J., Vilarino, M., Okamura, D., Soto, D.A., Zhong, C., Sakurai, M., Sampaio, R.V., Suzuki, K., Izpisua Belmonte, J.C. and Ross, P.J., 2018. Efficient derivation of stable primed pluripotent embryonic stem cells from bovine blastocysts. Proceedings of the National Academy of Sciences of the United States of America 115: 2090-2095. https://doi.org/10.1073/pnas.1716161115.

Bremer, S., Pellizzer, C., Hoffmann, S., Seidle, T. and Hartung, T., 2007. The development of new concepts for assessing reproductive toxicity applicable to large scale toxicological programmes. Current Pharmaceutical Design 13: 3047-3058. https://doi.org/10.2174/138161207782110462.

Brown, E.S., Jacobs, A. and Fitzpatrick, S., 2012. Reproductive and developmental toxicity testing: from *in vivo* to *in vitro*. ALTEX 29: 333-339. https://doi.org/10.14573/altex.2012.3.333.

Caloni, F., Fossati, P., Anadón, A. and Bertero, A., 2020. Beauvericin: The beauty and the beast. Environmental Toxicology and Pharmacology 75: 103349. https://doi.org/10.1016/j.etap.2020.103349.

Cao, H., Zhi, Y., Xu, H., Fang, H. and Jia, X., 2019. Zearalenone causes embryotoxicity and induces oxidative stress and apoptosis in differentiated human embryonic stem cells. Toxicology *In Vitro* 54: 243-250. https://doi.org/10.1016/j.tiv.2018.09.020.

Dell'Aquila, M.E., Asif, S., Temerario, L., Mastrorocco, A., Marzano, G., Martino, N.A., Lacalandra, G.M., Roelen, B.A., Carluccio, A., Robbe, D. and Minervini, F., 2021. Ochratoxin A affects oocyte maturation and subsequent embryo developmental dynamics in the juvenile sheep model. Mycotoxin Research 37: 23-37. https://doi.org/10.1007/s12550-020-00410-y.

de Oliveira Filho, J.W.G., Islam, M.T., Ali, E.S., Uddin, S.J., Santos, J.V.O., de Alencar, M.V.O.B., Júnior, A.L.G., Paz, M.F.C.J., de Brito, M.D.R.M., E Sousa, J.M.C., Shaw, S., de Medeiros, M.D.G.F., Dantas, S.M.M.M., Rolim, H.M.L., Ferreira, P.M.P., Kamal, M.A., Pieczynska,

M.D., Das, N., Gupta, V.K., Mocan, A., Dos Santos Andrade, T.J.A., Singh, B.N., Mishra, S.K., Atanasov, A.G. and Melo-Cavalcante, A.A.C., 2017. A comprehensive review on biological properties of citrinin. Food and Chemical Toxicology 110: 130-141. https://doi.org/10.1016/j.fct.2017.10.002.

Erceg, S., Mateo, E.M., Zipancic, I., Rodríguez Jiménez, F.J., Pérez Aragó, M.A., Jiménez, M., Soria, J.M. and Garcia-Esparza, M.Á., 2019. Assessment of toxic effects of Ochratoxin A in human embryonic stem cells. Toxins 11: 217. https://doi.org/10.3390/toxins11040217.

Fan, H., Ren, Z., Xu, C., Wang, H., Wu, Z., Rehman, Z.U., Wu, S., Sun, M.A. and Bao, W., 2021. Chromatin accessibility and transcriptomic alterations in murine ovarian granulosa cells upon Deoxynivalenol exposure. Cells 10: 2818. https://doi.org/10.3390/cells10112818

Fan, H., Wang, S., Wang, H., Sun, M., Wu, S. and Bao, W., 2021. Melatonin ameliorates the toxicity induced by Deoxynivalenol in murine ovary granulosa cells by antioxidative and anti-inflammatory effects. Antioxidants 10: 1045. https://doi.org/10.3390/antiox10071045.

Fang, H., Wu, Y., Guo, J., Rong, J., Ma, L., Zhao, Z., Zuo, D. and Peng, S., 2012. T-2 toxin induces apoptosis in differentiated murine embryonic stem cells through reactive oxygen species-mediated mitochondrial pathway. Apoptosis 17: 895-907. https://doi.org/10.1007/s10495-012-0724-3.

Gandolfi, F. and Brevini, T.A., 2010. RFD Award Lecture 2009. *In vitro* maturation of farm animal oocytes: a useful tool for investigating the mechanisms leading to full-term development. Reproduction, Fertility and Development 22: 495-507. https://doi.org/10.1071/RD09151.

Gautier, J., Richoz, J., Welti, D.H., Markovic, J., Gremaud, E., Guengerich, F.P. and Turesky, R.J., 2001. Metabolism of ochratoxin A: absence of formation of genotoxic derivatives by human and rat enzymes. Chemical Research in Toxicology 14: 34-45. https://doi.org/10.1021/tx000070j.

Genschow, E., Spielmann, H., Scholz, G., Pohl, I., Seiler, A., Clemann, N., Bremer, S. and Becker, K., 2004. Validation of the embryonic stem cell test in the international ECVAM validation study on three *in vitro* embryotoxicity tests. Alternatives to Laboratory Animals 32: 209-244. https://doi.org/10.1177/026119290403200305.

Gerez, J.R., Camacho, T., Brunaldi Marutani, V.H., Nascimento de Matos, R.L., Hohmann, M.S., Verri Júnior, W.A. and Bracarense, A.P.F.R.L., 2021. Ovarian toxicity by fusariotoxins in pigs: Does it imply in oxidative stress? Theriogenology 165: 84-91. https://doi.org/10.1016/j.theriogenology.2021.02.003.

Gerri, C., McCarthy, A., Alanis-Lobato, G., Demtschenko, A., Bruneau, A., Loubersac, S., Fogarty, N.M.E., Hampshire, D., Elder, K., Snell, P., Christie, L., David, L., Van de Velde, H., Fouladi-Nashta, A.A. and Niakan, K.K., 2020. Initiation of a conserved trophectoderm program in human, cow and mouse embryos. Nature 587: 443-447. https://doi.org/10.1038/s41586-020-2759-x.

Groebe, K., Hayess, K., Klemm-Manns, M., Schwall, G., Wozny, W., Steemans, M., Peters, A.K., Sastri, C., Jaeckel, P., Stegmann, W., Zengerling, H., Schopf, R., Poznanovic, S., Stummann, T.C., Seiler, A., Spielmann, H. and Schrattenholz, A., 2010. Protein biomarkers for *in vitro* testing of embryotoxicity. Journal of Proteome Research 9: 5727-5738. https://doi.org/10.1021/pr100514e.

Haschek, W.M. and Haliburton, J.C., 1986. *Fusarium moniliforme* and zearalenone toxicoses in domestic animals: a review. In: Richards, J.L. and Thurston, J.R. (eds.) Diagnosis of Mycotoxicoses, 1st Edition. Martins Nijhoff, Dordrecht, p. 213-235. https://doi.org/10.1007/978-94-009-4235-6_20.

Hayashi, K., 2019. *In vitro* reconstitution of germ cell development. Biology of Reproduction 101: 567-578. https://doi.org/10.1093/biolre/ioz111.

Hyttel, P., 2011. Electron microscopy of mammalian oocyte development, maturation and fertilization. In: Tosti, E. and Boni, R. (eds) Oocyte Maturation and Fertilization: A Long History for a Short Event, Bentham Science Publishers, p. 1-37. https://doi.org/10.2174/978160805182311101 01.

Hou, Y.J., Zhu, C.C., Xu, Y.X., Cui, X.S., Kim, N.H. and Sun, S.C., 2015. Zearalenone exposure affects mouse oocyte meiotic maturation and granulosa cell proliferation. Environmental Toxicology 30: 1226-1233. https://doi.org/10.1002/tox.21995.

Huang, C.H., Wang, F.T. and Chan, W.H., 2019. Prevention of ochratoxin A induced oxidative stress-mediated apoptotic processes and impairment of embryonic development in mouse blastocysts by liquiritigenin. Environmental Toxicology 34: 573-584. https://doi.org/10.1002/tox.22724.

Huang, F.J. and Chan, W.H., 2016. Effects of Ochratoxin A on mouse oocyte maturation and fertilization, and apoptosis during fetal development. Environmental Toxicology 31: 724-735. https://doi.org/10.1002/tox.22085.

Jain, A. K., Singh, D., Dubey, K., Maurya, R., Mittal, S. and Pandey, A.K., 2018. Models and methods for in vitro toxicity. In Dhawan, A. and Kwon, S. (eds.) In vitro toxicology. Academic Press, New York, NJ, p. 45- 65. https://doi.org/10.1016/B978-0-12-804667-8.00003-1.

Jia, H., Jia, C., An, Q., Cheng, Y., Jiang, X., Xu, Y., Zhao, R., Peng, W., Zhang, Y. and Su, J., 2020. Ochratoxin A exposure causes meiotic failure and oocyte deterioration in mice. Theriogenology 148: 236-248. https://doi.org/10.1016/j.theriogenology.2019.11.005.

Jiang, W.J., Liu, W., Li, Y.H., Jiang, H., Xu, Y.N. and Kim, N.H., 2022. Citrinin impairs pig oocyte maturation by inducing oxidative stress and apoptosis. Toxicon 2022 205: 84-90. https://doi.org/10.1016/j.toxicon.2021.11.124.

Kagawa, H., Javali, A., Khoei, H.H., Sommer, T.M., Sestini, G., Novatchkova, M., Scholte Op Reimer, Y., Castel, G., Bruneau, A., Maenhoudt, N., Lammers, J., Loubersac, S., Freour, T., Vankelecom, H., David, L. and Rivron, N., 2022. Human blastoids model blastocyst development and implantation. Nature 601: 600-605. https://doi.org/10.1038/s41586-021-04267-8.

Kinoshita, M., Kobayashi, T., Planells, B., Klisch, D., Spindlow, D., Masaki, H., Bornelöv, S., Stirparo, G.G., Matsunari, H., Uchikura, A., Lamas-Toranzo, I., Nichols, J., Nakauchi, H., Nagashima, H., Alberio, R. and Smith, A., 2021. Pluripotent stem cells related to embryonic disc exhibit common self-renewal requirements in diverse livestock species. Development 148: dev199901. https://doi.org/10.1242/dev.199901.

Kuijk, E.W., van Tol, L.T., Van de Velde, H., Wubbolts, R., Welling, M., Geijsen, N. and Roelen, B.A., 2012. The roles of FGF and MAP kinase signaling in the segregation of the epiblast and hypoblast cell lineages in bovine and human embryos. Development 139: 871-882. https://doi.org/10.1242/dev.071688.

Kuiper, G.G., Lemmen, J.G., Carlsson, B., Corton, J.C., Safe, S.H., van der Saag, P.T., van der Burg, B. and Gustafsson, J.A., 1998. Interaction of estrogenic chemicals and phytoestrogens with estrogen receptor beta. Endocrinology 139: 4252-4263. https://doi.org/10.1210/endo.139.10.6216.

Lan, M., Han, J., Pan, M.H., Wan, X., Pan, Z.N. and Sun, S.C., 2018. Melatonin protects against defects induced by deoxynivalenol during mouse oocyte maturation. Journal of Pineal Research 65: e12477. https://doi.org/10.1111/jpi.12477.

Lan, M., Zhang, Y., Wan, X., Pan, M.H., Xu, Y. and Sun, S.C., 2020. Melatonin ameliorates ochratoxin A-induced oxidative stress and apoptosis in porcine oocytes. Environmental Pollution 256: 113374. https://doi.org/10.1016/j.envpol.2019.113374.

Lawson, K.A., Dunn, N.R., Roelen, B.A., Zeinstra, L.M., Davis, A.M., Wright, C.V., Korving, J.P. and Hogan, B.L., 1999. Bmp4 is required for the generation of primordial germ cells in the mouse embryo. Genes & Development 13: 424-436. https://doi.org/10.1101/gad.13.4.424.

Leoni, G.G., Palmerini, M.G., Satta, V., Succu, S., Pasciu, V., Zinellu, A., Carru, C., Macchiarelli, G., Nottola, S.A., Naitana, S. and Berlinguer, F., 2015. Differences in the kinetic of the first meiotic division and in active mitochondrial distribution between prepubertal and adult oocytes mirror differences in their developmental competence in a sheep model. PLoS ONE 10: e0124911. https://doi.org/10.1371/journal.pone.0124911.

Li, N., Liu, X.L., Zhang, F.L., Tian, Y., Zhu, M., Meng, L.Y., Dyce, P.W., Shen, W. and Li, L., 2020. Whole-transcriptome analysis of the toxic effects of zearalenone exposure on ceRNA networks in porcine granulosa cells. Environmental Pollution. 261: 114007. https://doi.org/10.1016/j.envpol.2020.114007.

Liu, X.L., Wu, R.Y., Sun, X.F., Cheng, S.F., Zhang, R.Q., Zhang, T.Y., Zhang, X.F., Zhao, Y., Shen, W. and Li, L., 2018. Mycotoxin zearalenone exposure impairs genomic stability of swine follicular granulosa cells in vitro. International Journal of Biological Sciences 14: 294-305. https://doi.org/10.7150/ijbs.23898.

Lorenzetti, S., Altieri, I., Arabi, S., Balduzzi, D., Bechi, N., Cordelli, E., Galli, C., Ietta, F., Modina, S.C., Narciso, L., Pacchierotti, F., Villani, P., Galli, A., Lazzari, G., Luciano, A.M., Paulesu, L., Spanò, M. and Mantovani, A., 2011. Innovative non-animal testing strategies for reproductive toxicology: the contribution of Italian partners within the EU project. ReProTect Annali dell'Istituto Superiore di Sanità 47: 429-444. https://doi.org/10.4415/ANN_11_04_16.

Lu, Y., Zang, Y., Liu, J.Q., Zou, P., Jia, L., Su, Y.T., Sun, Y.R. and Sun, S.C., 2018. Comparison of the toxic effects of different mycotoxins on porcine and mouse oocyte meiosis. PeerJ 6: e5111. https://doi.org/10.7717/peerj.5111.

Luo, C., Zuñiga, J., Edison, E., Palla, S., Dong, W. and Parker-Thornburg, J., 2011. Superovulation strategies for 6 commonly used mouse strains. Journal of the American Association for Laboratory Animal Science. 50: 471-478.

Madeja, Z.E., Pawlak, P. and Piliszek, A., 2019. Beyond the mouse: non-rodent animal models for study of early mammalian development and biomedical research. The International Journal of Developmental Biology 63: 187-201. https://doi.org/10.1387/ijdb.180414ap.

Malekinejad, H., Schoevers, E.J., Daemen, I.J., Zijlstra, C., Colenbrander, B., Fink-Gremmels, J. and Roelen, B.A., 2007. Exposure of oocytes to the *Fusarium* toxins zearalenone and deoxynivalenol causes aneuploidy and abnormal embryo development in pigs. Biology of Reproduction 77: 840-847. https://doi.org/10.1095/biolreprod.107.062711.

Mastrorocco, A., Martino, N.A., Marzano, G., Lacalandra, G.M., Ciani, E., Roelen, B.A.J., Dell'Aquila, M.E. and Minervini, F., 2019. The mycotoxin beauvericin induces oocyte mitochondrial dysfunction and affects embryo development in the juvenile sheep. Molecular Reproduction and Development 86: 1430-1443. https://doi.org/10.1002/mrd.23256.

Mastrorocco, A., Ciani, E., Nicassio, L., Roelen, B.A.J., Minervini, F. and Dell'Aquila, M.E., 2021. Beauvericin alters the expression of genes coding for key proteins of the mitochondrial chain in ovine cumulus-oocyte complexes. Mycotoxin Research 37:1-9. https://doi.org/10.1007/s12550-020-00409-5.

McLaren, A., 1998. Germ cells and germ cell transplantation. The International Journal of Developmental Biology 42: 855-860.

Medvedova, M., Kolesarova, A., Capcarova, M., Labuda, R., Sirotkin, A.V., Kovacik, J. and Bulla, J., 2011. The effect of deoxynivalenol on the secretion activity, proliferation and apoptosis of porcine ovarian granulosa cells *in vitro*. Journal of Environmental Science and Health, Part B 46: 213-219. https://doi.org/10.1080/03601234.2011.540205.

Merton, J.S., Knijn, H.M., Flapper, H., Dotinga, F., Roelen, B.A., Vos, P.L. and Mullaart, E., 2013. Cysteamine supplementation during *in vitro* maturation of slaughterhouse- and opu-derived bovine oocytes improves embryonic development without affecting cryotolerance, pregnancy rate, and calf characteristics. Theriogenology 80: 365-371. https://doi.org/10.1016/j.theriogenology.2013.04.025.

Minervini, F., Dell'Aquila, M.E., Maritato, F., Minoia, P. and Visconti, A., 2001. Toxic effects of the mycotoxin zearalenone and its derivatives on *in vitro* maturation of bovine oocytes and 17 beta-estradiol levels in mural granulosa cell cultures. Toxicology *In Vitro* 15: 489-495. https://doi.org/10.1016/s0887-2333(01)00068-6.

Minervini, F. and Dell'Aquila, M.E., 2008. Zearalenone and reproductive function in farm animals. International Journal of Molecular Sciences 9: 2570-2584. https://doi.org/10.3390/ijms9122570.

Nitsche, K.S., Müller, I., Malcomber, S., Carmichael, P.L. and Bouwmeester, H., 2022. Implementing organ-on-chip in a next-generation risk assessment of chemicals: a review. Archives of Toxicology 96: 711-741. https://doi.org/10.1007/s00204-022-03234-0.

Osweiler, G.D., 1986. Occurrence and clinical manifestations of trichothecene toxicoses and zearalenone toxicoses. In: Richard, J.L. and Thurston, J.R. (eds.), Diagnosis of mycotoxicoses. Martinus Nijhoff, Dordrecht, p. 31–42.

Palma, G.A., Clement-Sengewald, A. and Krefft, H., 1993. *In vitro* production of cattle embryos from calf oocytes. Theriogenology 39: 278.

Park, H.J., Yang, S.G. and Koo, D.B., 2022. SESN2/NRF2 signaling activates as a direct downstream regulator of the PERK pathway against endoplasmic reticulum stress to improve the *in vitro* maturation of porcine oocytes. Free Radical Biology and Medicine 178: 413-427. https://doi.org/10.1016/j.freeradbiomed.2021.12.258.

Partridge, E.A., Davey, M.G., Hornick, M.A., McGovern, P.E., Mejaddam, A.Y., Vrecenak, J.D., Mesas-Burgos, C., Olive, A., Caskey, R.C., Weiland, T.R., Han, J., Schupper, A.J., Connelly,

J.T., Dysart, K.C., Rychik, J., Hedrick, H.L., Peranteau, W.H. and Flake, A.W., 2017. An extra-uterine system to physiologically support the extreme premature lamb. Nature Communications. 8: 15112. https://doi.org/10.1038/ncomms15112.

Peraica, M., Richter, D. and Rašić, D., 2014. Mycotoxicoses in children. Arhiv za higijenu rada i toksikologiju 65: 347-363. https://doi.org/10.2478/10004-1254-65-2014-2557.

Pereira Daoud, A.M., Popovic, M., Dondorp, W.J., Trani Bustos, M., Bredenoord, A.L., Chuva de Sousa Lopes, S.M., van den Brink, S.C., Roelen, B.A.J., de Wert, G.M.W.R. and Heindryckx, B., 2020. Modelling human embryogenesis: embryo-like structures spark ethical and policy debate. Human Reproduction Update 26: 779-798. https://doi.org/10.1093/humupd/dmaa027.

Piersma, A.H., 2013. Innovations in testing strategies in reproductive toxicology. Methods in Molecular Biology 94: 327-341. https://doi.org/10.1007/978-1-62703-131-8_25.

Rivron, N.C., Frias-Aldeguer, J., Vrij, E.J., Boisset, J.C., Korving, J., Vivié, J., Truckenmüller, R.K., van Oudenaarden, A., van Blitterswijk, C.A. and Geijsen, N., 2018. Blastocyst-like structures generated solely from stem cells. Nature 557: 106-111. https://doi.org/10.1038/s41586-018-0051-0.

Roelen, B.A.J., 2019. Bovine oocyte maturation: acquisition of developmental competence. Reproduction, Fertility and Development 32: 98-103. https://doi.org/10.1071/RD19255.

Roode, M., Blair, K., Snell, P., Elder, K., Marchant, S., Smith, A. and Nichols, J., 2012. Human hypoblast formation is not dependent on FGF signalling. Developmental Biology 361: 358-363. https://doi.org/10.1016/j.ydbio.2011.10.030.

Saitou, M., 2021. Mammalian germ cell development: from mechanism to *in vitro* reconstitution. Stem Cell Reports 16: 669-80. https://doi.org/10.1016/j.stemcr.2021.01.008.

Saitou, M. and Hayashi, K., 2021. Mammalian *in vitro* gametogenesis. Science 374: eaaz6830. https://doi.org/10.1126/science.aaz6830.

Sambuu, R., Takagi, M., Namula, Z., Otoi, T., Shiga, S., Rodrigues Dos Santos, R. and Fink-Gremmels, J., 2011. Effects of exposure to zearalenone on porcine oocytes and sperm during maturation and fertilization *in vitro*. Journal of Reproduction and Development 57: 547-550. https://doi.org/10.1262/jrd.11-026h.

Sanchez, F. and Smitz, J., 2012. Molecular control of oogenesis. Biochimica et Biophysica Acta 1822: 1896-1912. https://doi.org/10.1016/j.bbadis.2012.05.013.

Santos, R.R., Schoevers, E.J., Wu, X., Roelen, B.A.J. and Fink-Gremmels, J., 2015. The protective effect of follicular fluid against the emerging mycotoxins alternariol and beauvericin. World Mycotoxin Journal 8: 445-450. https://doi.org/10.3920/WMJ2014.1829.

Schoevers, E.J., Fink-Gremmels, J., Colenbrander, B. and Roelen, B.A., 2010. Porcine oocytes are most vulnerable to the mycotoxin deoxynivalenol during formation of the meiotic spindle. Theriogenology 74: 968-978. https://doi.org/10.1016/j.theriogenology.2010.04.026.

Schoevers, E.J., Santos, R.R., Fink-Gremmels, J. and Roelen, B.A.J., 2016. Toxicity of beauvericin on porcine oocyte maturation and preimplantation embryo development. Reproductive Toxicology 65: 159-169. https://doi.org/10.1016/j.reprotox.2016.07.017.

Schoevers, E.J., Santos, R.R. and Roelen, B.A.J., 2020. Alternariol disturbs oocyte maturation and preimplantation development. Mycotoxin Research 36: 93-101. https://doi.org/10.1007/s12550-019-00372-w.

Schoevers, E.J., Santos, R.R. and Roelen, B.A.J., 2021. Susceptibility of oocytes from gilts and sows to beauvericin and deoxynivalenol and its relationship with oxidative stress. Toxins 13: 260. https://doi.org/10.3390/toxins13040260.

Somoskoi, B., Kovacs, M. and Cseh, S., 2016a. T-2 mycotoxin slows down the development of mouse blastocysts, decreases their blastomere number and increases chromatin damage. Acta Veterinaria Hungarica 64: 390-400. https://doi.org/10.1556/004.2016.037.

Somoskoi, B., Kovacs, M. and Cseh, S., 2016b. Effects of T-2 mycotoxin on in vitro development and chromatin status of mouse embryos in preimplantation stages. Toxicology and Industrial Health 32: 1260-1265. https://doi.org/10.1177/0748233714555394.

Somoskői, B., Kovács, M. and Cseh, S., 2018. Effects of T-2 and Fumonisin B1 combined treatment on in vitro mouse embryo development and blastocyst quality. Toxicology and Industrial Health 34: 353-360. https://doi.org/10.1177/0748233718764039.

Spielmann, H., 2005. Predicting the risk of developmental toxicity from *in vitro* assays. Toxicology and Applied Pharmacology 207: 375-380. https://doi.org/10.1016/j.taap.2005.01.049.

Spielmann, H., 2009. The way forward in reproductive/developmental toxicity testing. Alternatives to Laboratory Animals 37: 641-656. https://doi.org/10.1177/026119290903700609.

Sun, M.H., Li, X.H., Xu, Y., Xu, Y., Pan, Z.N. and Sun, S.C., 2020. Citrinin exposure disrupts organelle distribution and functions in mouse oocytes. Environmental Research. 185: 109476. https://doi.org/10.1016/j.envres.2020.109476.

Takagi, M., Mukai, S., Kuriyagawa, T., Takagaki, K., Uno, S., Kokushi, E., Otoi, T., Budiyanto, A., Shirasuna, K., Miyamoto, A., Kawamura, O., Okamoto, K. and Deguchi, E., 2008. Detection of zearalenone and its metabolites in naturally contaminated follicular fluids by using LC/MS/MS and *in vitro* effects of zearalenone on oocyte maturation in cattle. Reproductive Toxicology 26: 164-169. https://doi.org/10.1016/j.reprotox.2008.08.006.

Takahashi, K., Tanabe, K., Ohnuki, M., Narita, M., Ichisaka, T., Tomoda, K. and Yamanaka, S., 2007. Induction of pluripotent stem cells from adult human fibroblasts by defined factors. Cell 131: 861-872. https://doi.org/10.1016/j.cell.2007.11.019.

Tervit, H., 1996. Laparoscopy/laparotomy oocyte recovery and juvenile breeding. Animal Reproduction Science 42: 227–38. https://doi.org/10.1016/0378-4320(96)01533-3.

Tian, X., Wang, F., He, C., Zhang, L., DunXian, T., Russel, J.L., Jing, X., PengJun, J. and GuoShi, L., 2014. Beneficial effects of melatonin on bovine oocytes maturation: a mechanistic approach. Journal of Pineal Research 57: 239-247. https://doi.org/10.1111/jpi.12163.

Tiemann, U., Viergutz, T., Jonas, L. and Schneider, F., 2003. Influence of the mycotoxins alpha- and beta-zearalenol and deoxynivalenol on the cell cycle of cultured porcine endometrial cells. Reproductive Toxicology 17: 209-218. https://doi.org/10.1016/s0890-6238(02)00141-7.

Torres-Rovira, L., Succu, S., Pasciu, V., Manca, M.E., Gonzalez-Bulnes, A., Leoni, G.G., Pennino, M.G., Spezzigu, A., Gallus, M., Dattena, M., Monniaux, D., Naitana, S. and Berlinguer, F., 2016. Postnatal pituitary and follicular activation: A revisited hypothesis in a sheep model. Reproduction 151: 215-225. https://doi.org/10.1530/REP-15-0316.

Wang, Y., Xing, C.H., Chen, S. and Sun, S.C., 2022. Zearalenone exposure impairs organelle function during porcine oocyte meiotic maturation. Theriogenology. 177: 22-28. https://doi.org/10.1016/j.theriogenology.2021.10.008.

Wu, J., Tu, D., Yuan, L.Y., Yuan, H. and Wen, L.X., 2013. T-2 toxin exposure induces apoptosis in rat ovarian granulosa cells through oxidative stress. Environmental Toxicology and Pharmacology 36: 493-500. https://doi.org/10.1016/j.etap.2013.03.017.

Wu, J., Tu, D., Yuan, L.Y., Yi, J.E. and Tian, Y., 2015. T-2 toxin regulates steroid hormone secretion of rat ovarian granulosa cells through cAMP-PKA pathway. Toxicology Letters 232: 573-579. https://doi.org/10.1016/j.toxlet.2014.12.016.

Xu, Y., Zhang, K.H., Sun, M.H., Lan, M., Wan, X., Zhang, Y. and Sun, S.C., 2019. Protective effects of melatonin against zearalenone toxicity on porcine embryos *in vitro*. Frontiers in Pharmacology 10: 327. https://doi.org/10.3389/fphar.2019.00327.

Xue, R., Li, S., Zou, H., Ji, D., Lv, M., Zhou, P., Wei, Z., Zhang, Z. and Cao, Y., 2021. Melatonin alleviates deoxynivalenol-induced apoptosis of human granulosa cells by reducing mutually accentuated FOXO1 and ER stress. Biology of Reproduction 105: 554-566. https://doi.org/10.1093/biolre/ioab084.

Yamanaka, Y., Lanner, F. and Rossant, J., 2010. FGF signal-dependent segregation of primitive endoderm and epiblast in the mouse blastocyst. Development 137: 715-724. https://doi.org/10.1242/dev.043471.

Yoshimatsu, S., Nakajima, M., Iguchi, A., Sanosaka, T., Sato, T., Nakamura, M., Nakajima, R., Arai, E., Ishikawa, M., Imaizumi, K., Watanabe, H., Okahara, J., Noce, T., Takeda, Y., Sasaki, E., Behr, R., Edamura, K., Shiozawa, S. and Okano, H., 2021. Non-viral induction of transgene-free iPSCs from somatic fibroblasts of multiple mammalian species. Stem Cell Reports 16: 754-770. https://doi.org/10.1016/j.stemcr.2021.03.002.

Yu, B., van Tol, H.T.A., Oei, C.H.Y., Stout, T.A.E. and Roelen, B.A.J., 2021. Lysophosphatidic acid accelerates bovine *in vitro*-produced blastocyst formation through the Hippo/YAP pathway. International Journal of Molecular Sciences 22: 5915. https://doi.org/10.3390/ijms22115915.

Zhang, F.L., Li, N., Wang, H., Ma, J.M., Shen, W. and Li, L., 2019. Zearalenone exposure induces the apoptosis of porcine granulosa cells and changes long noncoding RNA expression to promote antiapoptosis by activating the JAK2-STAT3 pathway. Journal of Agricultural and Food Chemistry 67: 12117-12118. https://doi.org/10.1021/acs.jafc.9b05189.

Zhang, G.L., Song, J.L., Zhou, Y., Zhang, R.Q., Cheng, S.F., Sun, X.F., Qin, G.Q., Shen, W. and Li, L., 2018. Differentiation of sow and mouse ovarian granulosa cells exposed to zearalenone *in vitro* using RNA-seq gene expression. Toxicology and Applied Pharmacology 350, 78-90. https://doi.org/10.1016/j.taap.2018.05.003.

Zhang, T.Y., Kong, L., Hao, J.X., Wang, H., Yan, Z.H., Sun, X.F. and Shen, W., 2020. Effects of Ochratoxin A exposure on DNA damage in porcine granulosa cells *in vitro*. Toxicology Letters 330: 167-175. https://doi.org/10.1016/j.toxlet.2020.05.011.

Zhang, Y., Han, J., Zhu, C.C., Tang, F., Cui, X.S., Kim, N.H. and Sun, S.C., 2016. Exposure to HT-2 toxin causes oxidative stress induced apoptosis/autophagy in porcine oocytes. Scientific Reports 6: 33904. https://doi.org/10.1038/srep33904.

Zhang, Y., Jia, R.X., Pan, M.H., Lu, Y., Cui, X.S., Kim, N.H. and Sun, S.C., 2017. HT-2 toxin affects development of porcine parthenotes by altering DNA and histone methylation in oocytes matured *in vitro*. Theriogenology 103, 110-116. https://doi.org/10.1016/j.theriogenology.2017.07.052.

Zhu, C.C., Zhang, Y., Duan, X., Han, J. and Sun, S.C., 2016. Toxic effects of HT-2 toxin on mouse oocytes and its possible mechanisms. Archives of Toxicology 90, 1495-1505. https://doi.org/10.1007/s00204-015-1560-3.

CHAPTER 15

Methods to decrease the risk of mycotoxin contamination

G. Brodal[1], A. Gallo[2] and R.R. Santos[3]*

[1]NIBIO (Norwegian Institute of Bioeconomy Research), Ås, Norway
[2]Department of Animal Science, Food and Nutrition (DIANA), Facoltà di Scienze Agrarie, Alimentari e Ambientali, Università Cattolica del Sacro Cuore, Piacenza, Italy
[3]SFR (Schothorst Feed Research), Lelystad, the Netherlands
**correspondence: guro.brodal@nibio.no*

1 Introduction

As discussed in the previous chapters, mycotoxins are produced by fungal species (moulds) within several genera as shown in Table 15.1. These filamentous fungi can infect or are associated with crop plants in the field during the growing season. Some of the most important mycotoxin-producing fungal species belong to the genus *Fusarium*, which can cause devastating diseases such as Fusarium Head Blight (FHB), also called Fusarium Ear Blight (FEB) and scab in small grain cereals, and Fusarium Ear Rot (FER) and Gibberella Ear Rot (GER) in maize. Moreover, *Aspergillus flavus* can cause ear or kernel rot in maize. (For more details, see Chapter 2). Moreover, species of the fungal genera *Claviceps* (causing ergot in heads of e.g., rye and several grasses) and *Alternaria* (often occurring as a saprophyte or sometimes causing disease in cereal heads) are also known to produce mycotoxins. Species of *Penicillium* and *Aspergillus* can cause grain deterioration and produce mycotoxins after harvest/during storage. Reduction in grain quality occurs due to development of disease and mycotoxins in the cereal heads/grain.

The grain can become contaminated to an extent that makes it less suitable for feed and food and can impact animal production. To reduce animal health risks, many countries have established maximum and guidance levels for mycotoxin content in grain and products for animal feed (Chapter 17). To limit as much as possible the harmful effects of mycotoxins, all possible practices to prevent their occurrence in feedstuffs and consequently in the diet should be considered. This implies adopting the best possible techniques, from pre-harvest to inclusion in the animal diet, following the HACCP system (Hazard Analysis Critical Control Point). Although dietary supplementation with mycotoxin decontaminants is the most well-known

 | DOI: 10.1163/9789004712720_16

TABLE 15.1 Important mycotoxins produced by fungal species within the genus *Fusarium* and by the genera *Alternaria, Aspergillus, Claviceps* and *Penicillium*. DON = deoxynivalenol, NIV = nivalenol, T-2/HT-2 = T-2 and HT-2 toxins, ZEN = zearalenone, FUM = fumonisins, ENNs = enniatins, BEA = beauvericin, MON = moniliformin, AOH/AME = alternariol and alternariol monomethyl ether, AFs = aflatoxins, EAs = ergot alkaloids, OTA = ochratoxin A.

Fungi	**Mycotoxins**											
	DON	**NIV**	**T2/ HT2**	**ZEN**	**FUM**	**ENNs**	**BEA**	**MON**	**AOH/ AME**	**AFs**	**EAs**	**OTA**
F. avenaceum						x	x	x				
F. culmorum	x	x		x								
F. graminearum	x	x		x								
F. langsethiae			x									
F. poae		x					x	x				
F. sporotrichioides			x				x	x				
F. tricinctum						x	x	x				
F. verticillioides (syn *F. moniliforme*)					x							
Alternaria spp.									x			
Aspergillus spp.										x		x
Claviceps spp.											x	
Penicillium spp.												x

measure for decreasing animal exposure to mycotoxins (Chapter 16), several measures/methods both pre-harvest, during harvesting/storage, and during processing procedures can reduce mycotoxin contamination of grains. Most mycotoxins in cereal grains are quite stable and are difficult to get rid of with e.g. thermal processes. Varying effects have been reported during extrusion and pelleting, etc. (Bullerman and Bianchini, 2007; Kabak, 2009; Kuchenbuch et al., 2018). It is therefore important to avoid/prevent as much as possible the development of mycotoxins where they arise, i.e. in the field and during storage. In a comprehensive review of factors influencing mycotoxin content during production of cereal-based feed, Hoffmans et al. (2022) stated that the prevention of mycotoxin production had a greater impact than the reduction during processing of mycotoxins already present in feed materials.

Methods against mycotoxins in cereals and feedstuff include agricultural practice, that contributes to the least possible infection by FHB/FEB and FER/GER and mycotoxin production in the field, and harvesting, handling, drying and storage routines that prevent the development of fungi and mycotoxins in grain after harvest, as well as sampling and analyses at elevators and feed mills to avoid grain lots with too high mycotoxin content from being used in feed production. Moreover, processing will reduce the mycotoxins in grain and feedstuff.

This chapter gives an overview of methods for reducing the risk of and decreasing contamination by mycotoxins in cereal grains for feed manufacturing and in maize silage. The main focus is on small grain cereals and maize.

2 Pre-harvest prevention

Disease resistance, cultural practice and plant protection measures can reduce the risk of mycotoxin contamination of crops. Implementation of risk-reducing measures in the field requires knowledge of the environmental and agronomic conditions that affect the infection by mycotoxigenic fungi and mycotoxin production. After a brief description of *Fusarium* epidemilogy and background for the increased mycotoxin occurrences in cereal grains recent years, this section gives a survey of the agricultural practices/cropping factors known to reduce the risk of mycotoxin development, primarily mycotoxins produced by *Fusarium* species.

Available inoculum (conidia and air blown ascospores) from infected plant debris, alternative hosts, *Fusarium* seedling blights and foot rot, along with moist and warm weather conditions, which are favourable for fungal growth and spore dissemination, can result in severe *Fusarium* infection (Figure 15.1) and the development of various mycotoxins in the grain when cereal plants are at the heading, flowering, and maturation stages (Parry et al., 1995; Miller, 2008; Shaner, 2003). In addition to infection of maize via silks, *Fusarium* also infect via insect damaged kernels (Munkvold, 2003). Hot and dry conditions are favourable for spore dispersal of *Aspergillus* from contaminated plants, plant debris, and soil and transmission by air or insects to maize where the fungus may produce AFs (Abbas et al., 2009).

FHB and mycotoxin occurrences, in particular DON, have increased in cereal-growing regions of the world over the last decades (Leonard and Bushnell, 2003). *Fusarium graminearum*, considered the major DON producer, was ranked as number four among the top ten most economically important plant pathogens (Dean et al., 2012). The main reasons for the increase are largely considered to be changed tillage practices, with the application of reduced/conservation tillage instead of deep ploughing, and at the same time more frequent production of wheat after wheat and more widespread growing of maize (McMullen et al., 2012). In addition, some regions have had significantly more rain and slightly higher average temperatures in the growing season than previously. This has resulted in increased inoculum potential and disease pressure, as for example in Norway, where more than 100% increased FHB frequencies in cereal grain have been recorded on average for the years 2000 until 2014, compared to the previous 30 years (Brodal et al., 2016a).

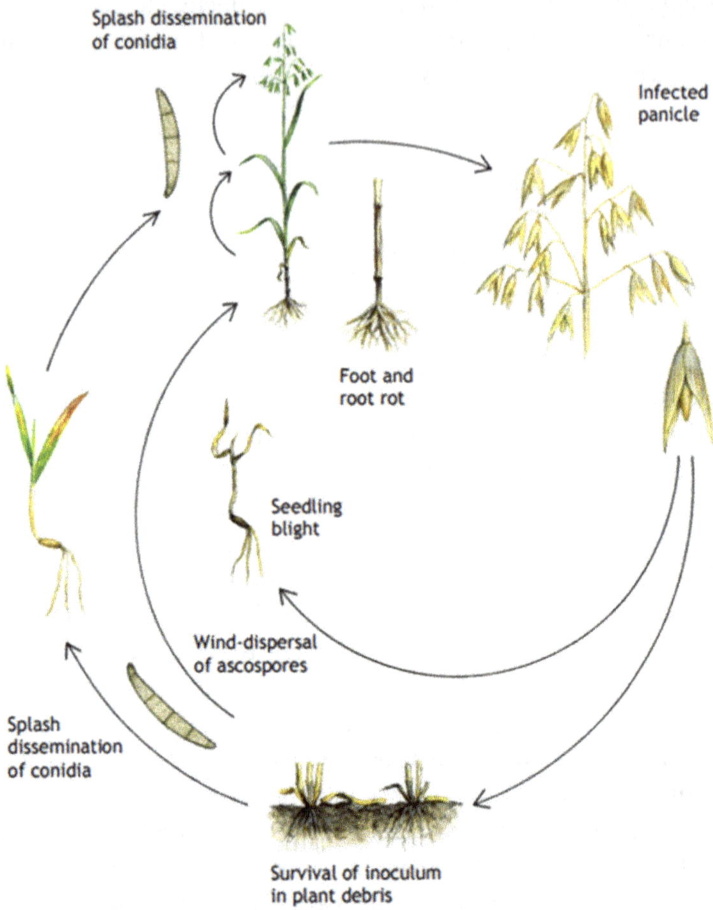

FIGURE 15.1 Disease cycle of *Fusarium* head blight in oats caused by *Fusarium graminearum* (drawing by Hermod Karlsen, in Brodal et al., 2009)

2.1 *Choice of plant variety and resistance breeding*

The growing of cereal varieties or breeding lines resistant to toxigenic fungi and the development of mycotoxins is the most sustainable and cost-effective method for reducing the risk of mycotoxin contamination in grains. Some differences in susceptibility to *Fusarium* infection and subsequent mycotoxin contamination among cereal species and varieties have been reported; however, there are not many highly resistant varieties available. Different resistance mechanisms to FBH have been described, including resistance to accumulation of DON (Miller et al., 1985; Mesterhazy, 1995). Resistance to FHB, particularly *F. graminearum*, which is the most important species causing FHB, is quantitative and based on several quantitative trait loci (QTLs) with minor effects (Buerstmayr et al., 2020; Sørensen et al., 2024), which makes resistance breeding complicated. Moreover, the selection of a variety based on mycotoxin risk is challenging because often a multi-mycotoxin contamination occurs, and a variety with some resistance to a fungal species and its mycotoxin may be susceptible to another species able to produce other mycotoxins. A Norwegian study ranking spring oats varieties showed that the variety Belinda was the most susceptible to DON contamination (800 μg/kg grain), whereas the variety Odal usually had the half DON levels (400 μg/kg grain) but was the most susceptible to T-2/HT-2 contamination (400 μg/kg grain) (Hofgaard et al., 2022a). Other studies have also observed a poor association between the levels of DON and T-2/HT-2, in oats (e.g. Edwards et al., 2009c) and in durum wheat (Aureli et al., 2015), indicating different environmental requirements for the development of their producers, which further complicates the breeding for resistance to these mycotoxins. However, the identification of many QTLs from comprehensive research and breeding programmes, the finding of genetic variation for FHB resistance, and molecular breeding methods and tools will hopefully in the near future result in cereal varieties with better resistance to FHB and mycotoxins (e.g. Huang et al., 2021; Michel et al., 2024; Morcia et al., 2024; Miedaner et al., 2024).

The selection of a plant variety must also consider geographical region, soil type, and climatic conditions. Munkvold (2014) pointed out the importance of selecting maize hybrids that are adapted to local conditions for at least two reasons: (i) their maturity rating will be appropriate and (ii) they will be less influenced by abiotic stress conditions that can predispose plants to mycotoxin contamination. Planting hybrids outside their adapted range may lead to higher mycotoxin contamination risk. In field experiments earlier maturing maize hybrids were shown to have a lower risk for mycotoxin contamination than later hybrids (Blandino et al., 2009a). Miedaner and Juroszek (2021) emphasised the need for increased efforts in breeding for resistance to mycotoxigenic fungi and insects in maize to meet the challenges of global warming and increased maize cultivation. The challenges behind and reasons for the rather slow progress in breeding for resistance to the major mycotoxigenic fungi (*F. graminearum*, *F. verticillioides* and *A. flavus*) and resistance to mycotoxin contamination in maize, were recently reviewed by Mesterhazy (2024). Several approaches and methods were discussed, and with prospects for hybrids with better resistance to the three main agents, it was indicated that the mycotoxin losses might be reduced by around 50% or more.

So far, the lack of cereal varieties with good resistance to *Fusarium* species and their mycotoxins means that other measures are very much needed.

2.2 *Crop rotation and tillage*

Crop rotation, i.e. growing of cereals after a crop that is not a host of *Fusarium*, can reduce the amount of *Fusarium* inoculum and the risk of mycotoxin contamination of grains (e.g. Parry et al., 1995; McMullen et al., 2012). *Fusarium* and *Aspergillus* species survive in crop residues, and infected crop residues are the most important inoculum source for *Fusarium* infections in cereal crops. Often the disease pressure is great if the crop residue is left on the soil surface, especially if cereals are grown in the same field for various seasons, and especially after maize (e.g. Abbas et al., 2009; McMullen et al., 2012).

The level of contamination in the field is affected by the crops used in a rotation system. Several studies have shown significantly lower levels of DON and ZEN in wheat grain when grown after other crops than maize (e.g. Beyer et al., 2006; Schaafsma et al., 2005; Edwards and Jennings, 2018). Drakopoulos et al. (2021) found higher amounts of *F. graminearum* in grain and straw and higher levels of DON in grains of barley after maize as pre-crop. In China, rice and maize as pre-crops for wheat resulted in higher DON contamination in wheat grain than cotton as a pre-crop (Selvaraj et al., 2015). In another study the DON level in wheat was 50% higher when the previous crop was maize compared to soybeans, probably because of contaminated crop residues left in the soil, combined with conservation tillage, which consists of leaving all crop residues on the surface of the soil (Dill-Macky and Jones, 2000). Whereas residues of soybeans are basically stem fragments, the remains from maize crops are formed by slow-degrading stalks, husk, and cob pieces, which may increase the survival of inoculum-borne residues. Lower T-2/HT-2 was detected when oat or barley was grown after a non-cereal crop compared to a cereal as pre-crop (Edwards, 2017; Orlando, 2010).

Soybean, oilseed Brassicas, legumes/pulses, sugar beet etc., considered poor hosts for *F. graminearum* with low inoculum potential in the residues, are recommended as break crops for cereals (e.g. Dill-Macky and Jones, 2000; Wenda-Piesik et al., 2017). Rasiukevičiūtė et al. (2018) reported that several non-cereal crops were able to maintain inoculum of *F. graminearum*, however, in rather low incidences, and the use of these crops as pre-crops was considered safe in cereal rotations. Recently, *Fusarium* spp. (mostly *F. avenaceum*, but also *F. graminearum* and other *Fusarium* species) have been isolated quite frequently from faba bean grains in Norway (Brodal and Aamot, unpublished). In addition, to clarify whether faba bean grains can contain harmful mycotoxins, further studies are needed to evaluate the potential of the crop as inoculum source for FHB and mycotoxin development when grown as pre-crop of cereals.

Several studies have reported that burying infected crop residues by ploughing reduces the disease inoculum (e.g. Dill-Macky and Jones, 2000; Edwards and Jennings, 2018). However, varying impacts of tillage on the reduction of FHB and mycotoxins have been reported, as reviewed by McMullen et al. (2012). In a study on the effect of tillage on the development of *Fusarium* and mycotoxins in oats, Hofgaard et al. (2022b) detected lower levels of *Fusarium langsethiae* and T-2/HT-2 in oat grains from ploughed plots compared to harrowed plots. Likewise, Kaukoranta et al. (2019) observed lower concentrations of T-2/HT-2 in oats with ploughing than without ploughing, and the toxins tended to increase with increasing cereal intensity. Lower levels of NIV and T-2/HT-2 were detected in oat samples from ploughed fields compared to samples from fields with reduced tillage (Schöneberg et al., 2018). Supronienė et al. (2012) reported reduced T-2/HT-2 and tendency to increased DON concentrations in winter wheat after ploughing. Hofgaard et al. (2022b) observed no impact of tillage on the content of DON, ENNs or BEA in the harvested oat grain. In wheat, Hofgaard et al. (2020) detected lower *Fusarium* inoculum potential after ploughing than after harrowing, but this did not result in a significant difference in the level of *Fusarium* and mycotoxins in the harvested grain. Drakopoulos et al. (2021) found increased incidence of *F. graminearum* and DON in the grain and straw of barley to be associated with reduced tillage. In an eight-year survey of *Fusarium* and mycotoxins in Swiss wheat grain, Vogelsang et al. (2019) found high levels of *F. graminearum* and DON and ZEN in samples from fields with the previous crop maize, reduced tillage, varieties with poor FHB resistance and strobilurin-based fungicides. Other previous crops and/or ploughing significantly reduced the DON and ZEN content. In contrast, high levels of *F. poae* and NIV were observed in wheat samples from ploughed fields and canola as pre-crop (Vogelsang et al., 2019). In maize, Herrera et al. (2023) reported that the concentrations of FUM and DON were not affected by the different soil tillage practices (conventional and no tillage).

Regardless of agronomic cropping measures, long-distance dispersal of ascospores of *F. graminearum* from other fields may still cause fungal infection and mycotoxin development (Schmale et al., 2006). Also, airborne conidia spores of *A. flavus* and *Fusarium* spp. from maize debris in a region might 'override' the residue management by tillage and crop rotation resulting in poor inoculum reductions (Munkvold, 2014).

2.3 *Seed, sowing date, sowing density*

High-quality seeds for sowing are essential for successful plant production. Seeds free from mycotoxigenic pathogens can reduce the risk of introducing such pathogens to 'healthy' areas where measures to reduce mycotoxin contamination are implemented. However, healthy seed cannot avoid infections from wind-borne ascospores. Although *Fusarium*-infected seeds might cause seedling blights, root and crown rots and poor plant establishment, and increase the risk of FHB and mycotoxin contamination of harvested grain, inoculum from *Fusarium*-infected seeds are considered less important than inoculum from plant residues in causing FHB epidemics (Shaner, 2003). However, *Fusarium proliferatum*-infected wheat seeds have been shown to systemically colonise wheat plants and develop FB_1 and BEA in grains (Guo et al., 2016), and fungicide seed treatment has been shown to reduce FHB compared to untreated seeds (Teich and Hamilton, 1985).

Changing the sowing date might influence the fungal infection and mycotoxin contamination of cereal grains which depend on the co-occurrence of a susceptible plant developmental stage, favourable environmental conditions, and sometimes vector activity. In maize production, earlier sowing dates have resulted in a lower risk for AFs, FUM and DON (Abbas et al., 2007; Blandino et al., 2009a; Krnjaja et al., 2022) and field experiments in spring barley showed less DON contamination of grain after early sowing than after later sowing (Choo et al., 2014). However, annual weather fluctuations can influence this aspect.

Sowing density can affect the risk of mycotoxins in at least two ways: (i) excessive plant densities can stress plants, increasing their susceptibility; and (ii) higher density sowing may create high-density crop canopies, which hold moisture and create a more favourable environment for fungal infection, causing greater levels of DON (Blandino et al., 2008a).

2.4 *Irrigation*

Drought stress can make crop plants vulnerable to fungal infections when moisture arises. To reduce the risk of FHB and mycotoxin production it is therefore important to avoid irrigation during flowering and maturation stages, especially for susceptible varieties (Cowger et al., 2009; Gautam and Dill-Macky, 2012; Kharbikar et al., 2015). Herrera et al. (2023) detected higher DON levels in maize grain from sprinkler-irrigated fields compared to flood irrigation, whereas the levels of FUM were not influenced by the irrigation system. Moreover, they detected higher DON and FUM levels when sprinkler irrigation was performed during night compared to day irrigation. However, irrigation to avoid drought stress reduced AFs in maize (Jones et al., 1981).

2.5 *Fertilisation*

Several studies have indicated that fertilisation does not greatly influence *Fusarium* infection and mycotoxin production in cereals, or they have shown inconsistent results. Although both nitrogen (N) deficient crops as well as very high N rates have shown elevated mycotoxin contaminations in the harvested grain, no significant effects were found on FHB, DON, ZEN and NIV in wheat grain by varying N rates (Yoshida et al., 2008; Krnjaja et al., 2015). However, the highest level of N under high inoculum pressure resulted in elevated levels of DON (Krnjaja et al., 2015). Lemmens et al. (2004) observed increased DON contamination in wheat grains by increasing N rates from 0 to 80 kg/ha, however, higher N levels (up to 160 kg/ha) did not result in a further increase in DON levels. Podolska et al. (2017) detected higher DON and ENNs in wheat grain of some of the included varieties when using 200 kg N/ha compared with 120 kg N/ha. However, no consistent trend between the N levels was observed for the other mycotoxins among the wheat varieties in the experiment. In barley, FHB and DON and ENNB were significantly lower in the grain from plots with N fertilisation (40, 80 and 140 kg/ha) than in grain from plots without N fertilisation (Hofer et al., 2016), indicating a possible strengthening of plants and increased defence against pathogens when N status is suitable. Moderate levels of N application have resulted in lower levels of FUM, AFs, and OTA in maize grain compared to grain from the non-fertilised control, whereas high N levels increased FUM, DON, OTA and ZEN (Blandino et al., 2008b). However, in another study, moderate levels of N in combination with boron were shown to increase FB_2 production in maize (Hassegawa et al., 2008). Some studies have shown reduced risk of AFs with optimum rates of N and phosphorus fertilisation (e.g. Abbas et al., 2009; Manoza et al., 2017). Scarpino et al. (2022) suggested N application in maize in accordance with the uptake to be optimal to prevent an elevated risk of contamination by mycotoxins. An 'alternative' fertiliser (potassium phosphite-based) claimed to improve the tolerance to *Fusarium* fungi and mycotoxin contamination in maize, showed promising control of FUM (up to 85% reduction of FB_1 and 100% reduction FB_2) in field experiments in Portugal (Simões et al., 2023).

2.6 *Fungicides*

Fungicide application is part of an integrated management strategy to optimise farming output. Fungicides are usually applied to decrease fungal diseases, but also to reduce the risk of mycotoxin production. The effects of fungicides have been variable. The type of fungicide (mode of action), the application scheme (timing), and the resistance level of the crop plant are important for the efficacy of the fungicide in controlling FHB and mycotoxin development.

At present, the most commonly used fungicides for controlling FHB and reducing DON risk are azoles such as tebuconazole, metconazole and prothioconazole; however, they may not provide satisfactory control. Reduction in FHB disease severity in wheat by these azoles has been found to be around 50 % and the reduction of DON in grain up to 45 % (Paul et al., 2008). Around 70-80% control of FHB and 63-73% control of DON in wheat compared to untreated control have been reported when using prothioconazole + tebuconazole in moderately resistant varieties (Moraes et al., 2022). Šíp et al. (2010) found 86% reduction in DON when combining variety resistance and fungicide (dimoxystrobin + epoxiconazole) treatment, whereas Mesterhazy et al. (2018) reported 91% reduction of DON after fungicide application (average several fungicides) at full flowering stage across varieties and years. Scarpino et al. (2015) evaluated the effect of prothioconazole on several mycotoxins in wheat and found a significantly reduction of ENNs, aurofusarin, MON, tentoxin and equisetin contents in all experiments (two seasons, three sites in north-west Italy, with medium FHB susceptible varieties), whereas DON, DON-3G and culmorin was reduced in the season with the highest disease pressure. A comprehensive study of prothioconazole-containing fungicides on mycotoxins in common and durum wheat revealed a significant reduction of DON, DON-3G, NIV, NIV-3G, culmorin and aurofusarin (Kleber et al., 2023).

Unlike fungicide treatment against e.g. leaf spot diseases in cereals, where treatment before heading and spike emergence is often needed, fungicide treatment to reduce the risk of FHB fungi and the development of mycotoxins is commonly performed around flowering/anthesis, which is the most susceptible stage. However, based on results from field experiments in wheat, Gonzales-Dominiques et al. (2021) suggested that the application of fungicides against FHB should be based on the time of *Fusarium* infection rather than on wheat growth stage. Several forecasting models have been developed for predicting potential FHB epidemics and DON contamination (e.g. De Wolf et al., 2003; Schaafsma and Hooker, 2007; Xu et al., 2013; Shah et al., 2019; Birr et al., 2019). They are mostly based on weather data (e.g. temperature, rain, relative humidity), and sometimes crop management options that influence the risk of FHB and DON (previous crop, tillage, host resistance levels) and the plant development stage. Some models are incorporated into decision support systems (e.g. Rossi et al., 2015) to help farmers and advisors determine whether the use of a fungicide is warranted. To find the optimal timing of fungicide application in oats in Norway, Hjelkrem et al. (2022) used functional data analysis to disclose weather conditions during the growth period associated with DON content in the harvested grain. They found low temperature and rainy and moist conditions prior to and during flowering of oats to be particularly important, in line with similar studies in wheat (e.g. De Wolf et al., 2003; Xu et al., 2013; Shah et al., 2019). The findings of Hjelkrem et al. (2022) are used on the digital Norwegian platform for prognosis, monitoring and decision support for integrated pest management (IPS) in agricultural crops, called VIPS (www.vips-landbruk.no), to advise farmers on the risk of DON development in oats and to consider whether application of a fungicide is justified. In barley, the optimal timing for fungicide application in closed-flowering varieties has been reported to be around the beginning of 'spent' anther extrusion and not during the flowering stage as in wheat and open-flowering barley (Yoshida and Nakajima, 2012). In winter barley, Cowger et al. (2023) found that rather late fungicide application, i.e. 6 days after full spike emergence had the best effect in reducing DON. Similarly, fungicide application 3 to 7 days after full heading was recommended for reduction of FHB and DON in spring barley (Friskop et al., 2023).

No or little effect has been observed on the content of T-2/HT-2 toxins in grain of oats and barley after the use of azoles or other types of fungicides (van der Fels-Klerx and Stratakou, 2010; Edwards and Anderson, 2011; Karron et al., 2020). A laboratory medium-based study indicated that T-2 toxin might increase as a response to low doses of the fungicides prochloraz or tebuconazole (Mateo et al., 2011). On the other hand, Somma et al. (2022) reported that azole fungicides were able to reduce *F. langsethiae* and *F. sporotrichioides* infections and related T-2 and HT-2 toxins by about 70-80% in durum wheat with natural infections under field conditions. Moreover, Abbas et al. (2009) mentioned that none of the fungicides tested against *A. flavus* reduced the AFs contamination in maize grain.

Other types of fungicides, such as strobilurins have shown no effect or in some cases even increased the DON contamination in wheat grain (Simpson et al., 2001; Feksa et al., 2019). Drakopoulos et al. (2021) observed higher DON and BEA in barley grains after application of strobilurin-based fungicides. Henriksen and Elen (2005) observed significantly higher *Fusarium* infection in grain of wheat and barley from early application (before flowering) of some fungicides, including a strobilurin compound, compared with grain from untreated plots. It might be speculated this could be due to the inhibition of microorganisms that would otherwise compete with *Fusarium* spp.

More recently a new fungicide, pydiflumetofen (a succinate dehydrogenase inhibitor) has been registered in several countries for use against several fungal diseases, including the management of FHB and DON. Pydiflumetofen has shown good effect against FHB in combination with propiconazole, reducing DON concentrations by 52-73% compared to non-treated control (Xia et al., 2021). A co-formulation of pydiflumetofen with prothioconazole showed a significant reduction of FHB and DON in winter wheat grain, more than pydiflumetofen separately, and with the high effect independent of application time from flag leaf fully emerged to late flowering (Edwards, 2022).

Comparing single and double application of fungicides in a field trial with wheat, Moraes et al. (2022) observed that although the double application (one at early anthesis and the second three or six days later) increased the costs, the gains in yield and grain weight, as well as the significant reduction in the levels of DON (34.1 mg/kg to < 2 mg/kg) and ZEN (7.4 to < 1.3 mg/kg) in the wheat, were able to outweigh the costs. In contrast, double application of fungicide gave no advantage in barley contamination with DON because this cereal is susceptible during a short period of time (Janssen et al., 2018).

2.7 *Weed control*

In addition to the survival on crop plant debris, several *Fusarium* spp. can colonise numerous weed species, often without showing any symptoms indicating their existence as endophytes (Postic et al., 2012; Supronienė et al., 2019). It was shown that *Fusarium* spp. isolated from weeds could infect wheat and thus be a source of FHB infection (Matelionienė et al., 2022). Moreover, lots of weeds will induce moisture, which is favourable for fungal growth and mycotoxin development. To prevent excessive moisture and to reduce weeds as an inoculum reservoir for mycotoxigenic fungi, weed management by herbicides and/or by ploughing is recommended. In maize, weed control by pre- and post-emergence herbicide application was shown to reduce the FUM and DON concentrations in the kernels by 47% compared to control (Blandino et al., 2007).

In a field study conducted in Norway by Henriksen and Elen (2005), the herbicide treatment during the growing season or glyphosate in autumn did not influence the level of *Fusarium* infection in the harvested grain of wheat, barley and oats. Nevertheless, Fernandez et al. (2009) reported a consistent association between previous glyphosate use and increased *Fusarium* infection of spikes in wheat and barley from several field experiments. Also, an earlier study indicated that glyphosate could increase the occurrence of *Fusarium* (Altman and Rovira, 1989).

2.8 *Insect control*

The control of insects such as European corn borer (ECB), *Ostrinia nubilalis*, and other moths during the growing season is one of the most important measures to reduce the risk of fungal infection and mycotoxin contamination in maize, such as FER and FUM and *Aspergillus* and AFs (Munkvold, 2014). By carrying fungal spores and by causing wounds and damage to the maize kernels the larvae of the moths facilitate the infection by mycotoxigenic fungi. The use of insecticides is widely used and aims to decrease insect damage to prevent fungal invasion. The most effective time point for application has been shown to be at the beginning of consistent ECB flight. With treatment at this time, Blandino et al. (2009b) reported on average 93% reduction in FUM contamination compared to untreated. An investigation on the effect of insecticides (pyrethroid and pyrethroid+diamide) on several mycotoxins showed a reduction in the incidence and severity of FER and FUM as well as other mycotoxins produced by *Fusarium* spp., including fusaric acid, fusaproliferin, fusarin C, BEA, bikaverin, MON and equisetin (Scarpino et al., 2018). However, the treatment did not reduce the content of DON and ZEN, or their modified or glycosylated metabolites.

Control of ECB in maize by growing hybrids containing Bt genes, where that is allowed, is considered the most effective measure to reduce the risk of fungal infection and mycotoxin contamination (Munkvold, 2014). Several studies have reported FUM reduction by growing Bt hybrids rather than conventional maize. E.g. in Iowa, USA, with moderate to high ECB populations, FUM were reduced by an average of 80% in Bt hybrids compared to conventional hybrids, whereas the reduction was 38% when ECB population was low (Munkvold, 2014).

2.9 *Biocontrol*

Biological control strategies as alternatives to synthetic fungicides in the management of plant diseases have recently received increased interest (Collinge et al., 2022). Several antagonistic microorganisms, such as bacteria, fungi, and yeasts (biocontrol agents/BCAs), based on suppression, mycoparasitism, competitions for space and nutrients, secretion of antibiotics and/or lytic enzymes, and induction of resistance in host plants have shown promising effects against *Fusarium* infection and mycotoxin production. Moreover, the use of natural compounds extracted from plants and other sources (botanicals, biopesticides) have also shown an effect against *Fusarium* and mycotoxins. Biocontrol agents and biopesticides can be applied to plant debris and on plants at flowering stage to suppress or limit the growth of toxigenic fungi.

Several studies on BCAs, such as species in the bacterial genera *Bacillus*, *Pseudomonas* and *Streptomyces*, and species in the fungal genera *Clonostachys*, *Cryptococcus* and *Trichoderma*, have proven to be antagonistic to *Fusarium* spp. and able to reduce mycotoxins in cereals, as reviewed by e.g. Shah et al. (2018), Sarrocco and Vannacci (2018), Powell and Vujanovic (2021), Habschied et al. (2021), Nešić et al.

(2021), and Chen et al. (2022). A recent study exploring the effect of a *Clonostachys rosea* strain in oats, when applied at anthesis, showed a reduction in both *F. graminearum* (79%) and DON level (80%) in mature kernels (Khairullina et al., 2023). Treatment of maize crop debris (stalk pieces) with *C. rosea* suppressed the perithecia development and ascospore discharge of *F. graminearum* from the infected stalks and thereby reduced the primary inoculum for FHB in subsequently grown wheat (Gimeno et al., 2020). In maize, strains of *Bacillus* were shown to suppress *A. flavus*, as a measure to control AFs (Vlajkov et al., 2021). Another study showed that endophytic *Bacillus* strains showed strong antagonism against FHB in wheat, with a reduction of around 80% (Kaul et al., 2022). Rojas et al (2020) studied the effect of naturally occurring fungal endophytes against *Fusarium* infection in wheat and found that some had biocontrol potential. In a study evaluating the efficacy of three commercial BCAs (based on *Streptomyces griseoviridis*, *Trichoderma asperellum* and *Pythium oligandrum*, respectively), all three showed high reduction of *F. graminearum* and DON in spikelets and straw of wheat (Pellan et al., 2021). Several microorganisms applied as seed treatments showed a potential to control seedling blight of maize (Koch et al., 2020).

Non-toxigenic strains of *A. flavus* were for more than 30 years ago shown to control development of AFs. Since then, several products based on non-toxigenic *A. flavus* strains have been developed and registered for biocontrol of AFs in maize and in other crops worldwide (reviewed by Sarrocco et al., 2019; Peles et al., 2021). The products are based on native strains selected in the area where they are going to be used. The application of non-toxigenic *A. flavus* has shown a reduction of AF contamination in maize by 73% (Savic et al., 2020).

Numerous phytochemicals have shown an effect against AFs (e.g. around 60% reduction) in maize (reviewed by Kumar et al., 2022). Phenolic compounds and essential oils extracted from plants have shown inhibitory activity against *F. graminearum* infections and DON and ZEN (Tian et al., 2016; Kalagatur et al., 2015; Drakopoulos et al., 2020a). A recent study with an extract of the medical plant *Zanthoxylum bungeanum* showed promising reduction of *F. graminearum* (up to 85%) and DON development (73%) in wheat heads (Abbas and Yli-Mattila, 2022).

Mulch layers of white mustard, Indian mustard or berseem clover used as a 'cut-and-carry biofumigation' were shown to decrease DON in wheat grain by up to 50%, 58% and 56%, respectively, and ZEN by 76%, 71% and 87%, respectively, compared with control (Drakopoulos et al., 2020b). Ashiq et al. (2022) investigated the potential of species within *Brassicaceae* used as cover crops to suppress *F. graminearum* and mycotoxins in maize by incorporation of the species in the soil. Maize was grown after *F. graminearum*-infected wheat crop as inoculum source. However, the severity of ear rot and stalk rot in maize was not reduced. Although the amounts of *F. graminearum* DNA in maize grain decreased by 58% upon incorporation of *Raphanus sativus* and DON was reduced by 50% after incorporation of *Brassica oleracea* compared to the control, the differences were not significant.

The main challenges with the use of BCAs and biopesticides are to develop formulations that are also highly effective in commercial use in the field (not only in the laboratory) and have a long shelf life with effect after storage.

2.10 *Organic farming*

The market in organic products has grown worldwide, and this includes feeding animals with organic feed. Many consumers prefer organic products because of the food safety interpretation based on the absence of synthetic pesticides or other contaminants, as well as mineral fertilisers during the growth of the grains and cereals. However, natural contaminants such as mycotoxins cannot be avoided.

After analysing 93 samples of organic cereals (wheat, oats, barley, spelt and rye) collected from local Italian markets, Juan et al. (2013) detected that 86% of the samples were multi-contaminated with mycotoxins, usually DON, NIV, ZEN, T-2/HT-2, and ENNs, and only 1% of the samples were contaminated with a single mycotoxin (ENNB1).

Comprehensive investigations in the UK detected no significant differences in levels of DON and ZEN in samples from organically vs. conventionally grown wheat, and no differences in DON and T-2/HT-2 concentrations in barley samples from the two growing systems, whereas the level of T-2/HT-2 was significantly lower in wheat and oats samples from organic production (Edwards, 2009abc). Brodal et al. (2016b) reviewed studies comparing the content of several mycotoxins in organically and conventionally grown wheat, barley, oats, and rye in temperate regions. Most studies found no significant differences in the levels of DON, ZEN, NIV and OTA in grain from the two farming systems, whereas the majority of the 12 comparisons of T-2/HT-2 reported lower levels in organically compared to conventionally produced grain. Organic farming combined with crop rotation resulted in decreased contamination of oats with T-2/HT-2, especially when oats were grown after a previous crop of grass (Kolawole et al., 2021). In Scotland, Daud et al. (2023) found significantly lower concentrations of T-2/HT-2, including conjugated glucoside forms, in organic compared to conventional oats. Bernhoft et al. (2022) performed a literature review on the mycotoxin levels in small-grain cereals cultivated in organic and conventional farming systems. In the selected 27 studies, 52% indicated that organic farming decreases DON contamination, 44% did not find any difference between the two systems (but as organic farms they were numerically less contaminated), and only one study (4%) indicated that organic grain is significantly more contaminated with mycotoxins than those from conventional farms. To exclude crop rotation as a confounding factor, these authors also compared the data from a 2-year field trial with wheat in Switzerland (Mäder et al., 2007; Griesshaber et al., 2004), demonstrating a significant decrease in the levels of DON in organic wheat (61 µg/kg) when compared with conventional one (82 µg/kg). These levels are, however, extremely low and represent food-quality wheat.

A more prominent decrease in contamination was described by Schneweis et al. (2005), after a 3-year trial whereby *Fusarium* mycotoxin levels were lower in organic (179 mg DON/kg and < 5 mg ZEN/kg) than in conventional (283 mg DON/kg and 28 mg ZEN/kg) wheat. Two studies on the FUM content in maize in organic compared to conventional farming (Pleadin et al., 2017; De Galarreta et al., 2015) did not indicate an effect of the farming system. There is no clear evidence that organic farming will increase or decrease the level of mycotoxin contamination. It is important to mention that many factors besides organic production may affect mycotoxin levels, since the agronomic practices may differ from region to region regarding plant variety, crop rotation, and tillage practices.

3 Harvest, storage and intervention at grain delivery

Harvest timing can have a major impact on the final levels of mycotoxins accumulated in the grain. The grain should be harvested as soon as possible after maturation and at appropriate moisture content. If necessary, the grain should be dried immediately to safe moisture levels that are lower than needed for fungal growth and mycotoxin production, which are around 18% for *Fusarium* spp. and 16% for *A. flavus*.

Delayed harvest of mature grain already infected by mycotoxigenic fungi may result in increased mycotoxin content, especially under moist weather conditions. A Swedish study based on more than 1400 cereal samples collected over 16 years found increased levels of DON and ZEN to be strongly associated with delayed harvest, whereas no effect of harvest date was observed on the levels of NIV or T-2/HT-2 toxins (Karlson et al., 2023). Kochiieru et al. (2021) reported higher levels of DON, ZEN and T-2 in wheat grain when harvest was delayed by 10 and/or 17 days after full maturity. Likewise, higher levels of DON and ZEN were observed in spring triticale when harvest was delayed, whereas no increase was detected in spring oats (Kochiieru et al., 2020). Compared to an average harvest day Edwards and Jennings (2018) found that four weeks delayed harvest of wheat increased the mean DON and ZEN concentrations in the grain 10- and 25-fold, respectively. Despite increased *Fusarium* spp. incidence level in the seed (from 6.9 to 13.9%), no difference in the DON content was observed after delayed harvest of barley grain (Xue et al., 2013). In maize, delayed harvest resulted in increased FUM in grain (Costa et al., 2018) and in silage (Mansfield et al., 2007). Jones et al. (1981) found reduced risk of AF contamination in maize with earlier harvests. Delayed harvest due to rainy weather can also result in increased risk of OTA, produced by *Penicillium* spp. and AOH/AME produced by *Alternaria* spp. (Li et al., 2001).

Before use it is important to ensure that all equipment for harvesting and transport are clean, dry, and without insects and signs of fungal contamination. To reduce the risk of mechanical damage to the grain during harvest, and the potential of mycotoxins to develop, the combine's cylinder speed and clearance should be adjusted to optimum functionality. The cutting height of the harvesting machines should be set to minimise soil contamination due to the widespread presence of *Fusarium* inoculum in the soil. Grain from lodged areas can have higher levels of fungal infections and mycotoxins and should be separated from healthier and better-quality grain.

Immediate storage of harvested grains will minimise the risk of mycotoxin contamination because delayed storage may expose the grain to temperature and humidity conditions that allow continued growth of toxigenic fungi and mycotoxin contamination (Munkvold, 2014). It is essential that raw materials are stored in suitable clean facilities free from contamination by fungi, insects and rodents. Magan et al., 2014, described 'Good postharvest storage practices for wheat grain', emphasising the importance of dry storage facilities, low grain moisture content and low grain temperature to ensure safe storage with minimal risk of continued growth of toxigenic fungi and OTA and DON production. The risk of DON production increases at moisture content > 20% and for OTA at 16-17% moisture content in grain. However, for long-term storage it is recommended to dry small grain cereals to 14-15% moisture content to prevent the growth of *Penicillium* and production of OTA and for maize 13% moisture to prevent growth of *Aspergillus* and development of AFs (Magan and Aldred, 2007).

Moreover, grain lots with different risks of mycotoxin contamination should not be mixed, and insects must be controlled as they can produce metabolic water resulting in 'hot spots'/pockets with increased moisture content and mycotoxin development in stored grain. Channaiah et al. (2014) referred to a simultaneous localisation and mapping (SLAM) strategy for management practices to prevent mycotoxin contamination during maize storage. The strategy includes sanitation, loading, aeration, and monitoring. Temperature and humidity monitoring will detect possible problems at an early stage when there is still time for corrective actions. Fleurat-Lessard (2017) reviewed integrated management strategies to prevent the development of grain spoilage mycotoxigenic fungi during storage, including monitoring and control of grain moisture and temperature, physical intervention and treatment methods and possible use of BCAs to prevent the development of fungi in grain bulks. Kumar et al. (2022) reported that essential oil from the plant *Toddalia asiatica* showed a promising effect against *Alternaria alternata* and its mycotoxins in stored wheat. Gaseous ozone can be used as an antimicrobial agent in grain storage to inactivate fungal spores and inhibit fungal growth. Some studies have reported that ozone can modify several mycotoxins, including AFs, FUM (FB_1) and DON (McKenzie et al., 1997; Sun et al., 2016).

Quantitative site-specific risk assessments or predictive models for mycotoxin accumulation during the growing season could contribute significantly to management efficiency (e.g., Battilani et al., 2003). Most of these models are tools for providing an early warning for the risk of unacceptable mycotoxin levels in grain at harvest but depend on within-season weather data (Hjelkrem et al., 2022).

A system for rapid screening of DON content in grain samples either immediately before or at delivery to the 'elevator'/'manufacturing plant' has been established in Norway (https://www.felleskjopet.no/alle-artikler/alle-artikler-planteproduksjon/artikler-korn/har-du-kontroll-paa-kornlageret/; https://kunde.kilde.no/Strand/kornogoljefroe/4/; https://www.fiska.no/kraftforkvalitet-korn-23). This makes it possible to direct the grain lots to the appropriate feed concentrate factory producing feed either for pigs/poultry, or for ruminants, or in worst case to outsource grain lots with excessive mycotoxin levels before entering.

4 Measures during primary processing of cereal grains

There are several measures that can be taken during processing to decrease feedstuff contamination before diet preparation. After harvest and before use, cereal grains are mechanically cleaned to remove impurities such as dirt, dust, weed seeds, straw pieces, chaff and small/lightweight/shrivelled, broken, and damaged kernels. As mycotoxins usually have highest concentrations in glumes, hulls/husks, and outer layers of the kernel (e.g. Cheli et al., 2013; Khabiti et al., 2014; Tibola et al., 2016; Meyer et al., 2022; Pietri et al., 2009), and small kernels have a high share of these structures, mycotoxin-contaminated kernels are often small and lightweight. The degree of penetration of fungal mycelium into the kernel and development of mycotoxins depends on the time of fungal infection in relation to heading, flowering and maturation stages of the plants. Moreover, mycotoxins are often accumulated in the grain dust (Straumfors et al., 2015; Colovic et al., 2019a; Limay-Rios and Schaafsma, 2021). Cleaning procedures, such as aspiration, brushing, sieving and sorting, including optical sorting, based on size, weight/density, shape, and colour are important steps for removing mycotoxin contaminated kernels/reducing the mycotoxin contamination of grain batches. These procedures have the advantage of decreasing mycotoxin contamination without affecting nutrient levels.

Sometimes mycotoxin-contaminated cereal grains have a normal size and weight (Seitz and Bechtel, 1985), and not all grain species will decrease in size and weight with the same intensity when infected by toxigenic fungi. For example, *Fusarium*-damaged kernels in wheat can undergo a 50% decrease in grain weight, whereas only 20% decrease has been observed in barley (Tekauz et al., 2000; Janssen et al., 2018). Hyperspectral imaging is considered to have potential for separating asymptomatic kernels despite high mycotoxin concentrations (Femenias et al., 2020). It has been shown that DON-containing healthy looking wheat kernels emitted light (fluorescence) when exposed to UV light and thus could be separated from kernels containing lower DON levels (Ochodzki and Goral, 2023).

As mycotoxins are often detected in the outer parts and layers of the kernels, dehulling and pearling/scouring are performed to remove husk/hulls and the outer layer of the grains. These procedures will remove the most contaminated parts of the grains before grinding/milling for feed and food production.

Several studies and reviews have shown that the above-mentioned procedures often decrease mycotoxin contamination in grain (e.g. Pettersson et al., 2008; Karlovsky et al., 2016; Peng et al., 2018; Colovic et al., 2019b; Schaarschmidt and Fauhl-Hassek, 2018, 2021; Brodal et al., 2020; Sipos et al., 2021; Leslie et al., 2021; Pascale et al., 2022; Hoffmans et al., 2022). In their comprehensive reviews, Schaarschmidt and Fauhl-Hassek (2018 and 2021) compiled studies that investigated the effects of cleaning, sieving, etc. of grain on mycotoxin concentrations in wheat and maize, respectively. For DON in wheat, where most of these investigations have been carried out, they found that some studies reported up to around 70-90% reduction in mycotoxin levels after cleaning, sieving and/or gravity separation (removing screenings) compared to the raw grain materials. Similar effects were also reported for NIV and ZEN in wheat, although only limited studies were available for these mycotoxins. Around 40% reduction in OTA was reported after cleaning, including scouring.

Mechanical cleaning can remove several different mycotoxins, especially in highly contaminated feedstuffs. For instance, after a highly DON-contaminated wheat batch (24 mg/kg) was subjected to simple mechanical cleaning (sieving, high air pressure and gravity separator), the cleaned fraction showed a 75% decrease of DON levels (6 mg/kg) and the rest (1% of the mass) accumulated 48 mg/kg DON (Santos RR; unpublished data). A multi-mycotoxin analysis showed that several other mycotoxins such as 3-ADON, 15-ADON, DON-3G, T-2 toxin, NIV, FUM, ZEN, OTA, ENNs, BEA, and MON, were present in the batch, especially in the rejected fraction (Figure 15.2). Similar findings of mycotoxins were reported in oats by Brodal et al. (2020).

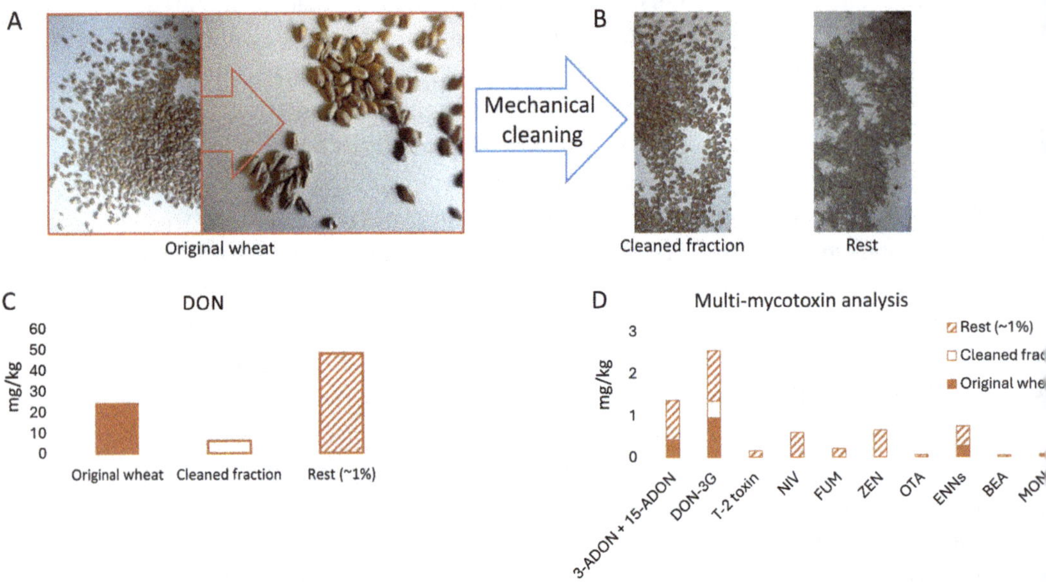

FIGURE 15.2 Representative images of a wheat sample before (A: original wheat) and after mechanical cleaning cleaned fraction and rest). In panel C, the levels of DON in the original wheat (24 mg/kg), cleaned fract (6.2 mg/kg) and rest fraction (48 mg/kg) are given, where the rest represents 1% of the mass loss during cleaning process. In panel D, a multi-mycotoxin analysis of the three fractions (original, cleaned, and r shows the presence of 3-ADON+15-ADON, DON-3G, T-2 toxin, NIV, FUM, ZEN, OTA, ENNs, BEA, MON, especially in the rejected fraction, i.e. the rest. Source: unpublished data; Santos RR.

In maize, many studies on the reduction of AFs, FUM, OTA, DON, and ZEN have been carried out, but also the reduction of other toxins, e.g. NIV, MON, T-2, HT-2, and AOH has been studied (Schaarschmidt and Fauhl-Hassek, 2021). In their comprehensive review Schaarschmidt and Fauhl-Hassek (2021) reported that several studies found reductions in AFs in the range 30-100% when using sieving, brushing, scouring, gravity separation and aspiration. However, they also reported that several studies found no or low reduction of AFs in maize batches after different cleaning operations. In a recent study, Pascale et al. (2022) found a reduction of FUM from 27 to 67%, of DON from 36 to 67%, and of ZEN from 67 to 87% by industrial scale sieving and density separation of maize. Moreover, Schaarschmidt and Fauhl-Hassek (2021) commented that reductions of AFs in the endosperm by dry milling varied from only a few percent up to around 90%.

In Table 15.2 and Table 15.3, we give some examples of the effects on primary processing (cleaning, size sorting/sieving, dehulling, scouring) on mycotoxin concentrations in barley and oats, respectively. In barley (Table 15.2), Perkowski et al. (1998) found a DON reduction of 80% by removing the small kernel fraction by size sorting (sieve size 2.5 mm), however the small kernel fraction contained 50% of the grain. In another study, Perkowski et al. (2003), also found 80% reduction in DON with the same sieve size, whereas the small kernel fraction then contained only 13% of the sample weight. Hietaniemi et al. (2009) reported approximately 50% reduction of DON by laboratory sorting of barley samples using a sorter with 2 mm sieve. Trenholm et al. (1991) showed 40% DON reduction by dehulling with 13% mass loss, whereas Clear et al. (1997) observed 59%, 49% and 86% reduction in DON content after dehulling during harvest, after laboratory dehulling, and after laboratory pearling (with 40% of original weight removed), respectively, in hulless barley, i.e. the lemma and palea in such cultivars are adhering loosely to the caryopsis. They mentioned that in addition to reduced DON levels after removing hulls at harvest, the nutritional value of the grain for feed was also improved. House et al. (2003) reported that the initial DON content was reduced after 15 s of pearling, with a grain mass loss of 15%. Moreover, Khatibi et al. (2014) found a reduction in DON levels of 37% and 85% in milled barley materials after removing hull-tissues by dehulling/milling with a roller mill and a precision mill, respectively. However, the mass loss from precision milling (23%) was almost twice as high as with roller milling (12%). Size sorting of barley has been reported to reduce 68% of HT-2 toxin and 81% of T-2 toxin using a sorter with 2.5 mm sieve (Perkowski et al., 2003). Another study showed a 50% reduction of T-2/HT-2 toxins using a sorter with 2 mm sieve (Hietaniemi et al., 2009). Perkowski et al. 2003 also reported a reduction of NIV by 94% and ZEN by 85% after size sorting, whereas Trenholm et al. (1991) did not detect ZEN in the grain after dehulling.

TABLE 15.2 Examples of the reduction (percentage) in the level of mycotoxins (DON, T-2, HT-2, NIV, and ZEN) and mass loss (percentage) by cleaning, sorting, dehulling, etc. (treatment method) of barley grain. See references for details on treatment methods and mycotoxin levels.

Treatment method	Mycotoxin	Reduction (%)	Mass loss (%)
Size sorting (2.5 mm sieve) [a]	DON	80	50
Size sorting (2.5 mm sieve) [b]	DON	80	13
Size sorting (2 mm sieve) [c]	DON	50	N.R.
Dehulling [d]	DON	40	13
Dehulling (hulless cultivars, in laboratory) [e]	DON	49	N.R.
Dehulling (hulless cultivars, at harvest) [e]	DON	59	N.R.
Pearling [f]	DON	86	40
Abrasive type dehulling (pearling) [f]	DON	66	15
Roller mill/dehulling [g]	DON	37	12
Precision mill/dehulling [g]	DON	85	23
Size sorting (2 mm sieve) [c]	T-2/HT-2	50	N.R.
Size sorting (2.5 mm sieve) [b]	T-2 and HT-2	81 (T-2); 68 (HT-2)	13
Size sorting (2.5 mm sieve) [b]	NIV	94	13
Size sorting (2.5 mm sieve) [b]	ZEN	85	13
Dehulling [d]	ZEN	100	14

N.R.: Not reported

a Perkowski et al. (1998); b Perkowski et al. (2003); c Hietaniemi et al. (2009); d Trenholm et al. (1991); e Clear et al. (1997); f House et al. (2003); g Khatibi et al. (2014).

In oats (Table 15.3), a Norwegian study (Brodal et al., 2020) showed that by removing the small kernel fraction (sieve size 2.2 mm) the mean concentration of DON was reduced by 24% (from 191 to 145 µg/kg) in one series of samples compared to unsorted grain. In the same study, T-2/HT-2 toxins were reduced by 56% (from 745 to 328 µg/kg) in one series of samples and by 32% (from 178 to 121 µg/kg) in another series of samples. The mass losses were 15% and 21%, respectively in the two series. In the same study, ENNB was reduced by 44% (from 1059 to 594 µg/kg) in one sample series. In addition, despite low levels, their analyses also showed a trend towards reduced levels of ADON, NIV, ENNA, ENNA1, ENNB1, and BEA after removing the small kernel fraction. In Finland, Hietaniemi et al. (2009), reported around 30-40 % reduction in the level of DON and 30-35% reduction in T-2/HT-2 toxins by size sorting oat kernels (sieve size 2 mm). They obtained 67-91% reduction in DON and could not detect T-2/HT-2 after dehulling in addition to size sorting. As a large proportion of the mycotoxins in oats is in the husks/hulls and the seed coat, a considerable reduction in mycotoxin content can be obtained by dehulling. The reduction in the content of DON by dehulling has been reported to vary from 40% to more than 90% (e.g. Scudamore et al., 2007; Yan et al., 2010; Ivanova, 2017; Tittlemier et al., 2020; Meyer et al., 2022) and a substantial reduction in the content of

TABLE 15.3 Examples of the reduction (percentage) in the level of mycotoxins (DON, 3+15-ADON, 3-ADON, DON-3G, T-2, HT-2, NIV, ENNB) and mass loss (percentage) by cleaning, sorting, dehulling, etc. (treatment method) of oat grain. See references for details on treatment methods, and mycotoxin levels.

Treatment method	Mycotoxin	Reduction (%)	Mass loss (%)
Size sorting (2 mm sieve)[a]	DON	30 - 40	N.R.
Size sorting (2 mm sieve) and dehulling[a]	DON	67 - 91	N.R.
Size sorting (2.2 mm sieve)[b]	DON	24	21
Size sorting (2.2 mm sieve)[b]	ADON	27	21
Commercial cleaning and dehulling[c]	DON	88	29
Dehulling[d]	DON	55 - 86	N.R.
Dehulling[e]	DON	94	31 - 43
Dehulling[f]	DON	87	28
Dehulling[g]	DON	40 – 95 (72)	23 - 31
Commercial cleaning and dehulling[g]	DON	4 – 80 (61) [L] 2 – 91 (57) [S]	26
Dehulling[e]	3-ADON	91	31 - 43
Dehulling[f]	3-ADON	100	28
Dehulling[e]	DON-3G	83	31 - 43
Dehulling[f]	DON-3G	48	28
Size sorting (2.2 mm sieve)[b]	T-2/HT-2	32 - 56	15 - 21
Size sorting (2 mm sieve)[a]	T-2/HT-2	30 - 35	N.R.
Size sorting (2 mm sieve) and dehulling[a]	T-2/HT-2	100	N.R.
Commercial cleaning and dehulling[c]	T-2/HT-2	89 and 95	29
Dehulling[h]	T-2/HT-2	95 (75-98) 58-98	N.R.
Cleaning and dehulling[i]	T-2/HT-2	80 - 95	N.R.
Cleaning[j]	T-2/HT-2	0 - 100	N.R.
Cleaning and dehulling[j]	T-2/HT-2	94 - 100	N.R.
Cleaning and dehulling[k]	T-2/HT-2	82 and 88	N.R.
Dehulling[e]	T-2/HT-2	94 and 95	31 - 43
Dehulling[f]	HT-2	100	28
Dehulling[g]	T-2/HT-2	25 – 97 (84)	23 - 31
Commercial cleaning and dehulling[g]	T-2/HT-2	60 – 97 (85) [L] 15 – 93 (66) [S]	26
Commercial cleaning and dehulling[c]	NIV	> 98	29
Dehulling[f]	NIV	100	28
Size sorting (2.2 mm sieve)[b]	ENNB	44	15

[L] Large kernels; [S] Small kernels; N.R.: Not reported.

a Hietaniemi et al. (2009); b Brodal et al. (2020); c Scudamore et al. (2007); d Yan et al. (2010); e Ivanova et al. (2017); f Tittlemier et al. (2020); g Meyer et al. (2022); h Edwards (2007); i Petterson et al. (2008); j Schwake-Anduschus et al. (2010); k Pettersson et al. (2011).

3-ADON and DON-3G has also been reported after dehulling (Ivanova et al., 2017; Tittlemier et al., 2020). Dehulling has mostly shown a large reduction in the levels of T-2 + HT-2 in oat grains varying from around 80% up to 100% reduction (Scudamore et al., 2007; Pettersson et al., 2008; Schwake-Anduschus et al., 2010; Pettersson et al., 2011; Ivanova, 2017; Tittlemier et al., 2020; Meyer et al., 2022; Edwards, 2007). A high effect of commercial cleaning and dehulling on the NIV content (reduced by > 98%) was reported by Scudamore et al. (2007), and Tittlemier et al. (2020) did not detect NIV after dehulling.

Sclerotia of EAs, sometimes found in rye, can be removed by optical-electronic colour sorters (Miedaner and Geiger, 2015). There is not much information on the behaviour of *Alternaria* toxins during cleaning and processing procedures of cereals. However, Janić Hajnal et al. (2019) found during milling that the largest share of *Alternaria* toxins was concentrated in outer parts of the wheat kernels, as around 70 to 100% were detected in bran and short fractions. Without quantifying the effects, Bretträger et al. (2023) showed that removing black-coloured *Alternaria* infected barley kernels by optical sorting could reduce the levels of *Alternaria* mycotoxins.

Several studies have shown that the effects of primary processing can vary considerably among grain batches. The results depend on the different cleaning, sorting and dehulling technology and equipment, such as aspirators, size sorters, screening graders, gravity tables, optical sorters, dehullers, etc. Their usage setting and performance influence the level/amounts of materials removed by cleaning. In addition, the reducing effect has been reported to be higher in severely contaminated grain compared to less contaminated (e.g. Scudamore et al., 2007; Meyer et al., 2022). In general, milling does not affect the total amount of mycotoxins in a grain batch but causes a redistribution in which e.g. DON or OTA is accumulated in the bran fraction and lower levels are found in flour (Kochiieru et al., 2019; Khaneghah et al., 2018).

Concerning animal feed production, the clean fraction after dehulling (Ríos et al., 2009; Odukoya et al., 2024) and milling (Schaarschmidt and Fauhl-Hassek, 2021) may be applicable to a very small category of animals, e.g. those used as breeders, whereas the obtained screenings and by-products grains (e.g. hulls and bran) are largely used in the feed production for the other animal categories. Usually, the by-products of these processes will contain very high levels of mycotoxins because most of the mycotoxins are located in the outer layers of the kernels (Tibola et al., 2016). These authors showed an increase of 28% in DON contamination in the wheat bran compared to the original feedstuff. Therefore, their inclusion levels in the diets should be taken with caution not only because of nutrient composition, but also because they may increase exposure to mycotoxins.

Assuming that the rejected fraction has a very low economic value, it is necessary to calculate the advantageous effect of sorting on the decrease of mycotoxins levels and, in the end, on the increased animal performance and economic gain. Considering that animal feed is often contaminated with a mixture of mycotoxins, a reduction in contamination before diet production will improve feed quality and animal production performance. The dehulling process will concentrate the contaminants in the hulls and may result in a mass loss of 10-35% with DON reduction from 45-70% in wheat, and a mass loss of 15-40% with DON reduction from 66-90% in barley (Peng et al., 2018), whereas mechanical cleaning may result in lower mass loss (1-7%) (Pascale et al., 2022), but can reach 30% (Tibola et al., 2016). The mass loss will depend on the quality of the batch, whereby sieving a maize batch with a high number of kernels broken in small particles may result in a high mass loss (69%) (Trenholm et al., 1991), which must be considered when deciding to apply such a procedure.

Feed cost ranges from 15% (sows) to > 60% (fattening pigs) of the production costs in the pig industry (Pomar et al., 2009; Solà-Oriol and Gasa, 2017). Similar numbers are observed for other livestock species. Hence, any extra processing to improve feed quality will result in increased financial investment. Therefore, the selected methods must consider the level of contamination, the efficacy, and the lowest cost of producing a high-quality diet.

5 Methods for managing/reducing mycotoxins in silage

In general, mycotoxin contamination of agricultural products should be prevented or counteracted by using pre-harvest or post-harvest strategies. Moreover, the crop production environment has a profound influence on the risk of mycotoxin contamination. The simplest strategy is based on the prevention of mycotoxin formation in the field, even if this is not ever possible. Cultivation of resistant cultivars, correct sowing period, crop rotations, burial of residues, adequate N supply, and prevention of damage to the kernels are all useful strategies to prevent the development and proliferation of mycotoxins.

The contamination of silage is correlated with both manageable and non-manageable factors. Silage management factors that are under the control of the farmer are forage species and agricultural background, stage of maturity or moisture concentration at harvest, harvesting and ensiling methods, type of storage structure, use of silage additives, feed-out methods, feed bunk management, and diet formulation (Mahanna and Chase, 2003). As reported by Bernardes et al. (2018), non-manageable climate-related factors are common or specific to warm or cold regions and can adversely affect silage production and utilisation but also influence some of the controllable factors mentioned above, such as moisture concentration at harvest.

Before the chopping phase, proper harvest scheduling is important in order to control the level of mycotoxin contamination in forages. However, harvest should be planned to ensure optimum yield of crops, proper DM concentration and a good nutritive value. Furthermore, the cutting height of the harvesting machines should be set to minimise soil contamination due to the widespread presence of *Fusarium* spores in the soil (Jouany, 2007).

Concerning storage of forage as silage, a low oxygen concentration and augmentation of carbon dioxide are efficient in preventing mould development (Oude Elferink et al., 2000). Consequently, all ensiling stages, such as aerobic, fermentation, stable, feed-out or aerobic spoilage phases, should be controlled and optimised as much as possible to assure adequate conservation of ensiled crops preventing proliferation of undesirable microorganisms such as mycotoxigenic fungi (Dell'Orto et al., 2015; Oude Elferink et al., 2000). However, ensiling procedures are not standardised, and farmers use different procedures to ensile and store silages (Gallo et al., 2016a,b). Incorrect silo management conditions, such as inappropriate DM content of crop at harvest time, inadequate particle length, slow silo filling, imperfect mass sealing, poor mass compression, delay in mass pH drop, air penetration in ensiled mass or inappropriate unloading equipment and techniques, could compromise any of the above-mentioned ensiling phases, exposing silage to the risk of air penetration and consequent activity of aerobic undesirable spoilage microorganisms (Dunière et al., 2013; Muck et al., 2018).

5.1 *Ensiling*

When focusing on ensiling procedures, to minimise or prevent the growth of mycotoxins or undesirable microbes, forages should be harvested at the recommended DM concentration or maturity stage and silos should be filled rapidly after harvest, tightly packed to achieve recommended densities and completely sealed to preserve anaerobic conditions (Jouany, 2007). Low silage pH and anaerobic conditions during ensiling inhibit the growth of spoilage and toxigenic fungi reducing the production of mycotoxins (Pahlow et al., 2003). During ensiling, moulds are often found in poorly compacted areas such as the top layer or sidewalls of silos, which are frequently more prone to oxygen infiltration than the rest of the silo (Borreani and Tabacco, 2010; Gallo et al., 2016a,b). Several studies indicate that sealing silos using oxygen-barrier plastic films can reduce oxygen penetration, mould infestation, and spoilage of silage (Borreani and Tabacco, 2008). Other studies have shown that such films limit aflatoxin production (Amaral et al., 2010; Cavallarin et al., 2011).

In well-preserved silage, most of the fungal growth is restricted by anaerobic or microaerophilic conditions and by organic acids (Kung et al., 2018). When oxygen gains access to ensiled mass, due to inadequate sealing during storage or at feed-out, yeasts and acetic acid bacteria start to oxidise the preservative acids and residual water-soluble carbohydrates. Concomitant aerobic deterioration results in a rise of pH and temperature. These conditions allow fungi to proliferate (Borreani and Tabacco, 2010; Bernardes et al., 2012). The dispersion of oxygen into a silage mass is influenced by silo management in the area of compaction, sealing and unloading. Compaction is closely related to the DM content of the ensiled mass, so ensiling forage crops at optimal DM content (for grass 350–450 g/kg – in baled silaged an higher dry matter is allowed – and for maize 300–350 g/kg) is crucial (Wilkinson and Davies, 2013). If compaction of the ensiled material is important, it is necessary to seal silages quickly and carefully. Furthermore, the seal should remain undamaged during the whole ensiling period. This is confirmed by Petersson (1998) and Richard-Molard et al., (1980), who observed that *P. roqueforti* spores stored under airtight conditions for three months had a reduced germination ability. Thus, leaving silos sealed for at least three months should have a beneficial effect on prevention of fungal growth.

5.2 *Feed-out phase*

Aerobic deterioration of silage during the feed-out phase is a major problem for farm productivity and feed quality worldwide. In contrast to the surface losses occurring in storage, air has a greater effect during the feed-out phase because exposure of the silage to oxygen is inevitable once the silo is opened, and air can penetrate via the silage face, up to distances of 1 to 4 m, especially in the periphery of the silo (Pahlow et al., 2003; Vissers et al., 2007). The greater penetration on the periphery is due to the higher porosity of the silage in these areas, and movement of oxygen is proportional to porosity (Pitt and Muck, 1993). To reduce mould infestation or mycotoxin contamination during feed-out, aerobic stability enhancing measures should be employed, including maintaining a straight firm silo face, feeding out the silage at a minimum rate of 10 to 16 cm per day during the winter and 25 to 35 cm in the summer, and feeding the silage immediately after removal from the silo (Jouany, 2007; Borreani and Tabacco, 2010). The dynamic of bacterial, yeast, fungal communities and mycotoxin production during ensiling of alfalfa silage and after silage air exposure was studied (Gallo et al., 2021). In brief, the study was performed using the mini-silos technique that simulates the bunker silos environment. Mini-silos were opened at different ensiling times and samples were analysed for fermentation parameters and microbiological populations (colony forming units/g) after 0 (before ensiling), 1, 2, 3, 7 and 14 days of ensiling focusing on *Fusarium*, *Penicillium*, and *Aspergillus* spp. During ensiling (Figure 15.3a), both homofermenter LAB and heterofermentative bacteria appeared dominant whereas moulds, yeasts, *Enterobacteriaceae* were inhibited. No *Acetobacter* spp. was counted. *Aspergillus*, *Fusarium* spp. and other fungi increased only during the first days after closing the silo (i.e. 2 days of ensiling) and then decreased (Figure 15.3b). In contrast, *Penicillium* spp. remained stable during all the ensiling phases. Among mycotoxins, DON increased up to about 562 µg/kg DM with ensiling, whereas the other *Fusarium*-produced mycotoxins remained constant. When exposed to air, moulds, yeasts, *Enterobacteriaceae*, *Acetobacter*, *Fusarium* and *Penicillum* spp. grown with different dynamic behaviours and MPA concentration increased up to 2,423 µg/kg DM, whereas *Aspergillus* spp. decreased. These results showed that different fungal species have different developmental dynamics in the ensiled mass and during the feed-out and exposure phase of the silo to air (Figure 15.3c).

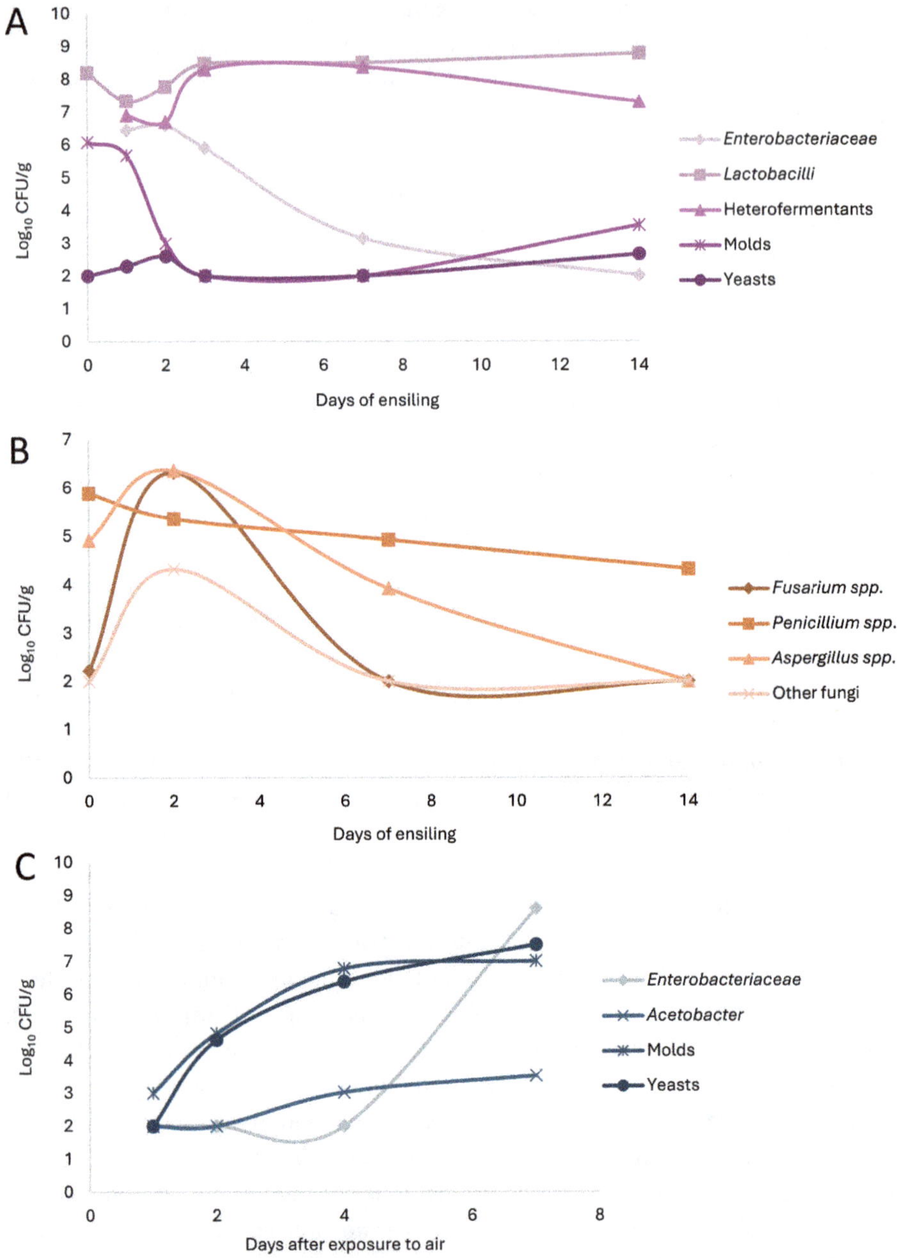

FIGURE 15.3 Dynamic evolution of bacterial, yeast and fungal communities during ensiling (A), and fungi dynamic evolution during ensiling (B) and feed-out time (C). Source: Gallo et al., 2021.

These results have to be take into account and analysed in depth despite the fact that the main pattern of aerobic deterioration has been known for approximately three decades (Wilkinson and Davies, 2013). When oxygen is introduced into the ensiled mass, aerobic microorganisms start to grow or re-grow, firstly respiring soluble substrates and then more complex compounds. Yeasts are generally the initiators of aerobic deterioration, consuming sugars and fermentation acids and raising silage temperature and pH (Pahlow et al., 2003). With an increase of pH, bacilli and other aerobic bacteria grow, increasing the temperature further. Finally, moulds complete the silage deterioration.

6 Conclusions

Mycotoxin contamination of cereal grains begins with infection of crop plants in the field by mycotoxigenic fungi, mainly *Fusarium*, and can develop further during improper storage, mainly by *Penicillium* and *Aspergillus*. Moreover, unsuitable conditions during harvesting and handling might enhance mycotoxin development. The grain can become contaminated to an extent that makes it less suitable for consumption by humans and livestock. As mycotoxins in grains are quite heat stable and difficult and costly to reduce, it is important to prevent their development in the field and in storage as much as possible. Several measures/methods both pre-harvest, during harvesting/storage and during processing procedures can decrease the risk of mycotoxin contaminations in food and feed.

Pre-harvest methods consist of Good Agricultural Practice (GAP) and Integrated Pest Management (IPM). Selection of a cereal variety with resistance to *Fusarium* and mycotoxin development is a sustainable and cost-effective method of reducing mycotoxin risk in grains. Unfortunately, not many highly resistant varieties are available; however, intensive breeding for better host resistance is ongoing in several countries. Crop rotation, i.e. growing of cereals after a crop that is not a host of *Fusarium*, e.g. legumes/pulses, brassica crops, potatoes, sugar beet, can reduce the amount of disease inoculum in the field and the risk of mycotoxin contamination. Maize is the riskiest pre-crop. Tillage practice that buries the infected crop residues, mainly deep ploughing, can reduce the disease inoculum, although some varying effects on mycotoxin development have been reported.

Moreover, the use of seeds free from *Fusarium* will reduce the risk of introducing the pathogen to 'healthy' areas. Weed management by herbicides and/or by ploughing is recommended to remove weeds that can be colonised by mycotoxigenic fungi and act as an inoculum reservoir. Weed control will reduce moisture in the canopy and prevent fungal growth and mycotoxin development. Regardless of reducing the inoculum potential in the field by crop rotation, tillage, weed control and use of healthy seeds, long-distance dispersal of ascospores of *F. graminearum* from other fields may still cause fungal infection and mycotoxin development. Also, airborne conidia spores of *A. flavus* and *Fusarium* spp. from maize debris in a region might 'override' the residue management.

Moderate N-fertilisation in accordance with the uptake in plants seems not to influence the mycotoxin risk, although dense crop canopies due to high N application has been shown to increase the risk of *Fusarium* growth and mycotoxin development. Under dry conditions irrigations can mitigate drought stress; however, it is important to avoid irrigation during flowering and maturation of the crop.

The use of insecticides to control ECB and other moths during the growing season is important to decrease insect damage (important entrance for fungal infection) and thereby reduce the risk of fungal infection and mycotoxin contamination in maize.

Fungicides are commonly used to control *Fusarium* and mycotoxin development. The type of fungicide (mode of action), the timing and rate, and the level of resistance of the crop plant are important for the efficacy of the treatment. The most efficient and used fungicides are azoles such as tebuconazole, metconazole and prothioconazole, however, they may not provide more than around 50% reduction in FHB disease severity and reduction of DON in grain. Moreover, little to no effect has been observed on the content of T2/HT2 toxins in grain of oats and barley after the use of azoles or other types of fungicides. A new promising fungicide, pydiflumetofen, has shown an improved effect against FHB and DON concentrations.

To strive for fungicidal use only when there is a risk of mycotoxin production, i.e. when *Fusarium* inoculum is present and the weather conditions are favourable for disease development, forecasting models have been developed and incorporated into online platforms to assist farmers in decisions on the need for fungicide applications.

While noting the beneficial effect of pesticides, e.g. fungicides, insecticides and herbicides, in crop protection, there must be consideration for residues that may accumulate and interact with mycotoxins, and be harmful for animals and human health (Guo et al., 2020).

Biological control agents (BCAs/antagonistic microorganisms) and natural compounds extracted from plants (botanical, biopesticides) can be applied to plant debris and on plants at flowering stage to suppress or limit the growth of mycotoxigenic fungi and mycotoxin development. However, it is a challenge to develop formulations that are highly effective in commercial use in the field (not only in the laboratory) and have a long shelf life with effect after storage.

There is no clear evidence that organic farming will increase or decrease the level of mycotoxin contamination. Several other factors, such as plant variety, crop rotation, and tillage practices may differ from region to region and may affect mycotoxin levels in other ways than just the farming system.

Delayed harvest of mature grain already infected by mycotoxigenic fungi may result in increased mycotoxin content, especially under moist weather conditions.

The grain should therefore be harvested as soon as possible after maturation and, if necessary, dried immediately to safe moisture levels to prevent (further) development of mycotoxins. Harvesting and transport equipment should be clean, dry, and without insects and signs of fungal contamination. To reduce the risk of mechanical damage to the grain during harvest, which increases the potential of mycotoxins to develop, the combine harvester should be set to optimum functionality. Grain from lodged areas can have higher levels of fungal infections and mycotoxins and should be separated from healthier and better-quality grain. Rapid screening of DON content in grain samples either immediately before or at delivery to the elevators and feed mills will enable outsourcing of grain lots with excessive levels and reduce the risk of unacceptable grain being used in feed production. Moreover, some forecasting models are designed to predict the risk of mycotoxins in harvested grain.

Good storage practices include dry and clean storage facilities, low grain moisture content and low grain temperature. To prevent temperature and humidity conditions that can result in continued growth of toxigenic fungi and mycotoxin contamination, the harvested grain should be dried and stored immediately after harvest. The risk of DON production increases when moisture content in grain is > 20% and for OTA at 16-17% moisture. For long-term storage it is recommended to dry grain to 14-15% moisture to prevent growth of *Penicillium* and production of OTA, and for maize 13% moisture to prevent growth of *Aspergillus* and development of AFs. Moreover, insects must be controlled in stored grain as they can produce metabolic water resulting in 'hot spots'/pockets with increased moisture content resulting in fungal growth and mycotoxin development.

There are several measures during processing to decrease feedstuff contamination before diet preparation. After harvest and before use, cereal grains are mechanically cleaned to remove impurities such as dirt, dust, weed seeds, straw pieces, chaff and small/lightweight/shrivelled, broken, and damaged kernels. Cleaning procedures, such as aspiration, brushing, sieving and sorting of grains, including optical sorting, based on size, weight/density, shape, and colour can significantly contribute to removal of the small, often mycotoxin-contaminated kernels, and reducing the mycotoxin contamination of grain batches. These procedures have the advantage of decreasing mycotoxin contamination without affecting nutrient levels. As mycotoxins are often located in the outer parts and layers of the kernels, dehulling and pearling/scouring are performed to remove husk/hulls and the outer layer of the grains. These procedures will remove the most contaminated parts of the grains before grinding/milling for feed and food production. Frequently discarded after cleaning grains, by-products including husks, hulls, and bran end up as sources of dietary fibre for farm animals (Pinotti et al 2016). As such, while preparing an animal diet, it is of the utmost importance to take their inclusion levels into account.

Despite different degrees of mycotoxin reduction by primary processing of cereal grains, overall results show that the level of mycotoxins can be significantly reduced by cleaning and removal of small kernels, and the quality can be further enhanced by removing husk/hulls and the outer layers of the kernels. These procedures might increase the number of grain lots that can be accepted as safe for food and feed.

By using appropriate methods for cutting, ensiling, and feed-out, it is possible to effectively control a certain level of contamination in the silage with mycotoxins. Scheduling the harvest is crucial in order to guarantee optimal DM concentration and prevent soil contamination during crop cutting. In order for ensiling to be effective, it is necessary to tightly compact the material and seal it using oxygen-impermeable plastic layers. This prevents the entry of oxygen, thereby generating an anaerobic environment that inhibits the formation of moulds. When silage is exposed to air, it causes the pH and temperature to rise, creating favourable circumstances for the growth of fungal species and the production of more mycotoxins. During the feed-out period, it is impossible to prevent oxygen from entering the silo. Hence, it is recommended to cut the silo at a pace of at least 10 to 16 cm per day in winter and 25 to 35 cm per day in summer.

Information on the management of mycotoxins in cereal crops has been listed in several national and international guidelines, including 'Commission Recommendation on the prevention and reduction of *Fusarium* toxins in cereals and cereal products' (EU 2006) and 'Code of practice for the prevention and reduction of mycotoxin contamination in cereals' (Codex Alimentarius 2003/2014). Nevertheless, it is not possible to completely avoid mycotoxin development if fungal inoculum is present and the weather conditions are favourable for fungal growth. However, a recent study showed that there might be a potential for better exploitation of opportunities for mycotoxin management at farm level among European farmers (Janssen et al., 2021). The study concluded that incentives, such as extra payments when wheat contains low levels of mycotoxins and testing of mycotoxin contamination for free, would influence farmers to change their agronomic approach to reduce the risk of *Fusarium* infection and mycotoxin development. Secundary fermentations of both aerobic and anaerobic migrorganisms such as Clostridia strains can then take place into stored vegetable mass (Mosconi et al., 2023).

References

Abbas, H.K., Shier, W.T. and Cartwright, R.D., 2007. Effects of temperature, rainfall and planting date on aflatoxin and fumonisin contamination in commercial Bt and Non-Bt corn hybrids in Arkansas. Phytoprotection 88: 41-50. https://doi.org/10.7202/018054ar.

Abbas, H.K., Wilkinson, J.R., Zablotowicz, R.M., Accinelli, C., Abel, C.A., Bruns, H.A. and Weaver, M.A., 2009. Ecology of *Aspergillus flavus*, regulation of aflatoxin production, and management strategies to reduce aflatoxin contamination on corn. Toxin Reviews 28: 142-153. https://doi.org/10.1080/15569540903081590.

Abbas, A., Yli-Mattila, T., 2022. Biocontrol of *Fusarium graminearum*, a causal agent of fusarium head blight of wheat, and deoxynivalenol accumulation: from *in vitro* to *in planta*. Toxins 14: 299. https://doi.org/10.3390/toxins14050299.

Altman, J. and Rovira, A.D., 1989. Herbicide-pathogen interactions in soil-borne root diseases. Canadian Journal of Plant Pathology 11: 166–172. https://doi.org/10.1080/07060668909501133.

Amaral, R.C., Nussio, L.G., Zopollatto, M., Goulart, R.S., Cabezas Garcia, E.H. and Toledo Filho, S.G., 2010. Lower temperature profile and less aflatoxin on corn silage covered by oxygen barrier plastic film. Advances in Animal Bioscience 1: 469–470. https://doi.org/10.1017/S2040470100000919.

Ashiq, S., Edwards, S., Watson, A. and Back, M., 2022. Biofumigation for the management of *Fusarium graminearum* in a wheat-maize rotation. Pathogens 11: 1427. https://doi.org/10.3390/pathogens11121427.

Aureli, G., Amoriello, T., Belocchi, A., D'Egidio, M. G., Fornara, M., Melloni, S. and Quaranta, F., 2015. Preliminary survey on the co-occurrence of DON and T2+HT2 toxins in durum wheat in Italy. Cereal Research Communications 43: 481-491. https://doi.org/10.1556/0806.43.2015.010.

Battilani, P., Rossi, V. and Pietri, A., 2003. Modelling *Fusarium verticillioides* infection and fumonisin synthesis in maize ears. Aspects of Applied Biology 68: 91–100. http://hdl.handle.net/10807/13720.

Bernardes, T.F., Nussio, L. and Amaral, R., 2012. Top spoilage losses in maize silage sealed with plastic films with different permeabilities to oxygen. Grass Forage Science 67: 34–42. https://doi.org/10.1111/j.1365-2494.2011.00823.x.

Bernardes, T.F., Daniel, J.L.P., Adegosan, A.T., McAllister, T.A., Drouin, P., Nussio, L.G., Huhtanen, O., Tremblay, G.F., Bélanger, G. and Cai, Y., 2018. Silage review: unique challenges of silages made in hot and cold regions. Journal of Dairy Science 101: 4001-4019. https://doi.org/10.3168/jds.2017-13703.

Bernhoft, A., Wang, J. and Leifert, C., 2022. Effect of organic and conventional cereal production methods on *Fusarium* Head Blight and mycotoxin contamination levels. Agronomy 12: 797. https://doi.org/10.3390/agronomy12040797.

Beyer, M., Klix, M.B., Klink, H. And Verreet, J.A., 2006. Quantifying the effects of previous crop, tillage, cultivar and triazole fungicides on the deoxynivalenol content of wheat grain — a review. Journal of Plant Disease and Protection 113: 241–246. https://doi.org/10.1007/BF03356188.

Birr, T., Verreet, J. A. and Klink, H., 2019. Prediction of deoxynivalenol and zearalenone in winter wheat grain in a maize-free crop rotation based on cultivar susceptibility and meteorological factors. Journal of Plant Diseases and Protection 126: 13-27. https://doi.org/10.1007/s41348-018-0198-9.

Blandino, M., Reyneri, A., Pascale, M., Haidukowski, M. and Saporiti, M., 2007. Weed control in maize to prevent mycotoxin contamination. Informatore Fitopatologico 57: 48-54.

Blandino, M., Reyneri, A. and Vanara, F., 2008a. Effect of plant density on toxicogenic fungal infection and mycotoxin contamination of maize kernels. Field Crops Research 106: 234-241. https://doi.org/10.1016/j.frc.2007.12.004.

Blandino, M., Reyneri, A and Vanara, F. 2008b. Influence of nitrogen fertilization on mycotoxin contamination of maize kernels. Crop Protection 27: 222-230. https://doi.org/10.1016/j.cropro.2007.05.008.

Blandino, M., Reyneri, A. and Vanara, F., 2009a. Effect of sowing time on toxigenic fungal infection and mycotoxin contamination of maize kernels. Journal of Phytopathology 157: 7-14. https://doi.org/10.1111/j.1439-0434.2008.01431.x.

Blandino, M., Reyneri, A., Vanara, F., Pascale, M., Haidukowski, M. and Campagna, C., 2009b. Management of fumonisin contamination in maize kernels through the timing of insecticide application against the European corn borer *Ostrinia nubilalis* Hübner. Food Additives and Contaminants 26: 1501-1514. https://doi.org/10.1080/02652030903207243.

Borreani, G. and Tabacco, E., 2008. Low permeability to oxygen of a new barrier film prevents butyric acid bacteria spore formation in farm corn silage. Journal of Dairy Science 91: 4272–4281. https://doi.org/10.3168/jds.2008-1151.

Borreani, G. and Tabacco, E., 2010. The relationship of silage temperature with the microbiological status of the face of corn silage bunkers. Journal of Dairy Science 93: 2620–2629. https://doi.org/10.3168/jds.2009-2919.

Bretträger, M., Scheibenzuber, S., Asam, S., Rychlik, M., Gastl, M. and Becker, T., 2023. Evolution of *Alternaria* toxins during the brewing process and the usability of optical sorting methods to reduce mycotoxin concentrations in beer. European Food Research and Technology 249: 1613–1626. https://doi.org/10.1007/s00217-023-04241-w.

Brodal, G., Henriksen, B. and Sundheim, L., 2009. Sjukdommer i korn, oljevekster og kjernebelgvekster [Diseases of cereals, oilseed crops and field legumes]. Pages 107-141 in: Bioforsk Fokus vol. 4 no. 4 Plantevern og plantehelse i økologisk landbruk. Bind 3 – Korn, oljevekster og kjernebelgvekster. Eds: Brandsæter LO, Mangerud K, Birkenes SM, Brodal G and Andersen A. Bioforsk, Ås, Norway. https://nibio.brage.unit.no/nibio-xmlui/handle/11250/2441679.

Brodal, G., Tangerås, H., and Henriksen, B. 2016a. Smitte av fusariose på såkorn i Norge gjennom 45 år. [Inoculum of *Fusarium* on cereal seeds in Norway during 45 years]. Jord- og Plantekultur 2016. NIBIO BOK Vol 2, Nr 1, 140-143. https://nibio.brage.unit.no/nibio-xmlui/handle/11250/2428749.

Brodal, G., Hofgaard, I.S., Eriksen, G.S., Bernhoft, A. and Sundheim, L., 2016b. Mycotoxins in organically versus conventionally produced cereal grains and some other crops in temperate regions. World Mycotoxin Journal 9: 755-770. https://doi.org/10.3920/WMJ2016.2040.

Brodal, G., Aamot, H.U., Almvik, M. and Hofgaard, I.S., 2020. Removal of small kernels reduces the content of *Fusarium* mycotoxins in oat grain. Toxins 12: 346. https://doi.org/10.3390/toxins12050346.

Buerstmayr, M., Steiner, B. and Buerstmayr, H., 2020. Breeding for Fusarium head blight resistance in wheat—Progress and challenges. Plant Breeding 139: 429–454. https://doi.org/10.1111/pbr.12797.

Bullerman, L.B. and Bianchini, A., 2007. Stability of mycotoxins during food processing. International Journal of Food Microbiology 119: 140-146. https://doi.org/10.1016/j.ijffoodmicro.2007.07.035.

Cavallarin, L., Tabacco, E., Antoniazzi, S. And Borreani, G., 2011. Aflatoxin accumulation in whole crop maize silage as a result of aerobic exposure. Journal of Science Food and Agriculture 91: 2419–2425. https://doi.org/10.1002/jsfa.4481.

Channaiah, L.H. and Maier, D.E., 2014. Best stored maize management practices for the prevention of mycotoxin contamination. In: Leslie, J.F. and Logrieco, A.F. (eds). Mycotoxin Reduction in Grain Chains. Wiley, Chapter 6. https://doi.org/10.1002/9781118832790.ch6.

Cheli, F., Pinotti, L., Rossi, L. and Dell'Orto, V., 2013. Effect of milling procedures on mycotoxin distribution in wheat fractions: a review. LWT – Food Science and Technology 54: 307-314. https://doi.org/10.1016/j.lwt.2013.05.040.

Chen, A. H. and Islam, T., 2022. An integrated pest management program for managing fusarium head blight disease in cereals. Journal of integrative Agriculture 21: 3434-3444. https://doi.org/10.1016/j.jia.2022.08.053.

Choo, T. M., Martin, R. A., Savard, M. E., & Blackwell, B. (2014). Effects of planting date and earliness on deoxynivalenol contamination in barley under natural epidemic conditions. Canadian journal of plant science 94: 1363-1371. https://doi.org/10.4141/cjps2013-369.

Clear, R.M., Patrick, S.K., Nowicki, T., Gaba, D., Edney, M. J. and Babb, J.C., 1997. The effect of hull removal and pearling on *Fusarium* species and trichothecenes in hulless barley. Canadian Journal of Plant Science 77:161-166. https://doi.org/10.4141/P96-014.

Collinge, D.B., Jensen, D.F., Rabiey, M., Sarrocco, S., Shaw, M.W. and Shaw, R.H., 2022. Biological control of plant diseases – what has been achieved and what is the direction? Plant Pathology 71: 1024-1047. https://doi.org/10.1111/ppa.13555.

Colovic, R., Puvaca, N., Cheli, F., Avantaggiato, G., Greco, D., Duragic, O., Kos, J. and Pinotti, L., 2019b. Decontamination of mycotoxin-contaminated feedstuffs and compound feed. Toxins 11: 617. https://doi.org/10.3390/toxins11110617.

Colovic, R., Vukmirovic, D., Pezo, L., Kos, J., Colovic, D., Bagi, F. and Memisi, N., 2019a. Corn grain brushing for deoxynivalenol reduction. Italian Journal of Food Science 31: 161-171. https://doi.org/10.14674/IJFS-1220.

Costa, R.V., Queiroz, V.A.V., Cota, L.V., Silva, D.D., Lanza, F.E., Almeida, R.E.M., Pereira, A.A., Alves, R.R. and Campos, L.J.M., 2018. Delaying harvest for naturally drying maize grain increases the risk of kernel rot and fumonisin contamination. Tropical Plant Pathology 43: 452-459. https://doi.org/10.1007/s40858-018-0234-0.

Cowger, C., Patton-Özkurt, J., Brown-Guedira, G. and Perugini, L., 2009. Post-anthesis moisture increased fusarium head blight and deoxynivalenol levels in North Carolina winter wheat. Phytopathology 99: 320–327. https://doi.org/10.1094/PHYTO-99-4-0320.

Cowger, C., Read, Q.D., Clark, L. And Dong, Y., 2023. Optimal timing of fungicide application to manage *Fusarium* head blight in winter barley. Plant Disease 107: 2054-2060. https://doi.org/10.1094/PDIS-01-23-0021-RE.

Daud, N., Currie, V., Duncan, G., Filipe, J.A.N., Yoshinari, T., Stoddart, G., Roberts, C. and Gratz, S.W., 2023. Free and modified mycotoxins in organic and conventional oats (*Avena sativa* L.) grown in Scotland. Toxins 15: 247. https://doi.org/10.3390/toxins15040247.

De Galarreta, J.I.R., Butron, A., Ortiz-Barredo, A., Malvar, R.A., Ordás, A., Landa, A. and Revilla, P., 2015. Mycotoxins in maize grains grown in organic and conventional agriculture. Food Control 52: 98-102. https://doi.org/10.1016/j.foodcont.2014.12.016.

De Wolf, E.D., Madden, L.V. and Lipps, P.E., 2003. Risk assessment models for wheat *Fusarium* head blight epidemics based on within-season weather data. Phytopathology 93: 428–435. https://doi.org/10.1094/PHYTO.2003.93.4.428.

Dean, R., Van Kan, J.A., Pretorius, Z.A., Hammond-Kosack, K.E., Di Pietro, A., Spanu, P.D., Rudd, J.J., Dickman, M., Kahmann, R., Ellis, J. and Foster, G.D., 2012. The Top 10 fungal pathogens in molecular plant pathology. Molecular Plant Pathology 13: 414-430. https://doi.org/10.1111/j.1364-3703.2011.00783.x.

Dell'Orto, V., Baldi, G., and Cheli, F., 2015. Mycotoxins in silage: Checkpoints for effective management and control. World Mycotoxin Journal 8: 603-617. https://doi.org/10.3920/WMJ2014.1866.

Dill-Macky, R., and Jones, R. K., 2000. The effect of previous crop residues and tillage on Fusarium head blight of wheat. Plant Disease 84: 71-76. https://doi.org/10.1094/PDIS.2000.84.1.71.

Drakopoulos, D., Meca, G., Torrijos, R., Marty, A., Kägi, A., Jenny, E., Forrer, H.R., Six, J. and Vogelgsang, S., 2020a. Control of *Fusarium graminearum* in wheat with mustard-based botanicals: From *in vitro* to *in planta*. Frontiers in Microbiology 11: 1595. https://doi.org/10.3389/fmicb.2020.01595.

Drakopoulos, D., Kägi, A., Gimeno, A., Six, J., Jenny, E., Forrer, H.R., Musa, T., Meca, G. and Vogelgsang, S., 2020b. Prevention of *Fusarium* head blight infection and mycotoxins in wheat with cut-and-carry biofumigation and botanicals. Field Crops Research 246: 107681. https://doi.org/10.1016/j.fcr.2019.107681.

Drakopoulos, D., Sulyok, M., Jenny, E., Kägi, A., Bänziger, I., Logrieco, A.F., Krska, R. and Vogelgsang, S., 2021. *Fusarium* head blight and associated mycotoxins in grains and straw of barley: Influence of agricultural practices. Agronomy 11: 801. https://doi.org/10.3390/agronomy11040801.

Dunière, L., Sindou, J., Chaucheyras-Durand, F., Chevallier, I. and Thévenot-Sergentet, D., 2013. Silage processing and strategies to prevent persistence of undesirable microorganisms. Animal Feed Science and Technology 182: 1–15. https://doi.org/10.1016/j.anifeedsci.2013.04.006.

Edwards, S.G., Barrier-Guillot, B., Clasen, P.E., Hietaniemi, V. and Petterson, H., 2009a. Emerging issues of HT-2 and T-2 toxins in European cereal production. World Mycotoxin Journal 2: 173-179. https://doi.org/10.3920/WMJ2008.1126.

Edwards, S.G., 2007. Investigation of *Fusarium* mycotoxins in UK barley and oat production. Project Report No.415, AHDB Cereals & Oilseeds: Stoneleigh, UK.

Edwards, S.G., 2009a. *Fusarium* mycotoxin content of UK organic and conventional wheat. Food Additives and Contaminants Part A 26: 496-506. https://doi.org/10.1080/02652030802530679.

Edwards, S.G., 2009b. *Fusarium* mycotoxin content of UK organic and conventional barley. Food Additives and Contaminants Part A 26: 1185-1190. https://doi.org/10.1080/02652030902919418.

Edwards, S.G., 2009c. *Fusarium* mycotoxin content of UK organic and conventional oats. Food Additives and Contaminants: Part A 26: 1063-1069. https://doi.org/10.1080/02652030902788953.

Edwards, S.G., 2022. Pydiflumetofen co-formulated with prothioconazole: A novel fungicide for *Fusarium* head blight and deoxynivalenol control. Toxins 14: 34. https://doi.org/10.3390/toxins14010034.

Edwards, S. and Anderson, E., 2011. Impact of agronomy on HT-2 and T-2 toxin content of oats. Plant Breeding and Seed Science 63: 49-57. https://doi.org/10.2478/v10129-011-0015-0.

Edwards, S.G., 2017. Impact of agronomic and climatic factors on the mycotoxin content of harvested oats in the United Kingdom. Food Additives and Contaminants: Part A 34: 2230-2241. https://doi.org/10.1080/19440049.2017.1372639.

Edwards, S.G. and Jennings, P., 2018. Impact of agronomic factors on *Fusarium* mycotoxins in harvested wheat. Food Additives and Contaminants: Part A 35: 2443-2454. https://doi.org/10.1080/19440049.2018.1543954.

Feksa, H.R., Do Couto, H.T.Z., Garozi, R., De Almeida, J.L., Gardiano, C.G. Tessmann, D.J., 2019. Pre-and post-infection application of strobilurin-triazole premixes and single fungicides for control of *Fusarium* head blight and deoxynivalenol mycotoxin in wheat. Crop Protection 117: 128-134. https://doi.org/10.1016/j.cropro.2018.12.003.

van der Fels-Klerx, H. and Stratakou, I. 2010. T-2 toxin and HT-2 toxin in grain and grain-based commodities in Europe: occurrence, factors affecting occurrence, co-occurrence and toxicological effects. World Mycotoxin Journal 3: 349-367. https://doi.org/10.3920/WMJ2010.1237.

Femenias, A., Gatius, F., Ramos, A. J., Sanchis, V. and Marin, S., 2020. Use of hyperspectral imaging as a tool for *Fusarium* and deoxynivalenol risk management in cereals: A review. Food Control 108: 106819. https://doi.org/10.1016/j.foodcont.2019.106819.

Fernandez, M.R., Zentner, R.P., Basnyat, P., Gehl, D., Selles, F., Huber, D., 2009. Glyphosate associations with cereal diseases caused by *Fusarium* spp. in the Canadian Prairies. European Journal of Agronomy. 31: 133-143. https://doi.org/10.1016/j.eja.2009.07.003.

Fleurat-Lessard, F., 2017. Integrated management of the risks of stored grain spoilage by seedborne fungi and contamination by storage mould mycotoxins–An update. Journal of Stored Products Research 71: 22-40. https://doi.org/10.1016/j.jspr.2016.10.002.

Friskop, A., Halvorson, J., Hansen, B., Meyer, S., Jordahl, J., Gautam, P., Chapara, V., Arens, A., Tjelde, T., Kalil, A., Fonseka, D., Schatz, B., Brueggeman, R., Baldwin, T., Gross, P., Deplazes, C. and Ransom, J., 2023. Effects of fungicides and cultivar resistance on Fusarium Head Blight and deoxynivalenol in spring barley from 2014 to 2019. Plant Health Progress 24: 16-23. https://doi.org/10.1094/PHP-05-22-0045-RS.

Gallo, A., Bertuzzi, T., Giuberti, G., Moschini, M., Bruschi, S., Cerioli, C. and Masoero, F., 2016a. New assessment based on the use of principal factor analysis to investigate corn silage quality from nutritional traits, fermentation end products and mycotoxins. Journal of Science Food and Agriculture 96: 437-448. https://doi.org/10.1002/jsfa.7109.

Gallo, A., Giuberti, G., Bruschi, S., Fortunati, P. and Masoero, F., 2016b. Use of principal factor analysis to generate a corn silage fermentative quality index to rank well- or poorly-preserved forages. Journal of Science Food and Agriculture 96: 1686-1696. https://doi.org/10.1002/jsfa.7272.

Gallo, A., Ghilardelli, F., Camardo leggieri, M., Fodil, S. and Battilani, P., 2021. Dynamic evolution of bacterial, yeast and fungal communities during ensiling of alfalfa silage and after exposure to air. MycoTWIN-MycoKey International Conference, Bari, 9 - 12 November 2021.

Gautam, P. and Dill-Macky, R., 2012. Free water can leach mycotoxins from *Fusarium*-infected wheat heads. Journal of Phytopathology 160: 484-490. https://doi.org/10.1111/j.1439-0434.2012.01928.x.

Gimeno, A., Kägi, A., Drakopoulos, D., Bänziger, I., Lehmann, E., Forrer, H.R., Keller, B. and Vogelgsang, S., 2020. From laboratory to the field: biological control of *Fusarium graminearum* on infected maize crop residues. Journal of Applied Microbiology 129: 680-694. https://doi.org/10.1111/jam.14634.

González-Domínguez, E., Meriggi, P., Ruggeri, M. and Rossi, V., 2021. Efficacy of fungicides against Fusarium head blight depends on the timing relative to infection rather than on wheat growth stage. Agronomy 11: 1549. https://doi.org/10.3390/agronomy11081549.

Griesshaber, D., Kuhn, F., Berger, U. and Oehme, M., 2004. Comparison of trichothecene contaminations in wheat cultivated by three different farming systems in Switzerland: Biodynamic, bioorganic and conventional. Mitteilungen aus Lebensmitteluntersuchung und Hygiene 95: 251–260.

Guo, Z. Q., Pfohl, K., Karlovsky, P., Dehne, H. W. and Altincicek, B., 2016. Fumonisin B-1 and beauvericin accumulation in wheat kernels after seedborne infection with *Fusarium proliferatum*. Agricultural Food Science 25: 138–145. https://doi.org/10.23986/afsci.55539.

Guo, H., Ji, J, Wang, J. and Sun, X., 2020. Co-contamination and interaction of fungal toxins and other environmental toxins. Trends in Food Science and Technology. 103: 162-178. https://doi.org/10.1016/j.tifs.2020.06.021.

Habschied, K., Krstanović, V., Zdunić, Z., Babić, J., Mastanjević, K. and Šarić, G.K., 2021. Mycotoxins biocontrol methods for healthier crops and stored products. Journal of Fungi 7: 348. https://doi.org/10.3390/jof7050348.

Hassegawa, R.H., Fonseca, H., Fancelli, A.L., da Silva, V.N., Schammass, E.A., Reis, T.A. and Correa, B., 2008. Influence of macro- and micronutrient fertilization on fungal contamination and fumonisin production in corn grains. Food Control 19: 36-43. https://doi.org/10.1016/j.foodcont.2007.01.006.

Henriksen, B. and Elen, O., 2005. Natural *Fusarium* grain infection level in wheat, barley and oat after early application of fungicides and herbicides. Journal of Phytopathology 153: 214–220. https://doi.org/10.1111/j.1439-0434.2005.00955.x.

Herrera, M., Cavero, J., Franco-Luesma, S., Álvaro-Fuentes, J., Ariño, A. and Lorán, S., 2023. Mycotoxins and crop yield in maize as affected by irrigation management and tillage practices. Agronomy 13: 798. https://doi.org/10.3390/agronomy13030798.

Hietaniemi V., Rämö S., Manninen P., Parikka P. and Hankomäki J., 2009. The effect of cleaning and de-hulling on the trichothecene content in oats and barley. NordForsk Mould and Mycotoxin Seminar, April 14-15 Uppsala, Sweden. http://fou02.planteforsk.no/NordforskNetworkMycotox/PDFs/PDFs%20Uppsala%202009/SariPresentation_April2009s.pdf.

Hjelkrem, A.G.R., Aamot, H.U., Lillemo, M., Sørensen, E.S., Brodal, G., Russenes, A.L., Edwards, S.G. and Hofgaard, I.S., 2022. Weather patterns associated with DON levels in Norwegian spring oat grain: a functional data approach. Plants 11: 73. https://doi.org/10.3390/plants11010073.

Hofer, K., Barmeier, G., Schmidhalter, U., Habler, K., Rychlik, M., Hückelhoven, R. and Hess, M., 2016. Effect of nitrogen fertilization on Fusarium Head Blight in spring barley. Crop Protection 88: 18-27. https://doi.org/10.1016/j.cropro.2016.05.007.

Hoffmans, Y., Schaarschmidt, S., Fauhl-Hassek, C. and van der Fels-Klerx, H.J., 2022. Factors during production of cereal-derived feed that influence mycotoxin contents. Toxins 25: 301. https://doir.org/10.3390/toxins14050301.

Hofgaard, I.S., Aamot, H.U., Seehusen, T., Riley, H., Dill-Macky, R., Holen, B.M. and Brodal, G., 2020. *Fusarium* and mycotoxin content of harvested grain was not related to tillage intensity in Norwegian spring wheat fields. World Mycotoxin Journal 13: 473-486. https://doir.org/10.3920/WMJ2020.2575.

Hofgaard, I.S., Brodal, G., Almvik, M., Lillemo, M., Russenes, A.L., Edwards, S.G. and Aamot, H.U., 2022a. Different resistance to DON versus HT2 + T2 producers in Nordic oat varieties. Toxins 14: 313. https://doi.org/10.3390/toxins14050313.

Hofgaard, I.S., Aamot, H.U., Seehusen, T., Holen, B.M., Riley, H., Dill-Macky, R., Edwards, S.G. and Brodal, G., 2022b. Reduced risk of oat grain contamination with *Fusarium langsethiae* and HT-2 and T-2 toxins with increasing tillage intensity. Pathogens 11: 1288. https://doi.org/10.3390/pathogens11111288.

House, J.D., Nyachoti, C.M. and Abramson, D., 2003. Deoxynivalenol removal from barley intended as swine feed through the use of an abrasive pearling procedure. Journal of Agricultural and Food Chemistry 51: 5172-5175. https://doi.org/10.1021/jf0342449.

Huang, Y., Yin, L., Sallam, A.H., Heinen, S., Li, L., Beaubien, K., Dill-Macky, R., Dong, Y., Steffenson, B.J., Smith, K.P. and Muehlbauer, G.J., 2021. Genetic dissection of a pericentromeric region of barley chromosome 6H associated with Fusarium head blight resistance, grain protein content and agronomic traits. Theoretical and Applied Genetics 134: 3963–3981. https://doi.org/10.1007/s00122-021-03941-9.

Ivanova, L., Sahlstrøm, S., Rud, I., Uhlig, S., Fæste, C.K., Eriksen, G.S. and Divon, H.H., 2017. Effect of primary processing on the distribution of free and modified *Fusarium* mycotoxins in naturally contaminated oats. World Mycotoxin Journal 10: 73-88. https://doi.org/10.3920/WMJ2016.2092.

Janić Hajnal, E., Mastilović, J., Bagi, F., Orčić, D., Budakov, D., Kos, J. and Savić, Z., 2019. Effect of wheat milling process on the distribution of *Alternaria* toxins. Toxins 11: 139. https://doi.org/10.3390/toxins11030139.

Janssen, E.M., Liu, C. and van der Fels-Klerx, H.J., 2018. *Fusarium* infection and trichothecenes in barley and its comparison with wheat. World Mycotoxin Journal 11: 33-46. https://doi.org/10.3920/WMJ2017.2255.

Janssen, E. M., Mourits, M. C., Oude Lansink, A. G. and Van Der Fels-Klerx, H. J., 2021. Incentives to stimulate European wheat farmers to adapt their fusarium species mycotoxin management. Toxins 13: 144. https://doi.org/10.3390/toxins13020144.

Jones, R.K., Duncan, H.E. and Hamilton, P.B., 1981. Planting date, harvest date, and irrigation effects on infection and aflatoxin production by *Aspergillus flavus* in field corn. Phytopathology 71: 810-816. https://doi.org/10.1094/Phyto-71-810.

Jouany, J.P., 2007. Methods for preventing, decontaminating and minimizing the toxicity of mycotoxins in feeds. Animal Feed Science and Technology 137: 342–362. https://doi.org/10.1116/j.anifeedsci.2007.006.009.

Juan, C., Ritieni, A. And Mañes, J., 2013. Occurrence of *Fusarium* mycotoxins in Italian cereal and cereal products from organic farming. Food Chemistry 141: 1747-1755. https://doi.org/10.1016/j.foodchem.2013.04.061.

Kabak, B., 2009. The fate of mycotoxins during thermal food processing. Science of Food and Agriculture. 89: 549-554. https://doi.org/10.1002/jsfa.3491.

Kalagatur, N.K., Mudili, V., Siddaiah, C., Gupta, V.K., Natarajan, G., Sreepathi, M.H., Vardhan, B.H. and Putcha, V.L., 2015. Antagonistic activity of *Ocimum sanctum* L. essential oil on growth and zearalenone production by *Fusarium graminearum* in maize grains. Frontiers in Microbiology 6: 892. https://doi.org/10.3389/fmicb.2015.00892.

Karlovsky, P., Suman, M., Berthiller, F., De Meester, J., Eisenbrand, G., Perrin, I., Oswald, I.P., Speijers, G., Chiodini, A., Recker, T. and Dussort, P., 2016. Impact of food processing and detoxification treatments on mycotoxin contamination. Mycotoxin Research 32: 179-205. https://doi.org/10.1007/s12550-016-0257-7.

Karlsson, I., Mellqvist, E. and Persson, P., 2023. Temporal and spatial dynamics of *Fusarium* spp. and mycotoxins in Swedish cereals during 16 years. Mycotoxin Research 39: 3-18. https://doi.org/10.1007/s12550-022-00469-9.

Karron, E., Runno-Paurson, E., Lõiveke, H., Islamov, B., Kütt, M.-L., Talve, T., Lauringson, E., Horak, H., Edesi, L. and Niinemets, Ülo., 2020. Application of widely used fungicides does not necessarily affect grain yield, and incidence of *Fusarium* spp. and mycotoxins DON, HT-2 and T-2 in spring barley in northern climates. Kvasny Prumysl 66: 215–223. https://doi.org/10.18832/kp2020.66.215.

Kaukoranta, T., Hietaniemi, V., Rämö, S., Koivisto, T. And Parikka, P., 2019. Contrasting responses of T-2, HT-2 and DON mycotoxins and *Fusarium* species in oat to climate, weather, tillage and cereal intensity. European Journal of Plant Pathology 155: 93–110. https://doi.org/10.1007/s10658-019-01752-9.

Kaul, N., Kashyap, P.L., Kumar, S., Singh, D. and Singh, G.P., 2022. Diversity and exploration of endophytic bacilli for the management of head scab (*Fusarium graminearum*) of wheat. Pathogens 11: 1088. https://doi.org/10.3390/pathogens11101088.

Khatibi, P.A., Berger, G., Wilson, J., Brooks, W.S., McMaster, N., Griffey, C.A., Hicks, K.B., Nghiem N.P. and Schmale III, D.G., 2014. A comparison of two milling strategies to reduce the mycotoxin deoxynivalenol in barley. Journal of Agricultural and Food Chemistry 62: 4204-4213. https://doi.org/10.1021/jf501208x.

Khairullina, A., Micic, N., Jørgensen, H.J.L., Bjarnholt, N., Bülow, L., Collinge, D.B. and Jensen, B., 2023. Biocontrol Effect of *Clonostachys rosea* on *Fusarium graminearum* Infection and Mycotoxin Detoxification in Oat (*Avena sativa*). Plants 12: 500. https://doi.org/10.3390/plants12030500.

Khaneghah, A.M., Fakhri, Y., Raeisi, S., Armoon, B., & Sant'Ana, A.S., 2018. Prevalence and concentration of ochratoxin A, zearalenone, deoxynivalenol and total aflatoxin in cereal-based products: A systematic review and meta-analysis. Food and chemical toxicology, 118: 830-848. https://doi.org/10.1016/j.fct.2018.06.037.

Kharbikar, L.L., Dickin, E.T. and Edwards, S.G., 2015. Impact of post-anthesis rainfall, fungicide and harvesting time on the concentration of deoxynivalenol and zearalenone in wheat. Food Additives and Contaminants Part A: Chemistry, Analysis, and Control 32: 2075–2085. https://doi.org/10.1080/19440049.2015.1084652.

Kleber, A., Gruber-Dorninger, C., Platzer, A., Payet, C. and Novak, B., 2023. Effect of fungicide treatment on multi-mycotoxin occurrence in French wheat during a 4-year period. Toxins 15: 443. https://doi.org/10.3390/toxins15070443.

Koch, E., Zink, P., Pfeiffer, T., von Galen, A., Linkies, A., Drechsel, J. and Birr, T., 2020. Artificial inoculation methods for testing microorganisms as control agents of seed-and soil-borne *Fusarium* seedling blight of maize. Journal of Plant Diseases and Protection 127: 883-893. https://doi.org/10.1007/s41348-020-00350-w.

Kochiieru, Y., Mankeviciene, A., Janaviciene, S., Jonaviciene, A. and Ceseviciene, J., 2019. The influence of milling and sifting processes on deoxynivalenol distribution in whole-wheat flour and its products. World Mycotoxin Journal 12: 133-140. https://doi.org/10.3920/WMJ2018.2404.

Kochiieru, Y., Mankevičienė, A., Cesevičienė, J., Semaškienė, R., Dabkevicius, Z. and Janavičienė, S., 2020. The influence of harvesting time and meteorological conditions on the occurrence of *Fusarium* species and mycotoxin contamination of spring cereals. Journal of Science of Food and Agriculture 100: 2999-3006. https://doi.org/10.1002/jsfa.10330.

Kochiieru, Y., Mankevičienė, A., Cesevičienė, J., Semaškienė, R., Ramanauskienė, J., Gorash, A., Janavičienė, S. and Venslovas, E., 2021. The Impact of Harvesting Time on *Fusarium* Mycotoxins in Spring Wheat Grain and Their Interaction with Grain Quality. Agronomy 11: 642. https://doi.org/10.3390/agronomy11040642.

Kolawole, O., de Ruyck, K., Greer, B., Meneely, J., Doohan, F., Danaher, M. And Elliott, C., 2021. Agronomic Factors influencing the scale of *Fusarium* mycotoxin contamination of oats. Journal of Fungi 7: 965. https://doi.org/10.3390/jof7110965.

Krnjaja, V., Mandić, V., Lević, J., Stanković, S., Petrović, T., Vasić, T. and Obradović, A., 2015. Influence of N-fertilization on *Fusarium* head blight and mycotoxin levels in winter wheat. Crop Protection 67: 251-256. http://dx.doi.org/10.1016/j.cropro.2014.11.001.

Krnjaja, V., Mandić, V., Bijelić, Z., Stanković, S., Obradović, A., Caro Petrović, V. and Gogić, M. 2022. Influence of sowing time on *Fusarium* and fumonisin contamination of maize grains and yield component traits. Agriculture 12: 1042. https://doi.org/10.3390/agriculture12071042.

Kuchenbuch, H.S., Becker, S., Schulz, M., Cramer, B. and Humpf, H.U., 2018. Thermal stability of T-2 and HT-2 toxins during biscuit- and crunchy muesli-making and roasting. Food Additives & Contaminants: Part A 35: 2158–2167. https://doi.org/10.1080/19440049.2018.1530456.

Kumar, P., Gupta, A., Mahato, D.K., Pandhi, S., Pandey, A.K., Kargwal, R., Mishra, S., Suhag, R., Sharma, N., Saurabh, V., Paul, V., Kumar, M., Selvakumar, R., Gamlath, S., Kamle, M., El Enshasy, H.A., Mokhtar, J.A. and Harakeh, S., 2022. Aflatoxins in cereals and cereal-based products: occurrence, toxicity, impact on human health, and their detoxification and management strategies. Toxins 14: 687. https://doi.org/10.3390/toxins14100687.

Kumar, P.S., Nattudurai, G., Islam, V.I.H. and Ignacimuthu, S., 2022b. The effects of some essential oils on Alternaria alternata, a post-harvest phyto-pathogenic fungus in wheat by disrupting ergosterol biosynthesis. Phytoparasitica 50: 513-525. https://doi.org/10.1007/s12600-021-00970-4

Kung Jr, L., Shaver, R.D., Grant, R.J. and Schmidt, R.J., 2018. Silage review: Interpretation of chemical, microbial, and organoleptic components of silages. Journal of Dairy Science 101: 4020-4033. https://doi.org/10.3168/jds.2017-13909.

Lemmens, M., Haim, K., Lew, H., and Ruckenbauer, P. 2004. The effect of nitrogen fertilization on Fusarium head blight development and deoxynivalenol contamination in wheat. Journal of Phytopathology, 152: 1-8. https://doi.org/10.1046/j.1439-0434.2003.00791.x

Leonard, K.J. and Bushnell, W.R., 2003. Fusarium head blight of wheat and barley. The American Phytopathological Society. St. Paul, Minnesota, USA. 512 pp.

Leslie, J. F., Moretti, A., Mesterházy, Á., Ameye, M., Audenaert, K., Singh, P. K., Forget, F.R., Chulze, S.N., Del Ponte, E.M., Chala, A., Battilani, P. and Logrieco, A.F., 2021. Key global actions for mycotoxin management in wheat and other small grains. Toxins 13: 725. https://doi.org/10.3390/toxins13100725.

Li, .Q., Toyazaki, N. and Yoshizawa, T., 2001.Production of *Alternaria* mycotoxins by *Alternaria alternata* isolated from weather-damaged wheat. Journal of Food Protection. 64: 567-571. https://doi.org/10.4315/0362-028X-64.4.567.

Limay-Rios, V. and Schaafsma, A.W., 2021. Relationship between mycotoxin content in winter wheat grain and aspirated dust collected during harvest and after storage. ACS Omega 6: 1857-1871. https://doi.org/10.1021/acsomega.0c04256.

Mäder, P., Hahn, D., Dubois, D., Gunst, L., Alföldi, T., Bergmann, H., Oehme, M., Amadò, R., Schneider, H., Graf, U., Velimirov, A., Fliessbach, A. and Niggli, U., 2007. Wheat quality in organic and conventional farming: Results of a 21 year field experiment. Journal of Science Food and Agriculture 87: 1826–1835. https://doi.org/10.1002/jsfa.2866.

Magan, N. and Aldred, D., 2007. Post-harvest control strategies: minimizing mycotoxins in the food chain. International Journal of Food Microbiology. 119: 131-139. https://doi.org/10.1016/j.ijfoodmicro.2007.07.034.

Magan, N., Aldred, D. and Baxter, E.S., 2014. Good postharvest storage practices for wheat grain In: Leslie, J.F. and Logrieco, A.F., Mycotoxin Reduction in Grain Chains Chapter 18. https://doi.org/10.1002/9781118832790.ch18.

Mahanna, B. and Chase, L.E., 2003. Practical applications and solutions to silage problems. In: Silage Science and Technology. Buxton, D.R., Muck, R.E. and Harrison, H (eds) 42: 855–895. https://doi.org/10.2134/agronmonogr42.c19.

Manoza, F.S., Mushong, A.A., Harvey, J., Wainaina, J., Wanjuki, I., Ngeno, R., Darnell, R., Gnonlonfin, B.G.J., and Massomo, S.M.S., 2017. Potential of using host plant resistance, nitrogen and phosphorus fertilizers for reduction of *Aspergillus flavus* colonization and aflatoxin accumulation in maize in Tanzania. Crop Protection. 93: 98-105. https://doi.org/10.1016/j.cropro.2016.11.021.

Mansfield, M.A., Archibald D.D., Jones A.D. and Kuldau G.A., 2007. Relationship of sphinganine analog mycotoxin contamination in maize silage to seasonal weather conditions and to agronomic and ensiling practices. Phytopathology 97: 504-511. https://doi.org/10.1094/PHYTO-97-4-0504.

Matelionienė, N., Supronienė, S., Shamshitov, A., Zavtrikovienė, E., Janavičienė, S. and Kadžienė, G., 2022. Weeds in cereal crop rotations may host *Fusarium* species that cause Fusarium Head blight and grain weight losses in wheat. Agronomy 12: 2741. https://doi.org/10.3390/agronomy12112741.

Mateo, E.M., Valle-Algarra, F.M., Mateo, R., Jiménez, M. and Magan, N., 2011. Effect of fenpropimorph, prochloraz and tebuconazole on growth and production of T-2 and HT-2 toxins by *Fusarium langsethiae* in oat-based medium. International Journal of Food Microbiology 151: 289-298. https://doi.org/10.1016/j.ijfoodmicro.2011.09.017.

McKenzie, K.S., Sarr, A.B., Mayura, K., Bailey, R.H., Miller, D.R., Rogers, T.D., Norred, W.P., Voss, K.A., Plattner, R.D., Kubena, L.F. and Phillips, T.D., 1997. Oxidative degradation and detoxification of mycotoxins using a novel source of ozone. Food Chemistry and Toxicology 35: 807-820. https://doi.org/10.1016/s0278-6915(97)00052-5.

McMullen, M., Bergstrom, G., De Wolf, E., Dill-Macky, R., Hershman, D., Shaner, G. and van Sanford, D., 2012. A unified effort to fight an enemy of wheat and barley: Fusarium Head Blight. Plant Disease 96: 1712-1728. https://doi.org/10.1094/PDIS-03-12-0291-FE.

Mesterhazy, A., 1995. Types and components of resistance to *Fusarium* head blight of wheat. Plant Breeding 114: 377-386. https://doi.org/10.1111/j.1439-0523.1995.tb00816.x.

Mesterházy, Á., Varga, M., Tóth, B., Kótai, C., Bartók, T., Véha, A., Acs, K., Vagvolgyi, C. and Lehoczki-Krsjak, S., 2018. Reduction of deoxynivalenol (DON) contamination by improved fungicide use in wheat. Part 1. Dependence on epidemic severity and resistance level in small plot tests with artificial inoculation. European Journal of Plant Pathology 151: 39-55. https://doi.org/10.1007/s10658-017-1350-2.

Mesterhazy, A., 2024. What is Fusarium Head Blight (FHB) resistance and what are its food safety risks in wheat? Problems and Solutions – A Review. Toxins 16: 31. https://doi.org/10.3390/toxins16010031.

Meyer, J. C., Birr, T., Hennies, I., Wessels, D. and Schwarz, K., 2022. Reduction of deoxynivalenol, T-2 and HT-2 toxins and associated *Fusarium* species during commercial and laboratory de-hulling of milling oats. Food Additives and Contaminants: Part A 39: 1163-1183. https://doi.org/10.1080/19440049.2022.2059576.

Michel, S., Steiner, B. And Buerstmayr, H., 2024. Efficiency of indirect selection for fusarium head blight resistance and mycotoxin accumulation in winter wheat (*Triticum aestivum* L.). Plant Breeding 143: 34-43. https://doi.org/10.1111/pbr.13136.

Miedaner, T. and Geiger, H.H., 2015. Biology, genetics, and management of ergot (*Claviceps* spp.) in rye, sorghum, and pearl millet. Toxins 7: 659-678. https://doi.org/10.3390/toxins7030659.

Miedaner, T. and Juroszek, P., 2021. Global warming and increasing maize cultivation demand comprehensive efforts in disease and insect resistance breeding in north-western Europe. Plant Pathology 70: 1032-1046. https://doi.org/10.1111/ppa.13365.

Miedaner, T., Flamm, C. And Oberforster, M., 2024. The importance of *Fusarium* head blight resistance in the cereal breeding industry: Case studies from Germany and Austria. Plant Breeding. 143: 44-58. https://doi.org/10.1111/pbr.13098.

Miller, J. D., Young, J. C., & Sampson, D. R., 1985. Deoxynivalenol and Fusarium head blight resistance in spring cereals. Journal of Phytopathology 113: 359-367. https://doi.org/10.1111/j.1439-0434.1985.tb04837.x.

Miller, J.D., 2008. Mycotoxins in small grains and maize: old problems, new challenges. Food Additives and Contaminants Part A, Chemistry, Analysis, Control, Exposure, and Risk Assessment 25: 219-230. https://doi.org/10.1080/02652030701744520.

Moraes, W.B., Madden, L.V. and Paul, P.A., 2022. Efficacy of genetic resistance and fungicide application against fusarium head blight and mycotoxins in wheat under persistent pre- and postanthesis moisture. 106: 2839-2855. https://doi.org/10.1094/PDIS-02-22-0263.

Morcia, C., Terzi, V., Ghizzoni, R., Carrara, I. and Gazzetti, K., 2024. Looking for *Fusarium* resistance in oats: an update. Agronomy 14: 505. https://doi.org/10.3390/agronomy14030505.

Mosconi M., Fontana A., Belloso Daza, M.V., Bassi, D. and Gallo, A., 2023. *Clostridium tyrobutyricum* occurrence in silages and cattle feed: Use of molecular and simulation data to optimize predictive models. Front. Microbiol. 14: 1118646. https://doi.org/10.3389/fmicb.2023.1118646

Muck, R., Nadeau, E.M.G., McAllister, T.A., Contreras-Govea, F.E., Santos, M.C. and Kung Jr., L., 2018. Silage review: recent advances and future uses of silage additives. Journal of Dairy Science 101: 3980-4000. https://doi.org/10.3168/jds.2017-13839.

Munkvold, G.P., 2003. Epidemiology of *Fusarium* diseases and their mycotoxins in maize ears. European Journal of plant pathology 109: 705-713. https://link.springer.com/article/10.1023/A:1026078324268.

Munkvold, G.P., 2014. Crop management practices to minimize the risk of mycotoxins contamination in temperate-zone maize. In: Mycotoxin Reduction in Grain Chains. Leslie, J.F. and Logrieco, A.F. (eds) Wiley, Chapter 5. https://doi.org/10.1002/9781118832790.ch5.

Nešić, K., Habschied, K. and Mastanjević, K., 2021. Possibilities for the biological control of mycotoxins in food and feed. Toxins 13: 198. https://doi.org/10.3390/toxins13030198.

Ochodzki, P. and Góral, T., 2023. Deoxynivalenol content in wheat kernels showing various levels of damage caused by *Fusarium culmorum*. Biuletyn Instytutu Hodowli i Aklimatyzacji Roślin 300: 67-75.

Odukoya, J.O., de Saeger, S., de Boevre, M., Adegoke, G.O., Devlieghere, F., Croubels, S., Antonissen, G., Odukoya, J.O. and Njobeh, P.B., 2024. Influence of traditional dehulling on mycotoxin reduction and GC-HRTOF-MS metabolites profile of fermented maize products. Heliyon 10: e23025. https://doi.org/10.1016/j.heliyon.2023.e23025.

Orlando, B., Barrier-Guillot, B., Gourdain, E. and Maumene, C., 2010. Identification of agronomic factors that influence the levels of T-2 and HT-2 toxins in barley grown in France. World Mycotoxin Journal 3: 169-174. https://doi.org/10.3920/WMJ2009.1191.

Oude Elferink, S.J.W.H., Driehuis, F., Gottschal, J. And Spoelstra, S., 2000. Silage fermentation processes and their manipulation. 161: 17–30. In: FAO Electronic Conference on Tropical Silage, FAO, Rome.

Pahlow, G., Muck, R.E., Driehuis, F., Oude Elferink, S.J.W.H. and Spoelstra, S.F., 2003. Microbiology of ensiling. On: Silage Science and Technology. Buxton, D.R., Muck, R.E. and Harrison, J.H. (eds) American Society of Agronomy, Inc., Crop Science Society of America Inc., Soil Science Society of America, Inc. Publications, Madison, WI. 31–93.

Parry, D.W., Jenkinson, P. and McLeod, L., 1995. *Fusarium* ear blight (scab) in small grain cereals: a review. Plant Pathology 44: 207-238. https://doi.org/10.1111/j.1365-3059.1995.tb02773.x.

Pascale, M., Logrieco, A.F., Lippolis, V., De Girolamo, A., Cervellieri, S., Lattanzio, V.M.T., Ciasca, B., Vega, A., Reichel, M., Graeber, M. and Slettengren, K., 2022. Industrial-scale cleaning solutions for the reduction of *Fusarium* toxins in maize. Toxins 14: 728. https://doi.org/10.3390/toxins14110728.

Paul, P.A., Lipps, P.E., Hershman, D.E., McMullen, M.P., Draper, M.A. and Madden, L.V., 2008. Efficacy of triazole-based fungicides for Fusarium Head Blight and deoxynivalenol control in wheat: a multivariate meta-analysis. Disease Control and Pest Management 98: 999-1011. https://doi.org/10.1094/PHYTO-98-9-0999.

Peles, F., Sipos, P., Kovács, S., Győri, Z., Pócsi, I. and Pusztahelyi, T., 2021. Biological control and mitigation of aflatoxin contamination in commodities. Toxins 13: 104. https://doi.org/10.3390/toxins13020104.

Pellan, L., Dieye, C.A.T., Durand, N., Fontana, A., Schorr-Galindo, S. and Strub, C., 2021. Biocontrol agents reduce progression and mycotoxin production of *Fusarium graminearum* in spikelets and straws of wheat. Toxins 13: 597. https://doi.org/10.3390/toxins13090597.

Peng, W.X., Marchal, L. and van der Poel, A.F.B., 2018. Strategies to prevent and reduce mycotoxins for compound feed manufacturing. Animal Feed Science and Technology 237: 129-153. https://doi.org/10.1016/j.anifeedsci.2018.01.017.

Perkowski, J., 1998. Distribution of deoxynivalenol in barley kernels infected by *Fusarium*. Food/Nahrung 42: 81-83. https://doi.org/10.1002/(SICI)1521-3803(199804)42:02<81::AID-FOOD81>3.0.CO;2-0.

Perkowski, J., Kiecana, I. and Kaczmarek, Z., 2003. Natural occurrence and distribution of *Fusarium* toxins in contaminated barley cultivars. European Journal of Plant Pathology 109: 331-339. https://link.springer.com/article/10.1023/A:1023547210060.

Petersson, S., 1998. Yeast/mold interactions during airtight storage of high-moisture feed grain. PhD thesis, Swedish University of Agricultural Sciences, Uppsala.

Pettersson, H., Börjesson, T., Persson, L., Lerenius, C., Berg, G. and Gustafsson, G., 2008. T-2 and HT-2 toxins in oats frown in Northern Europe. Cereal Research Communications 36: 591-592.

Pettersson, H., Brown, C., Hauk, J., Hoth, S., Meyer, J. and Wessels, D., 2011. Survey of T-2 and HT-2 toxins by LC-MS/MS in oats and oat products from European oat mills in 2005-2009. Food Additives and Contaminants Part B Surveillance 4: 110-115. https://doi.org/10.1080/19393210.2011.561933.

Pietri, A., Zanetti, M. and Bertuzzi, T., 2009. Distribution of aflatoxins and fumonisins in dry-milled maize fractions. Food Additives and Contaminants: Part A 26: 372-380. https://doi.org/10.1080/02652030802441513.

Pinotti, L., Ottoboni, M., Giromini, C., Dell'Orto, V. and Cheli, F., 2016. Mycotoxin contamination in the EU feed supply chain: A focus on cereal byproducts. Toxins, 8: 45. https://doi.org/10.3390/toxins8020045

Pitt, R.E. and Muck, R.E., 1993. A diffusion model of aerobic deterioration at the exposed face of bunker silos. Journal of Agricultural Engineering Research 55: 11–26. https://doi.org/10.1006/jaer.1993.1029.

Pleadin, J., Staver, M.M., Markov, K., Frece, J., Zadravec, M., Jaki, V., Krupic, I. and Vaheie, N., 2017. Mycotoxins in organic and conventional cereals and cereal products grown and marketed in Croatia. Mycotoxin Research 33: 219-227. https://doi.org/10.1007/s12550-017-0280-3.

Podolska, G., Bryla, M., Sulek, A., Waskiewicz, A., Szymczyk, K. and Jedrzejczak, R., 2017. Influence of the cultivar and nitrogen fertilisation level on the mycotoxin contamination in winter wheat. Quality Assurance and Safety of crops and Foods 9: 451-461. https://doi.org/10.3920/QAS2016.1064.

Pomar, C., Hauschild, L., Zhang, G.H., Pomar, J. and Lovatto, P.A., 2009. Applying precision feeding techniques in growing-finishing pig operations. Revista Brasileira de Zootecnia 38: 226-237. https://doi.org/10.1590/S1516-35982009001300023.

Postic, J., Cosic, J., Jurkovic, D., Vrandecic, K., Saleh, A.A. and Leslie, J.F., 2012. Diversity of *Fusarium* species isolated from weeds and plant debris in Croatia. Journal of Phytopathology 160: 76– 81. https://doi.org/10.1111/j.1439-0434.2011.01863.x.

Powell, A.J. and Vujanovic, V., 2021. Evolution of Fusarium Head Blight Management in Wheat: Scientific Perspectives on Biological Control Agents and Crop Genotypes Protocooperation. Applied Science 11: 8960. https://doi.org/10.3390/app11198960.

Rasiukevičiūtė, N., Supronienė, S., Kelpšienė, J., Švėgžda, P., Kadžienė, G., Šneideris, D., Ivaauskas, A. and Treikale, O., 2018. Susceptibility of non-cereal crops to *Fusarium graminearum* complex and their role within cereal crop rotation as a source of inoculum for Fusarium head blight. Spanish Journal of Agricultural Research 16: 1-12. https://doi.org/10.5424/sjar/2018164-13952.

Richard-Molard, D., Cahagnier, B. and Poisson, J., 1980. Wet grains storage under modified atmospheres: microbiological aspects. In: Controlled atmosphere storage of grains, Shejbal, J. (ed.) Elsevier Scientific Publishers, Amsterdam, 173– 182.

Ríos, G., Pinson-Gadais, L., Abecassis, J., Nakhia-Rozis, N. and Lullien-Pellerin, V., 2009. Assessment of dehulling efficiency to reduce deoxynivalenol and *Fusarium* level in durum wheat grains. Journal of Cereal Science 49: 387-392. https://doi.org/10.1016/j.jcs.2009.01.003.

Rojas, E.C., Sapkota, R., Jensen, B., Jorgensen, J.L., Henriksson, T., Jorgensen, L.N., Nicolaisen, M. and Collinge, D.B., 2020. *Fusarium* Head Blight modifies fungal endophytic communities during infection of wheat spikes. Microbial Ecology 79: 397–408. https://doi.org/10.1007/s00248-019-01426-3.

Rossi, V., Mansretta, V. and Ruggeri, M., 2015. A multicomponent decision support system to manage *Fusarium* head blight and mycotoxins in durum wheat. World Mycotoxin Journal 8: 629-640. https://doi.org/10.3920/WMJ2015.1881.

Sarrocco, S. and Vannacci, G., 2018. Preharvest application of beneficial fungi as a strategy to prevent postharvest mycotoxin contamination: A review. Crop Protection 100: 160-170. https://doi.org/10.1016/j.cropro.2017.11.013

Sarrocco, S., Mauro, A. and Battilani, P., 2019. Use of competitive filamentous fungi as an alternative approach for mycotoxin risk reduction in staple cereals: state of art and future perspectives. Toxins 11: 701. https://doi.org/10.3390/toxins11120701.

Savić, Z., Dudaš, T., Loc, M., Grahovac, M., Budakov, D., Jajić, I., Krstović, S., Barošević, T., Krska, R., Sulyok, M., Stojsin, V., Petres, M., Stankov, A., Vukotic, J. and Bagi, F., 2020. Biological control of aflatoxin in maize grown in Serbia. Toxins 12: 162. https://doi.org/10.3390/toxins12030162.

Scarpino, V., Reyneri, A., Sulyok, M., Krska, R. and Blandino, M., 2015. Effect of fungicide application to control Fusarium head blight and 20 *Fusarium* and *Alternaria* mycotoxins in winter wheat (*Triticum aestivum* L.). World Mycotoxin Journal 8: 499-510. https://doi.org/10.3920/WMJ2014.1814.

Scarpino, V., Reyneri, A., Sulyok, M., Krska, R. and Blandino, M., 2018. Impact of the insecticide application to maize cultivated in different environmental conditions on emerging mycotoxins. Field Crops Research 217: 188-198. https://doi.org/10.1016/j.fcr.2017.12.018.

Scarpino, V., Sulyok, M., Krska, R., Reyneri, A. and Blandino, M., 2022. The role of nitrogen fertilization on the occurrence of regulated, modified and emerging mycotoxins and fungal metabolites in maize kernels. Toxins 14: 448. https://doi.org/10.3390/toxins14070448.

Schaafsma, A.W. and Hooker, D.C., 2007. Climatic models to predict occurrence of *Fusarium* toxins in wheat and maize. International Journal of Food Microbiology 119: 116-125. https://doi.org/10.1016/j.ijfoodmicro.2007.08.006.

Schaafsma, A.W., Tamburic-Ilincic, L. and Hooker, D.C., 2005. Effect of previous crop, tillage, field size, adjacent crop, and sampling direction on airborne propagules of *Gibberella zeae/Fusarium graminearum*, fusarium head blight severity, and deoxynivalenol accumulation in winter wheat. Canadian journal of plant pathology 27: 217-224. https://www.tandfonline.com/doi/abs/10.1080/07060660509507219.

Schaarschmidt, S. and Fauhl-Hassek, C., 2018. The fate of mycotoxins during the processing of wheat for human consumption. Comprehensive Reviews in Food Science and Food Safety 17: 556-593. https://doi.org/10.1111/1541-4337.12338.

Schaarschmidt, S. and Fauhl-Hassek, C., 2021. The fate of mycotoxins during the primary food processing of maize. Food Control 121: 107651. https://doi.org/10.1016/j.foodcont.2020.107651.

Schmale III, D.G., Leslie, J.F., Zeller, K.A., Saleh, A.A., Shields, E.J. and Bergstrom, G.C., 2006. Genetic structure of atmospheric populations of *Gibberella zeae*. Phytopathology 96: 1021-1026. https://doi.org/10.1094/PHYTO-96-1021.

Schneweis, I., Meyer, K., Ritzmann, M., Hoffmann, P., Dempfle, L. and Bauer, J., 2005. Influence of organically or conventionally produced wheat on health, performance and mycotoxin residues in tissues and bile of growing pigs. Archives in Animal Nutrition 59: 155–163. https://doi.org/10.1080/17450390500147594.

Schöneberg, T., Jenny, E., Wettstein, F. E., Bucheli, T. D., Mascher, F., Bertossa, M., Musa, T., Seifert, K., Gräfenhan, T., Keller, B. and Vogelgsang, S., 2018. Occurrence of *Fusarium* species and mycotoxins in Swiss oats—Impact of cropping factors. European Journal of Agronomy 92: 123-132. https://doi.org/10.1016/j.eja.2017.09.004.

Schwake-Anduschus, C., Langenkämper, G., Unbehend, G., Dietrich, R., Märtlbauer, E. and Münzing, K., 2010. Occurrence of Fusarium T-2 and HT-2 toxins in oats from cultivar studies in Germany and degradation of the toxins during grain cleaning treatment and food processing. Food Additives and Contaminants 27: 1253-1260. https://doi.org/10.1080/19440049.2010.487499.

Scudamore, K.A., Baillie, H., Patel, S. and Edwards, S.G., 2007. Occurrence and fate of *Fusarium* mycotoxins during commercial processing of oats in the UK. Food additives and contaminants 24: 1374-1385. https://doi.org/10.1080/02652030701509972.

Seitz, L.M. and Bechtel, D.B., 1985. Chemical, physical, and microscopical studies of scab-infected hard red winter wheat. Journal of Agricultural and Food Chemistry 33: 373-377. https://doi.org/10.1021/jf00063a013.

Selvaraj, J.N., Zhao, Y., Sangare, L., Xing, F., Zhou, L., Wang, Y., Xue, X., Li, Y. and Liu, Y., 2015. Limited survey of deoxynivalenol in wheat from different crop rotation fields in Yangtze-Huaihe river basin region of China. Food Control 53: 151-155. https://doi.org/10.1016/j.foodcont.2015.01.021.

Shah, D.A., De Wolf, E.D., Paul, P.A. and Madden, L.V., 2019. Functional data analysis of weather variables linked to Fusarium head blight epidemics in the United States. Phytopathology 109: 96-110. https://doi.org/10.1094/PHYTO-11-17-0386-R.

Shah, L., Ali, A., Yahya, M., Zhu, Y., Wang, S., Si, H., Rahman, H. and Ma, C., 2018. Integrated control of *Fusarium* head blight and deoxynivalenol mycotoxin in wheat. Plant Pathology 67: 532-548. https://doi.org/10.1111/ppa.12785.

Shaner, G., 2003. Epidemiology of *Fusarium* head blight of small grain cereals in North America. In: K.J. Leonard and W.R. Bushnell (eds.), *Fusarium* head blight of wheat and barley. American Phytopathological Society Press, St. Paul, Minnesota, USA, pp 84-119.

Simões, D., Carbas, B., Soares, A., Freitas, A., Silva, A.S., Brites, C. and Andrade, E.D., 2023. Assessment of agricultural practices for controlling *Fusarium* and mycotoxins contamination on maize grains: Exploratory study in maize farms. Toxins 15: 136. https://doi.org/10.3390/toxins15020136.

Simpson, D. R., Weston, G. E., Turner, J. A., Jennings, P. and Nicholson, P., 2001. Differential control of head blight pathogens of wheat by fungicides and consequences for mycotoxin contamination of grain. European Journal of Plant Pathology 107: 421-431. https://link.springer.com/article/10.1023/A:1011225817707.

Šíp, V., Chrpová, J., Veškrna, O. and Bobková, L., 2010. The impact of cultivar resistance and fungicide treatment on mycotoxin content in grain and yield losses caused by Fusarium head blight in wheat. Czech Journal of Genetics and Plant Breeding 46: 21. https://doi.org/10.17221/93/2009-CJGPB.

Sipos, P., Peles, F., Brassó, D. L., Béri, B., Pusztahelyi, T., Pócsi, I., and Győri, Z., 2021. Physical and chemical methods for reduction in aflatoxin content of feed and food. Toxins, 13: 204. https://doi.org/10.3390/toxins13030204.

Solà-Oriol, D. and Gasa, J., 2017. Feeding strategies in pig production: sows and their piglets. Animal Feed Science and Technology 233: 34-52. https://doi.org/10.1016/j.anifeedsci.2016.07.018.

Somma, S., Scarpino, V., Quaranta, F., Logrieco, A. F., Reyneri, A., Blandino, M. and Moretti, A., 2022. Impact of fungicide application to control T-2 and HT-2 toxin contamination and related *Fusarium sporotrichioides* and *F. langsethiae* producing species in durum wheat. Crop Protection 159: 106020. https://doi.org/10.1016/j.cropro.2022.106020.

Sørensen, E.S., Windju, S., Jansen, C., Sonesson, A.K., Lillemo, M. and Alsheikh, M., 2024. Identification of haplotypes associated with resistance to *Fusarium graminearum* in spring oat (*Avena sativa* L.). Plant Breeding 143: 204-214. https://doi.org/10.1111/pbr.13156.

Straumfors, A., Uhlig, S., Eriksen, G.S., Heldal, K.K., Eduard, W., Krska, R. and Sulyok, M., 2015. Mycotoxins and other fungal metabolites in grain dust from Norwegian grain elevators and compound feed mills. World Mycotoxin Journal 8: 361-373. https://doi.org/10.3920/WMJ2014.1799.

Sun, C., Ji, J., Wu, S., Sun, C., Pi, F., Zhang, Y., Tang, L. and Sun, X., 2016. Saturated aqueous ozone degradation of deoxynivalenol and its application in contaminated grains. Food Control 69: 185-190. https://doi.org/10.1016/j.foodcont.2016.04.041.

Suproniené, S., Mankevičiené, A., Kadžiené, G., Kačergius, A., Feiza, V., Feiziené, D., Semaskiene, R., Dabkevicius, Z. and Tamošiūnas, K., 2012. The impact of tillage and fertilization on *Fusarium* infection and mycotoxin production in wheat grains. https://www.zemdirbyste-agriculture.lt/99(3)tomas/99_3_tomas_str5.pdf.

Suproniené, S., Kadziene, G., Irzykowski, W., Sneideris, D., Ivanauskas, A., Sakalauskas, S., Serbiak, P., Svegzda, P., Kelpsiene, J., Pranaitiene, S. and Jedryczka, M., 2019. Asymptomatic weeds are frequently colonised by pathogenic species of *Fusarium* in cereal-based crop rotations. Weed Research 59: 312-323. https://doi.org/10.1111/wre.12367.

Teich, A.H. and Hamilton, J.R., 1985. Effect of cultural practices, soil phosphorus, potassium, and pH on the incidence of Fusarium head blight and deoxynivalenol levels in wheat. Applied and Environmental Microbiology 49: 1429-1431. https://doi.org/10.1128/aem.49.6.1429-1431.1985.

Tekauz, A., McCallum, B. and Gilbert, J., 2000. *Fusarium* Head Blight of barley in western Canada. Canadian Journal of Plant Pathology 22: 9-16. https://doi.org/10.1080/07060660009501156.

Tian, Y., Tian, Y., Liu, N., Liao, Y., Sun, C., Wang, S. and Wu, A., 2016. Functional agents to biologically control deoxynivalenol contamination in cereal grains. Frontiers in Microbiology 7: 395. https://doi.org/10.3389/fmicb.2016.00395.

Tibola, C.S., Fernandes, J.M.C. and Guarienti, E.M., 2016. Effect of cleaning, sorting and milling processes in wheat mycotoxin content. Food Control 60: 174-179. https://doi.org/10.1016/j.foodcont.2015.07.031.

Tittlemier, S.A., Blagden, R., Chan, J., McMillan, T. L., Pleskach, K. and Izydorczyk, M.S., 2020. Effects of processing whole oats on the analysis and fate of mycotoxins and ergosterol. World mycotoxin journal 13: 45-56. https://brill.com/view/journals/wmj/13/1/article-p45_5.xml.

Trenholm, H.L., Charmley, L.L., Prelusky, D.B. and Warner, R.M., 1991. Two physical methods for the decontamination of four cereals contaminated with deoxynivalenol and zearalenone. Journal of Agriculture and Food Chemistry 39: 356–360. https://doi.org/10.1021/jf00002a026.

Vissers, M.M.M., Driehuis, F., Te Giffel, M.C., de Jong, P. and Lankveld, J.M.G., 2007. Concentrations of butyric acid bacteria spores in silage and relationships with aerobic deterioration. Journal of Dairy Science 90: 928–936. https://doi.org/10.3168/jds.S0022-0302(07)71576-X.

Vlajkov, V., Grahovac, M., Budakov, D., Loc, M., Pajčin, I., Milić, D., Novakovic, T. and Grahovac, J., 2021. Distribution, genetic diversity and biocontrol of aflatoxigenic *Aspergillus flavus* in Serbian maize fields. Toxins 13: 687. https://doi.org/10.3390/toxins13100687.

Vogelgsang, S., Beyer, M., Pasquali, M., Jenny, E., Musa, T., Bucheli, T. D., Wettstein, F.E. and Forrer, H.R., 2019. An eight-year survey of wheat shows distinctive effects of cropping factors on different Fusarium species and associated mycotoxins. European Journal of Agronomy 105: 62-77. https://doi.org/10.1016/j.eja.2019.01.002.

Wenda-Piesik, A., Lemańczyk, G., Twarużek, M., Błajet-Kosicka, A., Kazek, M. and Grajewski, J., 2017. Fusarium head blight incidence and detection of *Fusarium* toxins in wheat in relation to agronomic factors. European Journal of Plant Pathology 149: 515-531. https://doi.org/10.1007/s10658-017-1200-2.

Wilkinson, J.M., and Davies, D.R., 2013. The aerobic stability of silage: key findings and recent developments. Grass and Forage Science 68: 1–19. https://doi.org/10.1111/j.1365-2494.2012.00891.x.

Xia, R., Schaafsma, A.W., Limay-Rios, V. and Hooker, D.C., 2021. Effectiveness of a novel fungicide pydiflumetofen against *Fusarium* head blight and mycotoxin accumulation in winter wheat. World Mycotoxin Journal 14: 477-493. https://doi.org/10.3920/WMJ2020.2637.

Xu, X., Madden, L. V., Edwards, S. G., Doohan, F. M., Moretti, A., Hornok, L., Nicholson, P. and Ritieni, A., 2013. Developing logistic models to relate the accumulation of DON associated with Fusarium head blight to climatic conditions in Europe. European Journal of Plant Pathology 137: 689-706. https://link.springer.com/article/10.1007/s10658-013-0280-x.

Xue, A.G., Rowsell, J., Ho, K.M., Chen, Y., Chi, D.T., Manceur, A., Zhang, S. and Ren, C., 2013. Effect of harvest date on barley grain contamination with *Fusarium* spp. and deoxynivalenol in northeastern Ontario. Phytoprotection 93: 1-7. https://doi.org/10.7202/1015205ar.

Yan, W., Fregeau-Reid, J., Rioux, S., Pageau, D., Xue, A., Martin, R., Fedak, G., de Haan, B., Lajeunesse, J. and Savard, M., 2010. Response of oat genotypes to Fusarium head blight in Eastern Canada. Crop science 50: 134-142. https://doi.org/10.2135/cropsci2009.03.0129.

Yoshida, M., Nakajima, T. and Tonooka, T., 2008. Effect of nitrogen application at anthesis on Fusarium head blight and mycotoxin accumulation in breadmaking wheat in the western part of Japan. Journal of General Plant Pathology 74: 355–363. https://doi.org/10.1007/s10327-008-0109-1.

Yoshida, M. and Nakajima, T., 2012. Chemical control of Fusarium head blight and mycotoxin contamination in barley and wheat based on mycotoxin accumulation during grain development. Mycotoxins 62: 19-27. https://doi.org/10.2520/myco.62.19.

CHAPTER 16

Feed additives to mitigate mycotoxins in livestock

O. Kolawole[1]* *and A. Gallo*[2]

[1]*Institute for Global Food Security, Queen's University Belfast, UK*
[2]*Department of Animal Science, Food and Nutrition (DIANA), Facoltà di Scienze Agrarie, Alimentari e Ambientali, Università Cattolica del Sacro Cuore, Piacenza, Italy*
**correspondence: tobi.kolawole@wur.nl*

1 Introduction

Animal feed plays a key role in the global food supply chain. However, the contamination of feed with mycotoxins continues to present a huge challenge for livestock production due to their profound detrimental effects on animal health, welfare and productivity (Kolawole et al., 2020; Tolosa et al., 2021). As described in the previous chapters, exposure of farm animals to a wide range of types and concentrations of mycotoxins can result in various adverse health effects, with the symptoms and their severity depending on the type and concentration of mycotoxin, interactions with other mycotoxins, the exposure time and the age, sex, health status and stress levels of the animals concerned (Tolosa et al., 2021). Contamination of feed with mycotoxins can increase production costs due to increased veterinary care costs and reduced livestock productivity and increased environmental emissions due to wastage when animals refuse feed, meaning that additional feed must be produced to maintain livestock energy requirements (Kolawole et al., 2022). Previous large-scale global surveys of mycotoxin contamination of feed showed that feed and feed ingredients are frequently contaminated with more than one mycotoxin (Gruber-Dorninger et al., 2019). In addition, the presence and concentration of mycotoxins were found to depend on the geographical location, agricultural practices, food safety regulatory framework and, importantly, the climate (Gruber-Dorninger et al., 2019). Climate is the key factor influencing fungal growth and their biosynthesis of mycotoxins (Kolawole et al., 2021). Changes in climatic and therefore agronomic conditions will undoubtedly continue to drive elevated mycotoxin contamination of food and feed crops (Perrone et al., 2020).

To reduce the impact of mycotoxins on animal health and productivity, different management strategies have been developed, including the setting of maximum limits for mycotoxin residues in feed. Regulations have been established in many parts of the world for AFs, and guidance on maximum levels of the main mycotoxins that frequently contaminate feed and feed ingredients have been recommended

 | DOI: 10.1163/9789004712720_17

based on their potential to induce adverse effects in farm animals (Van Egmond et al., 2007). The European Union (EU) has the strictest mycotoxin guidelines globally (Chapter 17). As described in Chapter 15, a variety of mycotoxin decontamination methods before feed production have also been developed and widely practised by feed manufacturers (Čolović et al., 2019; Liu et al., 2022a). However, the use of feed additives is the most commonly used strategy among farmers and the feed industry, primarily for economic reasons (Boudergue et al., 2009; Kolawole et al., 2019). The extensive use of additives to mitigate the toxic effects of mycotoxins has prompted some countries to establish new recommendations regarding the use of additives to prevent potential adverse effects on animal health and performance from the additives themselves.

To date, tens of hundreds of additives for mycotoxin detoxification (or minimization of the exposure of farm animals to mycotoxins) have been developed and are available on the global market (Boudergue et al., 2009; Kolawole et al., 2019). The additives used for this purpose can be divided into three main groups based on their composition, mode of action and functionality: binders, modifiers and antioxidants (Figure 16.1). The market is rapidly expanding, with projected growth at a compound annual rate of 3.5% from 2019 (US $1.8 billion), reaching a value of US $3.7 billion by 2030. The poultry segment represents the largest increase and the aquatic animal segment is growing faster than all the others (Market Report, 2024). The following three major factors are expected to drive market growth: (1) increase in the demand for safe and nutrient-rich livestock feed to meet the global demand for livestock products; (2) the growing awareness and concerns about reduced animal feed efficiency and production due to elevated levels of mycotoxins in feed; and (3) the drive towards sustainable livestock production, i.e. there is a direct link between greenhouse gas emission intensities and livestock efficiency (Hidosa et al., 2017). The more efficient and productive the animal is, the lower the environmental impact of feed production will be. The main objectives of this chapter are (1) to provide a comprehensive overview of the distinct categories of substances that have been evaluated for their potential to reduce mycotoxins or mycotoxin-induced toxicity, and (2) to describe the anti-mycotoxin feed additive regulations and the criteria and requirements for new product approval in different countries.

2 Mycotoxin-binding additives

The aim of mycotoxin-binding additives is to prevent the absorption of mycotoxins from the GIT of livestock animals by adsorbing the toxins to their surfaces, to form a binder-mycotoxin complex, which is later excreted with the faeces. Mycotoxin binders can be divided into two sub-groups: inorganic (clay minerals) and organic (such as yeast cell wall, polymers and agricultural waste products). Their mycotoxin-binding efficiency and rate of adsorption are influenced by several physico-

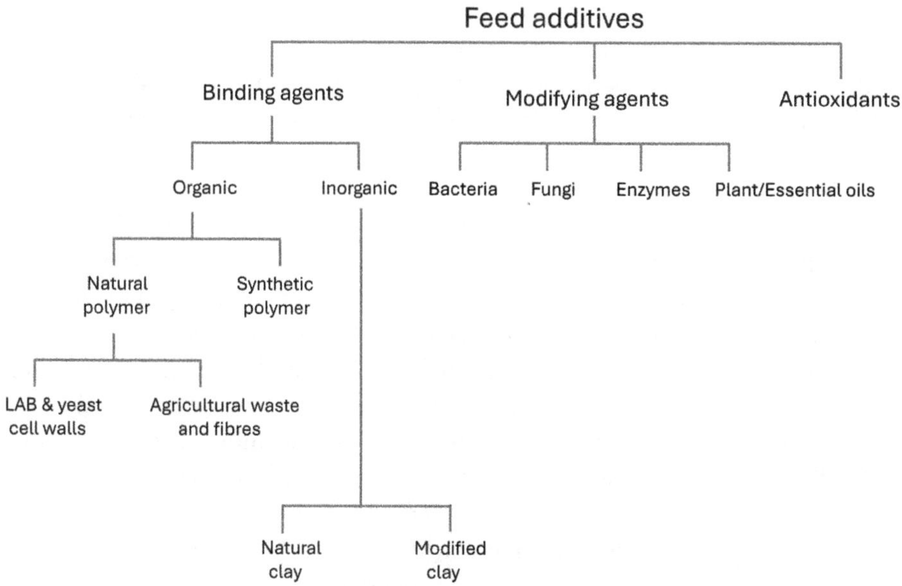

FIGURE 16.1 Categories/sub-categories of feed additives for reducing livestock exposure to mycotoxins

chemical properties including their size, porosity, total charge and cation exchange capacity (CEC) (Barrientos-Velázquez et al., 2016). In addition, their rate of mycotoxin adsorption is significantly influenced by the chemical properties of the toxin including its polarity, shape and size (Barrientos-Velázquez et al. 2016; Kolawole et al., 2019).

2.1 *Inorganic binders*

2.1.1 Natural clay minerals

Clays are fine graded natural rock and soil materials. Aluminosilicates are the largest and most important class of clay minerals. They are composed of silica, alumina and significant amounts of alkaline and alkaline earth ions (Andrade et al., 2011). Within the class of aluminosilicates, tectosilicates and phyllosilicates are the most important due to their wide range of applications. The basic structure of phyllosilicates is based on tetrahedral sheets (T), composed of individual tetrahedrons, and octahedral sheets (O) composed of individual octahedrons that share apical oxygen and hydroxyl anion groups with cations such as calcium and magnesium. The stacking of both sheets determines the chemistry and crystallography of each phyllosilicate. The T:O ratio is used to classify the layer structure of phyllosilicates into 1:1 (T–O) and 2:1 (T–O–T). Examples of phyllosilicates include montmorillonite, illites, bentonite and kaolinite. Tectosilicates have a three-dimensional framework structure in which all the four oxygens of the tetrahedron are shared with other tetrahedra. Examples of tectosilicates include zeolite, quartz and feldspar. Although phyllosilicates and tectosilicates are composed of O and T sheets as the predominant building blocks, the physicochemical properties of these minerals including their CEC, particle size, charge, polarity, expandability, pH and adsorption capacity, differ and depend on the exact structure and geographical origin (Andrade et al., 2011; Ito and Wagai, 2017).

Animal feeds were previously supplemented with clay minerals mainly as anti-caking agents to prevent lump formation. However, over recent decades, several researchers have demonstrated the ability of clay minerals to bind mycotoxins, particularly AFs (Afriyie-Gyawu et al., 2005). This phenomenon has led to the increasing use and exploration of new clay minerals for mycotoxin mitigation (Di Gregorio et al., 2014; Vila-Donat et al., 2018). Due to the negatively charged surface area and high CEC, some specific clay minerals can bind or adsorb mycotoxins to their interlayer spaces, external surfaces and edges. Several clay minerals with mycotoxin-binding properties have been patented; some of these are also commercially available as feed additives for livestock farmers and animal nutrition companies (He and Zhao, 2010). In fact, clay minerals constitute more than 90% of commercially available feed additives. The mechanisms of action for binding mycotoxins include chemisorption and ion exchange, i.e. binding between clay cations and mycotoxin carbonyl groups (Deng et al., 2010). In addition, the rate of adsorption depends on the origin and physicochemical properties of the clay and occurs mostly at low pH. Bentonite (mostly consisting of montmorillonite), the zeolite (clinoptilolite), activated carbon, diatomaceous earth and hydrated sodium calcium aluminosilicate are the most common commercially available clay minerals used for the alleviation of mycotoxin toxicities in farm animals.

Clay minerals are low cost and widely available. However, their application as additives for binding mycotoxins is limited because, as many studies have demonstrated, they bind AFs efficiently but they have little or no binding capacity for the other important mycotoxins (Di Gregorio et al., 2014; Vila-Donat et al., 2018; Kolawole et al., 2019). AFs are relatively hydrophilic with aromatic planar molecules; therefore, most clay minerals easily bind AFs by hydrogen bonding with the carbonyl oxygen and coordination via ion–dipole interactions. However, other mycotoxins including DON, ZEN, FUM, OTA, and T-2 toxin range from being moderately hydrophilic to highly hydrophobic compounds, so bind very weakly to natural clay minerals (Di Gregorio et al., 2014; Vila-Donat et al., 2018; Kolawole et al., 2019).

2.1.2 Modified clay minerals

To increase the affinity of clay minerals for other mycotoxins, organically modified clays are being developed by replacing the exchangeable inorganic cations present in the natural clays with high molecular weight cationic quaternary amines or polyhydroxy metal cations, resulting in increased hydrophobicity and enlargement of the interlayer spaces (Li et al., 2018). The common organic modifiers include octadecyltrimethylammonium chloride and benzyldimethylhexadecylammonium chloride. Modified clays can also be prepared by acid, alkali, oil and/or heat treatment of natural clays, leading to alteration of the surface properties including polarity, pH and CEC. These changes enable organically modified clays to possess high mycotoxin-sequestering abilities compared with natural clay minerals, with efficient binding of non-polar or weak polar mycotoxins (Wang et al., 2018; Li et al., 2018). The bentonite montmorillonite can be modified by using pillaring agents such as aluminium (Al) and iron (Fe), resulting in increased sorption area with high permanent porosity and high thermal stability (Gao et al., 2014), and such a technology has shown the ability to bind 35% DON (Zhang et al., 2021). Whilst modified clay minerals appear to be superior to natural clays in terms of their mycotoxin adsorption capacities, some of the modifying agents such as quaternary ammonium salts are toxic and may pose feed safety risks and have a negative environmental impact. Therefore, the labelling for the feed additive with modified clay should also mention the levels of possible contaminants.

2.2 *Organic binders*

2.2.1 Natural polymers

2.2.1.1 *Cell walls from yeast and lactic acid bacteria*

Yeast and lactic acid bacterial (LAB) cell walls are known to possess high-binding activity for a wide spectrum of mycotoxins, particularly for OTA, T-2 toxin and ZEN, when compared with inorganic clay binders. The cell wall of yeast (*Saccharomyces cerevisiae*) is composed of lipid, protein and polysaccharide fractions, with glucans and mannans being the two main constituents of the polysaccharide fraction. The mechanism of mycotoxin adsorption is mainly non-covalent, by hydrogen bonding and ionic or hydrophobic interactions (Liu et al., 2021; Luo et al., 2020). Many researchers have shown that β-D-glucans are the principal yeast cell wall component responsible for the binding of mycotoxins (Faucet-Marquis et al., 2014; Luo et al., 2020). Furthermore, the reticular organization of the β-D-glucans and the distribution between β-(1,6)-D-glucans and β-(1,3)-D-glucans both play significant roles in determining the capacity to adsorb mycotoxins (Faucet-Marquis et al., 2014; Luo et al., 2020).

The mycotoxin binding potential of LAB (mostly *Lactobacillus* species) is attributed to peptidoglycan, a unique and essential structural element in bacterial cell walls, and teichoic acid, a glycopolymer that is embedded within the peptidoglycan layers (Luo et al., 2020). Interestingly, modified or heat-treated cell walls of yeast and LAB offer higher rates of mycotoxin adsorption (Liu et al., 2022a). To summarize, the type of mycotoxin a cell wall product derived from yeast or LAB can adsorb and the efficacy of binding depend on the strain, growth conditions, pH and the number of binding sites determined by the percentage composition of cell wall components including mannoproteins and polysaccharides and the availability of negatively charged functional groups.

2.2.1.2 *Agricultural waste and plant fibre*

Sorption of mycotoxins by agricultural waste products and micronized plant fibres have recently emerged as a more favourable and low-cost detoxification strategy. Agricultural waste, including from the cultivation of grapes, oranges, bananas and almonds, contains insoluble fibres such as cellulose, hemicellulose, lignin and tannins that may facilitate the binding or inactivation of mycotoxins (Greco et al., 2019). Comparing the efficacy of several agricultural waste products to sequester mycotoxins, grape pomace, artichoke and almond hulls were found to bind AFs, ZEN and OTA effectively (Gambacorta et al., 2016). The efficacy of these materials was further evaluated *in vivo*. Results showed that grape pomace significantly reduced urinary biomarkers of AFs (67%) and ZEN (69%), with no significant reduction observed for OTA in pigs fed contaminated diet (Gambacorta et al., 2016). Other agricultural waste by-products, including durian peel, orange peel and banana peel, have also been shown to be effective for reducing mycotoxins *in vitro*.

Micronized insoluble plant fibres (less than 500 μm in size) can also adsorb mycotoxins from the GIT and ameliorate mycotoxin-induced toxic effects (Aoudia et al., 2009). Micronized fibres are obtained from different plant materials including cereals or legumes (wheat, barley, alfalfa, oats, pea hulls or green seaweed). Glucomannan, a water-soluble polysaccharide considered a dietary fibre, has also been shown to be a good mycotoxin adsorbent (Mohaghegh et al., 2017). It is commonly extracted from the roots and bark of plants. Some researchers have suggested that the hemicellulose component of cell walls could be responsible for mycotoxin adsorption. In addition to mycotoxin binding activity, plant fibres can promote gut health and microflora, if added to the diets at an optimal inclusion level.

2.2.2 Synthetic polymers

Synthetic polymers are large molecules composed of many monomers with a broad range of applications. They are produced by complexation of one or more functional monomer(s) in a solution by either covalent or non-covalent interactions, followed by a polymerization or copolymerization reaction with cross-linking monomers. Synthetic polymers including cholestyramine, polyvinylpyrrolidone, divinylbenzene-styrene and polyvinylpyrrolidone have been shown to possess high mycotoxin-binding capacity, especially for hydrophobic compounds. Furthermore, modified cellulosic polymers including hydroxypropyl methylcellulose, sodium carboxymethylcellulose, microcrystalline cellulose and molecularly imprinted polymers such as TMU95 have also been evaluated for their potential to bind mycotoxins (Solís-Cruz et al., 2017; Appell et al., 2007; Arak et al., 2019).

Taken together, there is substantial evidence in the literature to suggest that both inorganic and organic adsorbents can efficiently sequester mycotoxins. However, due to the variety of compounds, it is important to consider the safety and stability of these agents in the GIT and their effect on the palatability of animal feed. Furthermore, when applied in the diet, natural clays must be free of toxic metals or other contaminants.

3 Mycotoxin-modifying or biotransforming additives

Given the limitations of mycotoxin-adsorbing agents, particularly inorganic clays, the use of modifying agents of biological origin (fungi, bacteria, plants and extracted enzymes) has been shown to offer a promising alternative strategy to reduce animal exposure to mycotoxins. They have the capacity to degrade or modify the structure of mycotoxins to yield metabolites/degradation products that are non-toxic or at least less toxic when compared with the parent mycotoxins. Species of *Lactobacillus, Rhodococcus, Mycobacterium, Acinetobacter, Bacillus, Oenococcus, Bifidobacterium, Kluyveromyces, Rhodotorula, Kloeckera, Candida, Aspergillus, Penicillium, Rhizopus and Aureobasidium* genera isolated from different sources including soil, marine sources and fermented foods have been demonstrated to have the capacity to biotransform mycotoxins (Liu et al., 2022b; Ghadaksaz et al., 2022). In addition, pure cultures of microbes isolated from chicken and pig intestines have shown mycotoxin-degrading properties (Young et al., 2007; Yu et al., 2010). Many researchers have also identified enzymes extracted and purified from various microorganisms that degrade mycotoxins (Wang et al., 2019; Qin et al., 2021; Liu et al., 2022b). These include laccase enzyme purified from the mushroom *Trametes versicolor*, manganese peroxidase from the white-rot fungus *Phanerochaete spp.* and lactonohydrolase from the fungus *Candida rosea*. The most common mechanisms of mycotoxin detoxification include deacetylation, oxygenation, deepoxidation and glycosylation. In addition to microbes and enzymes, several researchers have also found some aqueous extracts and essential oils from plants including cinnamon, lavender, jojoba and peppermint that significantly reduce mycotoxin levels (Perczak et al., 2016, 2019; Xing et al., 2014). When supplementing diets with exogenous enzymes, a concern arises due to the exposure of these additives to temperatures above 60 °C because of the risk of their partial or total inactivation (Abdollahi et al., 2013). Therefore, this type of mycotoxin-biotransforming agents should be thermally stable to reach the desired efficacy. When enzymes are sensitive to high temperatures, they can be sprayed after pelleting, during the cooling stage, but it requires a specific line production in the feed mill. Importantly, the degradation of the parental mycotoxin should not result in more toxic metabolites.

4 Antioxidants

Antioxidants are added as feed supplements primarily to prolong the shelf life and sensory qualities by preventing rancidity and the oxidation of critical nutrients such as pigments and vitamins. Antioxidants do not bind or modify mycotoxins, but they can counteract the toxic effects of mycotoxins by reducing the oxidative stress and inflammatory signalling pathways triggered by mycotoxins. Oxidative stress is one of the major molecular mechanisms behind the toxic effects of mycotoxins (Shi et al., 2017; Graziani et al., 2015; Zhang et al., 2009). Exposure of farm animals to mycotoxins inhibits the expression of antioxidant enzymes including SOD and GPx resulting in the over-production of free radicals including ROS. When the production of ROS exceeds the antioxidant capacity of a cell, several intracellular reactions occur that promote cell death, including oxidative damage of DNA, proteins and membrane lipids (Shi et al., 2017; Graziani et al., 2015; Zhang et al., 2009). Many natural antioxidants have been demonstrated to act as free radical scavengers, thereby conferring beneficial, protective effects against the toxic effects induced by mycotoxins. Generally, most antioxidants activate the nuclear factor erythroid 2-related factor 2 (Nrf2)-associated genes leading to the release of a battery of detoxification and antioxidant enzymes, which protect the cells from mycotoxin-induced inflammation and oxidative damage (Kozieł et al., 2021). Many natural substances have been shown to possess antioxidant properties as superoxide anion scavengers and have been evaluated for their potential to protect against mycotoxin-induced toxicity: these include resveratrol and proanthocyanidin from grapes; silymarin from *Silybum*; cinnamaldehyde from cinnamon oil; ginsenosides from ginseng; vitamins A, C and E; some phenolic compounds including naringin, catechin, eugenol, quercetin, kaempferol, ellagic acid and curcumin; and the element selenium and zinc (da Silva et al., 2018).

5 Silage additives

Focusing on the most important component of dairy cow diets worldwide, silage safety is a critical topic because this stored-for-fermentation feed may contain several mycotoxins that are potentially hazardous to animal health, safety of milk or other animal food products (Driehuis et al., 2018; Gallo et al., 2015). In addition to regulations and recommendations regarding mycotoxin limits of contamination, many efforts to limit the production of mycotoxins in silage could be applied during the field, pre-harvest, harvest, and ensiling phases (Ogunade et al., 2018) (See also Chapter 15). Dalié et al. (2010) verified the LAB potential to mitigate mycotoxin contamination of silage *in vitro*. Several species belonging to *Lactococcus* and *Lactobacillus* genera (and *Pediococcus* and *Leuconostoc* genera) are potential agents to prevent or inhibit toxic fungal growth. Their antifungal activity was correlated with incubation period, temperature, and rapid decrease in pH. The highest production of antifungal metabolites by *Lc. lactis* subsp. *diacetylactis* was observed after three to four days of incubation. The peak yield of antimicrobial compounds produced by *Lc. lactis* subp. *diacetylactis* S 1-67-C was recorded at 25 °C. Furthermore, Piotrowska (2014) investigated that inoculating silage with *Lactobacillus brevis* and *L. plantarum* reduced OTA content by 16.9 to 35.0% when grown in de Man, Rogosa, and Sharpe broth and by 14.8–26.4% in PBS after 24 h of incubation. Inoculation of maize plants infested with southern rust using a mixture of *Pediococcus pentosaceus* and *Lactobacillus buchneri* at ensiling increased aerobic stability and prevented AF production (5,200 vs. 0 μg/kg) (Queiroz et al., 2012). In another study, *Lactobacillus buchneri* inoculation reduced AF concentration without any effects on concentrations of DON and ZEN in maize silage (Iglesias et al., 2005).

Inoculating maize silage with LAB, such as *Lactobacillus buchneri* and *Lactobacillus plantarum*, is a widely adopted strategy to improve the fermentation and increase biopreservation and aerobic stability of silage, because it allows for a delay in the growth of spoilage yeasts and moulds (Adesogan and Arriola., 2020; Muck et al., 2018). Antifungal effects of LAB are associated with their production of a plethora of metabolites such as organic acids (e.g., propionic acid and acetic acid), bacteriocins or phenolic compounds (phenyl lactic acid, n-decanoic and 3-hydroxydodecanoic acid) (Sadiq et al., 2019; Gallo et al., 2021). On the other hand, the antifungal effects of this practice are undefined, particularly in silage from high-moisture maize as it is an oxygen-rich microenvironment in which caked and clumpy areas develop visible green-grey mould indicative of mycotoxin production, including AFs and several *Fusarium*-produced mycotoxins (Gallo et al., 2015, 2021a). Application of inoculants containing obligate heterofermentative bacteria, such as *L. buchneri*, can increase the aerobic stability of silage (Muck et al., 2018). Despite the potential benefits of using LAB inoculants in silage, Ogunade et al. (2018) reported that very few studies have examined the use of mould-inhibiting chemical additives or microbial inoculants to reduce mycotoxin contamination. As previously mentioned, the LAB can also adsorb mycotoxins onto the polysaccharides and peptidoglycans in their cell walls (Lahtinen et al., 2004). The ability of LAB to adsorb mycotoxins is affected by their molecular structure, concentration, number and the physiological state of the cells (i.e., viable or dead), or pH (Ma et al., 2017; Sadiq et al., 2019).

Gallo et al. (2021) compared the effects of three commercial inoculants and six single strains of LAB on fermentation and levels of AFs and other mycotoxins in high-moisture maize silage. The storage time and the inoculants did not affect AF levels. No ZEN or TeA were detected in the mini silos, even if they contaminated the maize plants. Results indicate that different silage inoculants affected fermentation of high-moisture maize and its contamination by mycotoxins, although no inoculant influenced levels of AFs. Møller et al. (2021) evaluated the effect of LAB on the growth of *A. parasiticus* NRRL 2999 and its production of AF. They observed reductions of at least 50% in levels of AFB_1, OTA and ZEN following inoculation with *L. brevis* 2QB422. Guan et al. (2020) stated that the effect of inoculants is linked to the environmental conditions of a specific place; indeed, there are several strains of *L. plantarum* that are not suitable as inoculants in many countries of the world; they reported that using *L. rhamnosus* 753 could be an opportunity to prolong the aerobic stability and inhibit the accumulation of AFB_1 in hot and humid conditions. Wang et al. (2022) reported that the application of *Lentilactobacillus buchneri* and *Lactiplantibacillus plantarum* inoculants improved the aerobic stability of maize silage, delayed the onset of aerobic deterioration, and reduced the risk of mycotoxin production after silo opening, in particular that related to AFs, ZEN, FB_1 and FB_2, whereas no effects were observed for DON.

6 Criteria and requirements for new product approval

In Europe, the US and Brazil, regulations have been established to approve products with the capacity to reduce mycotoxins, although several claimed mycotoxin decontaminant products can be placed on the market. These products are commercially available and registered as emulsifiers, immunostimulants, sequestrants, anticaking agents and digestibility enhancers and given generic or catalogue names such as montmorillonite, bentonite and hydrated sodium calcium aluminosilicate. Generally, product efficacy data against mycotoxins effects are indirectly demonstrated by significant improvements in performance such as BW and FCR, as well as haematological and intestinal parameters.

6.1 *United States*

The US Food and Drug Administration (FDA) Centre for Veterinary Medicine (CVM) is responsible for the approval of new animal feed additives. Clay minerals including aluminosilicates, bentonites and diatomaceous earth are generally recognized as safe (GRAS) when used as anticaking agents in animal feed at a level not exceeding 2% and in accordance with feeding practices outlined in Code of Federal Regulations Title 21 Parts 582.2727 and 582.2729 (21 CFR 582.2727 and 582.2729) (FDA Part 582). However, these substances cannot be designated as GRAS when used in feed as mycotoxin binders or modifiers. To apply for the approval of a new feed additive, the company or manufacturer must submit a feed additive petition (FAP) to the FDA's CVM. The FAP must include satisfactory information to establish the safety and efficacy of the additive under the conditions of use detailed in the petition. Thereafter, the product will be scientifically evaluated by the CVM to verify the information submitted. A list of the general information required in an FAP submission is detailed in 21 CFR 571 and outlined below (FDA Part 571):

(1) The identity and composition of the additive including manufacturing methods and controls.
(2) The intended use, inclusion level and labelling (i.e. cautions, shelf life, directions for use)
(3) Data establishing the intended effect (physical, nutritional, or other technical effect)
(5) Safety evaluation (target animal and human food)
(6) Proposed tolerances for the feed additive, and
(7) Environmental assessment.

Upon successful verification, the FDA will issue a regulation concerning the feed additive and its use. The regulation will specify the types of animal feed that may contain the additive, the level of additive that may be used in the feed and the requirements for labelling the additive or for labelling the animal feed containing the additive. Following approval, use of the additive must be within the constraints of the established regulation. By August 2021, no product had been approved by the FDA for the binding, inactivation, modification or biotransformation of mycotoxins or the prevention of mycotoxin-induced toxicity. A list of additives permitted for use in US animal feed is available in 21 CFR 573 (FDA Part 573).

In January 2011, the Office of the Texas State Chemist (OTSC) published proposed rules for the use of aflatoxin-binding agents in customized formula feeds. The proposition was approved in Texas on 19 May 2011 (OTSC, 2011). The new regulation allows establishments or feed facilities to sell and include aflatoxin-binding agents in customized formula feeds. The guidelines for registration are defined in the Texas Agriculture Code §141.001(7). As the US FDA was not involved in this approval of AF binders, these substances are considered illegal under the Federal Food, Drug and Cosmetic Act. This implies that animal feeds containing AF binders are classified as adulterated once they enter interstate commerce outside Texas (OTSC, 2011). Consequently, the OTSC requires labelling to clearly state that the aflatoxin-binding product is for use in Texas only. A list of active feed facilities with a licence to use aflatoxin-binding agents in customized formula feeds in Texas is available on the Texas State Chemist's website (OTSC, 2022).

6.2 *Brazil*

In Brazil, the task force set up by the Ministry of Agriculture, Livestock and Food Supply in 2005 oversees the registration of anti-mycotoxin additives. Anti-mycotoxin additives must pass through a two-stage, *in vitro* and *in vivo*, process before they can be approved by the SAMITEC Institute, a research centre affiliated with the Federal University of Santa Maria, Laboratory of Mycotoxicological Analysis (LAMIC). In addition, products that have gained approval and been commercialized must be re-evaluated by LAMIC after two years, to check that the manufacturers are still marketing the same product composition that was originally approved. A press release published in 2009 indicated that more than 100 products have been submitted for evaluation, with only 30% being granted approval (Engormix, 2009). All the products approved were clay-based and highly specific for AFs.

6.3 *European Union*

In 2003, the European Commission (EC) regulation 1831/2003 established the guidelines regarding the authorization and use of additives in animal nutrition (EC, 2003, 2008). The commission in 2009 amended this regulation; a new functional group of feed additives, entitled 'substances for reduction of the contamination of feed by mycotoxins: substances that can suppress or reduce the absorption, promote the excretion of mycotoxins or modify their mode of action', was introduced. This additional functional group was a subgroup of the existing category of technological additives. The EU has the most stringent but explicit regulations regarding the evaluation and approval of additives for mycotoxin mitigation in farm animals. An application request for a new product or technological additive is received by the EC, which then seeks technical advice on the safety and efficacy of the product from the EFSA panel on Additives and Products or Substances used in Animal Feed (FEEDAP). The FEEDAP panel has published fundamental guidelines to help applicants prepare and present application dossiers for assessment and authorization in accordance with Article 5 of Regulation (EC) No 1831/2003 and its implementation rules (Regulation (EC) No 429/2008) (EC, 2008, 2012). In addition to the additive's identity (its name and the active substance), its characterization (physical, chemical and technological properties) and its conditions for use (inclusion level, target animal species and target mycotoxin), the FEEDAP panel requires scientific evidence of its safety and efficacy (EC, 2008, 2010, 2012). Importantly, the mode of action (i.e. inactivation, reduction of absorption, promotion of excretion or modification of mechanism of toxicity) must be clearly declared and demonstrated (EC, 2008, 2010).

In terms of product efficacy, the FEEDAP panel requires a minimum of three *in vivo* studies performed in at least three major species—poultry, monogastric mammal and ruminant—at two different locations. For additives intended to be used in fish, specific studies in fish, preferably salmonids, are required. The number of animals/replicates should be large enough to allow statistical evaluation of the results and to demonstrate the efficacy of the product at the lowest recommended dose. Collection of biological samples for the measurement of mycotoxins or their metabolites or biomarkers of exposure should be carried out over five days to demonstrate efficacy (EC, 2008, 2010). The EFSA has recommended endpoints to be measured for AFs, DON, ZEN, FUM, and OTA (Chapter 12, Table 2) (EC, 2008, 2010). However, recent studies have shown that mycotoxins undergo extensive phase I and II biotransformations (Gambacorta et al., 2013, 2016; Lauwers et al., 2019), so these biomarkers are not always optimal for each mycotoxin. Moreover, the recommended endpoints and biomarkers are too generic and not specific enough for each animal species. The optimal biomarkers to demonstrate the efficacy of a substance to reduce or modify mycotoxins in pig (plasma, urine, faeces) and broiler chickens (plasma and excreta), are outlined in Chapter 12 (Table 12.2).

Regarding the safety of the additive, for modifiers or biotransforming agents, the major metabolites and degradation products derived from the mycotoxin must be identified. Furthermore, the combined effects of both the additive and the degradation products or metabolites on the safety of the target animal and consumer must be evaluated in appropriate toxicology studies. For binding agents, their impact on the bioavailability of essential nutrients and veterinary substances must also be evaluated. The FEEDAP panel recommended the measurement of apparent digestibility of crude protein, zinc, retinyl or tocopheryl esters, thiamine, or pyridoxine and the addition of a coccidiostat to the feed for animal groups fed the highest recommended dose of the additive (EC, 2008, 2010). Authorized products must be submitted for re-evaluation after a 10-year period.

7 Other intervention: vaccination

Given the effects of mycotoxins on animal health, the possibility of vaccinating against mycotoxins was investigated. In particular, to date only vaccines against AFB_1 have been studied for ruminants, more precisely on lactating cows and heifers and not on calves. In 2011 Polonelli et al. conducted a study aimed at investigating the potential of anaphylatoxin B_1 ($AnAFB_1$) conjugated with keyhole limpet haemocyanin (KLH) as a vaccine ($AnAFB_1$-KLH) in controlling the carry-over of the AFB_1metabolite AFM_1 in cow's milk. $AnAFB_1$-KLH was used to immunize the cows, which enabled the onset of a good titre of anti-AFB_1 IgG (Abs) antibodies characterized by cross-reactivity with AFB_1, AFG_1 and AFG_2. The secretion of AFM_1 in the milk of cows fed continuously with AFB_1 is hindered by the presence of anti-AFB_1 antibodies produced as a result of vaccination. To further reduce the amount of AFM_1 transferred into cow's milk from cows exposed to AFB_1, a new trial was conducted comparing the immunogenicity of AnAFB1 covalently conjugated with KLH or with recombinant diphtheria toxin molecules (CRM197) and injected together with various adjuvants. Following vaccination, a reduction in the titre of anti-AFB_1 Ab was observed, whereas peripartum physical and metabolic stresses did not affect the concentration of anti-AFB_1 Ab (Giovati et al., 2014). In the same trial, the validity of the vaccine booster to consolidate and recover the Ab response against AFB_1 was also affirmed. Indeed, it was seen that time (and not the administration of AFB_1 in the diet) favours the disappearance of anti-AFB_1 Ab. The injection of KLH-conjugated $AnAFB_1$ mixed with Freund's complete (priming) and incomplete (booster) adjuvant was found to be effective in inducing the formation of anti-AFB_1 Ab antibodies. In particular, excellent results are obtained when vaccination is carried out prior to lactation, and the efficacy of the booster dose at the beginning of the immunization programme in reducing AFM_1 carry-over in milk was also found. By evaluating these results, it is therefore possible to ascertain the effectiveness of vaccination in reducing the risk of AFM_1-contaminated milk and cheese.

8 Conclusions

As depicted in Figure 16.1, a wide range of products to mitigate mycotoxins has been developed and evaluated, with more to come. This presents a huge challenge, particularly for farmers, the feed industry and animal nutritionists in terms of choosing the right product to ameliorate or reduce livestock exposure to mycotoxins to improve animal performance and increase profitability. The addition of products whose safety and efficacy have not been scientifically proven can induce potential adverse effects on animal health and productivity, or can elevate the costs of feed production without a return in animal performance.

Climate change, and its effects on agricultural conditions, is increasing the contamination of animal feed with multiple mycotoxins. Therefore, research should be devoted to the development of new innovative products that are low cost, environmentally friendly and nutritionally beneficial to the animal with broad-spectrum mycotoxin-detoxifying activity. It is important to note that the capacity of a substance to reduce mycotoxins cannot be measured effectively by *in vitro* studies alone, using buffers or simulated gastrointestinal models; animal feeding trials must also be conducted, using a complete and standard experimental design consisting of four treatments: (1) non-contaminated diet (negative control); (2) mycotoxin-contaminated diet (positive control); (3) non-contaminated diet with the additive; and (4) mycotoxin-contaminated diet with the additive. Additional treatments can be added to this experimental design, for example to test the effect of different levels of the decontaminant. In addition to demonstrating that the additive is safe, the efficiency of the additive to reduce mycotoxins or ameliorate toxicities induced by mycotoxins should be examined using specific and non-specific parameters. For non-specific parameters, in addition to BW, feed consumption and mortality rate data collected during the animal trial, blood samples should be analysed for haematological biochemical parameters including AST, ALT, ALP, LDH, protein, cholesterol, uric acid, IgA, haemoglobin and leukocytes, to evaluate any negative impacts of mycotoxins on the animals and the adverse or beneficial effects of the additive. The specific parameters to be measured to prove the efficacy of the additive include the quantification of mycotoxins and their metabolites in biological samples (milk, plasma, urine, or excreta). The pelleting process occurs during the production of most diets, which results in exposure to high temperatures. Therefore, feed additives with the ability to decontaminate mycotoxins should preferably be thermally stable. Silage inoculation with LAB has resulted in a positive effect when decreasing the mycotoxin levels of AFB_1 and OTA, and to a lesser extent *Fusarium* mycotoxins such as ZEN and FB_1 and FB_2. However, LAB strain efficacy may vary according to the geographical area. Finally, besides dietary supplementation with feed additives, vaccination is an extra intervention to support the mitigation of mycotoxin exposure. Some success has been demonstrated in cows, but only for AFB_1.

References

Abdollahi, M.R., Ravindran, V. and Svihus, B., 2013. Pelleting of broiler diets: an overview with emphasis on pellet quality and nutritional value. Animal Feed Science and Technology 179: 1-23. https://doi.org/10.1016/j.anifeedsci.2012.10.011.

Adesogan, A.T. and Arriola, K.G., 2020., Bacterial inoculants for optimizing silage quality and preservation and enhancing animal performance. https://ecommons.cornell.edu/server/api/core/bitstreams/f52be84f-3c95-4797-ae21-f2b83d2f9099/content.

Afriyie-Gyawu, E., Mackie, J., Dash, B., Wiles, M., Taylor, J., Huebner, H., Tang, L., Guan, H., Wang, J. and Phillips, T., 2005. Chronic toxicological evaluation of dietary NovaSil Clay in Sprague-Dawley rats. Food Additives and Contaminants 22: 259-269. https://doi.org/10.1080/02652030500110758.

Andrade, F., Al-Qureshi, H. and Hotza, D., 2011. Measuring the plasticity of clays: A review. Applied Clay Science 51: 1-7. https://doi.org/10.1016/j.clay.2010.10.028.

Aoudia, N., Callu, P., Grosjean, F. and Larondelle, Y., 2009. Effectiveness of mycotoxin sequestration activity of micronized wheat fibres on distribution of ochratoxin A in plasma, liver and kidney of piglets fed a naturally contaminated diet. Food and Chemical Toxicology 47: 1485-1489. https://doi.org/10.1016/j.fct.2009.03.033.

Appell, M., Kendra, D., Kim, E. and Maragos, C., 2007. Synthesis and evaluation of molecularly imprinted polymers as sorbents of moniliformin. Food Additives and Contaminants 24: 43-52. https://doi.org/10.1080/02652030600887586.

Arak, H., Karimi Torshizi, M., Hedayati, M. and Rahimi, S., 2019. The first in vivo application of synthetic polymers based on methacrylic acid as an aflatoxin sorbent in an animal model. Mycotoxin Research, 35: 293-307. https://doi.org/10.1007/s12550-019-00353-z.

Barrientos-Velázquez, A., Arteaga, S., Dixon, J. and Deng, Y., 2016. The effects of pH, pepsin, exchange cation, and vitamins on aflatoxin adsorption on smectite in simulated gastric fluids. Applied Clay Science 120: 17-23. https://doi.org/10.1016/j.clay.2015.11.014.

Boudergue, C., Burel, C., Dragacci, S., Favrot, M.C., Fremy, J.M., Massimi, C., Prigent, P., Debongnie, P., Pussemier, L., Boudra, H., Morgavi, D., Oswald, I., Perez, A. and Avantaggiato, G., 2009. Review of mycotoxin-detoxifying agents used as feed additives: mode of action, efficacy and feed/food safety. EFSA Journal 6: 22. https://doi.org/10.2903/sp.efsa.2009.EN-22.

Čolović, R., Puvača, N., Cheli, F., Avantaggiato, G., Greco, D., Đuragić, O., Kos, J. and Pinotti, L., 2019. Decontamination of mycotoxin-contaminated feedstuffs and compound feed. Toxins 11: 617. https://doi.org/10.3390/toxins11110617.

da Silva, E., Bracarense, A. and Oswald, I., 2018. Mycotoxins and oxidative stress: where are we? World Mycotoxin Journal 11: 113-134. https://doi.org/10.3920/WMJ2017.2267.

Dalié, D.K.D., Deschampes, A.M. and Richard-Forget, F., 2010. Lactic acid bacteria – Potential for control of mould growth and mycotoxins: A review. Food Control, 21: 370-380. https://doi.org/10.1016/j.foodcont.2009.07.011.

Deng, Y., Velázquez, A., Billes, F. and Dixon, J., 2010. Bonding mechanisms between aflatoxin B1 and smectite. Applied Clay Science 50: 92-98. https://doi.org/10.1016/j.clay.2010.07.008.

Di Gregorio, M., Neeff, D., Jager, A., Corassin, C., Carão, Á., Albuquerque, R., Azevedo, A. and Oliveira, C., 2014. Mineral adsorbents for prevention of mycotoxins in animal feeds. Toxin Reviews 33: 125-135. https://doi.org/10.3109/15569543.2014.905604.

Driehuis, F., Spanjer, M.C., Scholten, J.M. and Te Giffel, M.C., 2018. Review: Mycotoxins and mycotoxigenic fungi in silage: A review of the occurrence, detection, prevention and mitigation strategies. Journal of Applied Microbiology, 125: 907–921.

Engormix, 2009. LAMIC – New Online Service: Has your anti-mycotoxin additive been approved? Available at: https://en.engormix.com/MA-mycotoxins/news/lamic-new-online-service-t15106/p0.htm

European Commission (EC), 2003. Commission Regulation (EC) No 1831/2003 of the European parliament and of the council of 22 September 2003 on additives for use in animal nutrition. Available at: https://eur-lex.europa.eu/legal-content/EN/TXT/PDF/?uri=CELEX:32003R1831&from=EN

European Commission (EC), 2008. Commission Regulation (EC) No 429/2008 of 25 April 2008 on detailed rules for the implementation of Regulation (EC) No 1831/2003 of the European Parliament and of the Council as regards the preparation and the presentation of applications and the assessment and the authorisation of feed additives. Available at: http://extwprlegs1.fao.org/docs/pdf/eur79099.pdf

European Commission (EC), 2022. European Union Register of Feed Additives. Available at: https://ec.europa.eu/food/system/files/2022-01/animal-feed_additives_eu-register_1831-03.pdf

European Food Safety Authority (EFSA), 2010. Statement on the establishment of guidelines for the assessment of additives from the functional group 'substances for reduction of the contamination of feed by mycotoxins. Available at: https://efsa.onlinelibrary.wiley.com/doi/epdf/10.2903/j.efsa.2010.1693.

European Food Safety Authority (EFSA), 2012. Guidance for the preparation of dossiers for zootechnical additives. Available at: file:///C:/Users/3052903/Dropbox/Book%20Chapter/EFSA%20guidance.pdf

European Food Safety Authority (EFSA), 2017. Hazard analysis approaches small retail establishments and food donations: second scientific opinion. Available at: https://www.efsa.europa.eu/sites/default/files/wgs/biological-hazards/haccp2.pdf

Faucet-Marquis, V., Joannis-Cassan, C., Hadjeba-Medjdoub, K., Ballet, N. and Pfohl-Leszkowicz, A., 2014. Development of an in vitro method for the prediction of mycotoxin binding on yeast-based products: case of aflatoxin B1, zearalenone and ochratoxin A. Applied Microbiology and Biotechnology 98: 7583-7596. https://doi.org/10.1007/s00253-014-5917-y.

Food and Agriculture Organisation (FAO), 2006. Developing a HACCP plan Available at: https://apps.who.int/iris/bitstream/handle/10665/43598/9789241595032_eng.pdf?sequence=1

Food and Drug Administration (FDA). PART 571 - Food Additive Petitions. Available at: https://www.ecfr.gov/current/title-21/chapter-I/subchapter-E/part-571.

Food and Drug Administration (FDA). PART 573 - Food Additives Permitted in Feed and

Drinking Water of Animals. Available at: https://www.ecfr.gov/current/title-21/chapter-I/subchapter-E/part-573.

Food and Drug Administration (FDA). PART 582 - Substances generally recognized as safe. Available at: https://www.ecfr.gov/current/title-21/chapter-I/subchapter-E/part-582

Food Standard Agency (FSA), 2017. Hazard Analysis and Critical Control Point (HACCP). Available at: https://www.food.gov.uk/print/pdf/node/217.

Gallo, A., Giuberti, G., Frisvad, J.C., Bertuzzi, T. and Nielsen, K.F., 2015. Review on mycotoxin issues in ruminants: Occurrence in forages, effects of mycotoxin ingestion on health status and animal performance and practical strategies to counteract their negative effects. Toxins, 7: 3057–3111. https://doi.org/10.3390/toxins7083057.

Gallo, A., Fancello, F., Ghilardelli, F. Zara, S., Froldi, F. and Spanghero, M., 2021. Effects of several lactic acid bacteria inoculants on fermentation and mycotoxins in corn silage. Animal Feed Science and Technology, 277: 114962. https://doi.org/10.1016/j.anifeedsci.2021.114962.

Gambacorta, L., Pinton, P., Avantaggiato, G., Oswald, I. and Solfrizzo, M., 2016. Grape pomace, an agricultural byproduct reducing mycotoxin absorption: in vivo assessment in pig using urinary biomarkers. Journal of Agricultural and Food Chemistry 64: 6762-6771. https://doi.org/10.1021/acs.jafc.6b02146.

Gambacorta, S., Solfrizzo, H., Visconti, A., Powers, S., Cossalter, A., Pinton, P. and Oswald, I., 2013. Validation study on urinary biomarkers of exposure for aflatoxin B1, ochratoxin A, fumonisin B1, deoxynivalenol and zearalenone in piglets. World Mycotoxin Journal 6: 299-308. https://doi.org/10.3920/WMJ2013.1549.

Gao, Y., Li, W., Sun, H., Zheng, Z., Cui, X., Wang, H. and Meng, F., 2014. A facile *in situ* pillaring method—The synthesis of Al-pillared montmorillonite. Applied Clay Science 88–89: 228–232. https://doi.org/10.1016/j.clay.2013.12.004.

Ghadaksaz, A., Nodoushan, S., Sedighian, H., Behzadi, E. and Fooladi, A., 2022. Evaluation of the role of probiotics as a new strategy to eliminate microbial toxins: a review. Probiotics and Antimicrobial Proteins 14: 224-237. https://doi.org/10.1007/s12602-021-09893-2.

Giovati, L., Gallo, A., Masoero, F., Cerioli, C., Ciociola, T., Conti, S., Magliani, W. and Polonelli, L., 2014. Vaccination of heifers with anaflatoxin improves the reduction of aflatoxin B1 carry over in milk of lactating dairy cows. PLoS one 9: e94440. https://doi.org/10.1371/journal.pone.0094440.

Graziani, F., Pujol, A., Nicoletti, C., Pinton, P., Armand, L., Di Pasquale, E., Oswald, I., Perrier, J. and Maresca, M., 2015. The Food-associated ribotoxin deoxynivalenol modulates inducible NO synthase in human intestinal cell model. Toxicological Sciences 145: 372-382. https://doi.org/10.1093/toxsci/kfv058.

Greco, D., D'Ascanio, V., Santovito, E., Logrieco, A. and Avantaggiato, G., 2019. Comparative efficacy of agricultural by-products in sequestering mycotoxins. Journal of the Science of Food and Agriculture 99: 1623-1634. https://doi.org/10.1002/jsfa.9343.

Gruber-Dorninger, C., Jenkins, T. and Schatzmayr, G., 2019. Global Mycotoxin Occurrence in Feed: A Ten-Year Survey. Toxins 11: 375. https://doi.org/10.3390/toxins11070375.

Guan, H., Shuai, Y., Ran, Q.F., Yan, Y.H., Wang, X., Li D.D., Cai, Y., and Zhang, X., 2020. The microbiome and metabolome of Napier grass silages prepared with screened lactic acid bacteria during ensiling and aerobic exposure. Animal Feed Science and Technology 269: 114673. https://doi.org/10.1016/j.anifeedsci.2020.114673.

He, J. and Zhou, T., 2010. Patented techniques for detoxification of mycotoxins in feeds and food matrices. Recent Patents on Food, Nutrition and Agricolturee 2: 96-104. https://doi.org/10.2174/2212798411002020096.

Hidosa, D. and Guyo, M., 2017. Climate change effects on livestock feed resources: a review. Journal of Fisheries and Livestock Production 5: 259. https://doi.org/10.4172/2332-2608.1000259.

Iglesias, C.., Bach, A., Devant, M., Adelantado, C., Calvo, M.A., 2005. The effect of *Lactobacillus buchneri* inoculation on corn silages conservation. Scispace, 1: 611-613.

Ito, A. and Wagai, R., 2017. Global distribution of clay-size minerals on land surface for biogeochemical and climatological studies. Scientific Data 4: 170103. https://doi.org/10.1038/sdata.2017.103.

Kolawole, O., Graham, A., Donaldson, C., Owens, B., Abia, W., Meneely, J., Alcorn, M., Connolly, L. and Elliott, C., 2020. Low doses of mycotoxin mixtures below EU regulatory limits can negatively affect the performance of broiler chickens: a longitudinal study. Toxins 12: 433. https://doi.org/10.3390/toxins12070433.

Kolawole, O., Meneely, J., Greer, B., Chevallier, O., Jones, D.S., Connolly, L. and Elliott, C., 2019. Comparative *in vitro* assessment of a range of commercial feed additives with multiple mycotoxin binding claims. Toxins 11: 659. https://doi.org/10.3390/toxins11110659.

Kolawole, O., Meneely, J., Petchkongkaew, A. and Elliott, C., 2021. A review of mycotoxin biosynthetic pathways: associated genes and their expressions under the influence of climatic factors. Fungal Biology Reviews 37: 8-26. https://doi.org/10.1016/j.fbr.2021.04.003.

Kolawole, O., Siri-Anusornsak, W., Petchkongkaw, A., Meneeley, J. and Elliot, C., 2022. The efficacy of additives for the mitigation of aflatoxins in animal feed: a systematic review and network meta-analysis. Toxins 14: 707. https://doi.org/10.3390/toxins14100707.

Kozieł, M., Kowalska, K. and Piastowska-Ciesielska, A., 2021. Nrf2: a main responsive element in cells to mycotoxin-induced toxicity. Archives of Toxicology 95: 1521-1533. https://doi.org/10.1007/s00204-021-02995-4.

Lahtinen, S.J., Haskard, C.A., Ouwehand, A.C., Salminen, S.J., and Ahokas, J.T., 2004. Binding of aflatoxin B1 to cell wall components of *Lactobacillus rhamnosus* strain GG. Food Additives and Contaminants. 21: 158-164. https://doi.org/10.1080/02652030310001639521.

Lauwers, M., De Baere, S., Letor, B., Rychlik, M., Croubels, S. and Devreese, M., 2019. Multi LC-MS/MS and LC-HRMS methods for determination of 24 mycotoxins including major phase I and II biomarker metabolites in biological matrices from pigs and broiler chickens. Toxins 11: 171. https://doi.org/10.3390/toxins11030171.

Li, Y., Tian, G., Dong, G., Bai, S., Han, X., Liang, J., Meng, J. and Zhang, H., 2018. Research progress on the raw and modified montmorillonites as adsorbents for mycotoxins: A review. Applied Clay Science 163: 299-311. https://doi.org/10.1016/j.clay.2018.07.032

Liu, M., Zhao, L., Gong, G., Zhang, L., Shi, L., Dai, J., Han, Y., Wu, Y., Khalil, M. and Sun, L.,

2022a. Invited review: Remediation strategies for mycotoxin control in feed. Journal of Animal Science and Biotechnology 13: 19. https://doi.org/10.1186/s40104-021-00661-4.

Liu, L., Xie, M. and Wei, D., 2022b. Biological detoxification of mycotoxins: current status and future advances. International Journal of Molecular Sciences 23: 1064. https://doi.org/10.3390/ijms23031064.

Liu, Y., Wu, Q., Wu, X., Algharib, S., Gong, F., Hu, J., Luo, W., Zhou, M., Pan, Y., Yan, Y. and Wang, Y., 2021. Structure, preparation, modification, and bioactivities of β-glucan and mannan from yeast cell wall: A review. International Journal of Biological Macromolecules 173: 445-456. https://doi.org/10.1016/j.ijbiomac.2021.01.125.

Luo, Y., Liu, X., Yuan, L. and Li, J., 2020. Complicated interactions between bio-adsorbents and mycotoxins during mycotoxin adsorption: Current research and future prospects. Trends in Food Science and Technology 96: 127-134. https://doi.org/10.1016/j.tifs.2019.12.012.

Ma, Z.X., Amaro, F.X., Romero, J.J., Pereira, O.G., Jeong, K.C. and Adesogan, A.T., 2017. The capacity of silage inoculant bacteria to bind aflatoxin B1 in vitro and in artificially contaminated corn silage. Journal of Dairy Science, 100: 7198-7210. https://doi.org/10.3168/jds.2016-12370.

Market and Markets Report, 2024. Global feed mycotoxin binders and modifiers market analysis. Available at: https://www.researchreportsworld.com/enquiry/request-sample/19224847

Mohaghegh, A., Chamani, M., Shivazad, M., Sadeghi, A. and Afzali, N., 2017. Effect of esterified glucomannan on broilers exposed to natural mycotoxin-contaminated diets. Journal of Applied Animal Research 45: 285-291. https://doi.org/10.1080/09712119.2016.1174122.

Møller, C.O.D.A., Christensen, B.B. and Rattray, F.P., 2021. Modelling the biphasic growth of non-starter lactic acid bacteria on starter-lysate as a substrate. International Journal of Food and Microbiology, 337: 108937. https://doi.org/10.1016/j.ijfoodmicro.2020.108937.

Muck, R.E., Nadeau, E.M.G., McAllister, T.A., Contreras-Govea, F.E., Santos, M.C. and Kung Jr., L., 2018. Silage review: Recent advances and future uses of silage additives. Journal of Dairy Science, 101: 3980-4000. https://doi.org/10.3168/jds.2017-13839.

Office of the Texas State Chemist (OTSC), 2011. OTSC Quarterly Newsletter. Available at: https://otscweb.tamu.edu/Reports/PDF/NWL/2011/May-2011-NL.pdf

Office of the Texas State Chemist (OTSC), 2022. Active Feed Manufacturing Facilities. Available at https://otscweb.tamu.edu/Reports/ActiveFeedFacility.aspx

Ogunade, I.M., Jiang, Y., Cervantes, A.P., Kim, D.H., Oliveira, A.S., Vyas. D. and Adesogan, A.T., 2018. Effects of delayed ensiling and microbial inoculation on fermentation quality and aerobic stability of wilted silages of annual ryegrass and forage sorghum harvested at high moisture content. Grass Forage Science, 73: 206–216.

Perczak, A., Jus, K., Marchwinska, K., Gwiazdowska, D., Waskiewicz, A. and Golinski P., 2016. Degradation of zearalenone by essential oils under *in vitro* conditions. Frontiers in Microbiology 7: 2016. https://doi.org/10.3389/fmicb.2016.01224.

Perrone, G., Ferrara, M., Medina, A., Pascale, M. and Magan, N., 2020. Toxigenic fungi and mycotoxins in a climate change scenario: ecology, genomics, distribution, prediction and prevention of the risk. Microorganisms 8: 1496. https://doi.org/10.3390/microorganisms8101496.

Pietrowska, M., 2014. The adsorption of ochratoxin A by *Lactobacillus* species. Toxins, 6: 2826-2839.

Polonelli, L., Giovati, L., Magliani, W., Conti, S., Sforza, S., Calabretta, A., Casoli, C., Ronzi, P., Grilli, E., GAllo, A., Masoero, F. and Piva, G., 2011. Vaccination of lactating dairy cows for the prevention of aflatoxin B1 carry over in milk. PLoS one, 6: e26777. https://doi.org/10.1371/journal.pone.0026777.

Qin, X., Su, X., Tu, T., Zhang, J., Wang, X., Wang, Y., Wang, Y., Bai, Y., Yao, B., Luo, H. and Huang, H., 2021. Enzymatic degradation of multiple major mycotoxins by dye-decolorizing peroxidase from *Bacillus subtilis*. Toxins 13: 429. https://doi.org/10.3390/toxins13060429.

Queiroz, O.C.M., Adesogan, A.T., Arriola, K.G. and Queiroz, M.F.S., 2012. Effect of a dual-purpose inoculant on the quality and nutrient losses from corn silage produced in farm-scale silos. Journal of Dairy Science, 95: 3354-3362.

Sadiq, F.A., Yan, B., Tian, F., Zhao, J., Zhang, H. and Chen, W., 2019. Lactic acid bacteria as antifungal and anti-mycotoxigenic agents: a comprehensive review. Compreensive Reviews in Food Science and Food Safety, 18: 1403-1436. https://doi.org/10.1111/1541-4337.12481.

Shi, B., Su, Y., Chang, S., Sun, Y., Meng, X. and Shan, A., 2017. Vitamin C protects piglet liver against zearalenone-induced oxidative stress by modulating expression of nuclear receptors PXR and CAR and their target genes. Food and Function 8: 3675-3687. https://doi.org/10.1039/C7FO01301A.

Solís-Cruz, B., Hernández-Patlán, D., Beyssac, E., Latorre, J., Hernandez-Velasco, X., Merino-Guzman, R., Tellez, G. and López-Arellano, R., 2017. Evaluation of chitosan and cellulosic polymers as binding adsorbent materials to prevent aflatoxin B1, fumonisin B1, ochratoxin, trichothecene, deoxynivalenol, and zearalenone mycotoxicoses through an *in vitro* gastrointestinal model for poultry. Polymers 9: 529. https://doi.org/10.3390/polym9100529.

Tolosa, J., Rodríguez-Carrasco, Y., Ruiz, M. and Vila-Donat, P., 2021. Multi-mycotoxin occurrence in feed, metabolism and carry-over to animal-derived food products: A review. Food and Chemical Toxicology 158: 112661. https://doi.org/10.1016/j.fct.2021.112661.

van Egmond, H., Schothorst, R. and Jonker, M., 2007. Regulations relating to mycotoxins in food. Analytical and Bioanalytical Chemistry 389: 147-157. https://doi.org/10.1007/s00216-007-1317-69.

Vila-Donat, P., Marín, S., Sanchis, V. and Ramos, A., 2018. A review of the mycotoxin adsorbing agents, with an emphasis on their multi-binding capacity, for animal feed decontamination. Food and Chemical Toxicology 114: 246-259. https://doi.org/10.1016/j.fct.2018.02.044.

Wang, G., Miao, Y., Sun, Z. and Zheng, S., 2018. Simultaneous adsorption of aflatoxin B1 and zearalenone by mono- and di-alkyl cationic surfactants modified montmorillonites. Journal of Colloid and Interface Science 511: 67-76. https://doi.org/10.1016/j.jcis.2017.09.074.

Wang, X., Qin, X., Hao, Z., Luo, H., Yao, B. and Su, X., 2019. Degradation of four major mycotoxins by eight manganese peroxidases in presence of a dicarboxylic acid. Toxins 11: 566. https://doi.org/10.3390/toxins11100566.

Wu, B., Hu, Z., Wei, M., Yong, M., Niu, H., 2022. Effects of inoculation of *Lactiplantibacillus plantarum* and *Lentilactobacillus buchneri* on fermentation quality, aerobic stability, and microbial community dynamics of wilted *Leymus chinensis* silage. Frontiers in Microbiotechnology 13: 928731. https://doi.org/10.3389/fmicb.2022.928731.

Xing, F., Hua, H., Selvaraj, J.N., Yuan, Y., Zhao, Y. and Zhou, L., 2014. Degradation of fumonisin B1 by cinnamon essential oil. Food Control 38: 37-40. http://dx.doi.org/10.1016/j.foodcont.2013.09.045.

Young, J., Zhou, T., Yu, H., Zhu, H. and Gong, J., 2007. Degradation of trichothecene mycotoxins by chicken intestinal microbes. Food and Chemical Toxicology 45: 136-143. https://doi.org/10.1016/j.fct.2006.07.028.

Yu, H., Zhou, T., Gong, J., Young, C., Su, X., Li, X., Zhu, H., Tsao, R. and Yang, R., 2010. Isolation of deoxynivalenol-transforming bacteria from the chicken intestines using the approach of PCR-DGGE guided microbial selection. BMC Microbiology 10: 182. https://doi.org/10.1186/1471-2180-10-182.

Zhang, Q., Zhang, Y., Liu, S., Wu, Zhou, Q., Zhang, Y., Zheng, X., Han, Y., Xie, C. and Liu. N., 2021. Adsorption of deoxynivalenol by pillared montmorillonite. Food Chemistry 343: 128391. https://doi.org/10.1016/j.foodchem.2020.128391.

Zhang, X., Jiang, L., Geng, C., Cao, J. and Zhong, L., 2009. The role of oxidative stress in deoxynivalenol-induced DNA damage in HepG2 cells. Toxicon 54: 513-518. https://doi.org/10.1016/j.toxicon.2009.05.021.

CHAPTER 17

Hazard assessment of mycotoxins for farm and companion animals: a European perspective

*C. Terciolo, M. Garofalo, D. Payros, B. Willoquet, A. Pierron, P. Pinton and I.P. Oswald**

Toxalim (Research Centre in Food Toxicology, Université de Toulouse, INRAE, ENVT, INP- PURPAN, UPS, Toulouse, France
**correspondence: isabelle.oswald@inrae.fr*

1 Introduction

As described in Chapter 2, mycotoxins are secondary metabolites produced by filamentous fungal species. Several families of mycotoxins with more than 300 compounds are found in food and feed. Among the different food commodities contaminated, cereal grains are the main source of mycotoxin contamination. Because of their harmful effects and their common occurrence in feed, it is necessary to establish toxicological reference values not only for humans but also for the different farm and companion animal species exposed to these feed contaminants. At the European level, the establishment of these toxicological reference values is performed by the EFSA and published in various scientific opinions.

The EFSA opinions collect and scrutinize scientific information available in the public domain, in our case on the toxicity of mycotoxins. The animal hazard assessment process includes data on toxicokinetics as well as data on *in vivo* and *in vitro* toxicity. The farm and companion animals covered are ruminants (including small ruminants), pigs, poultry (broiler chickens, laying hens, turkeys and ducks), farmed fish, horses, farmed rabbits, dogs and cats. Based on the assessed toxicity data health, toxicological reference values are determined. The NOAEL or LOAEL are used as a reference point to characterize the chronic risks from exposure to mycotoxins.

This chapter gives an overview of the toxicological values determined for several mycotoxins (AFs, trichothecenes, ZEN, FUM, ochratoxins, BEA, ENNs, and EAs) on farm and domestic animal species. We especially summarize data obtained on pig (Table 17.1), poultry (Table 17.2), ruminants (Table 17.3), fish (Table 17.4) as well as horse, rabbit, cats and dog (Table 17.5).

 | DOI: 10.1163/9789004712720_18

2 Trichothecenes

DON belongs to the type B trichothecenes. It co-occurs with its acetylated (3-ADON and 15-ADON) and glucoside (DON-3G) derivatives. These three forms are expected to have the same *in vivo* toxicity as 3-ADON, and 15-ADON can be deacetylated and glucoside moiety of DON-3G cleaved in the intestinal tract. Therefore, a group toxicological reference value could be used for the sum of the three toxins (EFSA, 2017a). The available toxicity data allowed for the identification of an acute reference dose in pig, farmed mink and dogs. Chronic adverse effects were identified in many farm and companion animal species, and NOAEL/LOAEL could be identified for pigs, poultry species, ruminants (including small ruminants), farmed fish, cats dogs, horses and farmed rabbits (EFSA, 2017a; EFSA, 2023a) (Tables 17.1 to 17.5).

NIV also belongs to the type B trichothecenes. However, compared to DON, much less data are available concerning this toxin. No toxicity data have been identified for ruminants, rabbits, fish and companion animals but LOAEL were identified in pigs and poultry (EFSA, 2013; Tables 17.1 and 17.2).

T-2 and HT-2 toxins are members of the type A trichothecenes, characterized by an esterified group or free hydroxyl group at carbon 8. T-2 toxin is rapidly metabolized to a large number of products, HT-2 toxin being a major metabolite. The toxicological reference value was thus derived for the sum of T-2 and HT2 toxins (EFSA, 2011). Such reference values were identified for cattle, poultry, fish, rabbit and pig. This latter species was among the most susceptible to the effects of T-2 toxin, the most sensitive endpoints being immunological or haematological effects (Table 17.1).

3 Zearalenone

ZEN is a phenolic resorcylic acid lactone and exhibits oestrogen-like activity in farm animals by binding to the oestrogen receptors. In pigs, sheep and turkeys, the endpoint used for health-based guidance levels was mainly linked to the oestrogenic activity of the molecule, while others endpoints were used for broiler chickens and sheep (EFSA, 2017b). Of note, many different modified forms of ZEN have been identified with very different oestrogenic activity when compared to ZEN. To take that into account, relative potency factors were established based on *in vivo* uterotrophic assays performed in rodents; they were used to established a group toxicological reference value for ZEN and its modified forms in different animal species (EFSA, 2016).

4 Fumonisins

FUM are a large family of mycotoxins with structural similarities to sphingoid bases. FB_1 is the most documented toxin of this family, and FB_2 and FB_3 are also present in feed. The limited data on FB_2 and FB_3 suggest similar toxicity of the three toxins. The sum of FB_1, FB_2 and FB_3 was used to determine the reference values for farm animal species (EFSA, 2018; EFSA, 2022). Various endpoints were used to established these toxicological reference values. The very high sensitivity of horses and the low sensitivity of cattle should be noted (Tables 17.1 and 17.5).

5 Ochratoxins

Ochratoxins are a family of mycotoxins produced by fungi of the genera *Aspergillus* and *Penicillium*. Among the various members of this family, OTA is the most toxic and has been classified by the IARC as 'possibly carcinogenic to humans' (Group 2B). This toxin is also a potent nephrotoxic compound. The data in farm and companion animals are scarce and only one LOAEL on pig has been identified (EFSA, 2004a; 2023b; Table 17.1).

6 Ergot alkaloids

EAs are produced by species of *Claviceps* and are classified as indole alkaloids, derived from a tetracyclic ergoline ring system. More than 50 different EAs have been identified so far. Their structural similarity to dopamine, noradrenaline and serotonin is a factor in their toxicity and interrupts the neurotransmission. These alkaloids can also have negative effects on FI, growth, and reproduction but different factors can influence these negative outcomes (EFSA, 2012; 2024). Toxicological references values have been determined for both pigs and poultry (Tables 17.1 and 17.2).

7 Emerging mycotoxins

In addition to the above-mentioned toxins, for which regulation and/or recommendations exists in animal feed, many other secondary metabolites are produced by fungi. New mycotoxins called 'emerging mycotoxins' are neither regulated nor routinely detected. They have always existed and are now of growing interest. Among emerging mycotoxins, the most common are ENNs and BEA, both of them produced by *Fusarium* species (EFSA, 2014). Of the 29 species of ENNs known, ENNA, ENNA1, ENNB and ENNB1 are the most frequently detected in food and feed.

BEA and ENNs are cyclic hexadepsipeptides. They cause cell death at high concentrations through cell membrane permeabilization and destruction. At lower concentrations, they act by disrupting membranes and cell organelles (lysosomes, mitochondria) functions, leading to cell death by apoptosis.

For those emerging toxins, due to the paucity of data, toxicological reference values could only be established for poultry, based on zootechnical parameters (Table 17.2).

8 Aflatoxins

AFs are extremely potent carcinogens in all animal species, and most of the AF members are classified as 'carcinogenic to humans' by IARC (group 1). AFB_1 is the most abundant form of AFs and it targets the liver and increases the development of hepatic cancers; EU-regulated levels in feed (Directive 2002/32/EC) are given in Table 17.6. Whereas acute clinical intoxications are rarely seen in modern agricultural practice, sub-optimal weight gain, lower milk and egg production, as well as an increased susceptibility towards infectious diseases may lead to considerable economic losses in animal production due to AFs exposure. However, from the available data, NOAEL or LOAEL could not be defined for farm and companion animals (EFSA, 2004b).

TABLE 17.1 Toxicological references values determined by EFSA for pigs exposed to selected mycotoxins

Mycotoxin	NOAEL mg/kg feed *μg/kg BW*	LOAEL mg/kg feed *μg/kg BW*	Endpoint
DON	0.7 mg/kg feed		Reduced feed intake; reduced body weight gain; weight gain[a]
NIV		2.5 mg/kg feed *100 μg/kg BW*	Haematological alteration; reduced proliferation of lymphocytes from blood, spleen or thymus[b]
T2/HT-2		0.5 mg/kg feed *29 μg/kg BW*	Immunological or haematological effects[c]
ZEN	0.22 mg/kg feed *10.4 μg/kg BW*		Estrogenic effects[d]
FUM	1 mg/kg feed *40 μg/kg BW*		Effect on performances[e]
OTA	0.01 mg/kg feed	0.025 mg/kg feed	Decreased body weight[f]
EAs	0.6 mg/kg feed *1.74 μg/kg BW*	4.7 mg/kg feed	Increased relative weight of heart and spleen, decreased body weight gain and reduced feed intake[g]

a EFSA (2017a); b EFSA (2013); c EFSA (2011); d EFSA (2017b); e EFSA (2018); f EFSA (2023b); g EFSA (2024).
Note: no toxicological reference value was determined for pigs exposed to AFB_1, BEA, ENNB and ENNB1.

TABLE 17.2 Toxicological references values determined by EFSA for poultry exposed to selected mycotoxins

Mycotoxin	Poultry	NOAEL mg/kg feed *µg/kg BW*	LOAEL mg/kg feed *µg/kg BW*	Endpoint
DON	broiler chickens	0.6 mg/kg feed	1.9 mg/kg feed	Modifications of intestine morphology[a]
	laying hens	5 mg/kg feed		Zootechnical parameters (feed intake, egg fertility)[b]
	duck	7 mg/kg feed		Zootechnical parameters[b]
	turkey	0.6 mg/kg feed	1.7 mg/kg feed	Modifications of intestine morphology[a]
NIV	broiler chickens		3 mg/kg feed *360 µg/kg BW*	Alteration of the gizzard[c]
	laying hens		1 mg/kg feed *53 µg/kg BW*	Histological alteration of the kidney[c]
T-2/HT-2	broiler chickens		0.5 mg/kg feed *40 µg/kg BW*	Mucosal oral damage[d]
	laying hens		2 mg/kg feed *120 µg/kg BW*	Infertility and reduced egg production[d]
	duck		0.2 mg/kg feed *40 µg/kg BW*	Reduced body weight gain[d]
	turkey		0.5 mg/kg feed *48 µg/kg BW*	Mucosal oral damage[d]
ZEN	broiler chickens	50 mg/kg feed 7500 µg/kg BW		Reduced lymphocyte counts[e]
	turkey	100 mg/kg feed *9100 µg/kg BW*		Vent swelling[e]
FUM	Chickens	1 mg/kg feed *2600 µg/kg BW*	2.5 mg/kg feed	Modifications of intestine morphology[f]
	duck	8 mg/kg feed *400 µg/kg BW*		Alterations of liver enzymes[g]
	turkey	20 mg/kg feed *670 µg/kg BW*		Reduced feed intake; Increased Sa/So ratios in liver and kidney[g]
OTA	broiler chickens	0.03 mg/kg feed	0.1 mg/kg feed	Increased liver and kidney weight[h]
	laying hens	0.03 mg/kg feed	0.1 mg/kg feed	
EAs	broiler chickens	2.1 mg/kg feed		Reduced feed intake and body weight gain[i]
	laying hens	3.7 mg/kg feed	14.6 mg/kg feed	Reduced laying rate, daily egg mass, feed to egg mass ratio, eggshell weight, egg yolk colour and nitrogen and crude fat retention[i]
	ducks	0.2 mg/kg feed *47 µg/kg BW*	0.6 mg/kg feed *142 µg/kg BW*	Reduced feed intake and body weight gain[i]

TABLE 17.2 Toxicological references values determined by EFSA for poultry exposed to selected mycotoxins (*cont.*)

Mycotoxin	Poultry	NOAEL mg/kg feed *µg/kg BW*	LOAEL mg/kg feed *µg/kg BW*	Endpoint
ENNB	broiler chickens	12.72 mg/kg feed *763 µg/kg BW*		Live weights and feed intake[j]
	laying hens	11.23 mg/kg feed *674 µg/kg BW*		
ENNB1	broiler chickens	4.06 mg/kg feed *244 µg/kg BW*		Live weights and feed intake[j]
	laying hens	3.6 mg/kg feed *216 µg/kg BW*		
BEA	broiler chickens	12.6 mg/kg feed *1220 µg/kg BW*		Live weights and feed intake[j]
	laying hens	8.93 mg/kg feed *536 µg/kg BW*		
	turkey	2.48 mg/kg feed *136 µg/kg BW*		

a EFSA (2023a); b EFSA (2017a); c EFSA (2013); d EFSA (2011); e EFSA (2017b); f EFSA (2022); g EFSA (2018); h EFSA (2023b); i EFSA (2024); j EFSA (2014).
Note: no toxicological reference value was determined for poultry exposed to AFB_1.

TABLE 17.3 Toxicological references values determined by EFSA for ruminants exposed to selected mycotoxins

Mycotoxin	Ruminant	NOAEL mg/kg feed *µg/kg BW*	LOAEL mg/kg feed *µg/kg BW*	Endpoint
DON	dairy cow	5 mg/kg feed		Reduced feed intake, body weight gain, and weight gain[a]
	heifer	10 mg/kg feed		
	beef cattle, steers	18 mg/kg feed		
	sheep	16 mg/kg feed		
	goat	5 mg/kg feed		
T-2/HT-2	calves		*300 µg/kg BW*	Gastrointestinal lesions; alteration of serum proteins and haematology[b]
ZEN	sheep	1.5 mg kg feed *28 µg/kg BW*		Alteration of ovulation rates and lambing percentages[c]
FUM	cattle	31 mg/kg feed *600 µg/kg BW*		Increase in serum enzymes, cholesterol and bilirubin; decrease in lymphocyte blastogenesis[d]
EAs	dairy cow	0.1 mg/kg feed	0.3 mg/kg feed	Reduced feed intake, poor weight gain, hyperthermia, loss of tail switches and tips, early lameness and swelling of the feet, and reduced milk production.[e]
	beef cattle	0.1 mg/kg feed	0.2 mg/kg feed	
	sheep	0.3 mg/kg feed	0.9 mg/kg feed	Reduced body weight gain and related changes in the carcass characteristics.[e]

a EFSA (2017a); b EFSA (2011); c EFSA (2017b); d EFSA (2018); e EFSA (2024).
Note: no toxicological reference value was determined for ruminants exposed to AFB_1, NIV, OTA, BEA, ENNB and ENNB1

TABLE 17.4 Toxicological references values determined by EFSA for fish exposed to selected mycotoxins

Mycotoxin	NOAEL mg/kg feed *µg/kg BW*	LOAEL mg/kg feed *µg/kg BW*	Endpoint
DON	0.6 mg/kg feed		Reduced feed intake, body weight gain; lipid peroxidation; histopathological changes[a]
T-2/HT-2	0.63 mg/kg feed *13 µg/kg BW*		Reduced feed intake and growth; alteration of haematocrit[b]
ZEN	0.3 mg/kg feed *9 µg/kg BW*		Alteration of blood formula; lipid peroxidation[c]
FUM		10 mg/kg feed *0.5 µg/kg BW*	Reduced weight gain; neuronal apoptosis in brain[d]
OTA*	0.5 mg/kg feed		Zootechnical performance and hepatopancreatic effects[e]

a EFSA (2017a); b EFSA (2011); c EFSA (2017b); d EFSA (2018); e EFSA (2023b).
Note: no toxicological reference value was determined for fish exposed to AFB_1, NIV, EAs, BEA, ENNB and ENNB1.
*herbivorous fish.

TABLE 17.5 Toxicological references values determined by EFSA for rabbit, cat, dog and horse exposed to selected mycotoxins

Mycotoxin	Species	NOAEL mg/kg feed *µg/kg BW*	LOAEL mg/kg feed *µg/kg BW*	Endpoint
DON	rabbit	4 mg/kg feed		Reduced feed intake and body weight; alteration of the relative organ weights[a]
	cat	6 mg/kg feed		Reduced body weight gain[a]
	dog	4 mg/kg feed		Reduced body weight gain[a]
	horse	3.5 mg/kg feed	5.6 mg/kg feed	Reduced feed intake[b]
T-2/HT-2	rabbit	*100 µg/kg BW*		Alteration of drug metabolizing enzymes; microsomal oxidative damages[c]
FUM	rabbit		130 µg/kg feed	Decreased performance; alteration of plasma markers and blood formula; congestion of liver and kidney; impaired spermatogenesis, delayed puberty; increased Sa/So ratio[d]
	horses	1.0 mg/kg feed		Neurological effects[e]
OTA	rabbit	0.01 mg/kg feed	0.03 mg/kg feed	Decreased growth performance[f]
ZEN	dog		1.7 mg/kg feed *25 µg/kg BW*	Degenerative lesions of myometrium and endometrium; enlargement and atrophy of uterine glands; changes in blood formula; alteration of biochemical parameters[g]

a EFSA (2017a); b EFSA (2023a); c EFSA (2011); d EFSA (2018); e EFSA (2022); f EFSA (2023b); g EFSA (2017b).
Note: no toxicological reference value was determined for AFB_1, DON, NIV, T-2/HT-2, ZEN, FUM, OTA, EAs, BEA, ENNB and ENNB1, for the four animals' species considered.

TABLE 17.6 EU Regulatory limits of AFB_1 in animal feed

Animal category	Maximum content in mg/kg diet
Beef cattle, sheep, and goat	0.050
Dairy cattle	0.005
Calves and lambs	0.010
Pigs and poultry (except young animals)	0.020
Piglets and young poultry	0.010

Directive 2002/32/EC

9 Conclusions

In conclusion, mycotoxins are natural contaminants that are frequently present in animal feed. This review illustrates the scarcity of toxicological data, which are nevertheless necessary to determine toxicological reference values for farm and companion animal species. Indeed, with the exception of DON, it was not possible to define toxicological reference values for all the different animal species, notably ruminants (including small ruminants), pigs, poultry species, horses, farmed rabbits, farmed fish, cats and dogs. When characterizing risk from mycotoxin exposure, the identified NOAELs/LOAELs are compared to the dietary exposure estimates. For animals, since the reference points are obtained for the species considered, no safety factors are added during the risk assessment process. However, it should be noted that within an animal species, many factors might influence susceptibility, such as age, weight, sex, etc. For fish, most of the toxicological data are obtained in carp and extrapolated to all fish species. Mycotoxins often co-occur and it is important to take the combinations into consideration. So far, group reference values are determined for toxins within the same family (example DON and ZEN) but the co-exposure of mycotoxins from different families still need to be addressed (Grenier and Oswald, 2011; Alassane-Kpembi et al., 2017).

References

Alassane-Kpembi, I., Schatzmayr, G., Marin, D., Taranu, I., Puel, O. and Oswald, I.P., 2017. Mycotoxins co-contamination: Methodological aspects and biological relevance of combined toxicity studies. Critical Reviews in Food Science and Nutrition. 16: 3489. https://doi.org/10.1080/10408398.2016.1140632.

EFSA (EFSA Panel on Contaminants in the Food Chain). 2004a. Ochratoxin A as undesirable substance in animal feed. EFSA Journal 4: 365. https://doi.org/10.2903/j.efsa.2004.101.

EFSA (EFSA Panel on Contaminants in the Food Chain). 2004b. Aflatoxin B1 as undesirable substance in animal feed. EFSA Journal 2: 39. https://doi.org/10.2903/j.efsa.2004.39.

EFSA (EFSA Panel on Contaminants in the Food Chain). 2011. Presence of T-2 and HT-2 toxins in food and feed. EFSA Journal 9: 2481. https://doi.org/10.2903/j.efsa.2011.2481.

EFSA (EFSA Panel on Contaminants in the Food Chain). 2012. Ergot alkaloids in food and feed. EFSA Journal 10: 2798. https://doi.org/10.2903/j.efsa.2012.2798.

EFSA (EFSA Panel on Contaminants in the Food Chain). 2013. Risks for animal and public health related to the presence of nivalenol in food and feed. EFSA Journal 11: 3262. https://doi.org/10.2903/j.efsa.2013.3262.

EFSA (EFSA Panel on Contaminants in the Food Chain). 2014. Risks to human and animal health related to the presence of beauvericin and enniatins in food and feed. EFSA Journal 12: 3802. https://doi.org/10.2903/j.efsa.2014.3802.

EFSA (EFSA Panel on Contaminants in the Food Chain). 2016. Appropriateness to set a group health-based guidance value for zearalenone and its modified forms. EFSA Journal 14: 4425. https://doi.org/10.2903/j.efsa.2016.4425.

EFSA (EFSA Panel on Contaminants in the Food Chain). Knutsen, H.K., Alexander, J., Barregard, L., Bignami, M., Bruschweiler, B., Ceccatelli, S., Cottrill, B., Dinovi, M., Grasl-Kraupp, B., Hogstrand, C., Hoogenboom, L.R., Nebbia, C.S., Oswald, I.P., Petersen, A., Rose, M., Roudot, A.C., Schwerdtle, T., Vleminckx, C., Vollmer, G., Wallace, H., De Saeger, S., Eriksen, G.S., Farmer, P., Fremy, J.M., Gong, Y.Y., Meyer, K., Naegeli, H., Parent-Massin, D., Rietjens, I., van Egmond, H., Altieri, A., Eskola, M., Gergelova, P., Ramos Bordajandi, L., Benkova, B., Dorr, B., Gkrillas, A., Gustavsson, N., van Manen, M. and Edler, L., 2017a. Risks to human and animal health related to the presence of deoxynivalenol and its acetylated and modified forms in food and feed. EFSA Journal 15: 4718. https://doi.org/10.2903/j.efsa.2017.4718.

EFSA (EFSA Panel on Contaminants in the Food Chain), Knutsen, H.K., Alexander, J., Barregard, L., Bignami, M., Bruschweiler, B., Ceccatelli, S., Cottrill, B., Dinovi, M., Edler, L., Grasl-Kraupp, B., Hogstrand, C., Hoogenboom, L.R., Nebbia, C.S., Petersen, A., Rose, M., Roudot, A.C., Schwerdtle, T., Vleminckx, C., Vollmer, G., Wallace, H., Dall'Asta, C., Dänicke, S., Eriksen, G.S., Altieri, A., Roldan-Torres, R. and Oswald, I.P., 2017b. Risks for animal health related to the presence of zearalenone and its modified forms in feed. EFSA Journal 15: 4851. https://doi.org/10.2903/j.efsa.2017.4851.

EFSA (EFSA Panel on Contaminants in the Food Chain), Knutsen, H.K., Alexander, J., Barregard, L., Bignami, M., Bruschweiler, B., Ceccatelli, S., Cottrill, B., Dinovi, M., Edler, L., Grasl-Kraupp, B., Hogstrand, C., Hoogenboom, L.R., Nebbia, C.S., Petersen, A., Rose, M., Roudot, A.C., Schwerdtle, T., Vleminckx, C., Vollmer, G., Wallace, H., Dall'Asta, C., Eriksen, G.S., Taranu, I., Altieri, A., Roldan-Torres, R. and Oswald, I.P., 2018. Risks for animal health related to the presence of fumonisins, their modified forms and hidden forms in feed. EFSA Journal 16: 5242. https://doi.org/10.2903/j.efsa.2018.5242.

EFSA CONTAM Panel (EFSA Panel on Contaminants in the Food Chain), Schrenk, D., Bignami, M., Bodin, L., Chipman, J.K., del Mazo, J., Grasl-Kraupp, B., Hogstrand, C., Leblanc, J.C., Nielsen, E., Ntzani, E., Petersen, A., Sand, S., Schwerdtle, T., Vleminckx, C., Wallace, H., Dänicke, S., Nebbia, C.S., Oswald, I.P., Rovesti, E., Steinkellner, H. and Hoogenboom, L.R., 2022. Scientific Opinion on the assessment of informationas regards the toxicity of fumonisins for pigs, poultry and horses. EFSA Journal 20: 7534. https://doi.org/10.2903/j.efsa.2022.7534.

EFSA CONTAM Panel (EFSA Panel on Contaminants in the Food Chain), Schrenk, D., Bignami, M., Bodin, L., Chipman, J.K., del Mazo, J., Grasl-Kraupp, B., Hogstrand, C., Leblanc, J.C., Nielsen, E., Ntzani, E., Petersen, A., Sand, S., Schwerdtle, T., Vleminckx, C., Wallace, H., Dänicke, S., Nebbia, C.S., Oswald, I.P., Rovesti, E., Steinkellner, H. and Hoogenboom, R.L., 2023a. Scientific Opinion on the assessment of information as regards the toxicity of deoxynivalenol for horses and poultry. EFSA Journal 21: 7806. https://doi.org/10.2903/j.efsa.2023.

EFSA CONTAM Panel (EFSA Panel on Contaminants in the Food Chain), Schrenk, D., Bignami, M., Bodin, L., Chipman, J.K., del Mazo, J., Grasl-Kraupp, B., Hogstrand, C., Hoogenboom,L.R., Leblanc, J.C., Nebbia, C.S., Nielsen, E., Ntzani, E., Sand, S., Schwerdtle, T., Vleminckx, C.,Wallace, ., Gropp, J., Antonissen, G., Rychen, G., Ruiz, J.A.G., Innocenti, M.L., Rovesti, E. and Petersen, A., 2023b. Risks for animal health related to the presence of ochratoxin A (OTA) in feed. EFSA Journal 21 : 8375. https://doi.org/10.2903/j.efsa.2023.8375.

EFSA CONTAM Panel (EFSA Panel on Contaminants in the Food Chain), Schrenk, D., Bignami, M., Bodin, L., Chipman, J.K., del Mazo, J., Grasl-Kraupp, B., Hogstrand, C., Leblanc, J.C., Nebbia, C.S., Nielsen, E., Ntzani, E., Petersen, A., Sand, S., Schwerdtle, T., Vleminckx, C.,Wallace, H., Gropp, J., Mulder, P., Oswald, I.P., Woutersen, R., Ruiz, J.A.G., Rovesti, E. and Hoogenboom,L.R., 2024. Risks for animal health related to the presence of ergot alkaloids in feed. EFSA Journal 22: 8496. https://doi.org/10.2903/j.efsa.2024.8496.

Grenier, B. and Oswald, I.P., 2011. Mycotoxin co-contamination of foods and feeds: meta-analysis of publications describing toxicological interactions. World Mycotoxin Journal 4: 285. https://doi.org/10.3920/WMJ2011.1281.

CHAPTER 18

Climate change and mycotoxins

A.M. Gasperini[1*], *A. Medina*[2*] *and N. Magan*[2§]

[1]*MycoLab, Department of Clinical, Pharmaceutical and Biological Sciences, School of Life and Medical Sciences, University of Hertfordshire, Hatfield, UK*
[2]*Department of Environment and Agrifood at Cranfield University, Cranfield, UK*
[§]*In memoriam*
**correspondence: a.marcon-gasperini@herts.ac.uk; a.medinavaya@cranfield.ac.uk*

1 Introduction

Greenhouse gas emissions are the main factor in climate change. Such gases absorb infrared radiation emitted from the planet's surface increasing the atmospheric concentrations resulting in warmer temperatures. Carbon dioxide (CO_2) is the primary greenhouse gas emitted by human activities. The atmospheric CO_2 has increased by >40 % from 1800 to 2019 (National Academy of Sciences, 2020), and its concentration in the atmosphere has continuously increased from pre-industrial levels of 280 to 423 ppm in May 2022, and is expected to reach >900 ppm by the end of the century (NOAA, 2022).

Increases in CO_2 emissions coupled with substantial increases in global average surface temperatures and extreme wet/dry episodes on a regional basis have significant implications for regional shifts in climate (National Academy of Sciences, 2020). Indeed, many parts of the world are experiencing very unusual rainfall patterns and droughts resulting in increased fire risks and the destruction of land often used for staple food commodity production. Other effects of climate change include the melting of polar ice caps and increases in ocean levels and temperatures (EU Commission, 2022).

According to the United Nations (2019), the world population is estimated to reach 9.8 billion by 2050 and climate projections show a rise between +1.5 to +2.0 °C in global temperatures (IPCC, 2018). Human industrial activities suggest that currently global warming is increasing at a rate of 0.2 °C per decade (Allen et al., 2018). However, if global temperatures rise by >1.5 °C above pre-industrial levels, environmental changes could become irreversible. Extreme global warming could lead the current CO_2 reservoirs (forests and Arctic permafrost zones) to release excess carbon emissions resulting in further warming (Tollefson, 2022). In a scenario of a +1.5 °C increase in temperature, almost 147% of the world population will be exposed

 | DOI: 10.1163/9789004712720_19

to severe heat waves at least once every 5 years, and in a scenario with an increase of approx. +2 °C warming, the fraction of the global population affected becomes 3 times larger (37%), representing almost 1.7 billion people (Dosio et al., 2018).

Furthermore, climate change has implications for the future state of natural resources, as well as the resulting impacts on and constraints to agricultural production chains affecting both food security and safety (FAO, 2018). Taken together, the evidence available and the predictions suggest an overall negative effect on agricultural production, especially when significantly more food is required to feed an ever-increasing global population (FAO, 2017, 2015).

2 Overview of climate change impacts on crops and livestock

Agriculture is intrinsically sensitive to climate variability and changes either as a result of natural causes or human activities. Climate change is expected to directly influence crop production systems for food, feed, or forage, to affect livestock health, and to alter the pattern and balance in food production and trading, which today is done on a global basis (Godde et al., 2021).

Exposure of staple food plants to climate change has been shown to affect plant physiology, especially related to changes in number and spatial nature of stomata affecting photosynthesis and thus increasing sensitivity of plants to environmental stresses such as drought, waterlogging or heat, and plant diseases caused by pathogenic microorganisms (Mikkelsen et al., 2015). Such interactions between biotic and abiotic stresses under climate change conditions have a significant effect by decreasing yields and nutritional quality, which in turn impact the production of staple raw commodities on a global basis (Mikkelsen et al., 2015). Crop yields have been shown to decrease significantly at temperatures >30 °C (Boote et al., 2005; Schlenker and Roberts, 2009). Indeed, a decline of 3.8% in global maize production was reported based on climate warming in 1980 and 2008 (Lobell et al., 2011). Significant decreases in yield and quality of wheat, rice and soybeans have also been reported (Schauberger et al., 2017). A further negative outcome of elevated atmospheric CO_2 levels predicted by 2050 are associated with substantial declines in the zinc, iron, and protein content of staple crops such as wheat, rice, field peas and soybeans (FAO, 2016).

Projected changes in crop yield vary with alternative climate models and from one country to another in both size and magnitude. In general, yields were predicted to increase in Northern Europe, but decrease across Africa and South America (Ray et al., 2019; Rosenzweig and Parry, 1994). Throughout most tropical areas crop yields are more negatively affected than at higher latitudes, and with an increasing degree of climate change the impacts will be more severe (Wheeler and von Braun, 2013). For instance, the yields of major crops grown across Africa and South Asia under climate change may decline by 8% by 2050 (Knox et al., 2012). In India, increasing temperatures were associated with lower wheat yields from 1981 to 2009 (Gupta et al., 2017).

Moreover, yield variability significantly changed from 1961 to 2009 in the major crop-producing countries (Osborne and Wheeler, 2013). Year-to-year variations in yields of maize, soybean, rice, and wheat decreased in 19-33% of the global harvested area in 1981-2010. However, increased yield variability was shown for maize and soybean in Argentina and Northeast China, rice in Indonesia and Southern China, and wheat in Australia, France and Ukraine (Iizumi and Ramankutty, 2016).

In Australia declines in rainfall and rising daily maximum temperatures caused yield potential to decline by 27% from 1990 to 2015, according to simulations from 50 sites. However, elevated atmospheric CO_2 concentrations had a positive effect on wheat yields (Hochman et al., 2017). In contrast, high-temperature episodes combined with droughts during grain ripening (anthesis onwards) were found to have the most detrimental effects on wheat yields (Innes et al., 2015). The assessment of the impacts of higher temperature on yields of maize, pearl millet and sorghum in the West African Sudan Savanna showed that the +2.0 °C scenario had more negative impacts on yield than the +1.5 °C one. But the study found no evidence to suggest that yield variability would increase with the extra 0.5 °C of warming (Faye et al., 2018).

In summary, both the agrifood chains and those for livestock may not be very resilient to climate change (Figure 18.1). As feed nutritional quality is critical for livestock production, climate change will impact their development both directly and indirectly (Leggieri et al., 2021). Direct effects in terms of thermoregulation, metabolism, immune system functioning, and production would occur in warmer climate-change-related scenarios. In addition, indirect impacts from climate change include water shortages, food-borne diseases, and feed availability and quality (Cheng et al., 2022; Labanca et al., 2019; Hidosa and Guyo, 2017; Nardone et al., 2010). The feed availability for livestock can be affected by the decline in crop yields, nutritional value, and contamination with fungal toxins, mycotoxins. The presence of mycotoxins in feed and forage affects profits as result of decline in both animal productivity and health (Conte et al., 2020). Recent global surveys also suggest that feed maize is predominantly contaminated with multiple mycotoxins and that in some regions the predominant ones are changing (Raj et al., 2021). Additionally, extreme climate events are associated with negative effects on pastoral production (IPCC, 2019) since elevated CO_2, rising temperatures, and reduced precipitation could affect growth of essential forage crop species (Giridhar and Samireddypalle, 2015; Izaurralde et al., 2011). Exposure to elevated CO_2 has also been shown to increase the productivity of weeds which can then reduce yields of both food-related crops as well as forage (Vilà et al., 2021; Ziska and Dukes, 2010).

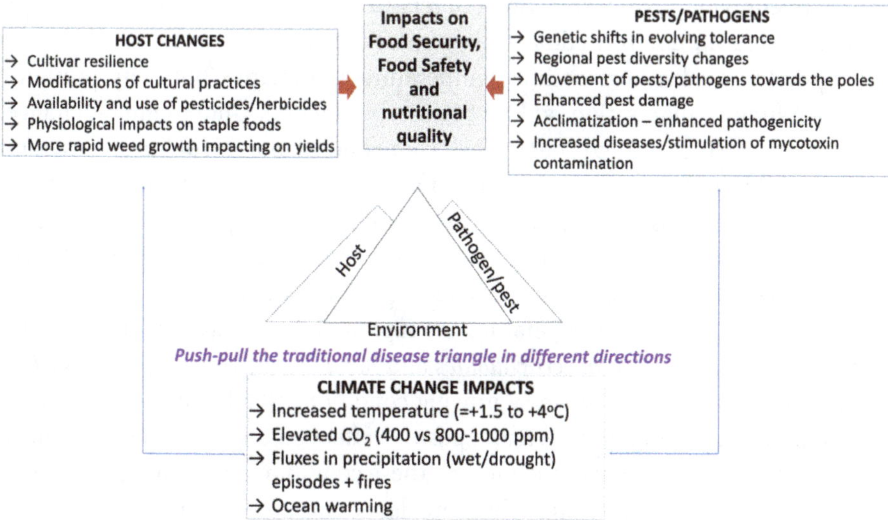

FIGURE 18.1 Climate change impacts the classic host changes and pest/pathogens, and environment triangle, which may occur due to the pressure of climate-change scenarios. Source: the authors.

3 Risks of mycotoxins in food and feed chains under climate change scenarios

Climate change is expected to increase the occurrence of mycotoxins in feed and food commodities worldwide (EFSA et al., 2020; FAO, 2020). The effects on mycotoxin production is complex and difficult to predict with complete accuracy because many of the biotic and abiotic factors will interact with each other (Dwivedi et al., 2013). Previously, interacting climate-change factors were cited as crucial factors for fungal colonisation and mycotoxin biosynthesis in crops (Magan et al., 2003). Regional studies have suggested that, certainly in Europe, changes in temperature of between +2 and +5 °C would significantly increase the areas where *Fusarium* mycotoxins such as DON in wheat may become more important and increase relative risks of contamination (Battilani et al., 2016). Studies by (Vaughan et al., 2016, 2014) showed that colonisation of maize by *Fusarium verticillioides* during silking was increased when maize plants were exposed to elevate CO_2. When there was an interaction between elevated CO_2 and drought stress, there was an increase in FB_1 contamination of the ripening maize cobs. A significant amount of research has focused on *Aspergillus flavus* and how interactions between elevated CO_2 (existing levels of 350-400 ppm vs. 650, 800 or 1000 ppm), increased temperature, and drought stress will influence growth and AFB_1 contamination of maize kernels. These studies, both *in vitro*, pre-harvest with ripening cobs and post-harvest with stored maize grain showed few effects on colonisation rates, but stimulated both regulatory and structural genes (*afl*R, *afl*D) involved in the biosynthesis of AFs as well as phenotypic toxin production (Garcia-Cela et al., 2021; Gilbert et al., 2018; Medina et al., 2014; 2015b; Medina et al., 2017b; Ponts, 2015).

Ecophysiological factors such as temperature and relative humidity primarily influence fungal colonisation and mycotoxin production. Under climate change, these factors are predicted to alter the behaviour and adaptation of fungal pests and pathogens (Moretti et al., 2019; Battilani et al., 2016; Paterson and Lima, 2010). Environmental stress has been shown to have significant consequences for secondary metabolite production, especially mycotoxins (Garcia-Cela et al., 2018; Fountain et al., 2016; Medina et al., 2015c; Ponts, 2015; Schmidt-Heydt et al., 2011).

Changes in climate and temperature can rapidly shift pathogen populations; for example, in maize, *Fusarium verticillioides* and *A. flavus* will outcompete *Fusarium graminearum* under drier, warmer, growing conditions (Reid et al., 1999; Szabo et al., 2018). Elevated temperatures had a positive effect on disease reduction, fungal biomass and DON accumulation, however these conditions could also promote infection by *A. flavus*, representing a far more dangerous problem (Damianidis et al., 2018; Dövényi-Nagy et al., 2020). The impact of climate-related abiotic factors on AFB_1 contamination has suggested that elevated temperatures (+4 to 5 °C above optimum), drought stress, and exposure to increased CO_2 levels (400 vs 1000 ppm) increased production of AFB_1 on maize (Garcia-Cela et al., 2021; Gasperini et al., 2019; Medina et al., 2014).

The ability to adapt to environmental changes demonstrated by mycotoxigenic fungi are related to the elevated level of fungal plasticity and the evolution of acclimatisation within the strains (Medina et al., 2015a). These plasticity traits differed among isolates, with values varying most in response to environmental extremes (Alster et al., 2021). However, generally microorganisms are considered to have a high degree of plasticity in order to reduce environmental threats to growth and propagation (Rong et al., 2019; Slepecky and Starmer, 2009).

4 Acclimatisation and evolution of resilience to climate change scenarios

There has been interest in the resilience of fungal pathogens, including mycotoxigenic species, when repeatedly exposed to climate-change-related abiotic factors. This is usually referred to as 'acclimatisation'. This has been defined as a process in which an individual organism adjusts to a change in its environment, allowing it to maintain fitness. This may be very important, especially if pathogenicity/virulence is modified or if mycotoxin production is increased. Váry et al. (2015) probably carried out the first studies that examined the acclimatisation of fungal pathogens of wheat in relation to elevated CO_2 levels (390 vs 780 ppm CO_2). Exposure of wheat plants to either 350 vs 650 ppm CO_2 in controlled plant growth chambers had effects on both crop physiology and disease severity. *Septoria tritici* and *Fusarium graminearum* sub-cultured for 10 and 20 generations in elevated CO_2 conditions, respectively, were found to change their infection and disease symptoms after such acclimatisation. Thus, the acclimatised strains of the two fungal pathogens had increased pathogenicity under such conditions, in terms of both disease symptoms and fungal biomass using molecular approaches which increased. Unfortunately, no analyses of concomitant contamination with type B trichothecenes was carried out. Recently, Peter Mshelia et al. (2020) acclimatised one strain each of *F. verticillioides* and *F. graminearum* for 10 generations on a milled maize-based media. However, this was only in relation to temperature with exposure to step-wise increases of 0.5 °C for each generation from 30 to 35 °C for the former, and from 30 to 33 °C for the latter species. No acclimatisation by exposure to elevated CO_2 was included. Subsequent exposure of these pathogens to changes in temperature x elevated CO_2 (400 vs 800 and 1200 ppm) x water activity (0.92, 0.95, 0.98 a_w) showed little effect on growth or FUM, and DON or ZEN production by *F. verticilioides* and *F. graminearum* respectively. Baazeem et al. (2022) examined the exposure of *A. flavus* strains to elevated CO_2 + elevated temperature (1000 ppm CO_2 + 37 °C) for 5 generations and compared growth and AFB_1 production. There was an effect on subsequent colonisation by one strain of pistachio nuts at different *aw* levels and a stimulation of AFB_1 for one strain but not the other. This suggests some differential effects on acclimatisation of strains of the same species that would have to be borne in mind.

5 Relationship between climate change factors and global shifts in impacts on mycotoxins

Hay et al. (2021) demonstrated that wheat plants were most susceptible to disease damage at warm temperatures and maize plants at cold temperatures. Elevated CO_2 (1000 ppm) was found to significantly increase the accumulation of DON. In wheat, despite reduced pathogen biomass and toxin accumulation at warm temperatures, *F. graminearum* was more aggressive with greater disease damage and toxin production per unit biomass. However, atmospheric CO_2 and temperature had no significant interactions, except for increasing DON accumulation in maize at lower temperatures and elevated CO_2. Disease outcomes are also pathogen strain specific, with complex interactions between host, strain, and growth conditions (Hay et al., 2021).

The geographical shift in the effects of climate change on fungal growth has also received attention. For example, when different time-frame scenarios were evaluated in the potential growth of *Alternaria* (far future > near future > current), it was observed that Poland's situation in the far future (2081–2100) will become similar to Spain's situation in the present time frame (1981–2000), showing a geographical shift in the problem (Perre et al., 2015). DON occurrence in triticale was favoured by local and regional agroclimatic factors and was amplified by extreme weather events (cold waves, tropical cyclones, heat waves) in Romania between 2012 and

2014. Extreme weather events revealed a higher contamination in Central Europe compared to south-eastern and eastern Europe (Gagiu et al., 2021).

Croatian maize samples had significant contamination with T-2 toxin indicating a *Fusarium* infection after heavy rainfall periods in 2010 (Pleadin et al., 2012). Furthermore, significantly higher concentrations of T-2/HT-2 toxin were found in oats related to the regional weather in Croatia. Higher precipitation, combined with extremely warm weather witnessed during 2018, resulted in conditions favouring the growth of *Fusarium* and the occurrence of T-2/HT-2 toxins, especially in the northern and eastern region (Kiš et al., 2021).

Additionally, higher temperatures due to climate warming could create new challenges related to alterations in insect populations (Wu et al., 2011). Insects are ectotherms, thus an increase in ambient temperature directly influences their metabolic rates, developmental rates, and activity patterns (Altermatt, 2010). This could lead to increased numbers of insects, increased damage to crops, higher occurrence of fungal contamination, and by extension, increased levels of mycotoxins (Medina et al., 2017a).

In the case of maize, the crop can become infected by several pathogens during the growing season, harvesting, drying and storage, resulting in significant losses. Pre-harvest pests represent an average of 35% of potential yield losses worldwide (Oerke, 2006), whereas post-harvest losses range between 14 and 36% (López-Castillo et al., 2018). However, the relationship between insect damage and mycotoxins in maize adds another dimension of complexity not found in wheat (Schaafsma and Hooker, 2007). There have been few if any studies examining the relationship between climate-change factors, pest damage, and *A. flavus* infection and how these impact on relative contamination levels.

6 Biocontrol as a strategy for mycotoxin mitigation under present and future climate scenarios

Biocontrol is considered to be one of the most promising technologies for sustainable agriculture and for reducing the use of chemical pesticides (De Cal et al., 2020; Tracy, 2014). Microbial BCAs represent an important approach towards this aim. However, it is important to take a holistic view for the successful development of BCAs which should include the impacts of climate change in relation to the different phases of screening, production and formulation of the best candidates (Magan and Medina, 2020).

While biocontrol has often been very effective for both disease and pest control in controlled environments such as glasshouses, there are still major hurdles related to its application outdoors where relative humidity and temperature fluxes mean that the resilience of BCAs is often compromised. The relative consistency is critical for significant substitution of existing chemical pesticides. Climate-change scenarios have almost never been considered in the screening for novel BCAs (Magan and Medina, 2020). Often, subsequent formulation of effective BCA strains again does not test for relative resilience under elevated CO_2 and increased temperature or drought stress. As a consequence, there are still hurdles on the road to the effective development of BCAs for field use under existing and future climate change regimes. Indeed, it has been pointed out that while most BCAs are genetically stable, their use has not been very successful due to their lack of resilience to climate change (Lahlali et al., 2022).

There are several reports of microbial strains from different genera being antagonistic to mycotoxigenic *Fusarium* spp. under laboratory and field conditions (Rojas et al., 2020; Timmusk et al., 2020). Naturally occurring endophytes were shown as a potential new source of biocontrol (Drakopoulos et al., 2020). However, they are often very slow growing and the targeting of mycotoxigenic pathogens to reduce mycotoxins has not been fully investigated. Different approaches indicate the potential to suppress FHB severity and mycotoxin accumulation by mycotoxin-degrading bacteria (Morimura et al., 2020) or detoxification of mycotoxin (ZEN) by fungal strains (Kosawang et al., 2014). Also, plant-based biopesticides were able to reduce FHB under controlled conditions, but the mycotoxin reduction was not consistent (Drakopoulos et al., 2020). Unfortunately, few if any of these studies identified strains that were resilient in the face of climate-change-related abiotic factors.

There have been significant successful studies of the use of atoxigenic strains of *A. flavus* for control of AFB_1 in both maize and groundnuts, especially in West and East Africa (Atehnkeng et al., 2008; Bandyopadhyay et al., 2016). However, while they have used local atoxigenic strains, these have not been examined for resilience to climate-change factors. Studies at Cranfield University have examined the impact of non-toxigenic *A. flavus* strains and other microbial BCAs on temporal control of AFB_1 and FB_1 in maize stored at different temperatures and water availabilities (Gasperini, 2019; Gasperini et al., 2019; Medina et al., 2017c; Mohale et al., 2013). It is especially important to understand how the potential biocontrol strains targeting AFB_1 control in maize may behave under interacting climate-change abiotic factors and whether they have the necessary resilience to reduce mycotoxin biosynthesis *in situ* (Gasperini et al., 2019).

Gasperini (2019) assessed the effects of simulated pest damage of kernels (0, 5, 15% damage) on the relative efficacy of atoxigenic strains of *A. flavus* for control of toxigenic strains in stored maize. The atoxigenic and toxigenic *A. flavus* strains were applied in a ratio 50:50 of conidia onto the maize kernels with different levels of damage and at different water availability conditions. Under climate-change exposure of increased temperature + CO_2 levels using conventional and GM maize with undamaged kernels or 5% of simulated damaged, the toxigenic *A. flavus* strain produced more AFB_1 when inoculated onto GM kernels (Figure 18.2). The study demonstrated that the efficacy of biocontrol was significantly affected under climate-change conditions (35 °C/1000 ppm CO_2) in both GM and non-GM maize.

There was also a reduction in the expression of one of the regulatory genes (*afl*R) involved in toxin biosynthesis due to the competitive nature of the atoxigenic strain biocontrol strain (Figure 18.3). However, when the CO_2 level was increased from 400 ppm to 1000 ppm + 5% kernel damage in the GM cultivar, the inhibition of *afl*R was very similar to the control (p=0.05) although the AFB_1 production was reduced in this treatment by 35 %. The full interaction of T (°C) × CO_2 × a_w × simulated pest damage had significant effects only in conventional maize. However, this effect did not occur when biocontrol was applied. The action of the atoxigenic strain is significant even in the increased CO_2 scenario, although overall efficacy was lower than in non-extreme conditions. This certainly suggests that it is important to consider climate-change factors when identifying resilient strains of candidate BCAs for use in the future (Magan, 2020; Magan and Medina, 2020).

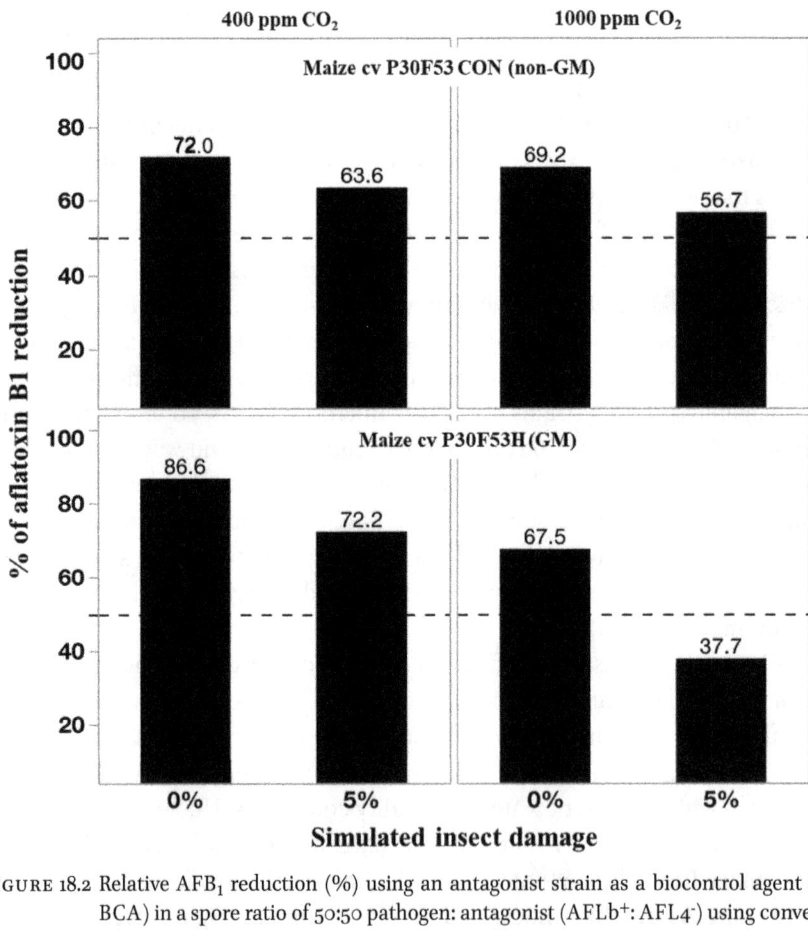

FIGURE 18.2 Relative AFB_1 reduction (%) using an antagonist strain as a biocontrol agent (AFL_4^-; BCA) in a spore ratio of 50:50 pathogen: antagonist ($AFLb^+$: $AFL4^-$) using conventional (P30F53 CON) and GM (P30F53 H®) maize with 0.98 a_w after 10 days at 35 °C in different CO_2 levels (400 and 1000 ppm) and simulated pest damage (0 and 5 %; Gasperini, 2019).

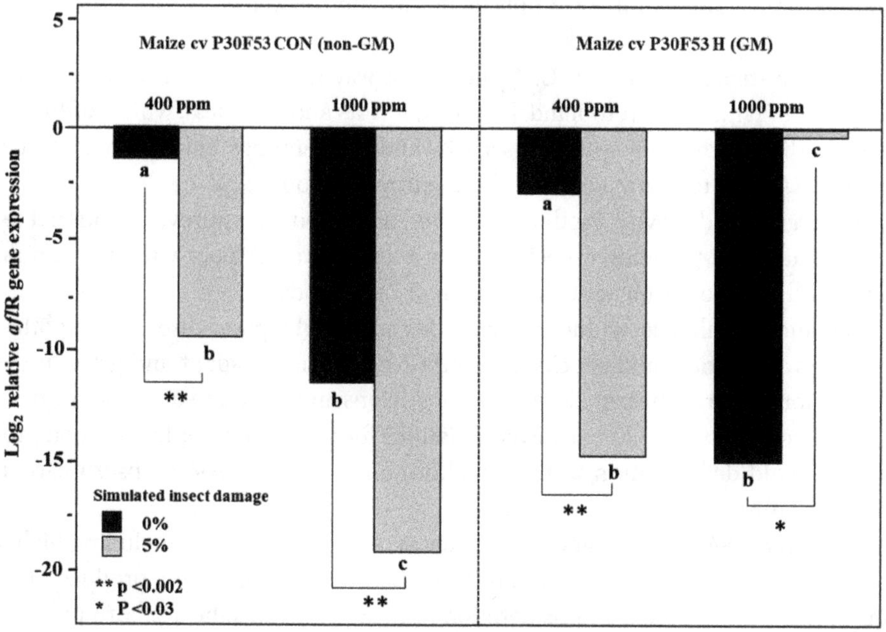

FIGURE 18.3 Effect of antagonist atoxigenic *A. flavus* strain against a toxigenic strain in a conidial ratio of 50:50 pathogen:antagonist (AFLb$^+$: AFL4$^-$) on relative expression of the regulatory gene *afl*R using conventional (P30F53 CON) and GM (P30F53 H*) maize at 0.98 a_w and stored for 10 days at 35 °C. The maize was exposed to existing CO_2 (400 ppm) or elevated (1000 ppm) levels in combination with simulated pest damage (0 and 5%) conditions. The expression was normalised for the control sample (AFLb$^+$) in each condition. *p*-values represent significant differences from the control; different letters indicate difference in the simulated damage and CO_2 levels (p<0.05; Gasperini, 2019).

7 Future challenges for feed safety considerations in agrifood production systems

Today, our food/feed production chains and supply systems are global in nature. Thus, fluxes in weather patterns, especially because of climate-change abiotic factors, are already resulting in significant impacts on food security on a global basis. This will be become even more pronounced in the future, especially in key global regions that are predominantly responsible for the production and transfer of staple foods and feed stocks across the globe. The impacts of climate change on the yield of cereals, considered to be between 5-15%, and of soya beans, up to 40%, will require significant changes in the development of new, more resilient cultivars. Perhaps gene editing will facilitate the more rapid development of staple crops that could be resilient under climate-change scenarios with conserved nutritional value and yields. However, the impact of different pests and diseases, which will change, represents a particular challenge. This would have to be considered in trying to minimise the vulnerability of supply chains for food as well as livestock productivity (EEA, 2021).

Climate-smart agriculture (CSA) may be a way forward, as it integrates management of landscapes (cropland, livestock, forests and fisheries), which addresses the interlinked challenges of food security and accelerating climate change (FAO, 2021). CSA aims to simultaneously achieve three outcomes:

Increased productivity: Produce more and better food to improve nutritional security and boost incomes, especially of 75% of the world's poor who live in rural areas and rely mainly on agriculture for their livelihoods.

Enhanced resilience: Reduce vulnerability to drought, pests, diseases, and other climate-related risks and shocks; and improve capacity to adapt and grow in the face of longer-term stresses like shortened seasons and erratic weather patterns.

Reduced emissions: Pursue lower emissions for each calorie or kilo of food produced, avoid deforestation from agriculture and identify ways to absorb carbon from the atmosphere.

However, CSA does not address the key aspects of feed safety including higher exposure of animals and populations to toxic residues or naturally produced toxins, including mycotoxins. Consideration of mycotoxins specifically, both pre- and post-harvest and their mitigation under climate change conditions, needs significantly more research. For example, it is not known if the diversity and movement of pests result in damage to crops in regions where they were not previously known and thus allow infection by mycotoxigenic fungi, increasing mycotoxin contamination. Or, if different mycotoxins become more important than those considered a priority at the present time in staple feedstuffs. At the present time, legislative limits are based on free mycotoxins only. It has been shown that perhaps up to 1/3 of mycotoxins may be bound to plant material (e.g., DON-3G) that can only be detected using sophisticated analytical means. It is important to investigate if, under climate change scenarios, the present legislative limits for specific mycotoxin limits in food/feed supply chains will become out of step with the reality due to changes in mycotoxin contamination.

References

Allen, M.R, Dube, O.P., Solecki, W., Aragón-Durand, F., Cramer, W., Humphreys, S., Kainuma, M., Kala, J., Mahowald, N., Mulugetta, Y., R. Perez, Wairiu, M., Zickfeld, K., 2018. Framing and Context, in: Masson-Delmotte, V., P. Zhai,H.-O. Pörtner, D. Roberts, J. Skea, P.R. Shukla, A. Pirani, W. Moufouma-Okia, C. Péan, R. Pidcock, S. Connors, J.B.R. Matthews, Y. Chen,X. Zhou, M.I. Gomis, E. Lonnoy, T. Maycock, M. Tignor, and T. Waterfield (Ed.), Global Warming of 1.5°C. An IPCC special report on the impacts of global warming of 1.5°c above pre-industrial levels and related global greenhouse gas emission pathways, in the context of strengthening the global response to the threat of climate change, sustainable development, and efforts to eradicate poverty. https://www.ipcc.ch/sr15.

Alster, C.J., Allison, S.D., Johnson, N.G., Glassman, S.I., Treseder, K.K., 2021. Phenotypic plasticity of fungal traits in response to moisture and temperature. ISME Communications 1: 43. https://doi.org/10.1038/s43705-021-00045-9.

Altermatt, F., 2010. Climatic warming increases voltinism in European butterflies and moths. Proceedings in Biological Sciences 277: 1281-1287. https://doi.org/10.1098/rspb.2009.1910.

Atehnkeng, J., Ojiambo, P.S., Ikotun, T., Sikora, R.A., Cotty, P.J., Bandyopadhyay, R., 2008. Evaluation of atoxigenic isolates of *Aspergillus flavus* as potential biocontrol agents for aflatoxin in maize. Food additives & contaminants Part A, Chemistry, analysis, control, exposure and risk assessment 25: 1264-1271. https://doi.org/10.1080/02652030802112635.

Baazeem, A., Medina, A., Magan, N., 2022. Effect of Acclimatization in elevated CO2 on growth and aflatoxin B1 production by *Aspergillus flavus* strains on pistachio nuts. Microorganisms 10: 49. https://doi.org/10.3390/microorganisms10010049.

Bandyopadhyay, R., Ortega-Beltran, A., Akande, A., Mutegi, C., Atehnkeng, J., Kaptoge, L., Senghor, A.L., Adhikari, B.N., Cotty, P.J., 2016. Biological control of aflatoxins in Africa: current status and potential challenges in the face of climate change. Special Issue: Mycotoxins in a changing world. https://doi.org/10.3920/WMJ2016.2130

Battilani, P., Toscano, P., Van der Fels-Klerx, H.J., Moretti, A., Camardo Leggieri, M., Brera, C., Rortais, A., Goumperis, T., Robinson, T., 2016. Aflatoxin B1 contamination in maize in Europe increases due to climate change. Scientific Reports 6: 24328. https://doi.org/10.1038/srep24328.

Boote, K., Allen, L., Prasad, P.V.V., Baker, J., Gesch, R., Snyder, A., Pan, D., Thomas, J., 2005. Elevated temperature and CO2 impacts on pollination, reproductive growth, and yield of several globally important crops. Journal of Agricultural Meteorology 60: 469–474. https://doi.org/10.2480/agrmet.469 .

Cheng, M., McCarl, B., Fei, C., 2022. Climate change and livestock production: A literature review. Atmosphere 13: 140. https://doi.org/10.3390/atmos13010140.

Conte, G., Fontanelli, M., Galli, F., Cotrozzi, L., Pagni, L., Pellegrini, E., 2020. Mycotoxins in feed and food and the role of ozone in their detoxification and degradation: an update. Toxins 12: 486. https://doi.org/10.3390/toxins12080486.

Damianidis, D., Ortiz, B.V., Bowen, K.L., Windham, Gary.L., Hoogenboom, G., Hagan, A., Knappenberger, T., Abbas, H.K., Scully, B.T., Mourtzinis, S., 2018. Minimum temperature, rainfall, and agronomic management impacts on corn grain aflatoxin contamination. Agronomy Journal 110: 1697-1708. https://doi.org/10.2134/agronj2017.11.0628.

De Cal, A., Guijarro, B., Larena, I., Melgarejo, P., 2020. Which biocontrol strategies best fit with other ipm system components?, in: De Cal, A., Melgarejo, P., Magan, N. (Eds.), How research can stimulate the development of commercial biological control against plant diseases, progress in biological control. Springer International Publishing 231–256. https://doi.org/10.1007/978-3-030-53238-3_14.

Dosio, A., Mentaschi, L., Fischer, E.M., Wyser, K., 2018. Extreme heat waves under 1.5 °C and 2 °C global warming. Environmental Research Letters 13: 054006. https://doi.org/10.1088/1748-9326/aab827

Dövényi-Nagy, T., Rácz, C., Molnár, K., Bakó, K., Szláma, Z., Jóźwiak, Á., Farkas, Z., Pócsi, I., Dobos, A.C., 2020. Pre-harvest modelling and mitigation of aflatoxins in maize in a changing climatic environment- A review. Toxins 12: 768. https://doi.org/10.3390/toxins12120768.

Drakopoulos, D., Meca, G., Torrijos, R., Marty, A., Kägi, A., Jenny, E., Forrer, H.-R., Six, J., Vogelgsang, S., 2020. Control of *Fusarium graminearum* in wheat with mustard-based botanicals: from *in vitro* to *in planta*. Frontiers in Microbiology 11: 1-15. https://doi.org/10.3389/fmicb.2020.01595.

Dwivedi, S., Sahrawat, K., Upadhyaya, H., Ortiz, R., 2013. Food, Nutrition and Agrobiodiversity Under Global Climate Change, in: Sparks, D.L. (Ed.), Advances in Agronomy. Academic Press, Elsevier Inc., 1 -128. https://doi.org/10.1016/B978-0-12-407686-0.00001-4.

EEA, 2021. Global climate change impacts and the supply of agricultural commodities to Europe — European Environment Agency. https://www.eea.europa.eu/publications/global-climate-change-impacts-and.

EU Commission, 2022. Consequences of climate change. https://ec.europa.eu/clima/climate-change/consequences-climate-change_en.

EFSA (European Food Safety Authority), Maggiore, A., Afonso, A., Barrucci, F., Sanctis, G.D., 2020. Climate change as a driver of emerging risks for food and feed safety, plant, animal health and nutritional quality. EFSA supporting publication 2020:146. https://doi.org/10.2903/sp.efsa.2020.EN-1881.

FAO, 2015. Climate change and food security: risks and responses. http://www.fao.org/3/a-i5188e.pdf.

FAO, 2016. The State of Food and Agriculture 2016. Climate change, agriculture and food security. https://www.fao.org/3/i6132e/i6132e.pdf.

FAO, 2017. The future of food and agriculture: trends and challenges. https://www.fao.org/3/i6583e/i6583e.pdf.

FAO, 2018. The future of food and agriculture – Alternative pathways to 2050. https://www.fao.org/3/I8429EN/i8429en.pdf.

FAO, 2020. Climate change: Unpacking the burden on food safety. https://doi.org/10.4060/ca8185en.

FAO, 2021. Climate-smart agriculture case studies 2021. https://doi.org/10.4060/cb5359en.

Faye, B., Webber, H., Naab, J.B., MacCarthy, D.S., Adam, M., Ewert, F., Lamers, J.P.A., Schleussner, C.-F., Ruane, A., Gessner, U., Hoogenboom, G., Boote, K., Shelia, V., Saeed, F., Wisser, D., Hadir, S., Laux, P., Gaiser, T., 2018. Impacts of 1.5 versus 2.0 °C on cereal yields in the West African Sudan Savanna. Environmental Research Letters 13: 034014. https://doi.org/10.1088/1748-9326/aaab40.

Fountain, J.C., Bajaj, P., Pandey, M., Nayak, S.N., Yang, L., Kumar, V., Jayale, A.S., Chitikineni, A., Zhuang, W., Scully, B.T., Lee, R.D., Kemerait, R.C., Varshney, R.K., Guo, B., 2016. Oxidative stress and carbon metabolism influence *Aspergillus flavus* transcriptome composition and secondary metabolite production. Scientific Reports 6: 38747. https://doi.org/10.1038/srep38747.

Gagiu, V., Mateescu, E., Dobre, A.A., Smeu, I., Cucu, M.E., Oprea, O.A., Alexandru, D., Iorga, E., Belc, N., 2021. Deoxynivalenol occurrence in triticale crops in romania during the 2012–2014 period with extreme weather events. Toxins 13: 456. https://doi.org/10.3390/toxins13070456.

Garcia-Cela, E., Kiaitsi, E., Medina, A., Sulyok, M., Krska, R., Magan, N., 2018. Interacting environmental stress factors affects targeted metabolomic profiles in stored natural wheat and that inoculated with *F. graminearum*. Toxins 10: 56. https://doi.org/10.3390/toxins10020056.

Garcia-Cela, E., Verheecke-Vaessen, C., Gutierrez-Poso, M., Kiaitsi, E., Gasperini, A.M., Magan, N., Medina, A., 2021. Unveiling the effect of interacting forecasted abiotic factors on the Aflatoxin B1 production kinetics by *Aspergillus flavus*. Fungal Biology 125: 89-94. https://doi.org/10.1016/j.funbio.2020.05.003.

Gasperini, A.M., 2019. Fungal diversity, pest damage and biocontrol of aflatoxins in GM and conventional Brazilian maize cultivars under existing and future climate change scenarios (Ph.D). Cranfield University, Cranfield, Beds, MK43 0AL, U.K.

Gasperini, A.M., Rodriguez-Sixtos, A., Verheecke-Vaessen, C., Garcia-Cela, E., Medina, A., Magan, N., 2019. Resilience of biocontrol for aflatoxin minimization strategies: climate change abiotic factors may affect control in non-gm and gm-maize cultivars. Frontiers in Microbiology 10: 2525. https://doi.org/10.3389/fmicb.2019.02525.

Gilbert, M., Medina, A., Mack, B., Lebar, M., Rodríguez, A., Bhatnagar, D., Magan, N., Obrian, G., Payne, G., Gilbert, M.K., Medina, A., Mack, B.M., Lebar, M.D., Rodríguez, A., Bhatnagar, D., Magan, N., Obrian, G., Payne, G., 2018. Carbon dioxide mediates the response to temperature and water activity levels in *Aspergillus flavus* during infection of maize kernels. Toxins 10: 5. https://doi.org/10.3390/toxins10010005.

Giridhar, K., Samireddypalle, A., 2015. Impact of Climate Change on Forage Availability for Livestock, in: Sejian, V., Gaughan, J., Baumgard, L., Prasad, C. (Eds.), Climate Change Impact on Livestock: Adaptation and Mitigation. Springer India, New Delhi, pp. 97–112. https://doi.org/10.1007/978-81-322-2265-1_7.

Godde, C.M., Mason-D'Croz, D., Mayberry, D.E., Thornton, P.K., Herrero, M., 2021. Impacts of climate change on the livestock food supply chain; a review of the evidence. Global Food Security 28: 100488. https://doi.org/10.1016/j.gfs.2020.100488.

Gupta, R., Somanathan, E., Dey, S., 2017. Global warming and local air pollution have reduced wheat yields in India. Climatic Change 140: 593–604. https://doi.org/10.1007/s10584-016-1878-8.

Hay, W.T., McCormick, S.P., Vaughan, M.M., 2021. Effects of atmospheric CO_2 and temperature on wheat and corn susceptibility to *Fusarium graminearum* and deoxynivalenol contamination. Plants 10: 2582. https://doi.org/10.3390/plants10122582.

Hidosa, D., Guyo, M., 2017. Climate change effects on livestock feed resources: a review. Journal of Fisheries and Livestock Production 5: 1-4. https://doi.org/10.4172/2332-2608.1000259.

Hochman, Z., Gobbett, D.L., Horan, H., 2017. Climate trends account for stalled wheat yields in Australia since 1990. Global Change Biology 23: 2071–2081. https://doi.org/10.1111/gcb.13604.

Iizumi, T., Ramankutty, N., 2016. Changes in yield variability of major crops for 1981–2010 explained by climate change. Environmental Research Letters 11: 034003. https://doi.org/10.1088/1748-9326/11/3/034003.

Innes, P.J., Tan, D.K.Y., Van Ogtrop, F., Amthor, J.S., 2015. Effects of high-temperature episodes on wheat yields in New South Wales, Australia. Agricultural and Forest Meteorology 208: 95–107. https://doi.org/10.1016/j.agrformet.2015.03.018.

IPCC, 2018. Summary for Policymakers: Global Warming of 1.5°C, in: V. Masson-Delmotte, P. Zhai, H.-O. Pörtner, D. Roberts, J. Skea, P.R. Shukla, A. Pirani, W. Moufouma-Okia, C. Péan, R. Pidcock, S. Connors, J.B.R. Matthews, Y. Chen, X. Zhou, M.I. Gomis, E. Lonnoy, T. Maycock, M. Tignor, T. Waterfield (Eds.), An IPCC Special report on the impacts of global warming of 1.5°C above pre-industrial levels and related global greenhouse gas emission pathways, in the context of strengthening the global response to the threat of climate change, sustainable development, and efforts to eradicate poverty. World Meteorological Organization, Switzerland. https://www.ipcc.ch/sr15/chapter/spm/.

IPCC, 2019. Summary for Policymakers, in: P.R. Shukla, J. Skea, E. Calvo Buendia, V. Masson-Delmotte, H.-O. Pörtner, D. Roberts, P. Zhai, R. Slade, S. Connors, R. van Diemen, M. Ferrat, E. Haughey, S. Luz, S. Neogi, M. Pathak, J. Petzold, J. Portugal Pereira, P. Vyas, E. Huntley, K. Kissick, M. Belkacemi, J. Malley (Eds.), Climate change and land: An IPCC special report on climate change, desertification, land degradation, sustainable land management, food security, and greenhouse gas fluxes in terrestrial ecosystems. Switzerland. https://www.ipcc.ch/site/assets/uploads/sites/4/2019/12/02_Summary-for-Policymakers_SPM.pdf.

Izaurralde, R.C., Thomson, A.M., Morgan, J.A., Fay, P.A., Polley, H.W., Hatfield, J.L., 2011. Climate Impacts on Agriculture: Implications for Forage and Rangeland Production. Agronomy journal 103: 371–381. https://doi.org/10.2134/agronj2010.0304.

Kiš, M., Vulić, A., Kudumija, N., Šarkanj, B., Jaki Tkalec, V., Aladić, K., Škrivanko, M., Furmeg, S., Pleadin, J., 2021. A two-year occurrence of *Fusarium* T-2 and HT-2 toxin in Croatian cereals relative of the regional weather. Toxins 13: 39. https://doi.org/10.3390/toxins13010039.

Knox, J., Hess, T., Daccache, A., Wheeler, T., 2012. Climate change impacts on crop productivity in Africa and South Asia. Environmental Research Letters 7: 034032. https://doi.org/10.1088/1748-9326/7/3/034032.

Kosawang, C., Karlsson, M., Vélëz, H., Rasmussen, P.H., Collinge, D.B., Jensen, B., Jensen, D.F., 2014. Zearalenone detoxification by zearalenone hydrolase is important for the antagonistic ability of Clonostachys rosea against mycotoxigenic *Fusarium graminearum*. Fungal Biology 118: 364-373.. https://doi.org/10.1016/j.funbio.2014.01.005.

Labanca, F., Raimondi, A., Fontanelli, M., Pisuttu, C., Rallo, G., Galli, F., Conte, G., Pellegrini, E., 2019. The effects of climate change on livestock production systems: the cases of mycotoxins in animal feed and animal heat stress. Agrochimica 2019: 99–106.

Lahlali, R., Ezrari, S., Radouane, N., Kenfaoui, J., Esmaeel, Q., El Hamss, H., Belabess, Z., Barka, E.A., 2022. Biological Control of Plant Pathogens: A Global Perspective. Microorganisms 10: 596. https://doi.org/10.3390/microorganisms10030596.

Leggieri, M.C., Toscano, P., Battilani, P., 2021. Predicted aflatoxin B_1 increase in Europe due to climate change: actions and reactions at global level. Toxins 13: 292. https://doi.org/10.3390/toxins13040292.

Lobell, D.B., Bänziger, M., Magorokosho, C., Vivek, B., 2011. Nonlinear heat effects on African maize as evidenced by historical yield trials. Nature Climate Change 1: 42–45. https://doi.org/10.1038/nclimate1043.

López-Castillo, L.M., Silva-Fernández, S.E., Winkler, R., Bergvinson, D.J., Arnason, J.T., García-Lara, S., 2018. Postharvest insect resistance in maize. Journal of Stored Products Research 77: 66–76. https://doi.org/10.1016/J.JSPR.2018.03.004.

Magan, N., 2020. Importance of Ecological Windows for Efficacy of Biocontrol Agents, in: De Cal, A., Melgarejo, P., Magan, N. (Eds.), How Research Can Stimulate the Development of Commercial Biological Control Against Plant Diseases, Progress in Biological Control. Springer International Publishing 1–14. https://doi.org/10.1007/978-3-030-53238-3_1.

Magan, N., Hope, R., Cairns, V., Aldred, D., 2003. Post-harvest fungal ecology: Impact of fungal growth and mycotoxin accumulation in stored grain. European Journal of Plant Pathology 109: 723–730. https://doi.org/10.1023/A:1026082425177.

Magan, N., Medina, A., 2020. Climate change and resilience of biological control agents, in: De Cal, A., Melgarejo, P., Magan, N. (Eds.), How research can stimulate the development of commercial biological control against plant diseases, progress in biological control. Springer International Publishing, 83–93. https://doi.org/10.1007/978-3-030-53238-3_6.

Medina, A., Akbar, A., Baazeem, A., Rodriguez, A., Magan, N., 2017a. Climate change, food security and mycotoxins: do we know enough? Fungal Biology Reviews 31: 143-154. https://doi.org/10.1016/J.FBR.2017.04.002.

Medina, A., Gilbert, M.K., Mack, B.M., OBrian, G.R., Rodríguez, A., Bhatnagar, D., Payne, G., Magan, N., 2017b. Interactions between water activity and temperature on the *Aspergillus flavus* transcriptome and aflatoxin B1 production. International Journal of Food Microbiology 256: 36–44. https://doi.org/10.1016/j.ijfoodmicro.2017.05.020.

Medina, A., Mohale, S., Samsudin, N.I.P., Rodriguez-Sixtos, A., Rodriguez, A., Magan, N., 2017c. Biocontrol of mycotoxins: dynamics and mechanisms of action. Current Opinion in Food Science 17: 41–48. https://doi.org/10.1016/J.COFS.2017.09.008.

Medina, A., Rodriguez, A., Magan, N., 2014. Effect of climate change on *Aspergillus flavus* and aflatoxin B_1 production. Front. Microbiol. 5. https://doi.org/10.3389/fmicb.2014.00348.

Medina, A., Rodríguez, A., Magan, N., 2015a. Climate change and mycotoxigenic fungi: impacts on mycotoxin production. Current Opinion in Food Science 5: 99–104. https://doi.org/10.1016/j.cofs.2015.11.002.

Medina, A., Rodriguez, A., Magan, N., 2015c. Changes in environmental factors driven by climate change: Effects on the ecophysiology of mycotoxigenic fungi. In: Climate Change and Mycotoxins 4: 71–90. https://doi.org/10.1515/9783110333619-006.

Medina, A., Rodríguez, Alicia, Sultan, Y., Magan, Nagan, 2015b. Climate change factors and

Aspergillus flavus: effects on gene expression, growth and aflatoxin production. World Mycotoxin Journal 8: 171–179. https://doi.org/10.3920/WMJ2014.1726.

Mikkelsen, B.L., Jørgensen, R.B., Lyngkjaer, M.F., 2015. Complex interplay of future climate levels of CO2, ozone and temperature on susceptibility to fungal diseases in barley. Plant Pathology 64: 319–327. https://doi.org/10.1111/ppa.12272.

Mohale, S., Magan, N., Medina, A., 2013. Comparison of growth, nutritional utilisation patterns, and niche overlap indices of toxigenic and atoxigenic *Aspergillus flavus* strains. Fungal Biology 117: 650–659. https://doi.org/10.1016/j.funbio.2013.07.002.

Moretti, A., Pascale, M., Logrieco, A.F., 2019. Mycotoxin risks under a climate change scenario in Europe. Trends in Food Science & Technology 84: 38-40. https://doi.org/10.1016/j.tifs.2018.03.008.

Morimura, H., Ito, M., Yoshida, S., Koitabashi, M., Tsushima, S., Camagna, M., Chiba, S., Takemoto, D., Kawakita, K., Sato, I., 2020. *In vitro* assessment of biocontrol Effects on *Fusarium* Head Blight and deoxynivalenol (DON) accumulation by DON-degrading bacteria. Toxins 12: 1-13. https://doi.org/10.3390/toxins12060399.

Nardone, A., Ronchi, B., Lacetera, N., Ranieri, M.S., Bernabucci, U., 2010. Effects of climate changes on animal production and sustainability of livestock systems. Livestock Science 130: 57–69. https://doi.org/10.1016/j.livsci.2010.02.011.

National Academy of Sciences, 2020. Climate Change - evidence and causes: Update 2020. National Academies Press, Washington, D.C. https://doi.org/10.17226/25733.

NOAA, 2022. Global climate summary for June 2022. https://www.climate.gov/news-features/features/global-climate-summary-june-2022.

Oerke, E.-C., 2006. Crop losses to pests. The Journal of Agricultural Science 144: 31. https://doi.org/10.1017/S0021859605005708.

Osborne, T.M., Wheeler, T.R., 2013. Evidence for a climate signal in trends of global crop yield variability over the past 50 years. Environmental Research Letters 8: 024001. https://doi.org/10.1088/1748-9326/8/2/024001.

Paterson, R.R.M., Lima, N., 2010. How will climate change affect mycotoxins in food? Food Research International 43: 1902–1914. https://doi.org/10.1016/j.foodres.2009.07.010.

Perre, E.V. de, Jacxsens, L., Liu, C., Devlieghere, F., Meulenaer, B.D., 2015. Climate impact on *Alternaria* moulds and their mycotoxins in fresh produce: The case of the tomato chain. Food Research International 68: 41–46. https://doi.org/10.1016/j.foodres.2014.10.014.

Peter Mshelia, L., Selamat, J., Iskandar Putra Samsudin, N., Rafii, M.Y., Abdul Mutalib, N.-A., Nordin, N., Berthiller, F., 2020. Effect of temperature, water activity and carbon dioxide on fungal growth and mycotoxin production of acclimatised isolates of *Fusarium verticillioides* and *F. graminearum*. Toxins 12: 478. https://doi.org/10.3390/toxins12080478.

Pleadin, J., Perši, N., Mitak, M., Zadravec, M., Sokolović, M., Vulić, A., Jaki, V., Brstilo, M., 2012. The natural occurrence of T-2 toxin and fumonisins in maize samples in Croatia. Bulletin of Environmental Contamination and Toxicology 88: 863–866. https://doi.org/10.1007/s00128-012-0559-1.

Ponts, N., 2015. Mycotoxins are a component of *Fusarium graminearum* stress-response system. Frontiers in Microbiology. 6: 2015. https://doi.org/10.3389/fmicb.2015.01234.

Raj, J., Farkaš, H., Ćujić, S., Jakovčević, Z., Magan, N., Čepela, R. and Vasiljević, M., 2021. Global mycotoxin survey of corn harvested in 19 countries. https://tinyurl.com/yju7kk6u.

Ray, D.K., West, P.C., Clark, M., Gerber, J.S., Prishchepov, A.V., Chatterjee, S., 2019. Climate change has likely already affected global food production. PLoS One 14: e0217148. https://doi.org/10.1371/journal.pone.0217148.

Reid, L.M., Nicol, R.W., Ouellet, T., Savard, M., Miller, J.D., Young, J.C., Stewart, D.W., Schaafsma, A.W., 1999. Interaction of *Fusarium graminearum* and *F. moniliforme* in maize ears: disease progress, fungal biomass, and mycotoxin accumulation. Phytopathology 89: 1028–1037. https://doi.org/10.1094/PHYTO.1999.89.11.1028.

Rojas, E.C., Jensen, B., Jørgensen, H.J.L., Latz, M.A.C., Esteban, P., Ding, Y., Collinge, D.B., 2020. Selection of fungal endophytes with biocontrol potential against Fusarium head blight in wheat. Biological Control 144: 104222. https://doi.org/10.1016/j.biocontrol.2020.104222.

Rong, M., Zheng, X., Ye, M., Bai, J., Xie, X., Jin, Y., He, X., 2019. Phenotypic plasticity of *Staphylococcus aureus* in liquid medium containing vancomycin. Frontiers in Microbiology 10: 809. https://doi.org/10.3389/fmicb.2019.00809.

Rosenzweig, C., Parry, M.L., 1994. Potential impact of climate change on world food supply. Nature 367: 133–138. https://doi.org/10.1038/367133a0.

Schaafsma, A.W., Hooker, D.C., 2007. Climatic models to predict occurrence of *Fusarium* toxins in wheat and maize. International Journal of Food Microbiology 119: 116–125. https://doi.org/10.1016/j.ijfoodmicro.2007.08.006.

Schauberger, B., Archontoulis, S., Arneth, A., Balkovic, J., Ciais, P., Deryng, D., Elliott, J., Folberth, C., Khabarov, N., Müller, C., Pugh, T.A.M., Rolinski, S., Schaphoff, S., Schmid, E., Wang, X., Schlenker, W., Frieler, K., 2017. Consistent negative response of US crops to high temperatures in observations and crop models. Nature Communications 8: 13931. https://doi.org/10.1038/ncomms13931.

Schlenker, W., Roberts, M.J., 2009. Nonlinear temperature effects indicate severe damages to U.S. crop yields under climate change. Proceedings of the National Academy of Sciences of the United States of America 106: 15594-15598. https://doi.org/10.1073/pnas.0906865106

Schmidt-Heydt, M., Parra, R., Geisen, R., Magan, N., 2011. Modelling the relationship between environmental factors, transcriptional genes and deoxynivalenol mycotoxin production by strains of two *Fusarium* species. Journal of the Royal Society Interface. 8: 117–126. https://doi.org/10.1098/rsif.2010.0131.

Slepecky, R.A., Starmer, W.T., 2009. Phenotypic plasticity in fungi: a review with observations on *Aureobasidium pullulans*. Mycologia 101: 823–832. https://doi.org/10.3852/08-197.

Szabo, B., Toth, B., Toth Toldine, E., Varga, M., Kovacs, N., Varga, J., Kocsube, S., Palagyi, A., Bagi, F., Budakov, D., Stojšin, V., Lazić, S., Bodroža-Solarov, M., Čolović, R., Bekavac, G., Purar, B., Jocković, D., Mesterházy, A., 2018. A new concept to secure food safety standards against *Fusarium* Species and *Aspergillus flavus* and their toxins in maize. Toxins 10: 372. https://doi.org/10.3390/toxins10090372.

Timmusk, S., Nevo, E., Ayele, F., Noe, S., Niinemets, Y., 2020. Fighting fusarium pathogens in the era of climate change: a conceptual approach. Pathogens 9: 1–14. https://doi.org/10.3390/pathogens9060419.

Tollefson, J., 2022. Climate change is hitting the planet faster than scientists originally thought. Nature d41586-022-00585–7. https://doi.org/10.1038/d41586-022-00585-7.

Tracy, E.F., 2014. The promise of biological control for sustainable agriculture: a stakeholder-based analysis. Journal of Science Policy and Governance 5: 2014.

United Nations, 2019. World Population Prospects 2019: Highlights, (ST/ESA/SER.A/423).

Váry, Z., Mullins, E., Mcelwain, J.C., Doohan, F.M., 2015. The severity of wheat diseases increases when plants and pathogens are acclimatized to elevated carbon dioxide. Global Change Biology 21: 2661–2669. https://doi.org/10.1111/gcb.12899.

Vaughan, M., Backhouse, D., Ponte, E.M.D., 2016. Climate change impacts on the ecology of *Fusarium graminearum* species complex and susceptibility of wheat to *Fusarium* head blight: a review. World Mycotoxin Journal 9: 685–700. https://doi.org/10.3920/WMJ2016.2053.

Vaughan, M.M., Huffaker, A., Schmelz, E.A., Dafoe, N.J., Christensen, S., Sims, J., Martins, V.F., Swerbilow, J., Romero, M., Alborn, H.T., Allen, L.H., Teal, P.E.A., 2014. Effects of elevated [CO2] on maize defence against mycotoxigenic Fusarium verticillioides. Plant, Cell and Environment 37: 2691–2706. https://doi.org/10.1111/pce.12337.

Vilà, M., Beaury, E.M., Blumenthal, D.M., Bradley, B.A., Early, R., Laginhas, B.B., Trillo, A., Dukes, J.S., Sorte, C.J.B., Ibáñez, I., 2021. Understanding the combined impacts of weeds and climate change on crops. Environmental Research Letters 16: 034043. https://doi.org/10.1088/1748-9326/abe14b.

Wheeler, T., von Braun, J., 2013. Climate Change Impacts on Global Food Security. Science 341: 508–513. https://doi.org/10.1126/science.1239402

Wu, F., Bhatnagar, D., Bui-Klimke, T., Carbone, I., Hellmich, R., Munkvold, G., Paul, P., Payne, G., Takle, E., 2011. Climate change impacts on mycotoxin risks in US maize. World Mycotoxin Journal 4: 79–93. https://doi.org/10.3920/WMJ2010.1246.

Ziska, L.H., Dukes, J.S., 2010. Weed Biology and Climate Change: Ziska/Weed Biology and Climate Change. Wiley-Blackwell, Oxford, UK. https://doi.org/10.1002/9780470958674.

CHAPTER 19

Perspectives regarding feed contaminated by mycotoxins: time for pragmatism

R.R. Santos

Schothorst Feed Research, Lelystad, the Netherlands
rsantos@schothorst.nl

Recognizing the risks of multi-mycotoxin contamination in feeds is the first step towards designing and applying action plans. These natural contaminants pose a threat to animal health and welfare, as well as being directly and indirectly associated with extra costs and economic losses.

In recent years, there have been increasing numbers of studies related to the effects of chronic exposure to diets naturally contaminated with mycotoxins. Farm animals, e.g., swine and poultry, under optimal hygienic and nutritional management will mostly experience subclinical effects causing decreased production performance, decreased vaccination efficacy, increased susceptibility to infectious diseases and, in some cases, increased use of antibiotics (Holanda and Kim, 2021; Okasha et al., 2024). Studies in ruminants are mostly focused on the carry-over of AFs, and few data are available for the effect of other mycotoxins, e.g. *Fusarium* and typical silage/haylage mycotoxins, on feeding behaviour, intestinal health, or milk cheesability (Gallo et al., 2022, 2024). Although aquafeed contamination with mycotoxins became a worldwide issue, reports are mostly focused on the accumulation of these toxins in edible tissues and human health risks, with very few studies addressing the effects of multi-mycotoxin exposure on fish growth performance and intestinal health (Bashorun et al., 2023; Koletsi et al., 2023; Boti et al., 2024). In the case of pets, they have been progressively fed a range of plant-based diets, either to provide texture and structure, or because they are subjected to changes related to the shift from animal-based to plant-based (vegan) diets (Liversidge et al., 2023). This has increased the risk of mycotoxin exposure and the number of chronic diseases that have been reported. Unfortunately, the negative impact of mycotoxins in pets, as well as in horses, is usually diagnosed in the clinic after acute intoxication, and the paucity of data related to chronic exposure results in an underestimation of the problem. Surveillance of animal diets for levels and types of mycotoxins is needed in order to safeguard feed safety and to provide awareness for veterinarians and

 | DOI: 10.1163/9789004712720_20

nutritionists. Biomonitoring, i.e. monitoring of mycotoxins in biological samples (fluids or excreta) of farm animals, is a method used for broadly assessing exposure (Muñoz-Solano and González-Peñas, 2023) and should be extended to companion animals. This approach does not replace toxicokinetic studies because of the inability to accomplish an adequate time point for sample collection in relation to individual feed intake of a group of animals. Moreover, the broad quantification of mycotoxins in serum or other biological samples will not entirely correlate with the amount of all the mycotoxins ingested but may include exposure through other routes (for example, via inhalation). However, it remains a relevant method for data collection, to generate knowledge on routine exposure to mycotoxins and their impact on the health and growth performance of a single animal or group of animals. When applying biomonitoring, the biotransformation of mycotoxins must be considered, as it is necessary to include their metabolites in the analysis. Furthermore, the most beneficial scenario for biomonitoring is the implementation of non-invasive and cost-effective analysis using excreta samples for the detection of mycotoxins, their metabolites or other biomarkers of exposure.

Co-exposure to multiple mycotoxins and their various modified forms, together with species-specific sensitivity to these toxins, makes the established health-based guidance value (HBGV) for individual mycotoxins challenging. To minimize risks, it is proposed to consider not only a single mycotoxin, but also its family, i.e. the modified forms added based on their toxic potencies. In the case of DON exposure, it is advised to calculate the sum of DON with its acetylated forms (3-ADON + 15-ADON) and DON-3G applying the same toxic potency for each compound. The relative potency of the 14 EAs is not well known. Therefore, the sum of 14 EAs is proposed also, considering equal potency. For ZEN, however, oestrogenic activity varies enormously among its modified forms, whereby β-ZEL is 5 times less potent and α-ZEL is 60 times more potent than ZEN. Therefore, the sum should apply the potency factor to each of these forms. The modified forms α- and β-ZEL were previously detected in maize, maize silage and animal feed samples (De Saeger et al., 2003; González Pereira et al., 2014), but these metabolites are often products of ZEN metabolism after consumption. Sows fed diets contaminated with ZEN can mostly transmit α-ZEL via colostrum and milk to lactating piglets (Benthem de Grave et al., 2021), whereas dairy cows commonly convert most ZEN into β-ZEL (Dänicke et al., 2014). A sustainable use of dietary ingredients that will certainly be contaminated with mycotoxins requires considering not only the order of less-to-more-sensitive species to the major contaminants, but animal age and production category. The expected response of an animal species after exposure to a single mycotoxin plus its family compounds can be a starting point. Subsequently, classes in major or minor risk must be identified per animal species, and, when possible, an order of sensitivity should be applied during decision-making. Generally, young animals are more sensitive than older ones to intestinal damage and immunosuppression caused by mycotoxins. Stressful conditions such as the weaning of piglets may disrupt the intestinal barrier, leading to a greater sensitivity to mycotoxins than grower-fattening pigs. Regarding poultry, an ingredient containing ZEN is best used to prepare diets for broiler chickens rather than for laying hens or broiler breeders. In general, caution is advised when incorporating by-products of cereals and grains, e.g. brans, hulls, and DDGS, in animal diets, as their mycotoxin levels are concentrated and can sometimes increase by a factor of three in comparison to the original ingredient.

Human health risk and global trade are part of the rationales behind more harmonized maximum regulated levels of mycotoxins in commodities destined for animal feed. International organizations such as FDA, EFSA, the Brazilian Surveillance Agency (ANVISA), and the South African Department of Agriculture, Forestry and Fisheries (DAFF) among others, established permitted levels of AFs in food and in feed because of the known carcinogenic effect. Maximum recommended levels of other mycotoxins, e.g. *Fusarium* mycotoxins, vary among continents and among countries on the same continent. Harmonizing all maximum levels of mycotoxins solely based on international trade is not feasible because some other multiple factors should be taken into account including global economic inequality, and pragmatic decisions should be taken based on the consequences in each geographic area. The toxicological DON reference value in poultry diet as determined by EFSA (0.6 mg/kg diet) is an ambitious target when considering food-feed competition and the tendency to an increased frequency of feed contamination with mycotoxins due to climate change (Kos et al., 2023). In high-income countries, dietary supplementation with feed additives to mitigate the negative impact of mycotoxins will probably be as common as the use of non-starch polysaccharide enzymes or phytases in animal diets. These feed additives that can counteract the negative impact of mycotoxins additionally appear as a tool for a sustainable use of feedstuffs, by reducing crop waste (Holanda and Kim, 2021). Food and feed safety, food insecurity and subsistence farming play an important role within populations from low-income countries, where there is still little dissemination of knowledge about animal feed production, processing, transport, storage, as well as risks, prevention and detection of mycotoxins (Jalilzadeh-Amin et al., 2023). In this case, educational programmes focused on the local reality and support for accessibility to analytical methods and solutions to counteract the mycotoxin impact should be offered beforehand.

Excluding crop yield losses and the health impact, in 2019 the cost of managing mycotoxins on the North American continent was estimated to be close to 5 billion dollars (FAO, 2019). Losses due to AFs alone in Indonesia, Philippines, and Thailand were approximately 0.9 billion dollars, of which 0.5 billion dollars were linked to human health effects (Schmale and Munkvold, 2014), thus excluding costs associated with livestock and companion animals. The costs associated with mycotoxins can be assessed by evaluating expenditures on veterinary services and drugs for mycotoxicosis treatment, investments in training and research initiatives, preventive strategies such as weather forecasting, optimal agricultural and storage practices, mycotoxin sampling and analysis in feed, inspections, crop price discounts, physical cleaning and sorting of cereal grains, and dietary supplementation with feed additives. However, costs are underestimated due to losses associated to impaired growth performance, chronic or opportunistic infectious diseases that elevate the use of veterinary drugs, including antibiotics, and vaccination failures resulting from mycotoxin exposure, making accurate calculation of these economic losses likely unattainable. Consequently, an effective application of existing knowledge and investment in solutions is required, which will be enhanced by establishing a dynamic network comprising farmers, breeders, practitioners, feed producers, industry stakeholders, policymakers, and scientists, thereby bridging fundamental and applied research.

References

Bashorun, A., Hassan, Z.U., Al-Yafei, M.A.A. and Jaoua, S., 2023. Fungal contamination and mycotoxins in aquafeed and tissues of aquaculture fishes and their biological control. Aquaculture 576: 739892. https://doi.org/10.1016/j.aquaculture.2023.739892.

Benthem de Grave, X., Saltzmann, J., Laurain, J., Rodriguez, M.A., Molist, F., Dänicke, S. and Santos, R.R., 2021. Transmission of zearalenone, deoxynivalenol, and their derivatives from sows to piglets during lactation. Toxins 13: 37. https://doi.org/10.3390/toxins13010037.

Boti, V., Martinaiou, P., Gkountouras, D. and Albanis, T., 2024. Target and suspect screening approaches for the identification of emerging and other contaminants in fish feeds using high resolution mass spectrometry. Environmental Research 251: 118739. https://doi.org/10.1016/j.envres.2024.118739.

Dänicke, S., Keese, C., Meyer, U., Starke, A., Kinoshita, A. and Rehage, J., 2014. Zearalenone (ZEN) metabolism and residue concentrations in physiological specimens of dairy cows exposed long-term to ZEN-contaminated diets differing in concentrate feed proportions. Archives of Animal Nutrition 68: 492-506. https://doi.org/10.1080/1745039X.973236.

De Saeger, S., Sibanda, L. and Van Peteghem, C., 2003. Analysis of zearalenone and -zearalenol in animal feed using high-performance liquid chromatography. Analytica Chimica Acta 487: 137-143. https://doi.org/10.1016/S0003-2670(03)00555-5.

Food and Agriculture Organization (FAO), Grain storage techniques – Evolution and trends in developing countries. Agricultural Services Bulletin No. 109.

Gallo, A., Catellani, A., Ghilardelli, F., Lapris, M. and Mastroeni, C., 2024. Review: Strategies and technologies in preventing regulated and emerging mycotoxin co-contamination in forage for safeguarding ruminant health. Animal 18: 101280. https://doi.org/10.1016/j.animal.2024.101280.

Gallo, A., Mosconi, M., Trevisi, E., and Santos, R.R., 2022. Adverse effects of fusarium toxins in Ruminants: a review of in vivo and in vitro studies. Dairy 3: 474–499. https://doi.org/10.3390/dairy3030035.

González Pereyra, M.L., Sulyok, M., Baralla, V., Dalcero, A.M., Krska, R., Chulze, S. and Cavaglieri, L.R., 2014. Evaluation of zearalenone, α-zearalenol, β-zearalenol, zearalenone 4-sulfate and β-zearalenol 4-glucoside levels during the ensiling process. World Mycotoxin Journal 7: 291-295. https://doi.org/10.3920/WMJ2013.1638.

Holanda, D.M. and Kim, S.W., 2021. Mycotoxin occurrence, toxicity, and detoxifying agents in pig production with an emphasis on deoxynivalenol. Toxins 23: 171. https://doi.org/10.3390/toxins13020171.

Jalilzadeh-Amin, G., Dalir-Naghadeh, B., Ahmadnejad-Asl-Gavgani, M., Fallah, A.A. and Mousavi Khaneghah, A., 2023. Prevalence and concentration of mycotoxins in animal feed in the Middle East and North Africa (MENA): a systematic review and meta-analysis. Toxins 15: 214. https://doi.org/10.3390/toxins15030214.

Koletsi, P., Wiegertjes, G.F., Graat, E.A.M., de Kool, M., Lyons, P. and Schrama, J.W., 2023. Individual and combined effects of deoxynivalenol (DON) with other Fusarium mycotoxins on rainbow trout (*Oncorhynchus mykiss*) growth performance and health. Mycotoxin Research 39: 405-420. https://doi.org/10.1007/s12550-023-00496-0.

Kos, J., Anić, M., Radić, B., Zadravec, M., Janić Hajnal, E. and Pleadin, J., 2023. Climate change—a global threat resulting in increasing mycotoxin occurrence. Foods 12: 2704. https://doi.org/10.3390/foods12142704.

Liversidge, B.D., Dodd, S.A.S., Adolphe, J.L., Gomez, D.E., Blois, S.L. and Verbrugghe, A., 2023. Extruded diet macronutrient digestibility: plant-based (vegan) vs. animal-based diets in client-owned healthy adult dogs and the impact of guardian compliance during in-home trials. Frontiers in Animal Science 4: 1288165. https://doi.org/10.3389/fanim.2023.1288165.

Muñoz-Solano, B. and González-Peñas, E., 2023.Biomonitoring of 19 mycotoxins in plasma from food-producing animals (cattle, poultry, pigs, and sheep). Toxins 15: 295. https://doi.org/10.3390/toxins15040295.

Okasha, H., Song, B. and Song, Z., 2024. Hidden hazards revealed: mycotoxins and their masked forms in poultry. Toxins 16: 137. https://doi.org/10.3390/toxins16030137.

Schmale, D.G. and Munkvold, G.P., 2014. Economic impact of mycotoxins. Mycotoxins in Crops: A Threat to Human and Domestic Animal Health, The American Phytopathological Society (APS).

Index